Human Development and Global Advancements through Information Communication Technologies:

New Initiatives

Susheel Chhabra
Lal Bahadur Shastri Institute of Management, Delhi, India

Hakikur Rahman
ICMS, Bangladesh

Senior Editorial Director:	Kristin Klinger
Director of Book Publications:	Julia Mosemann
Editorial Director:	Lindsay Johnston
Acquisitions Editor:	Erika Carter
Development Editor:	Myla Harty
Production Coordinator:	Jamie Snavely
Typesetters:	Jennifer Romanchak and Deanna Zombro
Cover Design:	Nick Newcomer

Published in the United States of America by
Information Science Reference (an imprint of IGI Global)
701 E. Chocolate Avenue
Hershey PA 17033
Tel: 717-533-8845
Fax: 717-533-8661
E-mail: cust@igi-global.com
Web site: http://www.igi-global.com

Library of Congress Cataloging-in-Publication Data
Human development and global advancements through information communication technologies : new initiatives / Susheel Chhabra and Hakikur Rahman, editors.
p. cm.
Includes bibliographical references and index.
Summary: "This book investigates the role of ICT in advancing economic, social, and political development as well as the potential setbacks experienced as a result of introducing new initiatives"--Provided by publisher.
ISBN 978-1-60960-497-4 (hbk.) -- ISBN 978-1-60960-498-1 (ebook) 1. Information technology--Economic aspects. 2. Information technology--Social aspects. I. Chhabra, Susheel. II. Rahman, Hakikur, 1957-
HC79.I55H86 2011
303.48'33--dc22
2011001619

British Cataloguing in Publication Data
A Cataloguing in Publication record for this book is available from the British Library.

All work contributed to this book is new, previously-unpublished material. The views expressed in this book are those of the authors, but not necessarily of the publisher.

Table of Contents

Detailed Table of Contents

Trends in international development cooperation point to the increasing networking of initiatives and programmes, facilitated by information and communications technology (ICT). This allows many more people and organizations from around the world to contribute to a given project, as with the case of online volunteers. There are various types of networks active in development cooperation, but network management needs to be incorporated by involved organizations in order to extract the expected benefits from their involvement. Network analysis practices will help determine if they are set up and managed appropriately.

The use of ICT in the developmental activities and the evolution of the telecentres across the world, for the deployment and the propagation of the ICT solutions, have been widely tested and accepted phenomena. In this paper, we examine the use of FOSS – Free/Open source software in those ICT solutions and categorize them into two types, in order to emphasize the difference between those solutions that are specifically developed to address the need of a community and solutions that utilize the generic FOSS and hence play, only a meager role in the solution. We discuss about our experience in developing a FOSS-based ICT solution that is specifically built to address the needs of a community information system. We illustrate our three-tiered multi-stakeholder model of development, deployment and sustained usage. We also describe our experience in the process of development and deployment of our solution in various locations of India, and the benefits obtained through the model by emphasizing the synergies between our model and the FOSS mode of ICT solution development.

The digital divide is nothing else than the reflection of the social divide in the digital world. The use of ICT for human development does offer opportunities to reduce the social divide for individual beings or communities; yet there exists a series of obstacles to overcome. The very existence of an infrastructure for connectivity is only the first obstacle, although it often receives an exclusive focus, due to the lack of an holistic approach which gives an essential part to digital and information literacy. Telecommunications, hardware and software are predictable prerequisites; however, the true pillars of human-focused information societies are education, ethics, and participation, interacting together as a systemic process. As long as decision makers are not ready to consider these evidences, and keep on favoring a mere technological vision, we will suffer from the most dangerous divide in terms of impact: the paradigmatic divide.

The transition to knowledge-intensive customer-centric enterprise is important, but never easy. Reorganizing people is likely to face critical structural and cultural change issues related to people. Addressing these issues is essential for the continued success of customer-value-building services and products. In light of today's competitive business environments and changing power of customers, organizations need to be able to deal with people-based issues in order to secure high quality customer service and long-life and profitable customer relationship. The article presents a recommended solution to deal with people change management in competitive business environments, viz. to 'reorganize people' in a customer-centric networked organization. 'Reorganization of people' is operationally defined by three sub-interventions: a) reconfiguring structure, b) reshaping culture, and c) rehabilitating people.

The emergence of Internet has revolutionalized the way businesses are conducted. The impact of e-commerce is pervasive, both on companies and society as a whole. It has the potential to impact the pace of economic development and in turn influence the process of human development at the global level. However, the growth in e-commerce is being impaired by the issue of trust in the buyer-seller relationship which is arising due to the virtual nature of e-commerce environment. The Online trading environment is constrained by a number of factors including web interface that in turn influences user experience. This paper identifies various dimensions of web interface that have the potential to influence trust in e-commerce. The empirical evidence presented in the paper is based on a survey of the web interfaces of 65 Indian B2B e-exchanges.

Learning is considered as one of the potential tool to empower a community. Over the past three decades, technology mediated learning has been recognized as an alternate channel replacing/ supporting/ strengthening the traditional forms of education in various forms, especially with the advent of interactive and collaborative learning. Open and distance learning (ODL) emerges as a potential means of enhancing educational access. On the other hand, open educational resources (OER) emerge as a potential material of this new paradigm of knowledge acquisition process. However, the organizational learning at the peripheries and capacity development at the grass roots remain almost unattended, despite recognized global efforts under many bottom-up empowerment sequences. Social components at large within the transitional and developing economies remain outside the enclosure of universal access to information and thus access to knowledge has always been constricted to equitably compete with the global knowledge economy. Despite challenges in designing and implementing collaborative learning techniques and technologies, this chapter would like to emphasize on introducing collaborative learning at community level and improve the knowledge capacity at the grass roots for their empowerment. This chapter, further, investigates the relationship of collaborative learning towards improved e-governance. In the main thrust section, the chapter goes through various channels of collaborative learning, methods that could be adopted during the implementation and technologies that could be availed during the dissemination phases of collaborative learning. Later on a few cases are being included, and before the conclusions, the chapter puts forward a few future research issues in the aspect of collaborative learning for empowerment of communities.

Many governmental institutions and other organizations have started to provide their customers with access to their documents by electronic means. This alters the way of interaction between authorities and citizens considerably. Hence, it is worthwhile to look at both the chances and the risks that this process of change implies for disabled citizens. Due to different laws or legal directives e.g. governmental authorities have a particular responsibility to consider also the needs of disabled persons. Therefore, they need to apply appropriate techniques for these groups to avoid an "Accessibility Divide". This discussion is built on the observation that governmental and other customer oriented processes are mostly based on the exchange of forms between authorities and citizens. Authors state that such processes can be distinguished into three scenarios, with the use of paper as means of transport on the one end and complete electronic treatment at the other end. For each scenario there exist tools to improve accessibility for people with certain disabilities. These tools include standard technologies like improved Web access by magnifying characters, assistive technologies like document cameras, and more sophisticated approaches like integrated solutions for handling forms and government processes.This article focuses on approaches that provide access to governmental processes for people with visual impairments, elderly people, illiterates, or immigrants. Additionally, it sees a chance to enable electronic processes in developing countries where the citizens have less experience in handling IT-based processes. The main part of the chapter describes an approach to combine scanned images of paper-based forms containing textual information and text-to-speech synthesis yielding an audio-visual document representation. It exploits standard document formats based on XML and web service technology to achieve independency from software and hardware platforms. This is also helpful for conventional governmental processes because people within the group of interest stated above often also have problems to access non-digitized information, for instance when they have to read announcements within public administration offices.

Despite the popularity, potency and perfection of electronic government (e-Government), it is yet somehow remain in uncharted territory for many countries in terms of implementing e-governance at the local government level. However, technology possess the potential for improving the way government works, and utilizing the newly evolved technology many countries have been engaged themselves for improving the way their citizens work. Local and national governments are trying to realize this potential by finding ways to implement novel technology in spearheading its utilization to achieve the best services for their citizens. They include awareness raising, knowledge acquisition, social networking, and mobilization in planning, developing, implementing, and evaluating local e-government initiatives to draw inferences and recommendations learning from local e-government pioneers across the globe. This chapter has tried to draw a line of reference by put forwarding the importance of local e-government organizational structure, and their supremacies in terms of utilization of information and communication technology (ICT). Along this context, the chapter has attempted to synthesize a few prospective local e-government scenarios, focus on their adaptation of ICT, puts forward recommendations to improve local e-government for better utilization of information services.

Diffusion of information and communication technologies is a global phenomenon. In spite of rapid globalization there are considerable differences between nations in terms of the adoption and usage of new technologies. Several studies exploring causal factors including national cultures of information and communication technology adoption have been carried out. The focus of this paper is slightly different from other studies in this area. Rather than concentrating on the individual information technology an overall e-Government readiness is the focus. This research conducted an analysis of the impact

This chapter explores information communication technologies (ICTs) (e.g., computer-mediated communication) and the implications for use in language learning and second language learning (L2). Further, the chapter presents culture and new trends in ICTs for L2 learning. Specific modality, challenges, and issues for future considerations in L2 learning are discussed. The paper argues for the need to understand culture and contextual appropriateness of L2 learning in ICT environments. Finally the paper contends that ICTs are best relegated as a supplemental role or tools, rather than as an outright substitute for traditional L2 learning and curricula.

The aim of this chapter is to illustrate a technology-based approach for promoting and diffusing Calabrian cultural heritage of the ancient Magna Graecia period (VIII cent. B.D.- I cent. A.D.) in a global perspective. To achieve this goal, the chapter focuses on the use of 3D technologies, as well as on virtual and augmented reality, with emphasis on the stereoscopic Virtual Theatre. These innovative tools support the creation of a global vision of the fragmentary archaeological Calabrian heritage, as well as the possibility to play with the virtual findings as in a videogame, by choosing what to explore and the contents to access. Moreover, these technologies exploit the entertaining component of the systems in order to provide personalized and interactive educational contents.

This essay presents a case study of Information Technology (IT) education as a contributor to economic and human development in rural Nigeria. The case of Summit Computers suggests that for developing countries to benefit from advances in IT, the following factors are of great importance and can be enhanced by IT education initiatives: convenience, affordability, emphasis on participation and empowerment of local users, encouragement of entrepreneurship, and building awareness among potential users. Additionally, careful attention should be given to how IT training can meet local employment and other needs are important factors in rural communities in developing countries such as Nigeria.

As Indian economy gets integrated to the global economy and strives to improve in terms of human development indicators, a special role exists for information and communication technologies (ICT) in this process. The strategic metamorphoses and the resultant expansion of ICT linked telecommunication services in India have favorably influenced the effort to accelerate the pace of human development by enabling equality in access to information, creation of employment, improving the quality of life, better livelihood opportunities in rural areas, growth of agriculture, impetus to business development, environmental management and many more. After the initiation of economic planning in India, telecom services were assumed to be natural monopoly and were provided by one entity without competition. The government launched ambitious ICT infrastructure initiatives, radically changing its communication policy framework. The resultant growth of ICT services in India has led to significant improvement in human development levels. It has led to a reduction in information asymmetry between the rich and the poor, improvement in telecom density and ICT accessibility in rural areas, fostering inclusive growth, providing better access to market information to people in remote and rural areas, facilitating technological leapfrogging, enhancing business networking and offering new opportunities from the perspective of human development.

This paper examines the characteristics of internet that motivate faculty members of Imam Muhammad Bin Saud University (IMSU) in Saudi Arabia to utilize the Internet in their research and instructional activities. The framework of the study was the attributes of innovations offered by Rogers. A modified instrument was adopted to collect the data and measure the attributes. The result revealed that the majority of IMSU faulty members used the Internet for research and academic activities twice a month or less, indicating a low Internet adoption rate. Multiple regression analysis showed that all attributes of innovation individually predicted Internet adoption. The combination of all attributes indicated the model could predict Internet adoption among faculty.

The emergence of inter-organizational system has facilitated easy and fast flow of information among the trading partners. This has affected the business relations among the trading parties involved. Though the inter-organizational systems have helped a lot in improving the business relations, the vulnerability

Preface

IMPORTANCE OF HUMAN DEVELOPMENT DOCUMENTATIONS

History of human development documentations by the United Nations goes back to 1990s when they first launched the Human Development Report (HDR) with the unified goal of placing human at the center of the development process in terms of economic debate, policy and advocacy. Despite its massiveness, the simplicity of the goal enabled peoples to enter into the development phenomena from each strata of the society. As mentioned by the HDR[1], it is "bring about the people, by the people, and for the people…". Later on many other parameters evolved around the ranking, measuring and evaluating human development indices among nations. Themes have circled around concepts and measurement, people's participation, poverty eradication, globalization, human right, democracy, MDGs[2], cultural liberty, climate change, human mobility and development, but making new technologies to work for human development remained as a major area of documentation (UNDP, 2000; 2001a; 2002; 2003; 2004; 2007; 2009). Though the focus has been implied to overall society development, but human aspects comprise a major portion of the commitment that suffices the importance of human development documentations by lead institution like the World Bank since 1978 through their World Development Reports, and their themes were also encompass the basic element of human development such as, poverty, health, environment, economy, agriculture and knowledge.

In essence human development became a development paradigm which integrates beyond the rise or fall of a nation's economy. It tends towards an enabling environment within nation and among nations where human were given to act at their full potentiality to be more productive. Eventually the development processes are expanding towards the peoples' choice leading to their elevated livelihoods and sustained economic growth (UNDP, 2009). Researchers and academics across countries and continents are working relentlessly to document progress of human wealth in terms of livelihood upgradation, socio-economic uplift and knowledge acquisition. Furthermore, apart from the UN or UN agencies, since the inception of organizational movements to engage in human development documentation numerous researches have been carried out by institutions like OECD, G8, G20, EU, EC, and other national, regional and international organizations.

ROLE OF ICT ON HUMAN DEVELOPMENT

In the realm of the newly evolved information communications and technologies (ICTs) human development has taken radical turn by making ICT as an enabler of development. ICT has been recognized as one of the most powerful force in the contemporary world to achieve widespread social and economic development (DOI, 2001). Like the United Nations, the World Economic Forum also acknowledges the role of ICT as the critical enabler of sustainable socio-economic growth and the important element for effective regional cooperation in creating larger market dimension (Mansion, n.d.), especially reaching out to the grass roots communities (Rahman, 2007a).

The spectacular growth of ICTs is by far in many ways is driving the global economy now and there are empirical evidences that ICT diffusion has a significant positive impact on GDP growth (Bedi, 1999; Pohjola, 2000; Oliner & Sichel, 2000; Baliamoune, 2002). Furthermore, similar to other technological innovations, the development of ICTs offers numerous opportunities for the achievement of global sustainability (WRI, 2004; UNDP, 2005). With its crosscutting nature, ICT allows applications in diversified ways (Kuhndt, Von Geibler & Herrndorf, 2006). Ranging from environmental eco-system and biodiversity management to entrepreneurship, urban-planning and transportation management to government and governance system to food security and poverty eradication and in all other facets of development ICT is predominant and human endeavors are at the locus.

At the same time, ICT has became an integral part of the development strategies in advanced and less advanced economies with great potential for social transformation by enhancing access to people, service, information and other technologies (Dutton et al., 2004). ICT approaches has been transformed to match with the demand of the communities and descended nearer to the people becoming more pro-poor or for the people at large (Rahman, 2007a). Uses of ICT has actively promoted for socio-economic development, job creation, rural development and poverty reduction (Nandi, 2002; Malhotra, et al., 2007). Moreover, ICT has the capacity to empower the grass roots utilizing socio-economic services by widening access to micro finance (Cecchni & Scott, 2003). Pervasive and ever increasing impact of ICTs on Entrepreneurships and Institutions, on Competitive Strategies and Innovation, on Financial and Utility Services, Employment, Education, Regional and Spatial Development, and Poverty Reduction has given it a new dimension of human development perception (Hanna, 2003).

FUTURE ISSUES AND RESEARCHES

It is natural that new technologies should be drivers of the future generation. However, just by looking at the new technologies, future prediction is difficult, and forecasting their impact on human society is more difficult. Comprehensive research approaches are needed to be developed with strategic ICT applications by looking at the global perspective (Rahman, 2009). Furthermore, the use of ICTs in development processes is not new. But, at the beginning of this millennium in 2000 they assumed a new standing, when the United Nations and G8 group of industrialized countries flagged ICTD (ICT for Development) as a global development priority. Since then, the understanding of ICTD as a core development issue has been rapidly evolving (UNDP, 2001b). Moreover, the new world economy characterized by globalization and an increasing emphasis on knowledge has also emerged over the past decade. Within the 'new economy', knowledge has become the crucial driver of economic growth. The economic, social and political eco-systems where future development will take place have, therefore, also changed. All countries, either rich or poor, must now reconsider their approach to development to incorporate this new reality (Shorrocks, 2003).

At the same time, the world economy has recently been changed. A new world economy has emerged over the last decade with two long-run broad trends, as the globalization and advances in ICT, but eventually they have converged. This has to be noted that the emerging 'new economy' is significantly different to the traditional 'old economy', as knowledge has replaced traditional productivity inputs, such as labor and natural resources, as the primary ingredient for economic growth (Clarke, 2003). This emerging phenomenon has placed human at the core of the future development processes. Hence, the future of different nations will depend on how they respond to the new economy through utilizing ICTs for human development.

WHERE THE BOOK STANDS

Among these perspectives and manifestations, documentation on human development and global advancement through ICT is not a challenge but necessity and this book through compilation of many multifaceted acts, activities and applications from the various corners of the globe has given a new dimension of acceptability, readability and research context. This book will be a guidebook in understanding the base line issues of ICT utilizations for human development and at the same time will lead to initiate future researches through various facets of human development paradigms.

ORGANIZATION OF THE BOOK

The book has been divided into four sections, Concepts and Perceptions; Culture and Learning; E-Government and E-Governance; and Applications and Cases. E-Government and E-Governance has four chapters, while other sections have five chapters each.

Information & Communication Technologies through Internet & networking infrastructure have facilitated the cooperation for development initiatives all over the world. Chapter 1 has described the possibility of cooperation and generation of value by multifaceted positive change in the knowledge sharing attitudes of individuals and organizations. The author has suggested the effective management of network through network analysis practices which can help us to reap the desired benefits.

The use of Free/Open source software for ICT solutions has significantly contributed to the developmental activities. Chapter 2 has discussed in detail authors' experience in development of a software solution and how it can play a symbolic role. Authors have also suggested some other solutions that are being specifically developed to address community information needs. The authors have recommended three-tier multi-stakeholder model of development, deployment and usage for community development.

The use of ICTs for human development offer opportunities to reduce the social digital divide for individuals as well as communities. Chapter 3 has emphasized the need to focus on basic pillars of social equality such as education, ethics, participation, interacting together as a systemic process. As long as decision makers are not ready to consider these evidences, and keep on favoring a mere technological vision, the society will suffer from the most dangerous divide in terms of impact: the paradigmatic divide and any resemblance to characters, projects, or policies in real life are quite intentional.

The ICT-based interventions has helped to reduce digital divide and facilitated economic development leading to equality in access to information, creation of employment, improving the quality of life, better livelihood opportunities in rural areas, growth of agriculture, impetus to business development, and environmental management. Chapter 4 has described as how strategic metamorphoses and the resultant expansion of ICT linked telecommunication services assumed to be natural monopoly in India and were provided by one entity without competition. The government launched ambitious ICT infrastructure initiatives, radically changing its communication policy framework. The resultant growth of ICT services in India has led to significant improvement in human development levels.

The adoption of Internet and ICT-based research initiatives are required to be examined in detail to facilitate value-based usage of such services. Chapter 5 examines the characteristics of internet that motivate faculty members of Imam Muhammad Bin Saud University (IMSU) in Saudi Arabia to utilize the Internet in their research and instructional activities. The framework of the study was the attributes of innovations offered by Rogers. A modified instrument was adopted to collect the data and measure the attributes. The result revealed that the majority of IMSU faulty members used the Internet for research and academic activities twice a month or less, indicating a low Internet adoption rate. Multiple regression analysis showed that all attributes of innovation individually predicted Internet adoption. The combination of all attributes indicated the model could predict Internet adoption among faculty.

Keeping in view the ubiquitous learning environment on the web interfaces, it has become important to develop tools for collaborative learning to empower communities. Chapter 6 has elaborated the need to use technology mediated learning as an alternative channel to strengthen the traditional forms of education. According to the author, the organizational learning at the peripheries and capacity development at the grass roots remain almost unattended, despite recognized global efforts under many bottom-up empowerment sequences. Social components at large within the transitional and developing economies remain outside the enclosure of universal access to information and thus access to knowledge has always been constricted to equitably compete with the global knowledge economy. The chapter has emphasized the need for introducing collaborative learning at community level and improves the knowledge capacity at the grass roots for their empowerment. The chapter, further, investigates the relationship of collaborative learning towards improved e-governance.

The online medium of instructions needs to be synchronized with the different cultures all over the world. Chapter 7 has described the implementation of Online Synchronous Learning (OSL) challenges to existing instruction technology theory because of the complexity of the digital age. This article describes the implementation of OSL for teaching English to foreign students from different cultures. The authors believe that the cultural historical Activity Theory is ideal for understanding OSL and its pedagogy. Through the lens of Activity Theory, this study takes close look at OSL courses and examines the socio-cultural factors affecting the success of the course as well as their complex relationships. Applying Activity Theory to analyze data collected over three years authors have developed a framework to help educators who intend to implement OSL from multiple cultural perspectives.

The ICT has transformed the way of understanding and translating the class sessions especially in different language settings. Chapter 8 has described in a case study as how a University class undertook a translation from Swedish to English in a keystroke logging environment and then replayed their translations in pairs while discussing their thought processes when undertaking the translations, and why they made particular choices and changes to their translations. Computer keystroke logging coupled with peer-based intervention assisted the students in discussing how they worked with their translations, and enabled them to see how their ideas relating to the translation developed as they worked with the text. The process showed that Computer Keystroke logging coupled with peer-based intervention has potential to (1) support student reflection and discussion around their translation tasks, and (2) enhance student motivation and enthusiasm for translation.

The computer mediated communication and learning environment is affected by the culture of the communicators. Chapter 9 explores computer-mediated communication (CMC) and information communication technology (ICT) use in language learning. More specifically, the chapter addresses the impact or implications of CMC tools for computer enhanced language learning. The author stresses the need to understand culture and contextual appropriateness of language, thus, it argues for communication technology to be used as a secondary resource rather than a primary tool for language learners. The discussion also addresses the dimensions of cultural variability with respect to language learning.

The Personal Digital Assistants, pocket PCs, and smart-phones etc. has facilitated to exploit and disseminate Cultural Heritage in 3 D object environments. Chapter 10 has illustrated as how Information and Communication Technologies (ICT) could be used to provide enriching learning experiences for different targets of users, especially young people. In fact, by the immersion in virtual museums or reconstructed worlds, users can build different paths of fruition interacting with 3D objects as in a videogame. The case study of Calabrian Magna Graecia (Italy) is presented, with particular reference to the projects "VirtualMuseum Net of Magna Graecia" and "NETConnect".

ICT-based government processes can facilitate smooth access to disabled citizens. Chapter 11 has observed that governmental processes mostly are based on forms. After distinguishing several processing scenarios, with the use of paper as means of transport on the one end and complete electronic treatment at the other end, they have suggested some important approaches that can provide access to governmental processes for people with visual impairments. One of the approaches described in this chapter is to associate synthesized speech to scanned images of printed documents, thus yielding an audio-visual document representation for disabled citizens.

The dream of ICT for human development can not be fulfilled unless local e-government set-ups are strengthened. Chapter 12 has stress upon the need to give importance to management of local e-governance in-line with national governments. This chapter has tried to draw a line of reference by putting forward the importance of local e-government organizational structure, and their supremacies in terms of utilization of ICT. Along this context, the chapter has attempted to synthesize a few prospective local e-government scenarios, focus on their adaptation of ICT, and suggested some recommendations to improve local e-government for offering better information services.

The e-governance readiness can not progress in equal dimensions all over the world due to differences in national & international cultures. Chapter 13 has focused on overall e-Government readiness instead of individual domains. The findings of the study is based on a research conducted on analysis of the impact of national culture has on e-Government readiness and its components for 62 countries. E-Government readiness assessment used in this study is based on the UN E-Government Survey 2008, while the national cultural dimensions were identified using Hofstede's model of cultural differences. The research model and hypotheses were formed and tested using correlation and regression analysis. The findings indicate that worldwide e-Government readiness and its components are related to culture.

The emergence of e-governance without a socio-cultural focus can destabilize the structure of societies as a whole. Chapter 14 has put forward the need to consider socio-cultural context as a framework to design e-government strategies and practices. The author has suggested that the information societies can be grouped in classes. It may help to analyze the classes' needs and possibilities and to formulate proper e-government agenda to be implemented. The real specificities and diversities among classes make the IS development multi-trajectory. In our diversified world, the effects will vary greatly, hence the need to develop a socio-cultural model for e-government readiness.

The smooth change management in knowledge-centric organizations has become important in view of socio-economic dynamics. The transition from existing traditional interventions to knowledge-intensive customer-centric enterprise requires critical structural and cultural change management issues. Chapter 15 has stressed the need to address these issues for the continued success of customer-value-building services and products. The chapter presents a recommended solution to deal with people change management in competitive business environments, viz. to 'reorganize people' in a customer-centric networked organization. 'Reorganization of people' is operationally defined by three sub-interventions: (a) reconfiguring structure, (b) reshaping culture, and (c) rehabilitating people.

The e-commerce especially the transactions carried out on the Internet are still grabbling with the trust issues. Since the impact of e-commerce is pervasive, both on companies and society as a whole, there is a need to develop a trust model. Chapter 16 has identified various dimensions of web interface that have the potential to influence trust in e-commerce. The empirical evidence presented in the chapter is based on a survey of the web interfaces of 65 Indian e-Marketplaces.

The expansion of ICT-based activities has led to the enhancement of usage and need to effective exploration of data in multiple data sources. Chapter 17 has illustrated as how to enhance service quality in a hospital using data mining. The improvement in service quality helps to create hygienic environment and enhance technical competence among staff members which generates value to patients. A weighting model is proposed to identify valid rules among large number of forwarded rules from various data sources. This model is applied to rank the rules based on patient perceived service parameters in a hospital. Results show that this weighting model is efficient. The model can be used effectively for determining the patient's perspective on hospital services like technical competence, reliability and hygiene conditions under a distributed environment.

The awareness and use of ICT education is required to be made available to rural community especially in developing countries since Information Technology education is associated with high cost and is not typically available outside urban areas. Chapter 18 has presented a case where the provision of IT education differs from the conventional emphasis on urban dwellers. The authors discuss the case of Summit Computers in a rural community in Nigeria. The analysis of the case suggests that for developing countries to benefit from advances in IT, awareness among the real users, convenience, affordability and consideration of how IT training can meet local needs and employment are important factors. Entrepreneurship, participation and empowerment of local users are also discussed as important factors that enhance the sustainability of IT education in rural communities.

The trust in ICT-based trading environment is required to be established to facilitate smooth B2B transactions. In Chapter 19 authors make an attempt to empirically examine the relationship between the levels of assurance with regard to deployment and implementation of relevant technology tools in addressing the identified technology related trust issues and ultimately enhancing the perceived level of trust in inter-organizational business relations. The empirical evidence presented in this article is based on a survey of 106 Indian companies using inter-organizational systems for managing their business relations.

CONCLUSION

The use of ICTs for human development has become a natural phenomenon over the years through application of technologies to improve socio-economic development of nations and communities. The ICT-based interventions has assisted to reduce social and digital divide, and facilitated socio-economic development leading to equality in access to information, creation of employment, improving the quality of life, better livelihood opportunities in rural areas, growth of agriculture, impetus to business development, and environmental management. Ranging from improved conceptual benchmarking; illustrations on various cases and applications on education, learning, and technologies; development aspects on e-government and e-governance; business, culture and human development contexts, this book will be able to serve as a practical guide to researchers, readers and learners. As the first book in this series, it is expected that it has established a significant benchmark in the aspect of utilization of ICTs for the human development.

REFERENCES

Baliamoune, M.N. (2002) *The New Economy and Developing Countries: Assessing the Role of ICT Diffusion*, Discussion Paper No. 2002/77, UNU/WIDER, Finland.

Bedi, A. S. (1999) 'The Role of Information and Communication Technologies in Economic Development: A Partial Survey'. Discussion Paper on Development Policy 7. Bonn: ZEF, Universitat Bonn.

Cecchini, Simone. (2004). Electronic Government and the Rural Poor: the Case of Gyandoot Research Note" *Information Technologies and International Development*, The Massachusetts Institute of Technology, 2(2): 65-75.

Clarke, M. (2003) e-development? Development and the New Economy, Policy Brief No.7, UNU World Institute for Development Economics Research (UNU-WIDER), Finland.

DOI (2001) Creating a Development Dynamics: Final Report of the Digital Opportunity Initiative. Accenture, Markle Foundation and UNDP.

Dutton, W.H., Gillett, S.E., McKnight, L.W. and Peltu, M. (2004) Bridging broadband Internet divides: reconfiguring access to enhance communicative power, *Journal of Information Technology*, 19(1): 28-38.

Hanna, N.K. (2003). Why National Strategies are needed for ICT-enabled Development, Staff Working Papers No. 3, Information Solutions Group (ISG), Bellevue, WA.

Malhotra, C., Chariar, C.V., Das, L.K. and Ilavarasan, P.V. (2007). ICT for Rural Development: An Inclusive Framework for e-Governance. In: G.P. Sahu (Ed.) *Adopting e-Governance*, GIFT Publishing, New Delhi, p. 216-226.

Kuhndt, M., Von Geibler, J. and Herrndorf, M. (2006) Assessing the ICT Sector Contributing to the Millennium Development Goals: Status quo analysis of sustainability information for the ICT sector, Wuppertal Report No. 3, Wuppertal Institute for Climate, Environment and Energy, Wuppertal, Germany.

Manson, H. (n.d.) The role of ICT in human development, Part One, two and three, Retrieved June 27, 2010 from http://www.mediatoolbox.co.za/pebble.asp?relid=2213.

Nandi, B. (2002) Role of Telecommunications in Developing Countries in the 21st century, *14th Biennial Conference* Seoul: International Telecommunications Society (ITS), 2002.

Oliner, S. D., and D. E. Sichel (2000) The Resurgence of Growth in the Late 1990s: Is Information Technology the Story?' *Journal of Economic Perspectives*, 14: 3-22.

Pohjola, Matti (2000) 'Information Technology and Economic Growth: A Cross-Country Analysis', in M. Pohjola (ed.), *Information Technology, Productivity, and Economic Growth: International Evidence and Implications for Economic Development*. New York: Oxford University Press, 242-56.

Rahman, H. (2007a) An Overview on Strategic ICT Implementations Toward Developing Knowledge Societies. In: Rahman, H. (Ed.) *Developing Successful ICT Strategies: Competitive Advantages in a Global Knowledge-Driven Society,* Idea Group Inc., USA, pp. 1-35.

Rahman, H. (2007b) Prologue. In: Rahman, H. (Ed.) *Selected Readings on Global Information Technology: Contemporary Applications*, Information Science Reference, USA, pp. xxv.

Rahman, H. (2009) Local E-Government Management: A Wider Window of E-Governance. In: *Handbook of Research on E-Government Readiness for Information and Service Exchange: Utilizing Progressive Information Communication Technologies,* Publisher: Information Science Publishing, USA, pp. 295-323.

Shorrocks, T. (2003) Forewords. In: Clarke, M. e-development? Development and the New Economy, Policy Brief No. 7, UNU World Institute for Development Economics Research (UNU-WIDER), Finland.

UNDP (2000) Human Development Report 2000: Human rights and human development, United Nations Development Programme (UNDP), New York, Oxford University Press.

UNDP (2001a) Human Development Report 2001: Making new technologies work for human development, UNDP, New York, Oxford University Press.

UNDP (2001b). *Essentials, Information Communications Technology for Development: Synthesis of Lessons Learned*, Evaluation Office No. 5, UNDP, New York, September 2001

UNDP (2002) Human Development Report 2002: Deepening democracy in a fragmented world, UNDP, New York, Oxford University Press.

UNDP (2003) Human Development Report 2003: Millennium Development Goals- A compact among nations to end human poverty, UNDP, New York, Oxford University Press.

UNDP (2004) Human Development Report 2004: Cultural liberty in today's diverse world, UNDP, New York, Oxford University Press.

UNDP (2005) Innovation: Applying Knowledge in Development, UN Millennium Project 2005, Task Force on Science, technology and Innovation, Sterling, USA.

UNDP (2007) Human Development Report 2007/2008: Fighting climate change- Human solidarity in a divided world, UNDP, New York.

UNDP (2009) Human Development Report 2009: Overcoming barriers- human mobility and development, UNDP, New York.

WRI (2004) Lessons from the field: An overview of the current use of information and communication technologies for development, World Research Institute.

ENDNOTES

1. http://hdr.undp.org/en/humandev/reports/
2. Millenium Development Goals

Chapter 1
Network Cooperation:
Development Cooperation in the Network Society

Manuel Acevedo
International Development Cooperation, Argentina

ABSTRACT

Trends in international development cooperation point to the increasing networking of initiatives and programmes, facilitated by information and communications technology (ICT). This allows many more people and organizations from around the world to contribute to a given project, as with the case of online volunteers. There are various types of networks active in development cooperation, but network management needs to be incorporated by involved organizations in order to extract the expected benefits from their involvement. Network analysis practices will help determine if they are set up and managed appropriately.

INTRODUCTION

Human Development in the Network Society

This article is rooted on two central paradigms, 'Human Development' and 'Network Society', and its aim is to explore one of its bridging drivers, namely the international development cooperation system (and its actors), that can contribute to how Human Development can be best advanced in the context of Network Society.

First the objective, which is Human Development. The concept was developed during the 80's by the Nobel laureate Amartya Sen (1999), and spread beyond academia as it becomes embraced by UNDP in the 90's, when it starts issuing Human Development Reports under the guidance of

DOI: 10.4018/978-1-60960-497-4.ch001

Mahbub ul Haq. Human Development is about expanding choices for people, so they can live a dignified life. A seemingly simple, yet highly powerful notion. Behind choices are freedoms, made possible by capacities and empowerment and capacities. Very importantly, Human Development provides the basis for a paradigm shift in development goals, moving fromc*s*. This last point has important implications in terms of information and communications technology (ICT) for Development, such as proactively seeking to open up opportunities and doing so by promoting local talent and capacities—in addition to directly helping to satisfy needs.

Secondly the context, namely the Network Society. The concept was developed by Manuel Castells, which he described as the social structure of the Information Age (1998a, 1998b). It is related to the more popular notions of 'Information Society' or 'Knowledge Societies' (Mansell & When, 1997), but more rigorously constructed and goes beyond the raw materials (information/knowledge) to infer the structural fabric of this stage in society. The Network Society paradigm characterizes new models of production, communication, organization and identity, all organized around and through networks. And while it is possible to speak of specific 'network societies', the combination of both economic/financial globalization and a widespread communications infrastructure provide meaning to the notion of a global Network Society.

In the Network Society, development may be viewed from the perspective of a higher-level connectedness, i.e. moving between inclusion/exclusion poles. Perhaps one of the most troubling consequences of the Network Society context is that exclusion from it amounts to a kind of absolute exclusion. Castells refers to a 'Fourth World' as an isolated and almost invisible realm outside the networks, not delimited necessarily by national boundaries, where people, institutions and entire social groups are connected and unconnected to the Network Society without their control. (1998b, p.335). It is the grand 'socio-economic-everything' exclusion, and while appearing in all countries[1] it certainly affects many more people in the under-developed South.

Therefore we should be asking about the types of elements and factors that favor the expansion of choices/freedoms (and thus inclusion) in a highly networked social context[2]. At this time, ICT is undoubtedly one of those elements to explore.

ICT4D 2.0

The field of ICT for Development (ICT4D) is arguably on the verge of a significant leap ahead. Reasons that point towards a much-accelerated absorption of ICT in development processes include (i) expansion in the ICT market covering less developed countries, (ii) an accumulation of moral imperatives (related to the impact of digital exclusion) and (iii) improved knowledge about what works and a sizeable number of experiences. And it is happening during an explosion of participation-oriented and network-friendly ICTs, grouped under the Web 2.0 label. Heeks (2008) appropriately terms this phase as ICT4D 2.0.

In his view, ICT4D 2.0 presents new opportunities for development processes and actors, as well as sector companies and professionals. It requires a new combination of expertise and world vision. Very importantly, it will be characterized by <u>innovation</u> of all sorts, not only technical but also process-related. By contrast, an earlier ICT4D 1.0 phase would be characterized by the popularization of the Web and its association with the Millennium Development Goals, a proliferation of pilot projects (few of which ever making the transition from the development cooperation 'laboratory' into regional or national scale) and little impact measured with inadequate methods.

ICT4D 2.0 would then be geared towards large scale implantation of the technologies to really finally exploit its much and often touted development impact. It is about putting these technologies to use to confront some of the most severe problems confronting mankind today, grouped in the concept of poverty that the South African writer and Nobel laureate Nadine Gordimer elegantly has described as "*the sum of all hungers*". And to do so with sober, systematic and realistic perspectives, in contrast with the exacerbated optimism and ad-hoc approaches exhibited at times during the ICT4D 1.0 phase.

For cooperation actors to become substantive, even relevant contributors to such a widespread, impact-oriented absorption of ICT, it is suggested that they together with the international system of development cooperation need to undergo significant changes. Moreover, such changes are also related to their adaptation to the Network Society paradigm, in line with the rationale of adequately placing Human Development in the same paradigm.

Therefore, the main argument proposed in this article is that development cooperation should itself transform towards networked cooperation models to best fulfil its purpose to stimulate and catalyze Human Development progress in the socio-economic and technological context of the Network Society. This transformation will include a substantial integration of ICT in the cooperation system/actors to make it effective—it is hard to think of productive networked activities where these technologies are not at the operational core. There is no reason why it should be any different for development cooperation; in fact one of the most recent wave of ICT benefits have to do precisely with collaborative work (the basis for cooperation of any kind). But the notion of network cooperation goes beyond the integration of ICT (which after all are tools serving specific ends), as the following section describes.

Characterizing Network Cooperation

During the process leading up to the first phase of the World Summit for the Information Society (WSIS)[3], the OECD held a forum on the integration of ICT into development programmes on March 2003[4]. In one of the keynote presentations, J.F. Rischard from the World Bank presented his views on the state of ICT mainstreaming, (2003) articulating three progressively higher degrees of integration, and giving cooperation agencies a grade. They are summarized below because they have a direct bearing on the topics hereby discussed, even if the level of ICT usage by these agencies has changed during these five years:

1. **Development solutions by themes or sectors:** He claimed that there had been significant advances and many existing experiences on which to draw, which could be replicated and multiplied. He gave the agencies a 'B', because despite the advancement, applications were often implemented by 'amateurs', applications failed to have tangible impact in the absence of national ICT strategies, and most development professionals were then not sufficiently aware of the role and possibilities of ICT.[5]
2. **Development programmes:** In his view, development cooperation was going through a silent revolution with respect to methods, improvements in the establishment of priorities, and improvements in the interface of the donors. However, it was still rare to find donor programmes and national strategies where ICT had found a place in. Few *Poverty Strategy Reduction Papers* had ICT components, and if so it was treated in peripheral fashion. The development community had not in general been a proponent for (or an actor in) ICT national strategies. Some reasons could be involved, including a certain degree of 'technophobia' in development circles, lack of resources/equipment dedicated to ICT in bilateral agencies, and an insufficient understanding of the knowledge-based economy. Rischard gave this category a 'C'.
3. **Advanced development thinking:** Rischard argued that true ICT mainstreaming in

advanced development schemes required more than thinking about ICTs—it required a change of mentality. He referred to the insufficient understanding of the knowledge-based economy in development circles, based on competitiveness, creation of opportunities and innovation. In turn, the emergence of such knowledge-based economies required in his view an overwhelmingly social vision of development. The supportive economic model would include not only economic structures and incentives, but a fortified educational system with access to ICT infrastructure, as well as collaboration among private sector, government and civil society to create vision, social cohesion and trust. Since this understanding was scarce in development agencies, the incentive to mainstream ICT was poor. Therefore, in this higher-up category of ICT mainstreaming, the agencies' efforts deserved a failing 'D' grade for him.

In essence, Rischard's point was that the development cooperation sector was far from adapting successfully to the Network Society environment. The problems were not so much in the instrumental incorporation of ICT but on the very vision of development held by cooperation agencies, and thus of how they conducted themselves, a view shared by other leading figures in development (Yunus 2004) and outside development circles (Castells, 2000[6]). Since 2003, the situation described in the first two points has improved, with more thematic ICT applications and more professionals in cooperation agencies aware of ICT's possibilities. However, the limitations shown in the last point, the 'higher order' development thinking, have not changed in any significant level.

The necessary vision for development, as proposed in the introduction, incorporates the Human Development paradigm in the context of the Network Society. The goals remain the same ones proposed by Amartya Sen, development as the increase of choices and freedoms. The international system of development cooperation, and not just individual agencies, needs to undergo significant transformation to properly respond to these changes in order to achieve satisfactory performance (some critics might even claim '*to remain relevant'*). Firstly, on its *architecture*, reticulated on variable geometries of nodes, links and systems, oriented towards collaboration and the use of knowledge. Secondly, with a *re-engineering* of cooperation practices, applying network dynamics and tools to projects, management, strategies and policies in order to reduce Sen's *unfreedoms* (*e.g.* poor health, lack of education, ineffectual governance, etc.) and to bring about greater opportunity. A significant informational and technological updating, what sometimes is referred as 'ICT mainstreaming' in cooperation (Annan, 2000) is an essential part of this re-engineering, but not sufficient to arrive at the for the comprehensive renovation. Always keeping in mind, at the risk of being reiterative, the *raison d'etre* for this renovation: to help achieve more opportunities and a better quality of life for the beneficiaries of cooperation actions.

To draw parallels that illustrate the nature of these changes, similar change processes have already advanced significantly in the private sector with e-business, public administrations with e-government or the educational sector (particularly universities) with e-learning. It is thus relevant to reflect over what an '*e-cooperation*' would look like, which by virtue of its structuring in and around networks, could be called 'network cooperation'. The following attributes would broadly characterize such a cooperation model:

- Incorporation of technological, institutional and human networks as *networks of production* in the *day-to-day* of cooperation (and not merely as contact networks). Networked operations would characterize

classic cooperation instruments such as projects or programmes.
- *The mainstreaming of ICT* into aid agencies' corporate tasks (internal dimension) as well as for field-related activities (external dimension) to improve the efficacy, understood as "effectiveness + efficiency" of cooperation actions. (Acevedo 2005)
- Emphasis on *knowledge management* (its generation, absorption and dissemination) as a constituent pillar of cooperation strategies and actions.
- Redesign of development project to become *networked projects*, where participation in this kind of project is not limited to its defined geographic space, but on spaces defined by information flows and knowledge generation, deliberately provoked and supported by the project management team.
- Promotion and channelling of *social capital* through networks, particularly through the increasingly popular web 2.0 applications. (Van Bavel et.al. 2004, Acevedo 2007,)
- Incorporation of *multi-stakeholder actions* for development, involving government, civil society, private sector, academia, etc. to leverage more numerous/diverse resources, as well as the sharing of responsibilities. The newer public-private alliances for development are examples of these.

Advocating a networked approach to cooperation is not new. Fukuda-Parr & Hill (2002) claim that networks among development practitioners and access to global knowledge systems can substitute for conventional models of technical cooperation. Nath (2000) proposed that knowledge networking need not be confined within the closed boundaries of information flows as it has the potential to evolve as an alternative institutional model for development promotion—suggesting that if traditional cooperation organizations are not good at managing knowledge, other actors will emerge to promote development.

In fact, cooperation practices are gradually becoming networked, most often on the side of civil society organizations. Many development actors form part of networks because it brings benefits to them, such as enhanced knowledge generation, increased potential for participation of people and entities in concrete activities, and improved productivity for certain joint actions. The use of Internet and specialized software (e.g. groupware), particularly with web 2.0 tools widely available, extends possibilities for already existing human/institutional networks as well as for the creation of new ones.

Networks of development practitioners across the globe are emerging, sharing relevant knowledge, information and experience from good/bad practices. As networks provide new modalities for information access, capacity building and knowledge acquisition, they help to overcome some of the failures of conventional development cooperation, like depending on donor-established channels for knowledge access and the faulty notion of the expert-counterpart model from North to South. Networks connect development actors in different sectors and project areas, fostering collaboration among individuals as well as among institutions. Properly devised and implemented, a development network provides its own source of support, and a superior one to that provided by a few designated experts.

Where significant added attention is required is in the treatment of various network-related aspects that can both facilitate (enable) and promote (publicize) advances for the reticulation of cooperation systems and agents. These aspects include network *typology*, *management* and *analysis*. These are examined below, but first we look at the design of development projects to illustrate immediate applications of a networked approach to cooperation.

The Development Project as a Network

Whereas knowledge networks are making notable inroads in development cooperation, the role of networks for delivering *products* in cooperation practices is much less advanced. Let us consider, for example, how development projects operate.

Most development projects are designed in ways that have not changed in decades. They seek the achievement of a set of objectives in a given physical location and time, and most of the participants are local staff, from the implementing organization together with some specifically hired to work in the project. The involvement of outside personnel (national or foreign) is minimal, often for training or evaluation purposes, including also a programme officer at the donor agency. The projects act in relative isolation, with sparse contact with other projects even within the same country or supported by the same donor agency. The results and knowledge gained in the projects are rarely applied outside of them.

By contrast, let us now illustrate how a car is made by any of the big multinationals in the automotive sector. A car is manufactured from thousands of components, produced in various countries (and with raw materials which may come from some additional ones). The engineering work is done at some central company labs, while assembly work can be carried out in factories in other places. Marketing, transport, sales and financing will increase the number of countries involved. There are numerous people involved along the way, and knowledge/experience serve as valuable assets which are carefully managed and exploited. The end product (the car displayed in the showroom) has proper levels of quality, price and safety which satisfies the consumer. The overall operation meets the objectives of the car maker: to sell cars and make a profit.

That car is the product of a highly networked process, involving the participation of a significant number of nodes around the world, using open and constant communications via multiple channels and following carefully designed specific procedures and protocols. ICT are utilized in every step and to connect the steps. The project, in comparison, involves few nodes in few countries, communication is intermittent, and activities take place in relative isolation. Use of ICT is far less pervasive. To be sure, there are development projects much more advanced that the ones depicted above, but they are a minority. Development projects today, by and large, are essentially un-networked.

It is possible to conceive a project as a network, designed over a set of networked nodes with well-defined individualized as well and collaborative tasks. These nodes channel the resources and carry out the necessary activities to achieve the project's objectives. By designing a project with network structures and dynamics in mind, it can allow for a significantly higher level of participation and inputs, many of them from outside the immediate vicinity of the project.

Figure 1 illustrates such a multi-faceted involvement of persons and organizations (whether in the project's central location or across the world) in a project network. Many of the participants would collaborate online, either as individuals (e.g. an online volunteer) or in teams (e.g. to prepare a publication). A team could be formed within a given organization (e.g. a group of technical specialists at a university), or set up as virtual teams across organizations (e.g. to design, host and manage the project's web site). The types of actors/nodes in this example include, besides the direct project staff (not shown in the sketch) [7]:

- individuals involved in a knowledge network (e.g. a given virtual community of practice) linked with the project themes;
- online volunteers participating on an individual basis, without institutional representation or affiliation (and 'indirect' volunteers that could support those 'direct' project OVs);

Figure 1. A network architecture for a development project

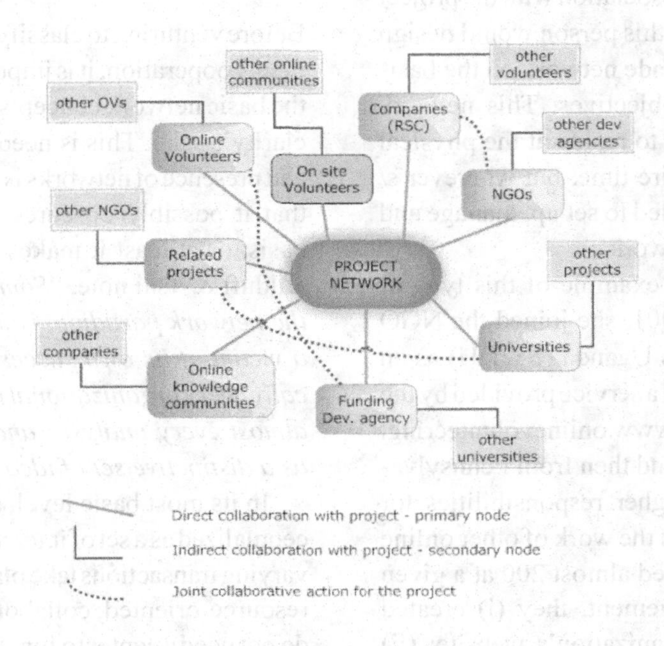

- staff/stakeholders from other related projects;
- members of one or more NGOs with thematic expertise or other direct interest in the project issues;
- corporate volunteers from one or more companies
- students/professors at one or more universities;
- onsite local volunteers, and
- staff from a different cooperation agency than the one(s) sponsoring the project.

Variable network geometries with innumerable configurations can thus be established, particularly as second-level collaborations emerge (shown in beige-colored rectangles in the figure). The key is to intentionally design and energize any such network, which means, among other things, that the project would have to be conceived with a network vision in mind.

Development projects around the world sometimes incorporate network functions, but in ad-hoc fashion and without explicit network approaches. Yet there are few impediments to designing and managing projects in this way. Lack of network expertise and traditional process inertia are the biggest obstacles to overcome. Much potential benefit can be yielded to a specific project (and thus to the people those projects are serving) in the form of substantially higher number of people involved, many likely in volunteer fashion. Not only they can bring in more knowledge, technical and even financial resources, but the social capital generated from this expanded 'project community' will contribute to the sustainability of its effects.

One of those impediments is that project directors usually have enough work in their hands, even with spatially/staff-wise compact projects, to consider juggling a significantly higher number of actors. This is particularly true when the project was not networked 'by design'. Development agencies could contribute to this shift in project architecture by incorporating 'network specialists' to their teams. In close association with the project director and other staff, this person would design and construct a tailor-made network on the basis of project needs and objectives. This network specialist may not need to reside at the physical project location the entire time, but wherever s/he would be better enabled to set up, manage and energize the project network

Laurie Moy was an example of this type of network manager. In 2001, she joined the NGO People with Disabilities Uganda (PWDU) as an online volunteer through a service provided by the UN Volunteers agency (www.onlinevolunteering.org). First from Texas, and then from Pennsylvania, Laurie assumed higher responsibilities for PWDU, and coordinated the work of other online volunteers, which reached almost 200 at a given time. Under her management, they (i) created and maintained the organization's website; (ii) published a monthly newsletter, (iii) developed project proposals; (iv) marketed the organization; and (v) raised funds (even putting together a benefit music CD with professional musicians). Laurie's husband was a professional therapist in a hospital, and through him she was able to involve health professionals and technicians from that hospital as well, i.e. the second level (and even third level) collaborations earlier mentioned. She later visited the NGO in Uganda, personally meeting its director, staff and beneficiaries, and establishing the library at PWDU during that visit. Her efforts made her a recipient of one of the first Online Volunteer of the Year Awards given by the UN since 2002[8]. She later established her own NGO, Pearls of Africa, (www.pearlsofafrica.org) also to benefit men, women and children in Uganda living with disabilities; and, as could be expected, she involved many online volunteers in Pearls of Africa as well.

Types of Networks for Development Cooperation

Before venturing to classify networks for development cooperation, it is important to briefly specify the basic network concepts used in this article, for clarity's sake. This is needed because the notion and presence of networks is so pervasive nowadays that it possibly obscures their meaning or real basis, or at least it makes them rather fuzzy. As Kilduff & Tsai note, "*Sometimes it appears that the network paradigm is in danger of becoming a victim of its own success—invoked by practically every organizational researcher, included in almost every analysis, and yet strangely absent as a distinctive set of ideas.*" (2008, p.9)

In its most basic level, a network can be conceptualized as a set of interconnected nodes, where varying transactions take place (communicational, resource-oriented, collaborative, etc.). Networks do not need a center to function properly, although they often exhibit one. Networks have higher efficiency than other organizational schemes (e.g. linear, hierarchical) because of their flexibility, modularity and agility.

Each node and connection can exhibit different characteristics. For example, the connections or links vary widely in intensity, from 'strong' links which indicate a frequent and productive relationship among nodes, to 'weak' links which essentially manifest mutual awareness (Wellman, 2001). Nodes can vary even more widely in terms of importance, responsibility and relationship. For example, a large node o set of closely space nodes can be considered an aggregator or hub

Networks have long existed as a form of organization and relation, in human society and much longer in nature (e.g. a bee hive). But in our days, in the Information Age, they have gained preponderant when enabled and powered by digital ICTs.

There are myriad ways to classify and distinguish among networks. In Moreno et. al. (2006) there is a comprehensive list of network types drawn from development bibliography. The present article only touches on a few categories, with the underlying common criteria that they are relevant for the comprehension of the functionality that a network can have for development organizations. For example, attributes of maturity or geographic location are not directly related to the kind of purpose that a network will have in a cooperation context.

The first classification useful for development cooperation is by social networks (among individuals) and organizational networks (among entities or organizations). Even if there is an ambiguous confluence space among the two, an organizational network is arguably more relevant for cooperation. Organizational networks have explicit purposes for the nodes to come together, a range of varying relations among the nodes, a structuring based on degrees of task decentralization and a relatively democratic decision-making. An organizational network often possesses a productive nature, or said a different way, it cannot be a mere set of contacts. These networks must produce something (e.g. a campaign, a project, reforms to a law), as a whole or through the multiple sub-networks that can be inscribed in them.

A second useful classification is morphological, separating networks essentially by structural complexity. A practical and easy way to characterize them is in terms of 'dimensions'. There are implications of morphology for development networks, essentially for their management purposes, as will be discussed ahead. (See Figure 2.)

A third classification is by functionality, i.e. what purposes they serve. The types identified are directly related to development cooperation and solidarity. Some appear more frequently than others (knowledge vs. online volunteer networks). For space limitations, they are only briefly described (omitting their structural models, operational methods and comparative advantages).

- *Institutional networks.* This is a generic term, similar to the earlier category of 'organizational network'. Most development/ cooperation networks will be under this umbrella. Their common denominator is

Figure 2. Morphological classification of networks by dimensions

1-D (linear); this topology does not differ much from non-networked structures. Its members are linked lineally, and communications must flow through the adjacent node before proceeding to the next one. The most common type would that depicted by the graph on the right, in a radial or star profile, where nodes are linked to a central node through which most of the information and transactions flow.	[9]
2D (planar); in planar networks, nodes can establish relationships with a nearby node (i.e. in terms of importance, not distance) or the center. This implies that some transactions could take place without involving the center. They present mesh geometries of varying complexity. Two such types are shown in the figures on the right.	
3-D (spatial); these networks present the highest degrees of freedom and flexibility in terms of relations, because any node can connect to any other one, regardless of importance. A way to visualize this network would be with the pipes of a refinery, which connect different items at different planes, sometimes going through various items and other times connecting two items directly. The figure on the right shows an approximation to this type of network. While the central node is still important, contrarily to 1-D and 2-D the majority of transactions at any given time will not pass through the center (in the sketch, signified by the red cube).	[10]

that they are established and maintained by one or more development organization. At a minimum, they share resources and information, and facilitate collaboration among its members. Some of the possible configurations under this broad category are:

- *among development actors*; for example, the bilateral cooperation agencies that form part of the OECD's Development Assistance Committee (DAC), or civil society networks formed to improve Third Sector participation in the complex WSIS process (McIver, 2003).
- *by individuals within a single entity*; cooperation organizations with a with a significant size, like some UN bodies (UNDP, World Bank, WHO), large NGOs/ Foundations (Oxfam, Rockefeller), or national agencies (like the German GTZ or the Spanish AECID) have a high volume of human resources, considering staff, collaborators and volunteers. They can therefore promote internal networks to help collaboration and contact among people who could be physically far apart. Usually, corporate intranets are used as the platforms. Examples are the 'virtual communities of practice' pioneered within the World Bank about a decade ago.
- *hybrid*, involving individuals that are outside the agency(ies) with others inside. An example is the UNDP Democratic Governance Network which involves outside governance specialists together with UNDP staff.
- *Knowledge (thematic) networks.* They are effective means of capacity building, and tend to be focused on some development topic of interests to its members (environment, microfinance, etc.). Often they are established by an organization (i.e. the aforementioned UNDP Governance network) but they can be outside organizations, using available Internet-based tools (discussion lists, Facebook, Yahoo Groups, etc.). The latter have valuable autonomy when discussing issues and making proposals, since they represent collective judgements outside direct organizational influence. The concept of knowledge networking can indeed go beyond capacity building into the more comprehensive notion of capacity development which aims to bring about changes in society as a whole (Brown, 2002).
- *Project networks.* They link development projects that have something in common (e.g. its subject matter or geographical domain). These networks can be made of projects from the same organization or from various agencies. Inadequate communications and collaboration among projects that could share resources means in practice to reduce the optimization of results in the cooperation sector, but it occurs rather extensively.
- *'Open Source' networks.* They are inspired on the ways of producing free/open source software (FOSS), like GNU/Linux, Apache, Mozilla Firefox, OpenOffice, etc. FOSS networked operational methodologies are effective and sophisticated, and can be good references for networked cooperation actions. Applied to development, these networks are composed of volunteers with a high level of technical knowledge and passion about a particular theme. They are oriented towards the generation of products (in this they differ from knowledge networks) and have well-defined structures that allows for the collaboration and decentralization of tasks for people is dispersed geographical locations. Although they are not much present in the cooperation sector yet, UNDP has expressed that the 'open-source' approach to development work is very promising (Denning, 2002).

- *Networked projects.* They were discussed earlier, and refer to development projects designed with network functions in mind. Because of their need to produce concrete results, these project could uptake some of the operational methods from free/open-source software, as alluded to in the previous point.
- *Diaspora (migrant) networks.* Diasporas are migrant communities whose members are immigrants residing in other countries and relatives/friends that remain in the home country. They tend to form well-structured networks, and generate social capital while pursuing common interests (e.g. lowering of the costs of remittances home, family joining rights in country of residence, etc.). They can include migrants of a single nationality (e.g. Peruvians in Spain, the U.S. or France), or various nationalities (e.g. Latin Americans in Spain). These networks are becoming relevant actors in so called 'co-development' schemes, and sometimes have direct economic effects by promoting active relationships among immigrant entrepreneurs with companies of their country of origin. For example, the UN ICT Task Force launched 'Digital Diaspora Networks' during the WSIS process (2003-2005) [11] to seek more opportunities for ICT entrepreneurs, trade and innovation in developing countries.

As can be readily drawn, actual cooperation networks combine some of these functional typologies. For example, the 1997 Nobel Peace Prize laureate 'International Campaign to Ban Landmines' (http://www.icbl.org), with over 1400 NGOs, incorporates elements of organizational as well as thematic networks, while its structure is reminding of a networked project. At any rate, for a detailed characterization of a cooperation network it is necessary to combine its description (morphological, functional, etc.) with its modus operandi, using network analysis methodologies described later on. But before discussing ways to analyze these cooperation networks, it is worthwhile to explore methods to manage them, which is treated next. (See Table 1.)

MANAGING DEVELOPMENT NETWORKS

Development and cooperation actors join networks, or structure themselves as networks, in order to obtain benefits, among which the ones below are often mentioned (UNSO 2000, Siochrú 2005, Mataix et.al. 2007, Moreno 2006):

- Shared use of resources (financial, knowledge, technological);
- Improved access to information;
- Enhanced collaboration potential;
- Orientation towards decentralization;

Table 1.

School Net Africa (www.schoolnetafrica.net)
As its name indicates, it is a networked organization. Its purpose is to improve the access, quality and efficiency of education with the use of ICT. Launched in 1999, it has become one of the main African educational networks, including national nodes from 31 countries. Through its national nodes it carries out projects to introduce ICT in groups of schools (mainly in secondary or high school). On the other hand, it functions like an organization that provides services, resources (information, advise) as well as educational contents. They involve teachers, students, politicians and pedagogues. *SNA promotes and facilitates the collaboration among peers, for example with teacher networks. Its supports the generation of African contents for and by the schools. It acts as a political agent when it participates in development fora and seeks to mobilize resources for its affiliates. In terms of typology, it has characteristics of a thematic network (about ICT for education), institutional (its nodes are various organizations) and of a project network (supporting national projects), with a high degree of cohesion and equivalence among its constituent nodes. In turn, some of the national nodes are networks themselves.*

- Mutual support and risk reduction;
- Higher credibility (mainly among smaller organizations);
- More structural flexibility (and operational agility); and
- More effective representativeness stemming from higher interlocution capacity.

These benefits do not occur spontaneously, and if so they will be limited and ad-hoc; in other words, networks do not self-manage. It is necessary then to employ specific network management styles. The challenges of network management are substantial because, with few exceptions,[12] it is a dimension of management with little tradition or scarcely in place. Network management among most development/cooperation actors is still in its early phases, and there is much to learn yet about how to work well in networked environments.

When a development organization makes a conscious effort to operate as network or become an active part of one, it will benefit by defining a corporate strategy to help manage the almost certain institutional changes needed as well as the emerging network itself. The strategy needs to be aimed at obtaining explicit results/objectives, and take stock of the existing organizational context. Four key elements of such a strategy are suggested, depicted in Figure 3 and briefly described below (Moreno et. al. 2006).

a. **Architecture:** a specific organizational network structure needs to be designed, one that identifies members (nodes, hubs) and their intended flows and relations (links). Such an organizational architecture should be conducive and coherent with the gradual emergence of a corporate networked culture favoring collaboration and horizontality.
b. **Processes:** they refer to the working procedures or methods to be implemented, or the modifications to existing ones, aimed at favoring operational reticulation. It should be applicable both internally (the organization as a network), e.g. how tasks and projects can be taken up by people across specific units; as well as externally (the organization as a node in a network), e.g. what resources it is willing to share and how. Key among such processes is knowledge management.
c. **Tools:** the instruments or resources needed to implement the strategy according to the

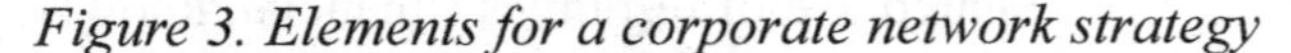

Figure 3. Elements for a corporate network strategy

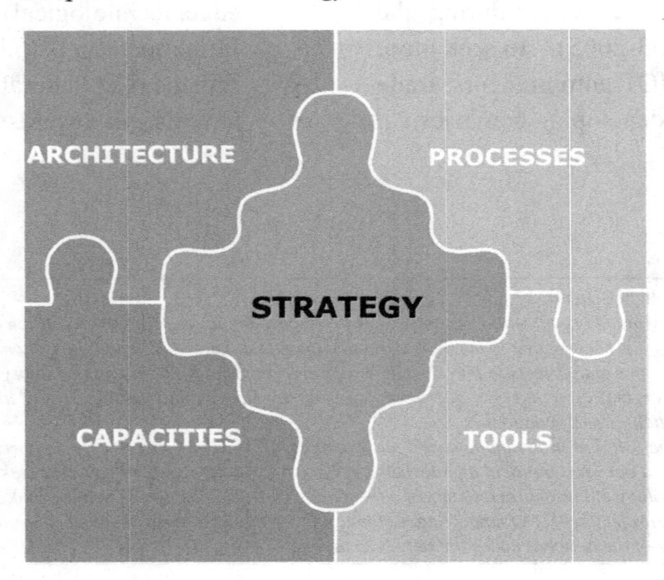

selected processes. Among the essential instruments for a successful networked strategy will certainly be the ICTs required, but there are others: financial, physical facilities, events, evaluations, etc.

d. **Human Capacity:** This element refers to the most essential organizational instrument, its human resources. Specific individual and collective capacity gaps relative to tools and processes should be identified and a plan drawn to address them. This can range from training on digital collaborative platforms (i.e. groupware) to team building exercises.

Special attention to so-called 'generative capacities' should be heeded (Moreno et al 2006). These are the capacities for (i) learning[13], (ii) systemic vision, (iii) new leadership styles, (iv) collaboration, and (v) feeding back to the organization. Networks are ideal environments to foster generative capacities, as long as they favor sharing and collective commitment. And in a feedback mode, such capacities also help to construct more creative and productive networks.

Effective network management requires and fosters participation, starting at the stage of determining its strategy. By the very nature of a network as a highly participatory context, the process of crafting its strategy should be open and participatory as well. Such an philosophy will not only result in a better strategy, but the process leading to it will be already a practical exercise in common decision making—a very useful skill when working in networks. Also, since there are not many 'guidebooks' available to safely follow in terms of setting up development networks, will be learned along the way through trial and error. In this regard, an open organizational attitude will help provide a more nurturing learning environment on which to support progressive networking practices.

Enabling vs. Representational Networks

Chosen strategies, operational culture or both can lead to important differences among management aspects of networks. One key difference relates to the ultimate purpose of a development network. Let us illustrate by examining the tendencies favored by two differing morphological types, 2-D (planar) and 3-D (spatial). We assume that the structure of a network is shaped by its objectives, so that a number of functional differences can be discerned between the two. Then we advance a hypothesis in terms of their relative adequacy for network cooperation approaches. (See Table 2.)

From these differences, it is suggested that planar network morphologies are more adequate for 'representational' networks, whereas spatial ones are more conducive to 'enabling' networks. A representational network serves mostly to advocate and negotiate on behalf of its members, advancing their commonly agreed interests (and defending them when threatened). On the other hand, an enabling network seeks to strengthen the capacities of its members to achieve their individual objectives, particularly their generative capacities, via collaborative tools and practices within as well as outside the network. Success for representational networks would thus be a function of joint actions undertaken by the network on behalf of the member nodes. In the case of enabling networks, success would rather result from the aggregation of collective activities undertaken by member nodes and supported by the network.

It can be argued that enabling 3-D networks offer more adequate environments in relation to the network cooperation approach hereby described than representational 2-D networks. One reason is that they are focused on strengthening each one member in relation to the member's own objectives, primarily by providing tools/methodologies that favor the free collaboration of the members within the network (and also outside of it. This

Table 2

2-D – planar geometry	3-D – spatial geometry
The central or principal node acts as the network **coordinator;** it largely determines which nodes will carry out particular functions/actions, and will know about these actions them in advance.	The main node (if there is one) acts as a network **dynamizer**, providing resources and tools to favor reticulated work among any of the nodes.
Procedures are very important: network operations are based primarily on a series of norms or protocol that give order and regulate the network's activities.	Network operations proceed in **ad-hoc** fashion, given the freedom and ease to establish productive relations among nodes, respecting a few basic institutional behaviour norms.
Planning for the network is very important, since the central node should orient resources and efforts towards their implementation.	**Periodic monitoring** is essential to know how the network is functioning, since it is not possible to plan all the possible collaborative activities among nodes.
The network prioritizes **access to information**, and specifically the central node fosters the availability of the information and provides access systems.	The network prioritizes **access to knowledge,** through the communication among nodes, the relationship with external entities and the systematization of information. The main node (if there is one) works on shared criteria for knowledge management, prioritizing the provision of tools/services that facilitate the effective and efficient managing of such knowledge.

generates a greater volume and range of collective, network-powered results. Another reason is that enabling networks take more advantage from the possibilities of networked collaboration, empowering the nodes to use reticulated work methods to suit their specific purposes. Moreover, it encourages members' initiative and responsive attitudes to the fast-changing environment of development cooperation, explicitly recognizing limitations on how well a coordinating or central node can effectively and efficiently orchestrate the collective capacities of the nodes. Finally, it takes more advantage of some of the differentiated traits of networks, such as relational freedom and flexibility—embedding hierarchical practices into reticulated structures reduces the latter's possibilities. In synthesis, it is recommendable to prime the creative and collaborative freedom inherent to development networks to take full advantage of their flexibility and agility, learning from them in a continuous action/feedback process.

Many development/cooperation networks exhibit both profiles, activating desired traits as needed, though normally one of them carries more weight, i.e. a network tends to be more representational than enabling, or viceversa. For example, a research study (Moreno, et.al. 2006) found that the *Spanish Confederation of Development NGOs* (*CONGDE*) acts as a network that more often resembles the representational approach, providing relevant information to its members, carrying out campaigns and acting as an important interlocutor to the Government. *SchoolNet Africa*, on the other hand, is an educational network whose priority is to strengthen the capacities to integrate ICT in education for its various nodes/members: teachers, schools and the national SchoolNet chapters. It acts mostly in an enabling way by providing services to its members, facilitating collaborative mechanisms for them to undertake joint actions, and opening opportunities for their participating in new collective projects.

Whether enabling networks better serve development purposes in general or only for particular occasions, it is worthwhile to consider how to transit from 2-D to 3-D network environments, as the latter is more complex[14] and less customary. It will depend to what degree the nodes actively participate, take ownership of their network environment and transform it in a source of resources and services to both 'reach and enrich' their objectives (Mataix et.al. 2007). Network management practices can take deliberate steps to foster such a 2-D to 3-D (or representational to enabling) transition through actions such as:

- Inducing flexibility in the modes of participation;
- Building generative capacities for effective networking;
- Applying knowledge management techniques for internal and continued reinforcement of reticulated practices; and
- Helping to establish networks for development projects.

Thus, if earlier we alluded to individual generative capacities as a strategic aspect of network strategies, we could now refer to a kind of 'collective' generative capacity to dynamize a network.

The best network management or strategy needs to be properly validated, otherwise they can remain attractive institutional exercises with no clear returns. The definitive measures of success will undoubtedly come from the results produced, both for the entities and in development terms. However, a previous step in assessing the success of a network, and possibly a decisive one, is to know whether we are indeed constructing and managing the kind of network we had in mind. For this, it is necessary to analyze the networks themselves.

Network Analysis

Network analysis practices can be used by an organization to assess and understand how it is really functioning as a network, or by a group of organizations in a network to understand how they are collectively doing. Most network analysis work has been directed to social networks, though some prominent authors also explore organizational networks from an analytical perspective. Castells is among them, examining the effect on organizational approaches derived from the influences of the digital revolution. This section merely touches upon possibilities made available by network analysis, as even a introductory treatment would extend beyond the possibilities of this article. But most importantly, it proposes that network analysis is not only useful but probably the only objective way to determine whether a development network is functioning as expected and what changes can be made, mainly in the relationships among the nodes, to improve its performance.

Network analysis methodologies serve to understand the relations among individuals, organizations and other social entities. It is based on the functional structure of the networks rather than on the attributes of its nodes. For example, for the case of a NGO manager the analysis would focus more on the patterns of social and professional relations (i.e. who does she interact with) than on personal characteristics of the staff like age, nationality, political orientation, etc.

According to Anhier and Katz (2005), there are two main categories of organizational networks that summarize their wide existing spectrum. One is that of 'simple networks', which examines the relations among nodes based on five parameters:

- **Cohesion:** characterizes the interconnection of social relations and their tendency to form areas of high relational density (hubs) where there are higher probabilities for links to exist or develop.
- Equivalence: describes to what extent the members of a network have similar relations with others, which helps to find zones or bands that facilitate the analysis by studying the relationships among those zones.
- **Prominence:** identifies the positioning of nodes in relation to others more or less prominent, which serves to visualize power relations.
- **Bridge:** identifies nodes that connect groups of nodes (or networks) not connected through other links or paths. Bridging nodes are important to understand infor-

mation flows and mobilization processes among groups.
- **Agency:** refers to situations in networks in which an actor observes the possibility of connecting empty spaces or nodes, which helps to characterize the enterprising role of some nodes in the establishment and interconnections of networks.

The other category proposed by Anhier and Katz is that of 'hypernetworks' or 'dual networks'. They imply the deduction of links through a coincidental participation in groups or events. For example, it can be assumed that between two NGOs participating in the same conference there is some type of link. The analysis of hypernetworks is appropriate in the case of relatively uncommitted affiliations to broad organizations (like a national federation of NGOs). This approach was used by Anhier and Katz to analyze NGO participation in the 2004 World Social Forum held in Mumbai, based on identifying which NGO got involved in which specific acts.

A different approach, better suited to assess the functional effectiveness of organizational networks comes from Arquilla y Ronfeldt (1999), and it appears well-suited to the development cooperation sector. It includes 5 different fields, which these authors hold should each be well formulated and integrated with the rest to achieve an effective network design in relation to organizational objectives:

- **Organizational:** its design or structure as an organization.
- **Narrative:** the story and messages it tells.
- **Doctrinal:** the strategies and collaborative methods.
- **Technological:** the information systems it uses.
- **Social:** the personal links that ensure loyalty and trust.

These two approaches are different but complementary. The orientation of the analysis of Arquilla and Ronfeldt is more *operational*, responding to the question, "*does the network behave according to predetermined purposes?*" Anhier and Katz's is more *diagnostic*, addressing the question "*why does this network function the way it does?*"

For example, an organization can observe as a node the network(s) in which it participates from the perspective of the 3-D enabling network mentioned in the previous sections, in order to ascertain how it is helping it to access to new/fortified relationships, access knowledge and management resources, and implement new and complementary styles of work. Using Anhier and Katz' parameters, it would be asking itself questions such as:

- Cohesion: *"Is it interesting for us to approach groups of NGOs which are very proactive in their relationships?"*
- Equivalence: *"Are some of the needs of our organization shared and already satisfied done by others in the network, and are the steps to be taken clearly described?"*
- Prominence: *"Do we want to have a active role in decision-making in the network?"*
- Bridge: *"Do we have links to NGOs in other networks we could activate for the members of this network?"*
- Agency: *"Do we want to lead or push new actions within the network?"*

One of the reasons that network analysis is well suited for development cooperation is that it is particularly useful when describing civil society groups at the international level, such as the participation of civil society organizations in the WSIS multi-stakeholder process (McIver 2003) or the 2004 World Social Forum in Mumbai[15] (Anhier and Katz 2005). Civil society beyond the local context (national, regional, global) is a highly relational phenomenon with at least an

Table 3.

International Open Source Network (http://www.iosn.net/) The IOSN is a international center of excellence for free/open-source software (FOSS) in the Asia-Pacific region, founded by UNDP. It has a small secretariat based at the UNDP Regional Centre in Bangkok and three centres of excellence in Manila, Suva and Chennai. Its objective is to facilitate and network FOSS advocates and human resources in the region, because it believes that developing countries in the region can use affordable yet effective FOSS ICT solutions to bridge their digital divides. For that purpose, it provides access to information/knowledge resources, conducts studies, participates in international and national events, and organizes trainings and awareness building seminars. It is interesting to compare it, from a network analysis perspective, with the example of the SchoolNet Africa Network (SNA) mentioned earlier. IOSN functions more as a point of encounter, while SNA had a stable role as implementer/coordinator of concrete projects. (IOSN occasionally also runs projects). IOSN, just as SNA, provides information and knowledge resources on its topics of interest. It could be deduced that knowledge management is IOSN's operational backbone, since it essentially works to generate and transmit knowledge. It typology is complex, since it is part of a large organizational network (UNDP) with a strong thematic orientation, integrated by highly variable nodes (from individual experts to government agencies, even companies – thus it would show low levels of equivalence. However, it has a strong cohesive character (it creates specific hubs) and some of its nodes have a high degree of bridge and agency attributes.

implicit reticulated nature. In addition, this type of analysis pushes aside the so-called 'methodological nationalism'[16], the tendency of social sciences to stick to statistical and conceptual categories proper of the nation-state, in contradiction with the realities of a globalized world. Finally, it simplifies the complexity of social relations to reveal standards and underlying trends.

To be sure, network analysis techniques still need to advance significantly and be widely tested, not only to provide meaningful results but even more important to be practical and applicable to a majority of networks and networked organizations. For example, other factors need to be added to results of network analysis to round up the understanding of a given network. One refers to its level of *maturity*, i.e. if it is in an early, advanced or declining phase in its evolution. Another one is its *geographical* framework, which includes the places where the nodes are represented (its socio-politic environment at home) as well as where they exercise their solidarity (whether internationally, their own countries, etc.). Ultimately, what matters is that these methods contribute to help organizations in forming and managing proper operational networks and to improve their reticulated way of working (internally and externally).(See Table 3.)

CONCLUSION

Development cooperation is only a small parcel within the field of development. Its main effect, particularly in technical cooperation, should be to catalyze change at a small scale and support national actors in their adoption and implementation of national development policies. It concerns itself with mitigating poverty, the most pressing issue confronting humanity today, and advancing human development processes beyond poverty. For those reasons, and since they are mostly using public funds, it clearly needs to operate with the maximum possibly efficacy.

For development cooperation to try to make good on its stated purposes, it needs to use available resources (always insufficient) and implement the most effective methods. There are periodic waves of controversy about the value of international cooperation and debate on whether the volumes of financial aid have been properly used and resulted in the intended benefits—many factors play a part, often the debates are too superficial and politically-motivated.

But there is little doubt that under the hood of the international development institutions, there is a highly complex system of development cooperation which can use a good tune-up. In the age of nearly-ubiquitous communications it is not rare that various development projects in the same place are not aware of the others, or that different departments within the same aid agency do not know what the others are doing. To be sure, this sometimes occurs also within 'simply' national Public Administrations, so it is no wonder that when adding a multi-country dimension there are added constraints for collaborating, sharing information/knowledge or coordinating actions. The challenge for the tune-up resides in which parts and what connections to concentrate the work on.

In this article we proposed two main aspects to be considered in that tune-up. One refers to the overall engine, and it is that networked cooperation models can better serve to advance Human Development processes in the context of the Network Society. This relatively straight-forward idea is derived from observation and documentation (i.e. it is already happening), but its implementation appears rather difficult, for a combination of three main reasons. First because it is ICT-intensive, though that is not too hard to overcome, and in fact can help to bring about the benefits described when discussing the ICT4D 2.0 phase. Secondly, because it involves not only one's organization but others as well, so control is elusive. But thirdly, and possibly the most important reason, lies in the changes (institutional, working methods) involved in incorporating network dynamics. To assuage those difficulties, some basic notions about network management were mentioned. The article merely highlights and threads basic points related to networked methods for cooperation, it is beyond its scope to treat them in more depth.

The other point for the tune-up is more micro, referring only to a sub-system of the car, and with the caveat that there is little evidence yet to support it (i.e. it is limited to qualitative observation by the author). It is the idea that enabling networks are in general best suited to a networked approach to development cooperation, when compared to networks that are more representational or centralized in nature. This draws from the arguments that (i) network dynamics favor collaborative and creative freedom, (ii) the member organizations are individually strengthened (with 'generative capacities') for their existing objectives and thus feel better supported, and also that (iii) collaborative results/effects have tendencies to increase with the size of the network, somehow related to how the value of a network grows as per its number of nodes - assuming it is properly managed.

On a final note, successfully achieving development targets such as those set by the MDGs or at national level requires a healthy combination of elements. Neither financial resources nor the proper political will can by themselves successfully respond to poverty and developmental challenges. The participation of lots of people and many organizations (NGOs, companies, universities, etc.) is required, i.e. the social capital that binds societies together. And today networks are the new greenhouses of social capital, nurturing knowledge, empowering people for action and organizing their collaboration. That is why properly tending for development networks can have a profound and widespread effect. Cooperation actors like bilateral aid agencies will be well advised, besides incorporating network approaches themselves as we advocated in the article, to increase direct investments and support for development networks around the world. It is a new and promising avenue for institutional strengthening.

REFERENCES

Acevedo, M. (2005). Las TIC en la Cooperación al Desarrollo. In *La Sociedad de la Información en el Siglo XXI: Un Requisito para el desarrollo – Vol II: reflexiones y conocimiento compartido* (pp. 44-66). Madrid: State Secretariat for Telecommunications and the Information Society, Ministry of Industry, Turismo y Comercio, Spain. ISBN 84-96275-09-4.

Acevedo, M. (2007). Network capital: an expression of social capital in the Network Society. *The Journal of Community Informatics* [Online], *3(2)*. Retrieved 18 Nov 2007 from <http://www.ci-journal.net/index.php/ciej/article/view/267/317>

Anheir, H., & Katz, H. (2005). Enfoques reticulares de la Sociedad Civil Global. In F. Holland, H. Anheir, M. Glasius, & M. Kaldor (Eds.), *Sociedad Civil Global 2004/2005* (pp. 221-238). Translated by José Luis González (original title: *Global Civil Society 2004-2005*). Barcelona: Icaria Editorial. ISBN: 84-7426-823-0.

Annan, K. (2000). *We the Peoples: The Role of the United Nations in the 21st Century. Millennium Report*. New York: United Nations Dept. of Public Information.

Arquilla, J., & Ronfeldt, D. F. (1999). *The emergence of noopolitik: toward an American information strategy*. Santa Monica, California: Rand Corp.

Brown, S. (2002). Introduction: rethinking capacity development for today's challenges. In Browne, S. (ed.), *Developing capacity through technical cooperation* (pp. 1-14). London: Earthscan Publications (for UNDP). ISBN 0-185383-969-99.

Castells, M. (1998a). *The rise of the Network Society (The Information Age: economy, society, culture; vol.1)*. Oxford: Blackwell Publishers. ISBN 0631221409.

Castells, M. (1998b). *End of millennium (The Information Age: economy, society, culture; vol.3)*. Oxford: Blackwell Publishers. ISBN 1-55786-872-7.

Castells, M. (2000). *Information Technology and Global Development*. [en línea] New York: UN Economic and Social Council (ECOSOC). Keynote address, ECOSOC High level segment July 2000. Retrieved 7 June 2001 from <http://www.un.org/esa/coordination/ecosoc/itforum/castells.pdf >

Denning, S. (2002). Technical Cooperation and Knowledge Networks. In S. Fukuda-Parr, S., C. Lopes, & K. Malik (Eds.), *Capacity for development: new solutions to old problems* (pp. 229-244). New York: Earthscan Publications. ISBN 1-85383-919-1.

Fukuda-Parr, S., & Hill, R. (2002). The Network Age: creating new models of technical cooperation. In S. Fukuda-Parr, S., C. Lopes, & K. Malik (Eds.), *Capacity for development: new solutions to old problems* (pp. 185-201). New York: Earthscan Publications. ISBN 1-85383-919-1.

Heeks, R. (2008, June). ICT4D 2.0: The next phase of applying ICT for international development. *IEEE Computer* (pp.26-33), June 2008. IEEE Computer Society

Holmen, H. (2002). *NGOs, networking, and problems of representation*. [Online]. Linköpings University and ICER, July 2002. Retrieved 18 Nov 2007 from <http://www.icer.it/docs/wp2002/holmen33-02.pdf>

ICASO. (2002). *HIV/AIDS networking guide*. [Online]. Canada: International Council of Aids Service Organizations. Retrieved 7 April 2007 from <http://www.icaso.org/publications/NetworkingGuide_EN.pdf >

Kilduff, M., & Tsai, W. (2008). *Social Networks and Organizations*. London: SAGE Publications. (reprinted 2008, 1st published in 2003). ISBN 978-07619-6957-0.

Mansell, R., & Wehn, U. (1998). *Knowledge Societies: information technology for sustainable development*. New York: Oxford University Press (for United Nations Commission on Science and Technology for Development). ISBN 0198294107.

Mataix, C., Moreno, A., & Acevedo, M. (2007). Estructuras en red: diseño y modelos para el Tercer Sector. Madrid: UNED (Spanish National Distance University), Fundación Luis Vives - *Module 8, 2007-2008 course on Strategic Management and Management Skills for Non-Profits Organizations*.

McIver, W. J., Jr. (2003). *The need for tools to support greater collaboration between transnational NGOs: implications for transnational civil cociety networking.* [Online]. State University of New York at Albany. Retrieved 20 Jul 2005 from <http://www.ssrc.org/programs/itic/publications/knowledge_report/memos/mcivermemo.pdf>

Moreno, A., Acevedo, M., & Mataix, C. (2006). *Redes 2.0 La articulación de las ONGD en España.* Madrid: Coordinadora de ONG para el Desarrollo-España (CONGDE).

Nath, V. (2000). *Knowledge Networking for Sustainable Development.* [Online]. Knownet Initiative, www.knownet.org. Retrieved 14 November 2001 from <http://www.cddc.vt.edu/knownet/articles/exchanges-ict.html >

Rischard, J. F. (2003). Integrating ICT in development programs. [Online]. *Keynote speech, Joint OECD/UN/World Bank Global Forum: Integrating ICT in Development Programmes*. Paris: OECD, 4-5 March 2003. Retrieved 14 Jun 2003 from www.oecd.org/dac/ictcd/docs/otherdocs/Forum_0303_summary.pdf

Sen, A. (1999). *Development as Freedom*. London: Oxford University Press.

Siochrú, S. Ó., & Girard, B. (2005). *Community-based Networks and Innovative Technologies: New models to serve and empower the poor*. New York: UNDP, 'Making ICT Work for the Poor' Series.

UNDP. (2000). *Optimizing efforts. A practical guide to NGO networking.* [Online]. Office to Combat Desertification and Drought (UNSO), UNDP. [Retrieved 3 Mar 2004 from http://www.energyandenvironment.undp.org/undp/indexAction.cfm?module=Library&action=GetFile&DocumentID=5256

UNDP. (2001). *Human Development Report 2001: making new technologies work for Human Development*. New York: Oxford University Press.

Van Bavel, R., Punie, Y., & Tuomi, I. (2004, July). ICT-enabled changes in Social Capital. [Online]. *The IPTS Report,* 85. Retrieved 3 Feb 2006 from <http://www.jrc.es/home/report/english/articles/vol85/ict4e856.htm>

Wellman, B. (2001). *Living networked in a wired world: the persistence and transformation of community. Report to the Law Commission of Canada.* Toronto: Wellman Associates.

Yunus, M. (2008). *Un mundo sin pobreza.* Translated by Monserrat Asensio (original title: *Creating a World Without Poverty*). Barcelona: Editorial Paidós Ibérica. ISBN 84-493-128.3.

ENDNOTES

1. Not only in least developed countries like Niger, Haiti or Sierra Leone; a country like Argentina with a medium-high Human Development Index has nearly 30% of its population under the poverty line—that´s a third of its people in the Fourth World.
2. While the reverse questions would also be valid, ie. 'what type of networked social environment can enhance the expansion of choices?' that discussion belongs more to the political and perhaps even philosophical realm, and falls outside the scope of development cooperation work, which is where we are positioning the discussion.
3. Held in Geneva on December 2003; the second phase took place in Tunis, on November 2005.
4. http://www1.oecd.org/dac/ictcd/docs/otherdocs/Forum_0303_conclusions.pdf
5. In our opinion, this 'B' grade was generous in light of the level of knowledge and awareness about ICT among development

professionals at the time; even today, five years later, their ICT capacity is arguably rather limited.

6 Castells has called for an 'informational development' paradigm supported by a new technological Marshall Plan

7 Direct project staff refesr to those positions with sustained, long-term involvement paid out by project finances, like a project director, a technical coordinator, a training specialist, etc. Although according to the scheme presented, a volunteer involved over a long period and with important contributions could well be considered in the future as part of the project staff as well—then the differentiation would be per paid staff and unpaid staff.

8 http://www.onlinevolunteering.org/stories/ovyear_list.php?yr=2002

9 The diagrams for 1-D and 2-D networks are taken from ICASO (2002), and used with permission.

10 The diagram for 3-D networks is taken from Moreno et.al. (2006), and used with permission.

11 http://www.unicttaskforce.org/stakeholders/ddn.html

12 As in the case of free/open source software networks, or in large consulting firms (like (Accenture, PriceWaterhouseCoopers, etc.).

13 Especifically, about 'learning to learn', ie. how to prepare oneself to learn better and from different methods/sources (including self-teaching).

14 This is not to say that representational networks are simple or that their careful network management is not required. The limited effectiveness of civil society networks oriented mainly towards representativity has been shown repeteadly in the literature (e.g. Holmen, 2002; McIver 2003; Siochrú & Girard, 2005; Anhier & Katz, 2005)

15 One of the conclusions reached from this analysis was the fragmented nature of the civil society networks at the Mumbai WSF, exhibiting many small and unconnected newtorks even when sharing common interests and being physically present at the same place.

16 Attributed to Beck 2003 and Shaw 2003, in Anheir & Katz 2005, p.221

This work was previously published in Information Communication Technologies and Human Development Volume 1, Issue 1, edited by S. Chhabra and H. Rahman, pp. 1-21, copyright 2009 by IGI Publishing (an imprint of IGI Global)

Chapter 2
FOSS Solutions for Community Development

Balaji Rajendran
Centre for Development of Advanced Computing, India

Neelanarayanan Venkataraman
Centre for Development of Advanced Computing, India

ABSTRACT

The use of ICT solutions in developmental activities and the deployment of them in modern telecentres have been widely accepted phenomena. In this article, we examine the use of FOSS – Free/Open source software in ICT solutions and categorize them into two types: 'FOSS in' and 'FOSS for', in order to emphasize the difference between those solutions, where FOSS play a meager or symbolic role and those solutions that are specifically developed to address a community need. We discuss about our experience in developing a FOSS-based ICT solution that is specifically built to address the needs of a community information system. We illustrate our three-tiered multi-stakeholder model of development, deployment and usage. We also describe our experience in the process of development and pilot deployment of our solution in various locations of India, and emphasize on the synergies between our model and the FOSS mode of ICT solution development.

INTRODUCTION

FOSS – Free/Open Source Software refers to the software that can be run for any purpose, studied, modified and redistributed in the modified or unmodified forms. FOSS is also referred as FLOSS – Free/Libre/Open Source, and the term free in FOSS or FLOSS is used as in 'freedom', and is not concerned with the price of the software. The idea of free software was conceived as a movement by Richard Stallman in 1983 to give software freedom to computer users (Stallman, 2002).

The term Open source software, as defined by Open source initiative – OSI, (Michael Tiemann, 2006) was derived mainly from the criteria of free software, as described above. The term Open

DOI: 10.4018/978-1-60960-497-4.ch002

source goes beyond the obvious meaning of just being able to view the source of software, but also the rights to modify, distribute, and redistribute. Hence OSI prefers to treat the term Open source as a trademark, and use it to only to describe that software licensed under an OSI approved license. However, there exist some differences between the two terms, though they fundamentally mean the same thing – freedom to use, study, modify, and redistribute the software. Hence, in this article, we use the term FOSS to refer to both Free (as in Freedom) and Open source software.

FOSS is often developed in a public and collaborative manner; generally through voluntary contributions from many developers. Over the years, as more and more FOSS software have gained wide-spread acceptance, they are becoming a compelling alternative to commercial and proprietary software, that have been in many areas of software utilization.

The growth and adoption of FOSS is also changing the dynamics of the business of software development. It also ushered in the transition from closed systems to open systems that provide the freedom to use, distribute, modify and redistribute the modifications made to the software of the system. The availability of FOSS without licensing fees and with source code, coupled with the freedom, have been the main factors behind their widespread acceptance and adoption, even in governments of various nations across the world.

The benefits offered by the FOSS model is particularly useful for the developing countries around the world; and the ability to obtain FOSS without licensing fees has proven to be very beneficial to the users in these regions as this makes information and communications technology (ICT) more affordable to them (Nah, 2006). Also, the vast pool of developers involved in FOSS development has made possible to consider options of using such software for ICT-enabled solutions. Also, the principles behind FOSS make such solutions to be ideal for use in the process of community development. (Rajiv & Jay, 2003; Virginia Report, 2001).

TELECENTRES AND COMMUNITY DEVELOPMENT

A Telecentre is generally a public place, where people living in a geographical region get access to a variety of services, through ICT tools, including computers, Internet, the related software and other digital accessories. The main aim of these telecentres is to support the process of community development, through the use of ICT tools in economic, educational, and social developmental projects and activities for a given community (Katherine & Ricardo, 2001).

Telecentres exists in many countries of the world, though they are called by different names, such as Community Information Centre, Kiosks, Common Services Centre, Public Internet Access Centre etc… Telecentres also differ by the services they offer, the people or the region or the community they serve, and by their business or organizational models. Telecentres may be operated by governments, or through NGOs or through PPP – Public Private Partnership models.

Telecentres have been around the world for a long period of time, evolving continuously with the new technologies and with different models of establishment and sustainability. Telecentres are also viewed as a movement, to bring the benefits of the new digital and connected information society to those under-privileged communities. Case-histories from some of the early telecentres established in both the developed and developing worlds (Fuchs, R, 1998) have been studied, and their analysis indicated much of the commonalities exist in the case of applications, infrastructure, technological innovation etc… A study of the telecentres that have been established in the Latin-American and the Caribbean regions for the developmental activities and their need to promote them in order to bridge the gap between the rich and the poor had been reported, along with the business perspective (Proenza, F., Bastidas-Buch, R. & Montero, G, 2001).

Thus telecentres, across the world, have been serving as the window to offer ICT tools and services to the common man, since a long time now, and have matured enough to directly progress towards implementation without piloting (Harris, R, 2007). The telecentre movement has therefore evolved and manifested itself into the second generation – Telecentre 2.0.

Hence, a number of ICT projects targeted for disadvantaged groups of people have been receiving support and funding from many agencies in the Asia-Pacific region are meant to be delivered through these evolved telecentres. This can be seen through the growing number of telecentres in this region. According to United Nations report, 2007 (United Nations, 2007), 400,000 new Telecentres are needed to reach the rural population in 16 countries of the Asia-Pacific region. Also, the number of projects that are aimed to setup Telecentres in the Asia-Pacific region alone stands at 74.

FOSS IN TELECENTRES

Telecentres are only the places for access to ICT solutions. They need to be provisioned with appropriate services and content, so as to take the communities to the next level (E-GovReport, 2005). Hence, the growing number of telecentres therefore throws up ample opportunity and market space for the usage of the existing FOSS software and also for the development and deployment of new FOSS software solutions aimed at community development. The telecentres could also be deployed with community information systems by using FOSS software to address the needs of the members of a local community (Lane & Plant, 1999; Balaji & Ravikumar, 2007).

Recent studies, from the South-Asian and Sub-Saharan African regions show that the FOSS (Free/Open Source Software) applications combined with low-cost hardware have emerged as an intelligent solution for sustainable telecentres (Bajwa, 2007). A compilation of the case studies on the use of FOSS in telecentres across the world, have shown that FOSS model of development, and licensing has become the right approach to software in telecentres meant for community development (Nah, 2006).

Categorization of FOSS

In this article, we categorize the usage of FOSS in community development efforts, into two types as: 'FOSS in' and 'FOSS for', in order to emphasize the difference between the use of generic FOSS tools and the development and use of special FOSS tools that are especially meant for community development efforts.

By, 'FOSS in', we refer to the usage of generic FOSS solutions in community development efforts. This refer to the use of existing generalized FOSS solutions such as GNU/Linux, Open Office etc... in telecentres as well as in kiosks meant for delivering ICT solutions. The usage of these FOSS solutions is meant to address a particular generic need, and may not be explicitly contributing to the needs of a community, thus playing a passive role in the process of ICT-enabled development processes.

By, 'FOSS for', we refer to those software solutions that are specifically built for addressing a particular community-developmental need, mostly addressing a livelihood aspect of a community. However, these solutions can be generalized and customized to be applied to similar communities or for addressing similar community developmental needs.

This categorization is done in order to emphasize the difference in the software solutions. In the 'FOSS in' category, the software is just a tool and usually does not play a vital role in the overall ICT solution. However, in the latter 'FOSS for' category, the software plays a key role in the overall ICT-based solution. These 'FOSS for' tools may also be built on top of other FOSS solutions. It is in this 'FOSS for' that one needs to focus, in order to have effective solutions developed exclusively to cater to the ICT-based developmental projects, that can capture the various livelihood needs of a community, and can improve them.

ICT Development Efforts and FOSS Solutions

There have been a lot of ICT-based developmental efforts happening across the world. In this section, we list some of those efforts, with emphasis in the Asia-Pacific Region. We also mention the usage of FOSS in these efforts, and categorize them as either 'FOSS in' or 'FOSS for' efforts.

FOSS in ICT Development Efforts

This section details some of the FOSS being used in various ICT development efforts, particularly in some of the Asia-pacific countries. This section only gives few examples and does not aim to survey the entire use of FOSS in the sphere of ICT development efforts in all these countries. It should also be noted that there are a number of ICT related community projects happening in these countries (United Nations, 2007), but the usage of FOSS in these efforts is not clear and explicit in many cases.

The Grameen Cyber Society project ("Grameen Cyber Society", n.d.) is executed in sectoral collaboration with a number of agencies for the rural people of Bangladesh. The main focus is to ensure that the rural communities have access to digital based learning materials to provide education for the overall community development. It also focuses on rural community development and uses FOSS tools for these purposes.

Bhutan's E-post project (International Telecommunication Union, 2008) is aimed at enhancing the communication facilities in remote areas of the country. It is meant for the local communities who had limited access to other forms of communications. Anyone can, who has access to Internet, send message to anyone else in the country through the post office and the post office would deliver the message as a local post. A total of 38 post offices including five in remote locations were identified for the project. Some of them have been utilizing the VSAT technology. They plan to use Simputers in the upcoming phases of the project that uses FOSS software in their systems. The project is executed by International Telecommunication Union (ITU) and Universal Postal Union (UPU) along with Encore Software.

The University of South Pacific (USP) and APDIP-initiated International Open Source Network (IOSN) have initiated Foss related projects. Among them are, an online examination system, micro-payment system, and FOSS distribution system hosting many FOSS packages.

FOSS for ICT Development Efforts

Sahana is a FOSS-based disaster management system that grew out of the 2004 Asian Tsunami disaster. It aims to be an integrated set of pluggable, Web-based disaster management applications that provide solutions to large-scale humanitarian problems in the relief phase of a disaster. It has 8 mature modules that address common disaster coordination and collaboration problems. They are the Missing Person Registry, Organization Registry, Request/Pledge Management System, Shelter Registry, Inventory Management, Catalogue Management, Situation Awareness and Volunteer coordination.

Sahana uses a plug-in, modular architecture and it can be deployed with just a small sub-set of these modules as decided by the administrator of the Sahana installation. The project was initiated in Srilanka and has now grown to become a globally recognized project with deployments in many other disasters such as the Asian Quake in Pakistan (2005), Southern Leyte Mudslide Disaster in Philippines (2006) and the Jogjakarta Earthquake in Indonesia (2006).

Identifying and Controlling Weeds – OSCAR - The Open Source Simple Computer for Agriculture in Rural Areas (OSCAR) project was developed by the French Institute of Pondicherry (IFP), that has developed an prototype software for weed identification and control in rice and wheat crops in the Indo-Gangetic Plains. OSCAR has been developed using FOSS tools and it encourages contributions to the software from various groups and organizations involved in that region. These groups help enhance the system by cataloguing the weed species in their region and their respective control measures. The application has been pilot-tested in Bangladesh, India, Nepal and Pakistan.

Open Enrich is software for managing content developed for use in local ICT Telecentres. The software enables local groups to organize information resources from the Internet and their own local computer systems. Users then navigate and browse content offline or, where connectivity allows, online. Information is managed through a structure of categories and sub-categories of information. The structure is customizable on the basis of relevance to a given community. The system has been deployed for community use in locations such as the, Seelampur-Zaffrabad in Delhi (Sarita, 2006), and the customized version of the software is used in setting up portals such as Akshaya ("Akshaya",n.d.) e-kendra.

The AgriBazaar ("Agribazaar," n.d.) is a portal that acts as an E-Market place to enable buyers and sellers to exchange various agricultural products and produce. It provides the Supply Chain Management (SCM) system that includes e-Stock, e-Logistic, e-Plan, e-Make, e-Support, and e-Payment modules. It uses a number of FOSS tools for development of this portal (Usage of FOSS in AgriBazaar, 2004) It is joint effort between the Malaysian Institute of Microelectronics Systems and Agricultural Department of Malaysian agriculture industry by enabling the various stake holders (e.g. farmers, fisherman, breeders, producers, retailers, distributors, exporters and importers) to conduct their daily agriculture production trading online.

Action Applications ("Usage of Action Apps",n.d.) is a collaborative web publishing tool especially designed for NGOs, to enable their organizations to create and manage the content of their websites as well as to exchange information with other websites easily by themselves. This application is integrated with complete mail management systems. It uses FOSS tools such as the Apache Web server and MySQL database for the system. Action Apps was initiated by Association for Progressive Communications (APC) -Internet and ICTs for social justice and development. Action Apps has been used in many ICT community efforts and some of them could be found in ("Usage of Action Apps", n.d.).

ECKO - Empowering Communities through Knowledge (Balaji, Neelanarayana, Ponraj & Kailash, 2005; Balaji & Ravikumar, 2007) is a tool to establish E-communities or E-society in a closed geographical region. The system was developed by C-DAC (Centre for Development of Advanced Computing), India. This article details our experience in the development and deployment of this system in the next section.

OUR EXPERIENCE IN DEVELOPMENT AND DEPLOYMENT OF A FOSS SOLUTION

Knowledge flow and permeation through a community is critical for the betterment of a community and its people. Knowledge permeation is dependent on the availability of mechanisms for communication and knowledge sharing. Rural communities are characterized by geographical isolation, poor communication channels and non-availability of knowledge sharing platforms.

We had envisioned for creating a virtual knowledge sharing platform, where the creators, users and owners of knowledge are the members of that community, who can virtually be connected and share their knowledge, thus leading to empowerment. This resulted in the conception and development of our ICT-based solution ECKO - Empowering Communities through Knowledge.

ECKO is a framework for setting up a community information system (Balaji et al., 2005), catering to the needs of a community, and thereby fostering an E-community. A community information system is a holistic system that provides ICT-based features specifically tailored to the livelihood needs of a community, in order to enhance the productivity, and save the cost and time related with those activities. Thus ECKO helps in fostering the formation of E-communities in regions that are bounded geographically and culturally.

ECKO is packaged web-based solution that can be classified under the 'FOSS for' category. It utilizes and is built on other FOSS tools such as the Apache HTTPD Web server, PHP engine, and the MySQL database, to deliver a web-based ICT solution. However the system has been specifically built to address the community needs.

ECKO as a software tool, consists of many components for provisioning of various services specific to a community, such as market-related information of various agricultural products, local job information etc…, general services such as capturing and sharing of information, along with communication mechanisms like messaging, chat etc…, and Knowledge inference services, that organizes the various fragments of information, over time, for future analysis and to provide services for policy-makers.

ECKO and its Architecture

ECKO has been designed as a generic, extensible, process-driven, software-based framework to build and nurture generic E-communities. It provides a platform for creating, using, and sharing information among the members of E-Communities. The architecture of ECKO [Figure 1] depicts the various services at different planes catering to the Users, Managers and Administrators. This architecture depicts the organization of ECKO from the user and usage-centric view point.

The first block of User Services, target the end-users of the system. The second block of Application services, cater to the role of a manger, who helps in running the system continuously, by managing the users, publishing the contents, archiving them, knowledge tagging and other such operational activities. The System services block consists of administrative tools for Installation and Configuration, Operational Configuration, and other facilities like Backup and Restore that help in customizing the system for the user requirements and for regular maintenance of the system.

However, from a general user point of view, the services facilitated through ECKO can be classified into the following three categories.

1. Community Services
2. Communication Services
3. Information Services

Community Services are those features that facilitate major community activities like 'Community Buying', 'Community Selling' and other group activities, like Opinion polls, Discussion Forums etc... The primary aim of these services is to offer popular development solutions and techniques like micro-financing, tele-consulting, etc.., and enhanced through the usage of ICT. Communication Services are those features that facilitate interaction and communication among members through messaging, chat etc... The aim of these services is to speed up and ease the process of communication among members. Information Services are those features that facilitate the dissemination of information through News, Information Repositories (InfoBase), Commodity Prices, Local Weather, Local Bullion Prices etc... The aim of these services is to pool information from various sources to serve the user on an instantaneous basis. The information is later organized and archived into repositories for later

Figure 1. Architecture of ECKO

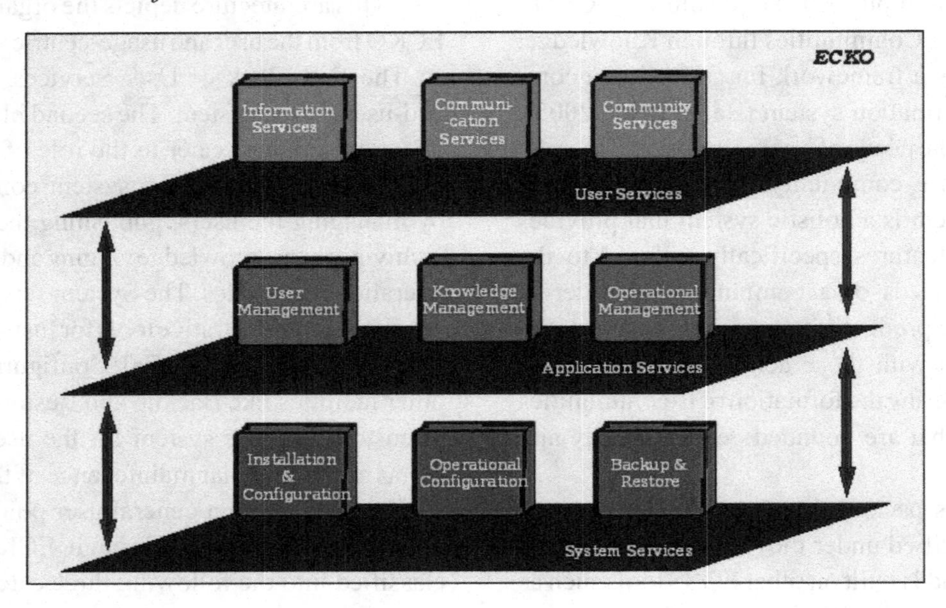

analysis and usage by the general users and policy makers.

The other salient features of ECKO are that it is highly customizable, and adaptable to the needs of a specific community. It also includes a localization framework that allows the system to be configured in any language. It has also been packaged with double-click easy-to-install, configure and management graphical interfaces.

The architecture of ECKO has therefore been designed to make it as a tool that emphasizes on bottom-up approach to local content generation, sharing, consumption, and management, with provisioning of communication services and other community services, depending on the needs of that particular community. As the flow of information increases, and the system grows, it can be used to tap the knowledge from several other similar communities, as well as serve as a tool for analysis for policy-makers.

Our Model of Deployment and Usage

The success of an ICT-based solution does not depend only on the technology or the software, but with the participation and involvement of the various stakeholders in the various processes, from design to implementation. Hence, we followed a three-tier multi-stakeholder model for the process of development, deployment, awareness creation, and sustained usage [Figure 2].

Our three-tier multi-stakeholder model consists of the ICT solution providers in one tier, various NGO's and self help groups who can take the ICT solutions to the field in the next tier and finally the people, whom the system would cater to, in the third tier (Subramanian, Balaji, & Ponraj, 2006).

For the usage and adoption of ICT-based solutions in under-served regions, awareness about the system needs to be created first. We, the ICT solution providers would train the NGO's and other social organizations by conducting awareness workshops and through field visits. Once these social organizations were trained, they in turn utilized their established kiosks for providing ICT services to the people. They therefore took the responsibility of creating the awareness, training the local people about ICT solution and its benefits to them. Thus they helped in creating

Figure 2. Three-tier model of development & deployment

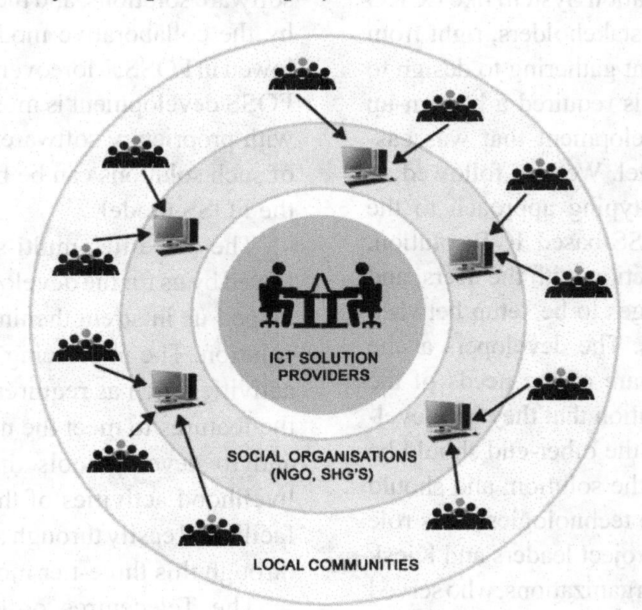

the much-required awareness among the people about ICT and technologies for their prosperity. They also sought opinions from the people about the usage of ICT solutions for the people's livelihood activities and guide them in using these techniques. Once trained and comfortable, the people could themselves utilize the system to meet their requirements. The people also use the services of the kiosk operator and may not get involved directly. The kiosk operator, with the support from the social service organizations gave the necessary support for this kind of users.

Hence this three-tier multi-stakeholder model was a fan-out model from the centre point to the perimeters of a community. The number of people involved and the reach of the solution expands as one goes to the boundaries of the model. The flow of information also gets filtered and customized to meet the targeted audience. For instance, the requirements from the users are spelt out through the kiosk operators of the social organizations, who can communicate it in a relatively technical manner to the ICT solution providers.

ECKO, as a community information system was deployed at various rural locations in India by the application of this three-tier multi-stakeholder model. Some of the locations were Dhan Foundation, at Melur in TamilNadu, AFARM, at Latur in Maharashtra, MSSRF Village Knowledge Centre in Pondicherry, SOHARD, at Alwar in Rajasthan etc… These organizations initially started by using the typical features of ECKO such as the News, Messaging and followed it by those activities that took the system to the people. The feedback from these organizations showed encouraging use over a period of time, progressing towards the demand for features that were impacting their livelihood activities directly.

Benefits of the Three-Tier Multi-Stakeholder Model

The three-tier Multi-stakeholder as illustrated in [Figure 2] served as a good model for development as well as for the deployment and outreach of our 'FOSS for' solution – ECKO.

A Community Information System like ECKO needs to involve various stakeholders, right from the process of requirement gathering to design to implementation. Also, this required a bottom-up approach to system development that was easily facilitated by the model. We also followed an iterative and rapid-prototyping approach to the development of this FOSS-based ICT solution, that required lot of interaction with the users, and that required virtual bridges to be setup between the developers and users. The developers at the one end, ought to be aware of the needs of the community, and ICT solution that they are developing; while the users at the other-end should be aware of the benefits of the solution, and should not be intimidated by the technologies. This role was well played by the Project leaders and Kiosk operators from the social organizations, who served as an intermediary between the developers and the users of the community, and therefore helped us in process of project management (Subramanian, Balaji, & NeelaNarayanan, 2007).

We also learnt that an ICT-based solution can start with generic services made accessible to the communities, but later have to adapt them to enhance the livelihood activities of the community. This would help the system to move from pilot deployment to sustained development. Our model was helpful in gathering the feedback regularly from the users of the system, and that resulted in the system evolving from a simple web-based communication system to a comprehensive community information system. Also, the role of the social organizations in providing valuable inputs collected through a continuous process from the users, served as a crucial input, as they were knowing the community members and their needs, much better, and also helped us in the process of localizing and customizing the system to the needs of their community.

WAY FORWARD

From our experience, we believe that a lot of sustained efforts are required to develop ICT-based software solutions, and they can be best developed by the collaborative model of development followed in FOSS. Moreover, the process followed in FOSS development is more open when compared with proprietary software, and also the life-cycle of such solutions can be better managed through the FOSS model.

The three-tier multi-stakeholder model followed by us for the development, and deployment helped us in strengthening the ICT-based FOSS solution. The interfacing with the users, and the activities such as requirement elicitation, tuning the features to meet the needs of the community and to develop tools that would enhance the livelihood activities of the community, were all facilitated easily through the social-organizations through this three-tier model.

The Telecentres or kiosks setup by these social organizations, helped us tremendously to directly deploy our FOSS based ICT solution. This model can be replicated in the growing number of telecentres (United Nations, 2007) along with the FOSS-based ICT solutions in order to attain sustainability.

Also, there are not many efforts from the commercial software providers to address community kind of social requirements in the rural regions, and we strongly believe that FOSS solutions applying similar kind of multi-tiered multi-stakeholder model for the process of development and deployment would be more suitable and ideal to cater to these efforts.

We also believe that the setting up of Community Information Systems in the telecentres or kiosks could foster the formation of E-communities in the regions, where they are located. As these Community Information Systems mature by enabling ICT-based approach to support the livelihood activities of a community, they can be networked with similar kind of E-communities in different regions to mutually cooperate and derive benefits from the cooperation. Such a means of federating the various Community Information Systems would result in clusters of Federated

Information Systems that would provide immense value not only to the own community members but also to the policy makers through the reflection of the needs and opinions of the people and would eventually usher the ICT revolution towards a equitable society.

A federation of such community information systems would however differ from the public Internet as it is wider in perspective and highly unorganized. The restriction of geographical boundaries in the formation of E-communities would help in the first propagating the benefits of ICT to all the members within a community, and that could be networked later. This restriction also brings coherence to the system in many factors such as the region, culture, occupation, polity etc… among the members of the community and also becomes easier to address the specific needs of a community. The federation of these empowered E-communities would then bring forth more meaningful collaboration paving the way for an empowered and equitable society.

CONCLUSION

We had discussed and categorized the role of the FOSS in ICT-enabled developmental activities. The 'FOSS for' ICT solutions, where the emphasis is on the active role of the ICT solution would serve the communities better than the 'FOSS in' ICT solutions, where there is a meager role for ICT.

We also discussed about the growing number of telecentres and their potential to serve as a means for establishing E-communities through FOSS based ICT-enabled solutions. We also discussed about the three-tier, multi-stakeholder model that could be followed for the development, deployment and sustained usage of similar FOSS based ICT-enabled solutions that could comprehensively serve the communities and function as an effective community information system leading to community development.

REFERENCES

AgriBazaar – Exchange for better price, project attempts to boost Malaysian agriculture by online trading. Retrieved December24, 2008 from http://www.agribazaar.com.my/.

Akshaya – Creating powerful e-Services to reach citizens, project addresses key issues in IT dissemination to masses. Retrieved December 24, 2008 from http://akshaya.kerala.nic.in

Bajwa, F. R. (2007). Telecentre Technology: The application of free and open source software. APDIP eNote 19, 2007. Retrieved December 24, 2008 from http://www.apdip.net/ apdipenote/ 19.pdf

Balaji, R., Neelanarayana, V., Ponraj, M., & Kailash, T. (2005). Establishment of Community Information Network in a Developing Nation. Proceedings from IEEE Tencon 2005: The International technical conference sponsored by IEEE Region, 10, 1- 6.

Balaji, R., & Ravikumar, B. (2007). Community computing for development. Proceedings from ISED 2000: International Conference on ICT solutions for Socio-economic Development, (pp. 9–16).

E-GovReport. (2005). *Global E-Government Readiness Report 2004: Towards Access and Opportunity, I-WAYS - The Journal of E-Government Policy and Regulation, 28(1)*. IOS Press.

Fuchs, R. (1998). Little Engines that Did: Case Histories from the Global Telecentre Movement. IDRC, 1998. Retrieved December 24, 2008 from http://www.idrc.ca/ en/ev- 10630-201- 1-DO_TOPIC.html.

Grameen Cyber Society. Reflecting on the pilot telecenter operation in rural bangladesh. Retrieved December 22, 2008 from http://siteresources. worldbank.org/ EDUCATION/ Resources/ 278200-1126210664195/ 1636971-1126210694253/ Grameen_Final_ Report.pdf

Harris, R. Telecentre 2.0: Beyond Piloting Telecentres. APDIP eNote 14, 2007. Retrieved December 24, 2008 from http://www.apdip.net/apdipenote/ 14.pdf

International Telecommunication Union. (2008). E-Services through post offices in bhutan, July 2008 http://www.itu.int/ ITU-D/tech/ RuralTelecom/ UPU_Bhutan.pdf.

Katherine, R., & Ricardo, G. (2001). Comparing Approaches: Telecentre Evaluation Experiences in Asia and Latin America. The Electronic Journal of Information Systems in Developing Countries (EJISDC 2001), 4(3). Retrieved November 24, 2007, from http://www.ejisdc.org.

Lane, C., & Plant, N. (1999). Community Computing in Rural Regeneration Networks, Cisc publication No 6, ISBN: 1860431666.

Nah, S. H. (2006). *Breaking Barriers: The potential of Free and Open Source software for sustainable human development. UNDP-APDIP ICT4D Series* (pp. 3–16). Elsevier Publications.

Proenza, F., Bastidas-Buch, R., & Montero, G. (2001). *Telecentres for Socio-Economic and rural Development in Latin America and the Caribbean*. Washington, D.C.

Rajiv, C. S., & Jay, P. K. (2003, Summer). Incorporating Societal Concerns into Communication Technologies. *IEEE Technology and Society Magazine*, 31–33.

Report, V. (2001). The Role of Institutions in the Design of Communication Technologies, Telecommunications Policy Research Conference, Alexandria, Virginia Report No: TPRC-2001-086: Rajiv, C. S. & Jay, P. K, 19 - 21.

Sahana – Sahana FOSS disaster management system (2007). Retrieved December 22, 2008 from http://en.wikipedia.org/ wiki/Sahana_FOSS_Disaster_Management_ System

Sarita, S (2006). eNRICH: Archiving and accessing local information, International Journal of Education and Development using Information and Communication Technology (IJEDICT), 2(1), 34-48.

Stallman, R. M. (2002). *Free Software, Free Society: Selected Essays of Richard M. Stallman, 2002*. GNU Press.

Subramanian, N., Balaji, R., & Neela Narayanan, N. (2007). Project management Approaches for developing Public Community Software Solutions. Proceedings from the Third International Conference on Project Management Leadership.

Subramanian, N., Balaji, R., & Ponraj, M. (2006). Model for Establishing Knowledge Platforms: A Case Study. *Proceedings from the International Conference on Digital Libraries, 2*, 559–566.

Tiemann, M. (2006-09-19). History of the OSI: Open Source Initiative. Retrieved December 22, 2008 from http://www.opensource. org/history; Retrieved on 2008-12-17.

United Nations. (2007). Report by United Nations Development Account Project on Knowledge Networks through ICT Access Points for Disadvantage Communities (2007), March 2007 Assessment of the Status of the Implementation and Use of ICT Access points in Asia and the Pacific, 12. Usage of Action Apps. Retrieved December 22, 2008 from http://apc.org/ actionapps/english/ general/slices.shtml.

Use of FOSS in AgriBazaar. (2004). Reterived 24 November, 2007 from http://r0.unctad.org/ ecommerce/event_docs/ fossem/azzman_ agribazzar.pdf, 15.

This work was previously published in Information Communication Technologies and Human Development, Volume 1, Issue 1, edited by Susheel Chhabra and Hakikur Rahman, pp. 22-32, copyright 2009 by IGI Publishing (an imprint of IGI Global)

Chapter 3
Digital Divide, Social Divide, Paradigmatic Divide

Daniel Pimienta
Networks & Development Foundation (FUNREDES), Dominican Republic

ABSTRACT

The digital divide is nothing else than the reflection of the social divide in the digital world. The use of ICT for human development does offer opportunities to reduce the social divide for individual beings or communities; yet there exists a series of obstacles to overcome. The very existence of an infrastructure for connectivity is only the first obstacle, although it often receives an exclusive focus, due to the lack of an holistic approach which gives an essential part to digital and information literacy. Telecommunications, hardware and software are predictable prerequisites; however, the true pillars of human-focused information societies are education, ethics, and participation, interacting together as a systemic process. As long as decision makers are not ready to consider these evidences, and keep on favoring a mere technological vision, we will suffer from the most dangerous divide in terms of impact: the paradigmatic divide. Any resemblance to characters, projects, or policies in real life is quite intentional.

INTRODUCTION

In 2000, the G8 initiated the "Dot Force", i.e. "Digital Opportunity Task Force" (United Nations, 2000) to raise international awareness on the subject. Since then, a concept has prevailed: the fight against the digital divide is a priority because ICT offers many possibilities for development for people, as well as for communities and countries. The concept of ICT for development (ICT4D) is now used by many international, regional and national organizations, and by all sectors (international, governmental, corporate, civil society, and the academy). We all share the same belief that the use of ICT for development holds very important promises.

DOI: 10.4018/978-1-60960-497-4.ch003

Yet, this is only a belief. Although it appears very credible, it remains a belief, because the highly remarkable lack of an impact evaluation has prevented the results of ICT4D projects in the last two decades from being clearly stated. The belief sometimes becomes myth or magic… for instance when one thinks that the mere fact of connecting a person to the Internet is going automatically to initiate a process causing this person to escape his/her situation of poverty.

In this article, my contention will be that this belief reveals a serious *lack of perspective*, ignoring that the digital divide is no more than the reflection of the social divide (Pimienta, 2002) in the virtual world. I will also demonstrate that there is an error of both focus and approach, which has very serious consequences when it comes from decision makers in the public arena. Field observers can see indeed that some projects seem to have positive impacts whereas others are never completed or do not have noticeable impacts. What are the criteria which allow the former to be distinguished from the latter? Can we identify the components that make it possible for public policies on the Information Society or ICT4D projects to produce positive impacts on the society?

I would like to make an hypothesis about the criteria, as well as to bring some elements to the analysis which are likely to sustain the hypothesis. The main hypothesis is that the crucial element is the approach; it matters more than being efficient in the ways policies are designed and projects are managed.

- An approach based on technology has every possibility of ending in failure for both policies and projects.
- An approach based on contents and applications will guarantee products but may fall short when it comes to the desired societal changes.
- An approach based on paradigm shift is the key to success in obtaining a positive societal impact.

My final objective is to show that maximum concentration is required on the education component that must support the policy or the project. The task related to digital and information literacy is both a priority, which is seldom fully considered in policies and projects, and an extraordinary challenge, considering the size it should have to reach the whole society. The greatest strategic element for the transformation towards information societies is education of the citizenship in the digital world as well as paradigm shifts linked to a new vision of society based on knowledge sharing. Be that as it may, the bottleneck lies in the decision makers' awareness and in the negative multiplying effects of their decisions when they have not adopted or not understood the correct approach (and its natural implications regarding the importance that multi-stakeholder partnership deserves).

These facts suggest that in addition of the social divide which lies behind the digital divide, there exists another divide which is not so clearly visible, that is not properly taken care of, and whose effect on the digital divide is still greater: the paradigmatic divide. This divide exists when decision makers in the field of the information society start from an erroneous approach, and keep on working within the logical framework of a previous societal paradigm, where society does not participate in the decision process.

ANTECEDENTS

This article compiles and synthesizes elements that come from several speeches delivered at international conferences on the information society in recent years (Pimienta & alt., 1993-2007). It is based on a series of concepts that were elaborated or collectively discussed (MISTICA Virtual

Community, 2002, 2004) among the Virtual Community of ICT4D actors in Latin America and the Caribbean (MISTICA, 1999-2006) between 1999 and 2006.

The very concept of *paradigmatic divide* rose very naturally during a presentation made by the author of this article (Pimienta, 2005). The newly formed concepts (see "Pimienta's Law", *infra*) were warmly welcomed by the audience which however was composed almost exclusively of academics, civil society actors, and staff from international institutions. Therefore, the message that had been designed for governmental decision makers could not reach these people, because they had decided to hold a parallel session, on the other side of the wall, in order to make decisions… while the other sectors were meeting to elaborate the criteria for well-focused decision making processes.

Since then, multi-stakeholder dialog has been being considered as an essential element of the policies for the information society. This tendency has been still greater since the World Summit on the Information Society (WSIS) (United Nations, International Telecommunication Union, 2003-2005). Nevertheless, the concept of multi-stakeholder approach keeps being perverted, in too many countries.

There are many ways to misuse the concept. The first and most frequent occurs when government representatives themselves select their interlocutors from the other sectors: thus, they eliminate those to whom they are not eager to listen. Such situations conspire against the essential pluralism, whereas in other cases civil society is exposed to a great contradiction, because the concept of representation is irrelevant in a framework of participative democracy. Another way to misuse the concept is quite common: it consists in organizing sham multi-stakeholder meetings, where the effective paradigm actually keeps being "top down", and where other sectors are supposed only to listen and approve governmental speeches so that government representatives can auto-attribute the seal of participation to themselves. Unfortunately, and in spite of its efforts, civil society has not managed to include the task of relevant evaluations in the priorities of regional agendas; such task would consist in evaluating all activities which are labeled as "multi-stakeholder partnership" with solid criteria. In any case, concrete cases when the official discourse contradicts the facts have been documented (FUNREDES, 2005).

A QUESTION OF INVESTMENT: DEFINITIONS

The analysis of ICT4D projects that do not produce impact in the field reveals that an obvious macroscopic cause is to be found in a bad distribution of the budgets among the main project headings.

The various headings can be seen in the following pyramid (See Figure 1).

"**Infrastructure**" refers to the devices that permit the signal to be transmitted (such as lines, microwaves, satellites), and to be carried (such as protocols for communication and routing devices), as well as the computer hardware and software that are involved in the transport of the information (operating systems, in the very broad sense, and communication protocols), reaching the users, whether through individual terminals or through terminals shared in a community (telecenters).

"**Infostructure**" refers to the contents and the applications that are located, are given access, and are executed above the infrastructure. It includes the software, the databases and the websites that are hosted in the computers which work as servers in the network. It is obvious that an *information* structure necessarily works closely linked with a *communication* structure. This leads to the concept of "commustructure"; for practical reasons, it is seen as part of the "infostructure", and the virtual communities are then treated as an integral part of this layer, next to the contents.

Figure 1. How to invest in ICT4D

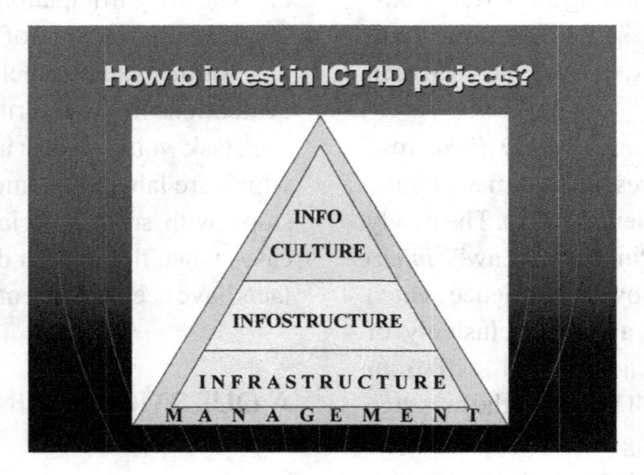

"**Infoculture**" refers to the sum of knowledge, methods, practices, and rules of good conduct that the people *possess* when they have gain ownership on the use of information and communication networks. The meaning of the word "infostructure" is not standardized yet in the specialized literature; there is even less standardization for the term "infoculture", and various definitions can then be found. What is required in order to acquire this culture (through an ownership process) is a **digital and information literacy** process (Wikipedia, 2007), as well as practice of uses which are relevant to the situation of these people. It is obvious that in a context of ICT4D where the paradigm shift is the essence of the change, cooperation and multi-stakeholder partnership are concepts that are parts of this layer.

The "**ownership**" is the learning process that leads people, groups or organizations, to have a control over ICT uses in coherence with their own environment. We can distinguish *technological ownership* and *social ownership*: the former occurs when the technology becomes transparent from its use, the latter when the technology becomes transparent from its social or economic function, hence when it becomes a mere tool. The process of ownership by people, groups and organizations who have not had the opportunity to reach such a close relation with ICT, due to their history or to their education, requires a specific **support** that combines education, practical skills, and uses which are meaningful within one's context. On the one hand, the size and the complexity of this support process cannot be underestimated; on the other hand, the rapid never-ending evolution of these technologies clearly raises the issue of lifelong learning.

The "**digital literacy**" process consists in equipping a given population with ICT concepts and methods, and putting the people in a situation in which they can make use of those technologies in order to get a real mastery (ownership) of them (particularly the use of a computer in a network context, it being clear that in a near future, the interface used by the network may well be a hybrid of a computer and a cell phone, maybe of digital TV as well). This should not be mistaken for office automation classes (i.e. courses oriented to the use of the computer office applications, generally on specific platforms). The acquired concepts must be independent of any platform, although practical training can take place on a given platform among the existing ones, for practical reasons. Training should be directed at ability to read/write using digital multimedia, the functionalities of the applications, as well as methods for good use and knowledge of the usages and practices of the environment. These kinds of programs are usually long and progressive, unlike office automation classes.

The "**information literacy**" process consists in providing the people with concepts and training in order for them to process data and transform them into information, knowledge and decisions. It includes methods to search and evaluate information, elements of information culture and its ethical aspects, as well as methodological and ethical aspects for communication in the digital world. These kinds of programs are usually very long and progressive, and they require an appropriate combination of theory and practice. If the practical part is missing, the efforts invested in literacy prove to be insufficient to obtain a real social ownership. As far as national programs are concerned, the support must involve all the sectors in the design of applications and contents that have a social and a national meaning. This terminology just begins to prevail, not without being sometimes rejected by some professionals in the education field, who fear a perversion of the (basic) concept of literacy. The expression "ICT training" is still very widely used, although some ICT professionals reject it because it carries an instrumental and oversimplified image and does not convey a correct idea of the complexity of the processes at stake. Education in ICT or in the digital world represents valid options to describe the concept. A consensus has been reached among the group of information professionals regarding the terminology of *information literacy*, i.e. Alfin, in Spanish (Alfin, 2006-2007) and the proclamation of the central part that this group should play (Prague Declaration, 2003) and the declaration of Toledo by the Spanish librarian groups (Alfin, 2006). That group does not always differentiate between digital literacy and information literacy and does include both concepts in the same terminology.

The "**management**" (of ICT projects) includes all the processes that, starting off with the very setting of the project (which includes user support and traffic management), ensure the organizational, financial and institutional sustainability of the project, and integrate, from its beginning and throughout the life of the project, the evaluation of results and impacts. It must be clear, however, that the processes of multistakeholder partnership, an essential ingredient of the success of an ICT4D project, require specific elements of management.

Multistakeholder partnership is a process in which citizens and economic actors actively participate in all the stages of national policies. That makes it possible for them to fully appropriate ICT for development, and brings consensual solutions to elaborate national policies embedded in national culture and structures. See for instance Bolivian Strategy of ICT4D (ETIC, 2005) the exemplary effort that was made in Bolivia towards a participation process –unfortunately, it seems that the process could not survive a change of government. For practical reasons, the effort may be initiated with key stakeholders (i.e. those who have legitimate and well-informed interests in the process), then expand, giving more capacity to those key stakeholders, and creating motivation to diffuse to the other citizens (Funredes, 2003-2005).

The term tends to progressively replace "public-private partnership" insofar as it conveys the idea of participation of all the groups (global governments, national governments, local governments, private sector and civil society, academics being sometimes distinguished from civil society) in a clearer and more precise way. The concept prevailed as one of the key elements of the World Summit on the Information Society (WSIS) and of its recommendations, because it seems obvious that the participation of all sectors is required to build a society (and it is precisely the point when we talk of Information Society)…

A MATTER OF INVESTMENT: PIMIENTA'S LAW

This four-layer frame allows to establish "**Pimienta's law**", which was coined in a presentation during the meeting mentioned above, as a provocative way to deliver a message from civil society to the governments which manage ICT4D projects (English readers should be aware of the fact that "Pimienta" is both the name of the author and the word for "pepper", in Spanish –hence an amusing pun on the spiciness of the theory). In short, the message is that fighting against the digital divide does not amount to the simple issue of connecting everybody: infrastructure in itself cannot represent an end, and must not receive all the attention

"Pimienta's law" enunciates that:

1. An ICT4D project of which the proportion of infrastructure exceeds 60% will in all probability end in serious trouble due to deficiencies in the other dimensions.
2. An ICT4D project of which the proportion of infrastructure exceeds 80% will in all probability end in a disaster…
3. An ICT4D project which receives almost 100% of its budget for infrastructure should be an object of scrutiny for the offices in charge of detecting and preventing corruption… a strong probability exists indeed that its purpose is to generate substantial commissions to buy equipment which will become obsolete in a few years, before it is properly used, because nothing has been planned for this use…

Three Focuses: Three Ways

Where does this "spicy" law come from? Is it from a thorough econometric analysis? Actually, there is no mathematics behind it, but only a simple observation of public management for the past few years in various places. How many computers, bought with foreign currencies, end up in the corner of a school, without being used (when they do not remain in their boxes) because no support has been planned to train the teachers? –just to mention an example! How many modern telecenters end up with equipment out of order because there was no consultation before selecting a technology which is not appropriate, and no planning for maintenance was made?

Although in case 3 (100% for infrastructure) the honesty of the people who manage such projects can be put in doubt, in most cases the error is based on a fundamental misunderstanding of the nature of ICT4D use and of the reality of the information society.

To make the point clearer, focuses, perceptions and approaches can be classified in 3 categories.

1. **ICT for the Sake of ICT: ICT as Ends in Themselves:** The technological approach assigns an excessive importance to infrastructures. The ground for this is strengthened by the explicit goal to install technology or by the implicit belief that a bottom-up process will automatically occur. For that reason the results of those projects boast the number of computers installed and the bandwidth… without too much concern for the real and effective use of the technology and still less for the social impact. This vision leads to a bad use of national (or international) resources and it is a smoke screen on the real needs of development. It is clear that this approach is less and less assumed today, because government discourse adjusts to international discourse; nevertheless, behind those pretty speeches, it is not difficult to find that the promised support is almost empty, and to perceive the reality of a disastrous approach for development.
2. **ICT for Development: ICT as Means Serving Development**: This approach focuses on the contents and the applications. In this approach, ICT are only tools which allow applications and contents, therefore **uses**, that do matter and that are going to be the leading elements of development. ICT are given the sole importance of being tools, sometimes with the belief that they can be

neutral at the economic and the cultural level regarding societal impacts. This vision clearly surpasses the technological vision, and it allows the development of applications and contents which support development. The approach sensitizes to the necessity of indicators that reflect uses rather than the technology; without a doubt, it is efficient in terms of development (because it integrates the essential components of support for the use). However, it is not sufficient regarding the essence of the paradigm shifts. That makes this approach too tolerant to the perversions of multi-stakeholder partnership processes. It is often naive when it tries to ignore the huge cultural and linguistic implications that come with technology and its standardized uses. Its very characteristics make this vision come naturally with the traditional world of international cooperation, through its bilateral and multilateral agencies, and it tends to ignore a major contradiction: talking of an information society the central element of which is relations in networks, without changing the very obsolete paradigm of international cooperation management for relations of these kinds (Jansen & Pimienta, 2007).

3. **ICT for Human Development: ICT as catalysts/facilitators of the paradigm shift**: This approach focuses on infoculture, cooperation, and participative processes. The ground for this approach is the conviction that ICT are more than simply tools serving development: they are the catalyzing and facilitating factor for deep changes that the society needs and that should occur **independently** of the existence or the presence of ICT (although obviously those changes would occur in a more natural and more effective way thanks to the proper use of ICT).

In this sense, the common thing between e-government, e-health and e-learning is not the weak "e" as in "electronic", but the potentially hidden "p", as in Process, Participation, and Paradigm.

It is not "the fault of ICT" that education must change to something more focused on the group, that the function of professors must shift from knowledge suppliers to facilitators of learning processes, that the relationships between the actors in the field of education must adopt a network configuration which needs to teach to learn rather than to focus on erudition… These changes are certainly required by the evolution of the society; thinking that the cause of the change lies in ICT represents a critical error of focus. ICT do not replace pedagogy; an education project integrating ICT cannot work if pedagogy is not integrated in the new framework. In many places, representative democracy is reaching its operative limits; introducing new modalities of participative democracy is necessary to restore credibility with citizens. That is not because ICT demands it -once again, thinking this way is a deep error of perspective. ICT clearly offers valuable resources for participative democracy as well as wonderful examples of participative collective construction (Wikipedia for example). However, once again, it plays its part when it accompanies a political will for change; it cannot replace this will.

Finally, we can safely state that although there are situations where organizational changes are required **without** ICT implications, on the contrary, ICT should never represent per se a valid justification to proceed to organizational changes (and a good organizational design should make provision to keep performing although more slowly in case of computer breakdown). Considered as a thermodynamic process, computerization of processes amplifies entropy indeed: a well-organized company which gets equipped with computers will have a still better organization; a weakly organized company which gets equipped with computers without previously paying attention to its organization model will have a still weaker organization, and will probably be in danger of going bankrupt.

Anyone can think of various other fields in which deep changes are required (e.g. health), analyze them and find out that although ICT are perfect tools to speed up the changes, they are neither the cause nor the justification for making them. Hence a principle everyone should share: if changes are led by the mere application of technology, without thinking in terms of organization, a failure can be foreseen. The following table can help determine where is the focus of a generic Information Society policy or where is the approach of a specific ICT4D project uses (See Table 1).

Some governments still do have an "ICT4ICT" approach: this can be observed in the deficient support to infrastructure investments, or in the fact that training support limited to classes on the use of some software applications is too often taken for real information literacy (although associated costs differ by a scale factor of 1 to 100).

The world of international cooperation has for some time now been in favor of a focus on ICT4D, and has begun using the vocabulary of ICT4HD. However, words do not always show the deep complexity of the concepts, especially about participation. It is equally true, though, that it is difficult and delicate for intergovernmental organizations to lecture their State members.

The civil society group involved in the Information Society matters has the clearest concepts, and makes a real effort to lead its interlocutors towards paradigm shift, although with obvious difficulties; this was noticeable in the WSIS process. Be that as it may, results vary, due to existing misunderstandings and to the lack of long-term impact evaluations in the field.

A PROCESS-ORIENTED VIEWPOINT ON THE DIGITAL DIVIDE

How is it possible to fight so fundamental a misunderstanding that decision makers make investments in the name of the struggle against the digital divide which actually do not fit priorities, and only take into consideration the technological aspect?

This section aims at both presenting a constructivist framework to understand the complexity of the digital divide and clearly demonstrating that providing an infrastructure is hardly one of the 10 commandments of the right to communication and knowledge…

We are now going to identify the various elements the digital divide is made of, thanks to 3 graphics, labeled "*The hurdle track from ICT to human development*" (Pimienta & Blanco, 2006). The whole process is seen as a resolution process.

Table 1. Information society policy approaches

APPROACHES FOCUS	ICT4ICT	ICT4D	ICT4HD
STARTING POINT	access	information	knowledge
EXTENT	specific	general	holistic
ECONOMY	consumption	use	production
EVALUATION	results	use	impacts
PROJECT MANAGEMENT	results	products	process
FLOW	top down	consultative	participative
GENERAL	technology	application/content	paradigm

Figure 2. The framework of the process

The sequence has its own logic, although it should be clear that obstacles do not always appear in real life in the same order as indicated here, and that support strategies can choose a different order (especially after obstacle #6).

Obstacle #1: Access/Infrastructure

The possibility one has to physically access ICT.

It is quite obvious that there is no way to give access without an adequate infrastructure. The connectivity between nodes is guaranteed by the Internet; what is left to public policies must be the final part, which unites the users and the network, whether at the individual level (i.e. an individual with his/her own computer) or at the collective level (i.e. a telecenter).

It is important to mention here the crucial issue of **accessibility**. The network and its applications absolutely must be designed in order to guarantee that disabled people can have full access, through adequate devices. For instance, a requirement would be that webpages should be designed so as to be scrolled and read by voice synthesizer software. Moreover, it is worth acknowledging that a website which follows accessibility guidelines (W3C, 1994-2006) is more friendly to all users, not only to people suffering from any kind of disability.

Obstacle #2: Access/Affordability

The balance between the price of access to the infrastructure and the financial capacities of the users.

Now, of what use is an infrastructure to me if I do not have the financial capability to pay the toll? The topic of "universal access" should be understood not only in terms of geographical cover (e.g. providing rural access), but also in terms of economic cover (providing low-income people with access).

That raises the issue of countries in the South (Pimienta, 1993) where people live in a situation caused by economic globalization: although wages are local (and generally speaking in an order of magnitude lower than in the North), prices are global, except that in many countries in the South, especially in Africa, charges for telecommunication services are even higher than in the North.

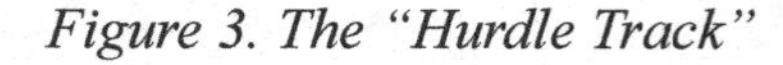

Figure 3. The "Hurdle Track"

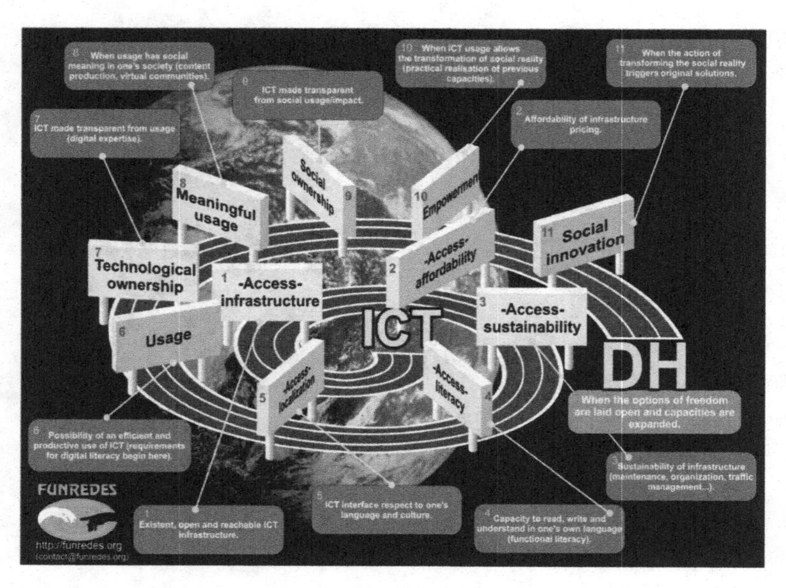

Obstacle #3: Access/Sustainability

The organization of access resources should be durable and its development should follow demand.

Now, of what use is financial capability to me if the resource is not managed professionally in order to make it work in the future, and to adapt it to a growing demand without response time becoming unbearable?

Sustainability is often a crucial problem for telecenters. In the best case, that is an organization issue. In the worst case, that is a financial issue: when external funding ends, the telecenter is not able to maintain financial balance between costs and income, or achieves it by charging a rate for its services which is higher than the market one.

Obstacle #4: Access/Basic Functional Literacy

Users should have the functional ability to read and write in order to have an adequate use.

Now, of what use is an access to information to me if I do not have the knowledge to decipher it, and to process it to produce new knowledge?

It is of course possible (and necessary) to design innovating interfaces so that illiterate people can communicate using oral language and/or very intuitive icons. However, it would be quite difficult for the use of ICT by these groups of people to reach a significant level in terms of development capacity. Let us be serious about priorities: before we imagine a digital literacy or an information literacy, let us begin with basics. Today, the first obstacle for ICT is still basic functional literacy. In the context of the development of mass media, this form of literacy cannot be limited to paper forms: it should consider new digital media (the first one being the screen), and take into account the fact that nowadays, the read/write ability must be designed in a way that integrates multimedia (because sounds and images which are integrated into the text tend to be both an integral part and an integrated part of communication).

Obstacle #5: Access/ Linguistic Localization

When using the system, people should be able to use their mother tongue.

Now, of what use is access to me if the system I use is not able to communicate in my mother tongue?

The *location of a language in the digital world* refers to the treatment of the characters written in this language, through electronic means (Diki-Kidiri, 2007). English or Spanish, more generally all languages sharing the same alphabet, are clearly localized. Let us not forget, though, the long lasting problems which were due to a codification standard (ASCII) which did not have enough bits: this made it possible for English to be perfectly localized, because the language does not use diacritics, but diacritical characters (such as ñ, é or ç) were ignored until the MIME protocol made the integration of extended ASCII possible.

However, the obstacle has not yet been overcome for many of the some 7,000 languages still in use. Unicode (Unicode, 1991-2007) is doing a great work in order to establish codes for different alphabets, yet there is a huge challenge for many languages which only exist in an oral form, and whose speakers have to agree on a written form to exist in the virtual world.

It should be underlined here that although the very existence of a written form and a code for electronic processing are basic components for localization, the requirements for a full use may well be more complex –keyboard, parser, translation software, etc. (Millán, 2000).

Obstacle #6: Actual Use

The possibility of making an effective and efficient use of ICT.

Now, of what use is an access in one's mother tongue if the person does not know how to make a relevant use of the technology?

The requirements to make effective and efficient use of ICT ask for being able to handle digital tools as well as to understand conceptual, methodological, and cultural issues in relation with the digital environment. This introduces the double concept of **digital literacy and information literacy**, which are as crucial in the North as in the South, and which can be considered as the main challenge of integrating a given nation into the Information Society.

Obstacle #7: Technological Ownership

Users should be skilful enough for the technology to be transparent for their personal use.

When people use tools skillfully enough so as not to be slowed down by technology, and so as to be able to create new uses answering their questions, they can concentrate on what they want to do rather than on how to do it. This ownership requires sophisticated capacities, which include being fluent in computer use and in editing software, as well as some expertise in searching for information whatever its digital form; that implies that the user must have come one step further than digital literacy, to a process of information literacy.

Obstacle #8: A Meaningful Use

Using ICT in a way which has a social meaning in the user's personal, professional, and community environments.

Again, this article does not deal with ICT use, but more specifically with ICT use for development. Hence, use is not restrained to playing games or communicating in an interpersonal way: using ICT makes it possible to meet some of the needs in the field of development. This includes the ability to produce content and/or to create virtual communities. The concept of "meaningful use" (a concept coined by our friend Ricardo Gómez in 2001, in Spanish, as "uso con sentido", see Gómez, 2001) implies that people evolve from being mere information consumers to producers of knowledge and social relations. That requires a fair level of information literacy, and enough digital literacy to bring complementary abilities to be a producer, of contents as well as of communities (i.e. to organize virtual communities).

Obstacle #9: Social Ownership

Users should be expert enough so that the technology is transparent from its social use.

This level requires an exact understanding of the societal impacts of ICT use and of the cultural (culture of network or information culture) and methodological aspects related to the medium. Those elements are part of the advanced level of information literacy (for a classification of the various capacities which are required for the literacy process, see Pimienta & Dhaussy, 1999 –please note, however, that the paper should certainly be updated, due to the numerous new features which have appeared since 1999 in the history of the Internet such as Web 2.0). It should be clear that the education-learning process is necessary but not sufficient: in order to reach this level, it is absolutely necessary that the acquired concepts be put into practice. At this stage, a coherent project of literacy should plan practice work and "real" deliveries as part of the curriculum –not only examples, drills and training exercises.

Obstacle #10: Empowerment

Preliminary note on "empowerment": the word stresses several important aspects at the same time: to gain capacities and the knowledge to use them to defend one's (social) causes, eventually to gain (social) power in the process.

Individuals or communities should be able to transform the social reality they live in through social ownership of ICT.

This item deals with putting into practice the acquired capacities individually and collectively. In principle, this level should be the one all the organized actors who are specialized in ICT4D have; they try to share with counterparts from civil society or communities working in other fields, for instance organizing workshops.

Obstacle #11: Social Innovation

The action of transformation should be likely to bring original solutions, designed by an individual or by the community.

"Underdeveloped people" do not exist! As an individual trained in Europe who has changed into a Latin American and Caribbean person in the last 20 years, the author is a privileged witness of a fact which is evident but often forgotten: underdevelopment is not a matter of people but of collective organization and institutionalization. And there is something more: the context of chronic difficulties in which people live in the South is a permanent engine of creativity. There is more daily creativity in a poor suburb in the South to face the permanent challenges than in a city of the North; however, the capacity to transform this creativity into innovation is limited by the lack of education and/or by the lack of "empowerment" (in the context of ICT as in any other context). For that reason it is crucial to overcome the previous obstacles, especially in the context of ICT, where adverse factors affect in smaller proportion (apart from the case of electrical power, which should be paid attention to no matter which ICT4D project in several countries), due to the virtual and global nature of the framework –which is one of the strong arguments to maintain a belief in the "digital opportunities".

The Finishing Line: Human Development

Options for individual and collective freedom become open to people (or to communities) and they have the capacity to take advantage of them.

The hypothesis could be confirmed in the future, when impact evaluation is given appropriate importance, and when criteria to analyze factors of success and of failure are agreed. By the way, the trend which consists in creating databases of "success stories" should be observed with a little sane skepticism: daily experience teaches us that

Table 2. A methodological table

INFRASTRUCTURE ACCESS
The possibility an individual or a group has to physically access ICT
FINANCIAL ACCESS
The balance between the price of access and the financial capacities of the person or the group of people
SUSTAINABLE ACCESS
The organization of access to resources should be durable and its development should follow demand
LITERACY ACCESS
Users should be able to read and write (in their mother tongue, obviously)
LOCATION ACCESS
Maternal languages should be used in interactions
USE
The possibility of making an effective (which reaches the set goal) and efficient (time-wise) use of ICT
TECHNOLOGY OWNERSHIP
Users should be skilful enough for the technology to be transparent for their personal use
MEANINGFUL USE
Using ICT in a way which has a social meaning in the user's personal, professional, and community environments
SOCIAL OWNERSHIP
Users should be expert enough that the technology be transparent from its social use.
EMPOWERMENT
Individuals or communities should be able to transform the social reality they live in through appropriating ICT
SOCIAL INNOVATION
The action of transformation should be likely to bring original solutions, designed by an individual or by the community.
HUMAN DEVELOPMENT
Options for individual and collective liberty should become open to people or to communities who are then able to take advantage of them

much more can be learnt from errors than from successes…

This hypothesis is that the people who have overcome the first 10 obstacles have exceptional opportunities for human development, and can bring change to their personal lives as well as the life of their communities. Among these people, the most creative ones can demonstrate the huge capacities for innovation that exist in the South.

Besides trying to offer pedagogical material to fight the paradigmatic divide, the "hurdle track" can be used as a table for systematizing methodology. It has indeed been used as such in one of its early versions, in order to identify the obstacles to overcome in considering the question of linguistic diversity on the Internet (Pimienta, 2006). It should be possible to use it in the same way with other issues, as has already been done in various workshops.

CONCLUSION

The main thrust of this article has been the issue of the importance of education, to reach a critical mass of citizens who can take part in the current transformation of societies, and who do not mistake the technology for the paradigm shifts at stake (or avoid being mistaken on that matter).

Figure 4. The three pillars of shared-knowledge societies

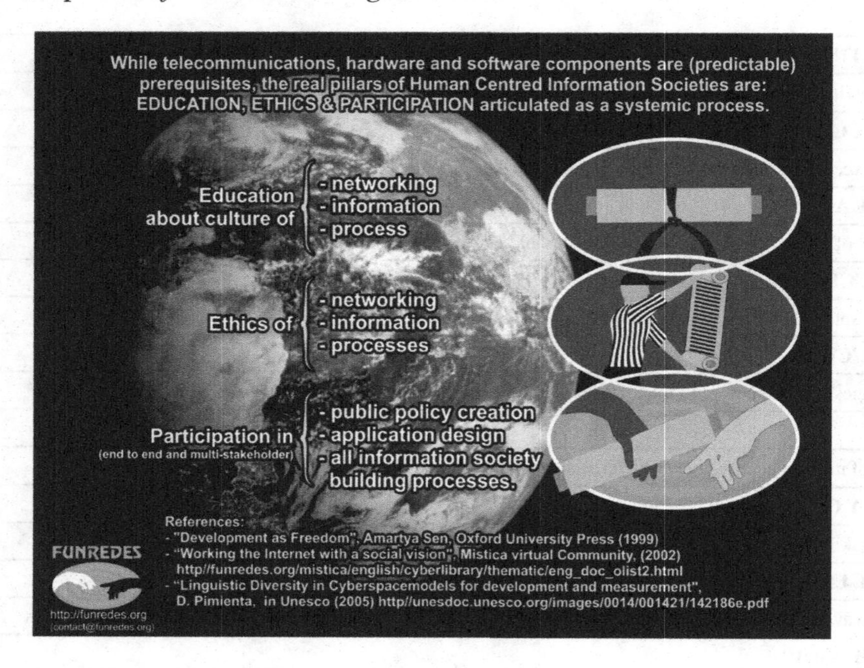

The recurring cry of the document has been the issue of the requirement for a real and organized multi-stakeholder partnership to build new social projects, a requirement too often misled by fake promises.

It should be clear that we are within a systemic process, where each component interacts with the others: the education process must be participative and education is necessary so that all actors of the society participate.

Ethics is the third pillar of an appropriate process for the building of shared-knowledge societies. About this new coined word, Adama Samassekou can be credited with the phrase "Sociétés des savoirs partagés", which can be translated into English as "shared-knowledge societies", bearing in mind that English does not distinguish between the concepts of "savoir" and of "connaissance". Samassekou is President of the Academy of African Languages and of the MAAYA Linguistic Diversity Network (http://maaya.org); he facilitated the first part of the WSIS process. He thus gave an answer, from civil society, to the limitations of the phrases "information society", "knowledge society", or "communication society". The plural form of "society" insists on the following point: no single model exists, and each nation must build its own, according to its culture and its history (see Ambrosi, A., Peugeot, V., Pimienta, D., eds., 2005). It has not been explicitly dealt with ethics in this article although it constantly appears as a watermark. Ethics consists, more precisely, in information ethics, communication ethics, and network ethics. In this context, ethics is subject to the same systemic conditions; that raises the multiple issue of the necessity of ethics in education, of education in ethics (Toro, 2000), of ethics in participative processes (Pimienta, 2007) and of participation in an ethical discourse which must explore new frontiers.

REFERENCES

W3C (1994-2006). *Web Accessibility Initiative*. Retrieved December 30, 2007, from http://www.w3.org/WAI

Alfin (2006-2007). *Blog sobre alfabetización informacional.* Retrieved December 30, 2007, from http://www.alfinred.org/blog

Alfin (2006). *Bibliotecas por el aprendizaje permanente, Declaración de Toledo sobre la alfabetización informacional.* Retrieved December 30, 2007, from http://www.lectores.info/formacion/file.php/38/Modulos/Documentos/Dec_Toledo.pdf

Ambrosi, A., Peugeot, V., & Pimienta, D. (Eds.). (2005). *Word Matters, Multicultural perspectives on information societies*, C&F Editions. Online version retrieved December 31, 2007, from http://www.vecam.org/article698.html?lang=en

Diki-Kidiri, M. (2007). Comment assurer la présence d'une langue dans le cyberespace? *UNESCO.* Retrieved December 31, 2007, from http://unesdoc.unesco.org/images/0014/001497/149786F.pdf

ETIC. (2005). *Estrategia Boliviana de TIC para el Desarrollo.* Retrieved February 12, 2008, from http://etic.bo

FUNREDES. (2005). *UNDP.DO supports Multi-stakeholder Partner Cheap.* Retrieved December 30, 2007, from http://funredes.org/undp.do

FUNREDES. (2003-2005). *Multistakeholder partnership methodology.* Retrieved December 30, 2007, from http://cmsi.funredes.org/inc/multistakeholder_en.htm

Gómez, R. (2001). *MISTICA: Re: Comunidades virtualizadas?* Retrieved December 31, 2007, from http://funredes.org/mistica/castellano/emec/pro/memoria6/0145.html

Illustrated version available in Spanish (2004). Retrieved December 29, 2007, from http://funredes.org/mistica/castellano/ciberoteca/tematica/trabajando.pdf

Jansen, S., & Pimienta, D. (2007). *Perspectivas de la Cooperación Sur-Sur (CSS) en el marco de las Sociedades de los Saberes Compartidos: Visión desde el terreno.* Retrieved December 30, 2007, from http://funredes.org/mistica/castellano/ciberoteca/tematica/css-si-final.pdf

Millán, J. A. (2000). *La lengua que era un tesoro: el negocio del español y como nos quedamos sin el.* Retrieved December 31, 2007, from http://jamillan.com/tesoro.htm

MISTICA. (1999-2006). *Methodology and Social Impact of Information and Communication Technologies in America.* Retrieved December 29, 2007, from http://funredes.org/mistica

Pimienta, D. (1993, August). *Research Networks in Developing Countries: Not exactly the same story!* Paper presented at INET'93, San Francisco, CA. Retrieved December 30, 2007, from http://funredes.org/english/publicaciones/index.php3/docid/31

Pimienta, D. (2002). *The Digital Divide: the same division of resources?* Retrieved December 29, 2007, from http://funredes.org/mistica/english/cyberlibrary/thematic/eng_doc_wsis1.html

Pimienta, D. (2005, June). *Una vision desde la sociedad civil.* Presentation made at the Regional Preparatory Conference of Latin America and the Caribbean for the Second Phase of the World Summit on the Information Society, Rio de Janeiro, Brazil. http://www.riocmsi.gov.br/english/cmsi

Pimienta, D. (2006). Measuring linguistic diversity on the Internet. *UNESCO.* Retrieved December 31, 2007, from http://unesdoc.unesco.org/images/0014/001421/142186e.pdf

Pimienta, D. (2007). At the Boundaries of Ethics and Cultures: Virtual Communities as an Open Ended Process Carrying the Will for Social Change (the "MISTICA" experience). In Capurro, R. & al. (Eds.), *Localizing the Internet. Ethical Issues in Intercultural Perspective* (pp. 205-229). München: Fink Verlag. Online version retrieved December 31, 2007, from http://funredes.org/mistica/english/cyberlibrary/thematic/icie/

Pimienta, D., & Blanco, A. (2007). *The Hurdle Track from ICT to Human Development, first published in English at Global Knowledge Partnership Beyond Tunis: Flightplan.* Retrieved December 30, 2007, from http://www.globalknowledge.org/gkpbeyondtunis/INDEX.CFM?menuid=33&parentid=30

Pimienta, D., & Dhaussy, C. (1999). *Users Training: A Crucial but Ignored Issue in Remote Collaborative Environments.* Retrieved December 31, 2007, from http://www.isoc.org/inet99/proceedings/posters/157/index.htm

Pimienta, D., & al. (1993-2007). *Presentaciones de Funredes (y de sus asociados en proyectos).* Retrieved December 29, 2007, from http://www.funredes.org/presentation/

Prague Declaration. (2003). *Towards an Information Literate Society.* Retrieved January 31, 2008, from http://www.nclis.gov/libinter/infolitconf&meet/post-infolitconf&meet/PragueDeclaration.pdf

Toro, J. B. (2000). *Educación para la democracia.* Retrieved December 31, 2007, from http://funredes.org/funredes/html/castellano/publicaciones/educdemo.html

Unicode (1991-2007). *Unicode Works.* Retrieved December 31, 2007, from http://www.unicode.org

United Nations. (2000). *United Nations Information and Communication Technologies Task Force.* Retrieved December 29, 2007, from http://www.unicttaskforce.org

United Nations, International Telecommunication Union. (2003-2005). *World Summit on the Information Society.* Retrieved December 30, 2007, from http://www.itu.int/wsis

Virtual Community, M. I. S. T. I. C. A. (2002). *Working the Internet with a Social Vision.* Retrieved December 29, 2007, from http://www.funredes.org/mistica/english/cyberlibrary/thematic/eng_doc_olist2.html

Wikipedia. (2007). *Information Literacy.* Retrieved December 30, 2007, from http://en.wikipedia.org/wiki/Information_literacy

This work was previously published in Information Communication Technologies and Human Development, Volume 1, Issue 1, edited by Susheel Chhabra and Hakikur Rahman, pp. 33-48, copyright 2009 by IGI Publishing (an imprint of IGI Global)

Chapter 4
Reorganizing People in Customer Knowledge Management Change

Minwir Al-Shammari
University of Bahrain, Kingdom of Bahrain

ABSTRACT

The transition to knowledge-intensive customer-centric enterprise is important, but never easy. Reorganizing people is likely to face critical structural and cultural change issues related to people. Addressing these issues is essential for the continued success of customer-value-building services and products. In light of today's competitive business environments and changing power of customers, organizations need to be able to deal with people-based issues in order to secure high quality customer service and long-life and profitable customer relationship. The chapter presents a recommended solution to deal with people change management in competitive business environments, viz. to 'reorganize people' in a customer-centric networked organization. 'Reorganization of people' is operationally defined by three sub-interventions: a) reconfiguring structure, b) reshaping culture, and c) rehabilitating people.

INTRODUCTION

Today's fast-changing business world is witnessing aggressive fluctuations, higher degrees of uncertainty, and fierce competition. The changing nature of business environments requires high organizational requirements as well as high involvement from people. The increasing dominance of knowledge as a basis for improving efficiency and effectiveness of organizations triggered many companies to find new ways of utilizing knowledge they have gained in devising or improving their business practices (Awad and Ghaziri, 2004). A knowledge-based customer-centric strategy is

DOI: 10.4018/978-1-60960-497-4.ch004

centered on the creation of DCC based on customer knowledge with the aim of creating sustainable competitive advantage (SCA) for the business. As the long-term objective of business competitive strategies is to build SCA, focus should be on 'difficult-to-imitate' resource-based capabilities (Salck et al., 2006). The competitive advantage of imitable resources is short-lived; it may soon be rapidly imitated by a capable competitor or made obsolete by an innovation of a rival.

This paper seeks to examine the role of people in the implementation of customer knowledge management (CKM) strategic change. The ability of an organization to compete in rapidly changing business environments is contingent upon its ability to develop competitive strategies that enable leverage of distinctive core competencies and delivery of value-adding products or services to customers. Once the knowledge-based customer-centric competitive strategies have been identified, a plan is developed to 'reorganize people' in order to enable the CKM change strategy. 'Reorganizing people' is used in this paper to refer to transformation of organizations from hierarchical to networked organizations, restructuring of units in which people operate into self-controlled teams and assignment of 'case managers', and changing the corporate culture and leadership style of the newly formed organizations.

Two basic perspectives are used to relate to the process of reorganizing people: structure and culture. It is true that sometimes terminologies are used in a vague or contradicting manner. As of the term 'reorganizing', it could mean different things to different people. For instance, Weiss (2001) offered a contribution which explained three approaches to reorganization: restructuring, reengineering, and rethinking. Restructuring involves the redesigning of organizational units through initiatives such as downsizing, reengineering refers to attempts to introduce dramatic change in business processes, whereas rethinking involves the redesign of thinking and mindset through initiatives such as the learning organization.

Role of People in CKM Change

People refer to human resources, such as front-line staff, support staff, business managers, as well as general managers, as well as knowledge workers who are involved in CKM activities. Knowable workers are those employees who can think or work with ideas. Knowledge workers add value to a company's products and services and have direct impact on the efficiency and productivity of the work process by capturing, applying, sharing, and disseminating their knowledge within the organization are called knowledge workers (Awad and Ghaziri, 2004). A knowledge worker is the 'product' of experience, values, processes, education, and the ability to be creative, innovative, and in tune with the culture of the company. Knowledge worker is the one who wants a challenge and to be on a winning team. Examples of knowledge workers are managers, lawyers, engineers, system analysts, strategic planners, market analysts, and accountants (Awad and Ghaziri, 2004). Other remaining categories of employees may be considered as support to knowledge workers.

People in Structural Change

Organizational system and its components can be analyzed using the analogy of human body. De Wit and Meyer (2004) offered a way to divide organizational systems into three parts: anatomy (structure), physiology (processes), and psychology (culture). Salaman and Asch (2003) classified organizations based on three components through which the capability of organizations is produced, i.e. organization structures, organization systems and processes, and organizational cultures.

Due to the dynamics of today's business environments, and the shift towards knowledge-based customer-centric organizations, the above classifications may not suffice to analyze organizational dynamics, and thus need to be expanded. In line with the proposed CKM framework, Figure 1 presents a seven sub-system classification of organizational systems that are analogous to human body as follows:

- **Analytical (mind):** knowledge, intelligence, and continuous learning
- **Physical (bones):** form (structure), technology, and other tangible resources
- **Physiological (flesh):** content
- **Managerial (heart):** planning, coordination, command, control, and adjustment processes
- **Informational (blood):** data and information
- **Psychological (soul):** people and culture
- **Social (social interactions):** relationships with the external environment, i.e. customers, suppliers, business partners, and alike

People in Cultural Change

When addressing the role of people in organizational structures, it is equally important to address existing corporate culture, the type of culture that the organization is trying to foster, and bridge the gap between the two by revamping the existing set of cultural values accordingly. Corporate culture has been recognized as a pervasive force influencing organizational effectiveness.

Deal and Kennedy (1982) have conceived culture, rather than structure, strategy, or politics, as the prime mover in organizations Cultural change programs start with identifying current shared organizational values and norms, and then proceed to identifying what the culture should be, and end with identifying the gap between the two and developing a plan to close it (Morgan & Studrdy, 2000)

Corporate culture plays an integral role in knowledge sharing among people and in a successful development of a CKM strategy. It helps in fostering or hindering information and knowledge flow in an organization and in promoting distinction in delivery of customer products or services. Although are considered important tools for disseminating information or knowledge within an organization, ICTs alone cannot secure efficient flow of information or knowledge, if corporate culture is not conducive to knowledge sharing among employees. People in the organization need to be capable, willing, and ready to share knowledge or provide a high quality product or service.

Besides, corporate culture plays a significant role in facilitating or hindering the delivery of products or services preferred by customers, and

Figure 1. Anatomy of knowledge-based customer-centric organizations

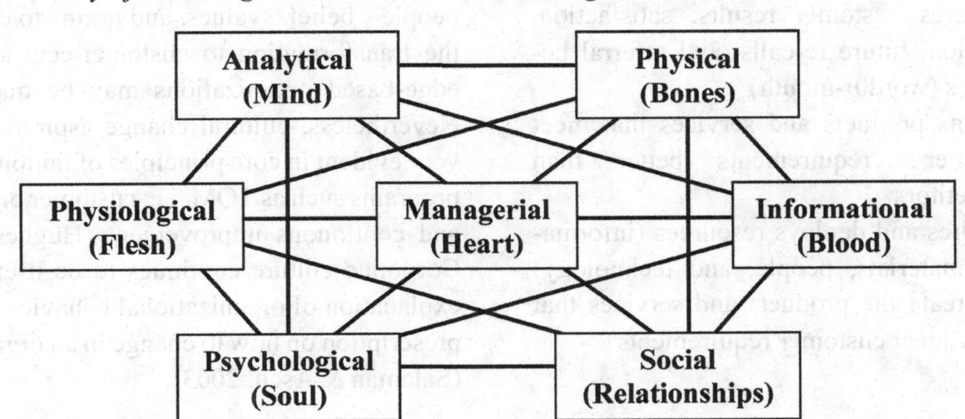

ultimately accomplishment of organizational effectiveness. Culture has got an influence on people's behavior and many aspects of organizational life starting from product or service planning and design, and development and ending up with marketing, sales, delivery, and customer service. As of the CKM change strategy, it is extremely essential for businesses to realize the importance of shifting their corporate culture from internally oriented to customer-oriented in order to be able to create SCA. Therefore, CKM change requires an analysis of the level of commonality of culture (breadth of widely shared beliefs, values, and norms) and plan for implementation of a cultural transformation program at the individual, team, and organizational levels. However, one should note here that cultural change programs that address all the three levels of depth of culture may require from two to five year implementation programs (Davenport, 1993).

Therefore, organizations need to transform to a customer-centric cooperative and knowledge sharing rather than competitive and knowledge hoarding culture. More specifically, an effective customer-centric corporate culture is the one that (Buttle, 2004):

- Identifies which customers to serve
- Understands customer's current and future requirements
- Obtains and shares customer knowledge across the enterprise
- Measures customer results: satisfaction, retention, future re-calls, and referral behaviors (word0f-mouth)
- Designs products and services that meet customer's requirements better than competitors
- Acquires and deploys resources (information, materials, people, and technology) that create the products and services that satisfy meet customer requirements
- Develops the strategies, processes, and structure that enable the company to satisfy customer needs

However, changes in culture rather than structure or technology, are the most difficult to undertake among various pillars of organizations. This difficulty is due to several factors such as:

- The enormous amount of effort and time that is required to create noticeable changes. Cultural change takes place through an ongoing socialization process that may take whole life span of employees.
- The feasibility and ethicality of organizational change to accomplish lasting and long-term change in beliefs, values, and norms of people is questionable (Salaman, 1997).
- The employees' resistance to organizational change, viz. business strategy, job design, organizational structure, business processes, and technology, which works in the opposite direction of the change program.

Although considered essential and having an influential impact on the success of customer-centric business transformation, organizational culture has been viewed as intellectually flawed and practically impossible (Ogbonna & Wilkinson, 2003). Furthermore, the feasibility of changing people's beliefs, values, and norms to cope with the transformation to customer-centric knowledge-based organizations may be questioned. Nevertheless, cultural change aspirations were very evident in core principles of famous change programs such as TQM, i.e. customer orientation and continuous improvement (Hughes, 2006). Corporate culture continues to be useful as an explanation of organizational behavior than as a prescription on how to change in an organization (Salaman & Asch, 2003).

Issues in Managing People

Reorganizing people in customer-centric teams carry with it a number of issues as follows:

Attracting People

Attracting the 'right' applicant to fill the required skill, knowledge, or behavioral characteristics gap would have its profound impact on the accomplishment of business strategies, goals and objectives, and ultimately on the success of its CKM strategy. Recruitment of people in customer-centric organizations is usually based not only on the applicant's qualifications, but also on the extent to which the job applicant's cultural values 'fit' with the required organizational culture, team-based work conditions, and the prospects of the applicants to add value, develop, and grow within the organization (Peppard & Rowland, 1995). But, the challenge that faces CKM strategies is that not all individuals are willing, capable, or ready to work in teams.

Managers alongside other personnel staff usually perform the selection process, yet there are increasing numbers of instances where teams play an important role in 'choosing their colleagues', or even customers choosing their service providers. For example, South West Airlines used its frequent flyers to select cabin crew (Heskett, et al., 1994). However, the challenge that remains facing customer-centric organizations is how to assess the potential of a person to add value to customers as well as to the company. The degree of 'fitness' with the required organizational culture is very difficult to assess, and remains a rather subjective and illusive quality trait. However, sometimes, motivation and intellectual ability may provide a sound guide for selection (Peppard & Rowland, 1995).

Developing People

Development of people refers her to a set of activities that are initiated by customer-centric companies on-the-job (i.e., job placement and job rotation) or off-the-job (i.e., training workshops and professional seminars, and conferences), with the aim of improving the value-adding contribution of human resources through:

- Acquisition of new customer-centric ICT and job-related skills and knowledge (i.e., marketing, sales, customer service). Examples are:
  - How to capture customer data? How to survey customers?
  - How to analyze customer's data?
  - How to profile or segment customers?
  - How to design a customized campaign program?
  - How to create customized product or service?
- Acquisition of customer-focused team-based behavioral characteristics and decision-making skills (soft skills): refers to values, beliefs, attitudes, and behaviors. Examples are trustworthiness, innovation and creativity, perfection, team spirit, and risk taking.
- Continuous customer-oriented learning and improvement: refers to learning how to design a new order fulfillment procedure that requires less time, effort, and money, and ultimately, pleases customers.

Having the right behavior with customers shall at the end be reflected on customer's purchasing behavior and decisions. For example, front-office staff cannot be polite, courteous, and committed to their clients, if their own internal organization behaves rudely and indifferently. The behavior of front office employees will have a direct bearing on perceptions, loyalty and retention of customers. Similarly the behavior towards suppliers will determine the perception of the company and play a part in motivating partnerships between companies. However, behaving well towards customers and people is not enough. Customers get fed up with the most attentive staff if they simply cannot actually do it, so skills of staff are also necessary (Peppard & Rowland, 1995).

In this regard, traditional organizations cope with major changes by assigning functional units to selected parts of the issue or problem- temporarily removing the burden from view. But for a new form of organization to be a true 'learning organization,' it must develop an atmosphere that is conducive to long-term rather than short-term benefits (Wysocki & DeMichiell, 1997). In new forms of organizations, the role of people has shifted from doing to improving the work. If 'improving' is becoming an equal to 'doing' for each employee, then 'learning' as well as 'performing' is becoming a key objective for the company as a whole (Peppard & Rowland, 1995).

Planning for training and development needs is based on gap analysis of current versus desired levels of skills, knowledge, and behavioral characteristics as well as of customer-centric teams. Skills, knowledge capabilities, and behavioral characteristics of customer-centric team members should be carefully planned, designed, and developed in order to help in delivering successful products or services, achieving business goals and objectives, and ultimately business strategies. Current and desired corporate cultural values and the degree of empowerment will have to be also considered when identifying the necessary behaviors, skills, and subsequent training and development needed expanding a team member's ability to secure a high quality customer service.

Achieving significant changes in a team member's capacity is required for achieving improvement in customer products or services, but is not enough. Self-controlled teams should also be expanded to self-learning teams. Continuous learning is essential to cope with continuous environmental changes. It is not enough for employees to only learn the 'how to do things', but they should also learn how to solve business problems, how to add value, and how to develop and maintain interpersonal relationships within and outside the organization. Therefore, customer-oriented development of human resources capacity to provide high quality customer service should take place on an ongoing basis that starts with the introduction of new staff members to a company and its teams, and continues throughout people's career life.

However, not all people are willing, ready, or capable to work in teams. Team-based work may not be successful all the time, even if we manage to select those who are capable of doing so, Sometimes working in teams may be unfruitful, unharmonious, and does not lead to the desired work synergy (the team's output is greater than that of individuals acting alone). Working in teams may pose a challenge to interpersonal relationships among team members, and may create problems such as personal clashes and conflicts, groupthink, and time consumption (Peppard & Rowland, 1995). The situation will be more aggravated when national cultural values in which organizations operate are intolerant of diversity, but supportive of uniformity that shuns personal differences and sees them as equal to personal hostilities.

The chronic issue that still faces organizations is the viability of the decision to change people's skills, values, beliefs, attitudes, and behaviors. To a large extent, developing team members' hard skills (i.e. structured work-related knowledge and skills) of people is easy. In contrast, developing team-members' soft skills (team-based work values, attitudes, and behaviors) is much harder to implement and reach fruitful results, especially in the short-run time horizon and with contradicting national cultural values. Similar difficulty exists in the attempt to introduce change in the culture of organizations, which are, by nature, enduring and deeply rooted. Supportive and visionary leadership plays a crucial role in successful structural and cultural change programs.

Maintaining People

What is important is not only to attract and develop the capacity of team members, but also to be able to keep them loyal, committed, and ultimately retained. Companies need to continue to provide attractive motivation, both intrinsic and extrinsic, in addition to management support. Compensation systems may be viewed by some people as extrinsic motivators, but may be viewed by others as 'hygiene' factors', which do not motivate workers, but rather, may lead to job dissatisfaction if badly designed (Herzberg, 1987).

It is not expected of teams to perform well from day number one of their forming. Usually team working evolves through four stages (Jassawalla & Sahittal, 1999): forming (acquaintance), storming (interactions), norming (acceptance), and performing (goal attainment). However, the challenge that may continue to face such organizations is how to evaluate performance of customer-centric team members' activities, how to align the pay scale with flexi-working hour systems, how to differentiate between high performers and average performers within the same team when applying team-based compensation, and how to improve people's loyalty, job satisfaction, retention, and ultimately job performance.

Traditional productivity measures are frequently inadequate, inaccurate, and may be inapplicable in customer-centric networked organizations. The traditional way of rewarding (e.g. factory floor workers, sales people) was piece rate based, where people are paid according to the number of 'pieces' they process? For example, how could one differentiate between two knowledge workers who provided same customer service? Is it by time taken? Is it customer's value of purchase? Is it customer retention rate? Is it number of customer complaints? Is it percentage of new customers acquired? Is it customer satisfaction?

Modern organizations, especially virtual ones, often face the challenge of supervising and evaluating a workforce that is geographically spread across the world, working in isolation from direct supervision, and working more in teams. Rather than working in a central office, many salespeople work remotely and rely on hand-held pen computers, cellular phones, and pagers to link them to customers and the head office. The nature of team-based work makes it hard to apportion individual-based rewards (Pearlson & Saunders, 2006). Therefore, direct employee supervision may need to be replaced by electronic tracking of employees' activities, such as the number of calls processed, e-mail messages sent, or time spent surfing the Web.

Virtual teams cannot be managed in the same way in more traditional teams. The differences in management control of performance activities are particularly pronounced. Monitoring behavior of virtual teams is likely to be more limited than in traditional teams, as behavior of virtual team members cannot be easily observed. Therefore, performance is more likely to be evaluated in terms of output than on displays of behavior (Pearlson & Saunders, 2006). Therefore, evaluation of employees may be partially conducted by using objective compensation systems that reward people for deliverables produced (i.e., a report produced by certain date) or targets achieved (i.e., sales quota), as opposed to subjective systems that emphasize factors such as 'attitude', feel, etc (Pearlson & Saunders, 2006). However, in CKM strategy, subjective performance aspects of the work, such as quality of service and interactions with customers, are considered as important as objective performance measures in creating and maintaining relationships with customers, and cannot be easily skipped

In networked organizations, there is no hierarchical and departmental status, but empowerment and an appreciation of the team as a whole, e.g., the name of every team member is shown on business cards and pamphlets (Peppard &

Rowland, 1995). As organizations migrate from traditional structures to new forms of organizations, so should their compensation systems. People in traditional organizations may consider 'low pay' as a cause of dissatisfaction, but may not consider 'high pay' as a cause of satisfaction. In contrast, members of customer-centric self-managed teams may consider job-related intrinsic factors, such as empowerment, team membership, management recognition, and self-actualization, as replacements to materialistic motivators, viz. salary increase.

Therefore, new forms of customer-centric competitive organizations should encourage:

- Team-based compensation
- Customer and quality focus evaluation: performance evaluation system is based on the contribution of team members to the well being of customers.
- Knowledge-sharing focus evaluation: knowledge-sharing behaviors need to be incorporated in performance appraisal systems, and rewarded through recognition, pay raise, and financial incentives.
- Continuous learning and value-adding customer offerings.

Managing Resistance to Change

Organizational transformation decisions, like downsizing, automation, or process revamping have got an inescapable cost of eliminating some positions such as low-level service or clerical jobs and even middle managers. In collectivist cultures, people openly criticize companies that lay-off people because they cut off their salaries. For instance in the Arab culture, people frequently repeat this saying 'hanging by the nick is better off than cutting-off means of living'. Whenever companies undertake major reengineering programs, people's resistance to change is expected to intensify especially in collectivist societies. In cultures known for their uncertainty avoidance, shunning off risk taking, and high fetish for conformity and passive stability, reengineering is viewed as a threat to people's job security. The challenge here is how to handle or cope with such resistance to change. Companies need to offer more educational and awareness programs before and throughout the change itself, and offer job placement advice service and post-termination support programs for 'victims' of the change program.

The shift from individual-based to team-based reward systems may be challenged by lack of cooperation among team members (Pearlson & Saunders, 2006). Organizations need also to be aware of the sensitivity of change to people in their organizations. For instance, compensation systems that try to devise new appropriate ways to provide rewards to team members may cerate negative reactions from employees. Another challenge is related to culture-sensitivity of some structural change decisions such as in compensation systems. For example, in national cultures with higher levels of individualism, many workers may prefer reward systems to be linked with the performance of individual employees, whereas same reward system may be counterproductive in a more collectivistic culture (Griffin, 2005).

Leading People

CKM change requires visionary, inspiring, and supportive leadership who can coach not boss. The new leadership role is to provide support and a clear strategic vision of the change program but should not be promising more than what can be realistically delivered. New leaders coach and sponsor rather than direct or give orders, and may not be the most senior in the team, but need to possess an admirable work-related knowledge. New leadership also needs to have a total rather than a partial view of the work (holistic that looks at the work as a one whole unit), a participative rather than authoritative style, a friendly rather than reserved attitude, and a customer rather than product orientation.

Leadership in CKM transformation programs is especially important to decide on the level of the program (i.e., operational, analytical, or strategic), prioritize the CKM program over other wide-scale organization programs, provide high-level ownership, support, and oversight of the project, and break down the business functional silo walls because CKM programs are cross functional in nature.

The CKM change programs need to be championed at CEO level. However, a lower level of change, i.e. operational CRM projects, needs champions at senior functional management level such as chief marketing officer or sales manager. Analytical CRM needs champions at lower levels yet. In general, CKM champions tend to reside at higher levels or at marketing, sales, or service functional levels. However, it should be noted that if ICT people, with limited business knowledge, champion CKM, there is a danger that it will be seen as an implementation of a pure ICT project, at the expense of its potential business benefits (Buttle, 2004).

Empowerment provides employees with intrinsic rewards and a higher moral status, but not all people have got the preference and ability for empowerment. Some people may feel uncomfortable with works that do not follow clear and structures rules and regulations. Such a preference is not purely an individual choice per se, rather, it could relate to national cultures that shun risk taking in favor of uncertainty avoidance. However, the challenge that faces organizations is how to decide on the appropriate level of empowerment provided to employees, and sometimes to customers or suppliers.

Empowerment needs to be advocated the same way should technology be advocated; it should be appropriate (Peppard & Rowland, 1995). The appropriate level of empowerment is based on two factors: the extent of organizational empowerment and people's preference and ability for empowerment. Organizational empowerment refers to the extent to which the organization defines systems and procedures that staff must work to. On the other hand, people's preference and ability for handling empowerment refers to the extent to which people are comfortable, motivated, and able to take the initiative to work without strict procedures (Clutterbuck et al., 1993).

Empowerment should be provided at its right level to the right people at the right time. The 'appropriate' level of organizational empowerment itself is usually contingent upon the situation. For instance, low empowerment is need in some situations, such as the rigid standards must be adhered to such as financial procedures and guidelines, whereas considerable discretion in meeting clients' needs may be needed in others, such as attending to an ad-hoc request of customers. However, the real challenge is when leadership needs to balance between employees' ability and preference for empowerment from one side, and customer satisfaction with quality of the service from the other side. For instance, nothing is more annoying to customer than when the person attending to their needs has to continually refer back up the hierarchy to obtain approval for a particular request. Disempowerment can lead to an extension of lead times, dissatisfied customers, and a general inability to innovate.

Managing Knowledge Workers

The need for knowledge workers in specific continues to grow as the importance of knowledge-based competition grows among business firms. However, managing knowledge workers usually pose many challenges to organizations. Knowledge workers often like to work independently, require extensive and highly specialized training, define performance based on terms recognized by other members of their profession rather than their organization, (Griffin, 2005).

Organizations nowadays face the challenge to attract, evaluate, compensate, and retain self-directed knowledge workers. Large companies compete for the attraction of knowledge workers, and work hard to retain them, but not every organization is willing to make the human capital investments necessary to take advantage of these jobs. A special importance is to be paid to knowledge workers' professional and soft skills, and the match with the requirements of the job. The knowledge worker is expected to possess both professional and soft skills. Professional skills relate to technical skills and abilities, whereas soft skills relate to s sense of cultural, political, and personal aspects of knowledge in the business. The personal aspects of knowledge include open, candid, and effective communication skills, a warm and pleasant personality that nurtures knowledge creation, manipulation, sharing, and application in a group setting, sensitivity to the political pressures in the department or organization in general (Awad & Ghaziri, 2004).

Measuring productivity of knowledge workers is not as simple as traditional piece rate performance evaluation (e.g., number of units sold, number of units produced, and number of customer served). Furthermore, performance of knowledge workers may fall below organizations' expectations. Several factors may limit knowledge worker's performance (Awad & Ghaziri, 2004):

- Time constraint: As there is always more work and less time to do, either quality level or completion time might lag behind targets. Motivation is also affected where urgency supercedes motivation.
- Working smarter and harder but accomplishing little in the short-run: limited time, effort, and manpower are often behind frustrating results.
- Doing work that the firm did not hire them to do.
- Heavy work demands invariably affect a knowledge worker attention span, motivation, and patience, regardless of pay or benefits.
- Dislike of ideals proposed by management, avoidance

Although knowledge workers are usually highly paid compared to other people in the same organization, however, they may monitor the going salary rate in the market, and if they find it higher, it may adversely affect their continuity on the job. Managing knowledge workers with control of corporate knowledge as the core asset of business requires a 'handling with care' approach. Carefully designed and customized systems for selecting, evaluating, and compensating knowledge workers help a lot in reducing their prospective mobility. Sometimes, leadership support and favorable corporate culture may create a noticeable impact on alleviating possible drainage of intellectual assets of organizations.

A PROPOSED MODEL FOR THE REORGANIZATION OF PEOPLE

The recommended solution to deal with people in CKM change is to 'reorganize people' in a customer-centric networked organization. 'Reorganization of people' is operationally defined by three sub-interventions: (a) redesigning structure, (b) reshaping culture, and (c) rehabilitating people.

Redesigning Structure

Turbulent business environments are creating complex problems for business organizations, which cannot be solved by traditional solutions. One of these solutions is organizational design. It is almost becoming a fact of life that there is no one best way to design organizations, as the best design is contingent upon many external as well as internal factors such as the organization's environment, goals, size, strategy, and technology (Bowditch & Buono, 2005).

Organizations adopt a flexible organizational design forms, i.e. the networked organization, or at another extreme point may even adopt a VC design. In between, some organizations, i.e. banks, may take a mediocre design choice by adopting

Figure 2. A recommended model for the reorganization of people

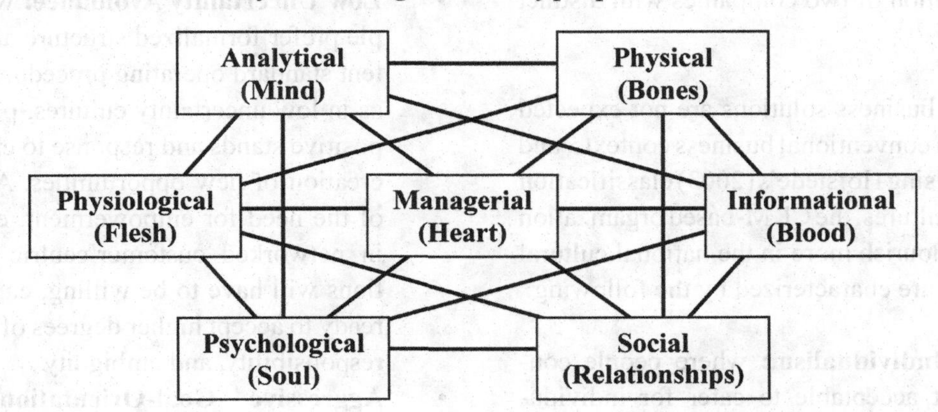

a hybrid design, which combines the features of both hierarchical and networked organizations, thus creating a mixed balance of centralization and decentralization. In such situations, it may well be appropriate if customer-facing front-office units such as sales, billing, and customer services follow a networked team-based form, whereas back-office supporting units, such as purchasing, human resources, accounting, and ICT services are kept under the functional and hierarchical structure.

Reshaping Culture

CKM organizational transformation requires not only changes in structure, but also nurturing knowledge-sharing customer-oriented culture. Knowledge sharing culture helps employees in handling customer complaints and converting these complaints from being a challenge to being an opportunity. The ability of employees to excel in handling customer complaints as opposed to their competitors would give them a SCA in terms of creating customer satisfaction and ultimately customer profitability.

The knowledge-sharing culture may be fostered through incorporating it as an element in both performance evaluation and pay and rewards systems, mentoring programs to senior members to transfer their knowledge, training programs in knowledge-sharing methods, and informal organizational gatherings and to improve interpersonal relationships among employees.

The structural and cultural changes would be more effective once they were compatible with the national cultural fabric in which an organization operates. National socio-cultural settings do have various profound impacts on product or service delivery, standards of business conduct, and ethics that the society is considering appropriate or inappropriate. For instance, consumer preferences for color, style, taste, and so on may change from one place to another.

As culture represents an integral part of customer-centric organizational transformation, therefore, management of changes in culture becomes very crucial for customer-oriented organizations in order to:

- Enable maintenance, innovation and development of the existing customer-oriented value-adding cultural values.
- Foster a new customer-oriented value-adding corporate culture that prevails throughout the organization and emphasizes values such as excellence, trust, respect, teamwork, and focus on achievement.
- Resolve or minimize conflicts between subcultures within different teams or units, especially in the case of merger between or

acquisition of two companies with distinct cultures.

Creative business solutions are not expected to flourish in conventional business contexts, and vice versa. Using Hofstede's (2003) classification of national cultures, the CKM-based organization is likely to flourish more in the national cultural settings that are characterized by the following:

- **High Individualism:** where people consider it acceptable to cater for individualized preferences of customers, and let these preferences take premium over those of masses of customers in the market. Providing customized product or service may be seen as unacceptable favoritism treatment somewhat of discriminatory nature that is based on purchasing power of customers. That is why it is quite possible to find consumers in some countries; for instance, who are willing to pay premium prices for tailor-made clothes, whereas consumers in other countries may be unwilling to pay that premium, and in turn, prefer to purchase ready-made ones.
- **High Power Tolerance:** Socio-cultural values also affect the way workers in a society feel about the importance of their jobs and organizations (Griffin, 2005). The role of superiors is changing from directing to sponsoring, coaching, guiding, and mentoring. The role of employees is also changing from receiving orders to being fully trusted with the power to do the job. Employees' empowerment enables the delivery of higher value to customers, but it works well in power tolerance social contexts, where less significance is attached to a person's position in the hierarchy, and control of power is no longer resides in the hands of superiors, but distributed and shared by all staff members.
- **Low Uncertainty Avoidance:** where people prefer formalized structure and consistent standard operating procedures, whereas in low uncertainty cultures, people take positive stands and response to change and creation of new opportunities. As a result of the need for empowerment, employees in networked customer-centric organizations will have to be willing, capable, and ready to accept higher degrees of authority, responsibility, and ambiguity.
- **Aggressive Goal-Orientation:** where people in this culture place a high value on the purchasing power and financial worth of customers, but this is done through building, maintaining, and expanding relationships with customers, and being concerned with their welfare.
- **Long-Term Time-Orientation:** organizations are supposed to be concerned with customer throughout their life cycle time. The suitable cultures for CKM strategic change are those that carry a mixture of short-term and long-term time orientations. In hybrid time outlook, people prefer delivery of products or services that provide more immediate rewards from customers at the early stage of customer's life cycle (acquisition), maintain customer relationships at the mid-term horizon in order to cast intermediate rewards, and work hard for many years to get more rewards on the long-run (expansion of profitability from customers).

Rehabilitating People

A rehabilitation of staff knowledge and attitude towards customers is essential in the development of a CKM-based organization. A comprehensive and customized staff rehabilitation program is essential to ensure continuation of superior quality in customer products and services and a high level of customer satisfaction and loyalty. Besides, effective rehabilitation of staff may be accomplished by changing the mind set of people as well as their paradigms, skills, and capabilities by informing and training them on customer-centric business environment and how to understand and meet customer requirements. What is needed in CKM based organizations is a complete change not only in *hard skills* (what they do and how) but in *soft skills* (how they interact with customers) as well, and organization's pay and reward system.

CONCLUSION

Creating customer-centric SCA from organizational changes require a flexible structure, outsourcing of non-core activities, empowerment of employees, a constant and reliable knowledge-sharing culture, and process-based teamwork. To be able to function effectively in rapidly dynamic and complex business environments, it is inevitable for forward-looking organizations to adapt to change, add value to customers, reward and capitalize on creative ideas and distinct capabilities, create new business opportunities, and develop an atmosphere that is conducive to continuous life-long systemic-based learning. Today's fast-changing business world is witnessing aggressive fluctuations, higher degrees of uncertainty, and fierce competition. The changing nature of business environments requires dynamic rather than static organizational forms.

The people component of organizations represents a major organizational pillar in facing today's changing business environment through creating distinctive core competencies. In terms of people-based structural changes, the evolution of the world's economy from an industrial-based to an information-based economy enabled the trend to shift from the functional and hierarchical to the flexible and networked organizations.

There is a clear trend nowadays to shift away from functional organizations based on individuals performing individual tasks to networked structures.

Traditional, multilevel functional hierarchies are rigid structures that depend heavily on rules, procedures, and vertical and lateral referral, which make these organizations intrinsically inflexible, inefficient, ineffective, and unfit for competition. The hierarchical structure must be adjusted to ensure flexibility, speed of service, and the integration among business functions

Networked organizations apply team-based incentives, and a well-designed reward and recognition system that helps reinforce the desired behavior of being customer-focused. Besides, networked organizations distribute authority and power to people through empowerment. Empowerment helps boosting employee's morale and improves customer satisfaction. Although few organizations have actually reached higher levels of customer-centric organizational design flexibility, many customer-centric companies are expected to move toward it. Structural changes in customer-centric organizations also require changes in corporate culture and leadership style. The corporate culture needs to be supportive of knowledge sharing not knowledge hoarding and distinction in customer service not execution of customer service. Leaders of customer-centric organizational transformation need to be visionary, inspiring, and supportive, and need to coach not boss.

Several issues could emerge while transforming to a customer-focused organization. Reorganizing people in teams requires reviewing and upgrading in people skills, values, attitudes, behaviors, and performance in order to secure provision of value-adding customer products or services. The product of mismatch between people's skills and requirements of the job is poor performance. The challenge here is and how to enhance people's skills, knowledge, motivation, and commitment throughout an ongoing learning program.

More migration is expected in the future from rigid and internally-focused towards flexible and externally oriented structures, and from unicultural to multicultural global organizations. In embarking upon these changes, managers must be able to draw a fine line between maintaining a culture that is functioning well and changing a culture that has become dysfunctional.

REFERENCES

Awad, E., & Ghaziri, H. (2004). *Knowledge Management*. Upper Saddle River, NJ: Prentice Hall.

Bowditch J.L. & Buono (2005). *A Primer on Organizational Behavior*. 6th edition, Hoboken: John Willey & Sons.

Buttle, F. (2004). *Customer Relationship Management: Concepts and Tools*. Oxford, England: Elsevier Publishing.

Davenport, T. (1993). *Process Innovation: Reengineering Work through Information Technology*. Boston: Harvard Business School Press.

De Wit, B., & Meyer, R. (2004). *Strategy: Process, Content and Context*. London: Thomson Learning.

Deal, T. E., & Kennedy, A. A. (1982). *Corporate Cultures: The Rights and Rituals of Corporate Life*. Reading, MA: Addison-Wesley.

Griffin, R. (2005). *Management* (8th ed.). Boston: Houghton Mifflin Company.

Herzberg, F. (1987). One more time: How do you motivate employees? *Harvard Business Review*, (January-February): 109–120.

Heskett, J., Jones, T., Earl, L., & Schlesinger, L. (1994). Putting the service-profit chain to work. *Harvard Business Review*, (March-April): 164–174.

Hofstede, G. (2003). Cultural constraints in management theories. In Reddding, G., & Stening, B. W. (Eds.), *Cross-cultural management* (*Vol. II*, pp. 61–74). Cheltenham: Edward Elgar Publishing.

Hughes, M. (2006). *Change Management: A Critical Perspective*. London: CIPD Publications.

Jassawalla, A. R., & Sahittal, H. C. (1999). Building collaborative new product teams. *Academy of Management Review, 13*(3), 50–60.

Morgan, G., & Sturdy, A. (2000). *Beyond Organizational Change: Structure, Discourse and Power in UK Financial Services*. London: Macmillan.

Ogbonna, E., & Wilkinson, B. (2003). The false promise of organizational culture change: a case study of middle managers in grocery retailing. *Journal of Management Studies, 40*(5), 1151–1178. doi:10.1111/1467-6486.00375

Pearlson, K., & Saunders, C. (2006). *Managing and Using Information Systems: A Strategic Approach* (3rd ed.). New York: John Wiley.

Peppard, J., & Ronald, P. (1995). *The Essence of Business Process Re-Engineering*. Upper Saddle River, NJ: Prentice-Hall.

Salaman, G. (1997). Culturing Production. In Du Gay, P. (Ed.), *Production of Culture/Cultures of Production*. London: Sage.

Salaman, G., & Asch, D. (2003). *Strategy and Capability: Sustaining Organizational Change*. Oxford: Blackwell.

Slack, N., Chambers, S., Johnston, R., & Betts, A. (2006). *Operations and Process Management: Principles and Practice for Strategic Impact*. Essex, England: Pearson Education.

Weiss, J. W. (2001). *Organizational Behavior and Change: Managing Diversity, Cross-Cultural Dynamics, and Ethics*. Cincinnati, Ohio: South-Western.

Wysocki, R. K., & DeMichiell, R. L. (1997). *Managing Across the Enterprise*. New York: John Wiley.

Chapter 5
Role of Web Interface in Building Trust in B2B E-Exchanges

Muneesh Kumar
University of Delhi South Campus, India

Mamta Sareen
University of Delhi, India

ABSTRACT

The emergence of Internet has revolutionized the way businesses are conducted. The impact of e-commerce is pervasive, both on companies and society as a whole. It has the potential to impact the pace of economic development and in turn influence the process of human development at the global level. However, the growth in e-commerce is being impaired by the issue of trust in the buyer-seller relationship which is arising due to the virtual nature of e-commerce environment. The Online trading environment is constrained by a number of factors including web interface that in turn influences user experience. This chapter identifies various dimensions of web interface that have the potential to influence trust in e-commerce. The empirical evidence presented in the chapter is based on a survey of the web interfaces of 65 Indian B2B e-exchanges.

INTRODUCTION

The proliferation of Internet technologies into business has fundamentally changed the inter-firm relationships. It has provided faster access and better knowledge of commodities and prices for meeting their sourcing and selling needs. The ability to exchange information in both directions between firms has created a relationship not previously possible. The impact of e-commerce is pervasive, both on companies and society as a whole. It is the first mass application of information and communication technologies in the movement towards digital economy. It has broken all man-

DOI: 10.4018/978-1-60960-497-4.ch005

made boundaries and provided an opportunity for both buyers and sellers to interact among themselves regardless of difference in language, society, culture and tradition. The rapid growth of e-commerce is now being related to economic development and is often been cited as a driver of economic growth. E-commerce is also been touted as a powerful medium through which less developed economies can exploit the potential of global markets. It, thus, has the potential to impact the pace of economic development and in turn influence the process of human development at the global level.

Use of web technologies in inter-organizational business transactions has opened up new business opportunities not only for the manufacturers but also for intermediaries. A number of models are being used to exploit these opportunities in the B2B e-commerce space. Electronic Exchanges (E-exchanges) represent one of these models of B2B e-commerce that provide a unique platform for both buyers and sellers to interact and transact online in a highly competitive environment.

E-exchanges are a logical extension of the ability of organizations to help their various processes like procurement, selling, etc. Sometimes e-exchanges are established by intermediaries who spot the opportunity; other times they represent collaboration among various manufacturers or service companies (Ranganathan C. 2003). They can be vertically focused on particular industries, or they can be horizontally focused to provide goods and support services across a wide variety of industries, or can be procurement focused, or wholesale or retail focused. E-exchanges can also be sell-side or buyer-side or neutral depending upon their attractiveness to buyers, sellers or both (Kalpan & Sawhney, 2000). To succeed E-exchanges need to attract a large number of buyers and sellers so as to create liquidity at all ends (Kalpan & Sawhney, 2000). However, the issue of trust is cited to be a legitimate concern in the growth of E-exchange (eMarket Service, 2009). Trust is often stated to play a key role in helping users to overcome perceptions of risk, uncertainty, anonymity offered by the virtual environment (Mcknight et al, 2002, Tan & Theo, 2001; Hoffman, 1999). Since, the transactions in this virtual environment are conducted through the 'veil' of web interface, trust becomes an important issue. The web interface acts like the only 'contact point' among the buyers and sellers. Hence, there is a need for the web interface to induce trust in online environment. The focus of the present chapter is to identify various trust inducing web dimensions that may enhance the effectiveness of web interface and there by help in inducing trust among the e-commerce players.

TRUST AND E-COMMERCE

Trust is defined as "the willingness of a party to be vulnerable to the actions of another party based on the expectations that the other party will perform a particular action important to the trustor, irrespective of the ability to monitor or control that other party" (Mayer, Davis & Schoorman 1995). In the context of e-commerce, trust may be regarded as a judgment made by the user, based on general experience learned from being a customer/seller and from the perception of a particular merchant. In other words, trust is also seen as a generalized expectancy that the word, promise, or written statement of another party can be relied on (Rotter, 1980).

To date, research on understanding online trust and e-commerce is limited (Grabner-Kräuter & Kaluscha 2003; Yoon 2002; Corritore et al. 2003; Kolsaker & Payne 2002). In their critical reviews of website and/or ecommerce trust, Corritore et al. (2003) and Grabner- Kräuter and Kaluscha (2003) argued that there is a lack a conceptual understanding of online trust and theoretical support for its role in online transactions and relationships. Without trust, businesses are unable to function (Reichheld et al. 2000). Jian, Bisantz, and Drury (2000) and Bailey et al. (2003) claim that trust not only plays a strong role in human-to-human interactions, but also plays a critical role in human-to-computer interactions.

LITERATURE REVIEW

A rich web interface may have a positive impact on trust in the faceless environment of e-commerce. Several studies like Fogg et al (2001); Lee and Kim & Moon, (2000); Neilsen, (1999, 2005), etc. reported evaluations of a list of design features that could potentially appear on the web interface to impact trust.

Smith and Spiers (2008) tested the robustness of Technology Acceptance Model (TAM) as applied to e-commerce adoption by both the senior and the net generation. They observed strong relationship between the website perceived usefulness and behavioral intention to use the web site. Similarly, ease of use was found to be strongly related to attitude towards using the website. They found that website usability to exert greatest influence on perceived ease of use of web sites. In another study Urban et al (2009) found that online trust extends beyond privacy and security and is closely related to website design.

Zhou X. (2005) asserted that poor quality of web interface, lack of proper content in the web sites, unintuitive navigation, etc. can diminish the trust in the concerned company in e-commerce activities. Bailey et al (2005) stressed that visual aesthetics and navigation quality of a web site help to assess its trustworthiness in e-commerce activities. Besides other causes, poor Web site design and long server down times was found to be held responsible for the failure of several e-commerce endeavors (Razi et al. 2004).

Ang and Lee (2000) stated that if the web site does not lead the buyer to believe that the seller is trustworthy, no business can be conducted. In other words, one key consideration in fostering online trust in e-commerce is to build a trust inducing web interface. Lohse and Spiller (1998) identified four interface design features that affect the effectiveness of the web interface. Their results indicated that features like effective navigation, detailed product descriptions, links, etc. affect the trust in e-commerce activity. Jarvenpa et al, 1999 stated that a web site with trust inducing features functions as a skillful sales person for that company and therefore moderates the disadvantages of an impersonal web site. It is believed that online buyers in e-commerce look for the presence of positive cues about a site's general trustworthiness, as well as for the absence of negative cues. Hence, the e-commerce players by carefully designing their site to set and meet user expectations can influence the trustworthiness of other players.

Arion et al. (1994) asserted that user interface is the point where trust is generated. They stated that trust is a dynamic process, initially based on faith due to the lack of evidence that seeks to reach a certain level of confidence, i.e., where there is conclusive evidence in favor of trusting behavior. In their consideration of computer-supported cooperative work (CSCW) systems, Arion et al. (1994) argued that development of trust in the human-computer interaction need to be supported by the infrastructure/system.

A web interface provides total "user experience"[1]. Hence, an effective web interface can make this contact point between the trading partners more meaningful and help in building up trust.[2] The effectiveness of the web interface may also be determined by factors such as the aesthetic appearance of the site, the content and the way the information is presented to the user. The web interface is not just how it looks; it is how easy it is to learn, how well it recedes into the sub consciousness of users, and how well it supports users' tasks.[3] Different authors have suggested various features for making the web interface more effective (Egger, 2003; Neilson, 1999; Wang et al 2007 and others). These features can be classified into three broad categories i.e. *appeal, content and usability.*

Appeal: It refers to 'attitude' component and the first impression a user gets when accessing a site for the first time Lindgaard (1999) stated that an immediate negative impression may well determine the subsequent perception of the site's quality and usability, whereas one may inherently judge a site by its first impression. Literature from psychology also stresses the important role of a party's first impression, as someone's confirmation bias would entail that all user actions will unconsciously seek to confirm the first impression rather than falsify it (Kahneman & Tversy, 1973; Good, 1988). Fogg et al. (2002) reported, in their large study about how people evaluate the credibility of websites that, almost 50% of all comments made by participants referred to graphic design. They therefore argue that, in the context of online credibility and trust, findings indicate that *looking* good is often interpreted as *being* good, credible and trustworthy. Hence, appeal has largely to do with the site's graphics design and layout. In addition, Demonstrating important clients or providing links towards company's various policies also instills trust among the users (Doney & Canon, 1997).

Usability: Usability is the measure of the quality of a user's experience when interacting with the web site Sweden Canada Link (2001) stated that usability is about making the visit to the website as effective as possible for the users. The focus of usability is on enabling users, whatever their interests and needs, by removing barriers and making the system as easy to use as possible. According to usability expert Jakob Nielsen, usability is a necessary condition for survival on the web. If a web site is difficult to use, people leave. If the homepage fails to clearly state what a company offers and what users can do on the site, people leave. Indeed, *visual design* is presented to the user passively, while the user actively needs to *navigate* the website in order to access relevant information. Usability is all the more important in the context of online shopping as it is known to be an important condition for the acceptance and adoption of new technologies. The Technology Acceptance Model (TAM), as defined by Davis (1989), holds that usefulness and ease of use are both strong predictors of trust. This model has also been explicitly used to relate trust and e-commerce by researchers like Gefen and Straub (2000), Pavlou (2001), amongst others. Navigation Design of a website is concerned with the browsing of the website with ease. Cyr (2008) argued that even if detailed information is put on the site the customer may be leaving the site if he finds it difficult to search for the information he wants. Harridge-March (2006) had argued that proper navigation helped the customer save time and overcome performance risks and therefore leads to trust. Yoon (2002) also empirically showed that navigation design results in trust.

Content: Websites contain information and serve as a medium that predominantly is used for the transfer of information (either technical or not). This plays an important role in the effectiveness of the site. Product information has historically been regarded as a critical element of the content of web interface[4]. A number of researchers have investigated the relationship between web interface and information structure (e.g., Gay et al., 1991: Radha & Murphy, 1992; Mohageg, 1992; Utting & Yankelovich, 1989), concluding that information structure is an essential element of an effective web interface. However, a web site may also contain other information such as detailed and relevant information about the company, its complete offline address, seals of approvals from various trusted third parties, etc. Green (1998) stated that in the e-commerce environment, information plays an important role as business audiences seek more information about products/services and the company. To be able to convey the information effectively, it is necessary to structure it properly. Correct and detailed descriptions of the products and services offered by the company helps the users to make informed decisions about their transactions. Features that reduce user costs, such as comparisons with competitive products, may also be seen as a sign of honesty and competence (Egger, 2002). In addition, the provision of related content, if relevant, can also be interpreted as the company truly understanding its customers' needs. The credibility of the information has also been observed to be very important about a company's ethical standards. According to Korgaonkar et al (2006) inclusion of features like option to communicate with the salesperson, reviews from other shoppers, third party evaluation and information exchange with online vendor act as antecedents to trusting intention. Ribbink et al (2004) argued that communication is part of e-quality and is an antecedent to satisfaction leading to trust. Similarly, Mukherjee and Nath (2003) also argued that timely communication has the capacity to generate trust by resolving disputes and ambiguities.

Various dimensions of the web interface address the appeal, usability and content features of an effective web interface. Various authors have identified a number of such dimensions (Kim 1998; Egger 2003; Wang et al 2006; etc.). Some of these dimensions are categorized as trust inducing dimensions. Kim (1998) had identified four interface design features which contribute to trustworthiness of the web site. Wang et al. (2007) also stressed on certain web dimensions to enhance the richness of web interface. He proposed four broad dimensions of web interface namely, graphic design dimension, Structure design dimension, Content design dimension and Social-cue design dimension that may influence trust. However, the various studies do not include features that are commonly found in any modern e-commerce web site and the e-commerce players are interested in. These include reference to the kind of security policy, privacy policy being followed by the company for online transactions, the technology related policies and procedures followed by the company to address the security, privacy etc. issues, the statements from well known customers, media excerpts, etc. Thus, there is a need for a more comprehensive model relating trust and web interface.

WEB INTERFACE AND TRUST MODEL

In order to incorporate the various gaps in the earlier model relating trust with web interface, the present chapter proposes five trust inducing dimensions of an effective web interface. These dimensions are: (a) User Interface Dimension; (b) Information Structure Dimension; (c) Information Content Dimension; (d) Demonstrability Dimension; and (e) Social-ability Dimension. The relationships between these dimensions and three features identified earlier are exhibited in the Figure 1.

User Interface Dimension: User Interface indicates the appearance and the tools available for accessing the information contained in the web site/portal. Constaine (1995) pointed out that interface is important aspect as the more intuitive the user interface is, the easier it is to use and trust. The website must be recognizable as from the organization. That is, it must be obvious that the look of the site – colors, logos, layout, etc. is consistent with other collateral from the organization. The various features offered by the web site that normally giving the first impression about the company lays the initial foundation of trust building process. Kim and Moon, (1999) reported that the overall color layout and graphical interface influences the trustworthiness of the web site. This dimension aids in addressing the appeal aspect of an effective web interface. Various features that may be useful in enhancing trust levels among the users include: (a) Home Page; (b) Graphics Interface; (c) Links; (d) Professionalism; and (e) Loading Time.

Information Structure Dimension: The structure dimension defines the overall organization and accessibility of displayed information on the web site. Ease of navigation has frequently been mentioned as a key to promote online trust (e.g., Cheskin/Sapient Report, 1999; Neilsen, 1998). In other words, users must be able to easily locate the information they seek on the web site.

Figure 1. Web interface and trust model

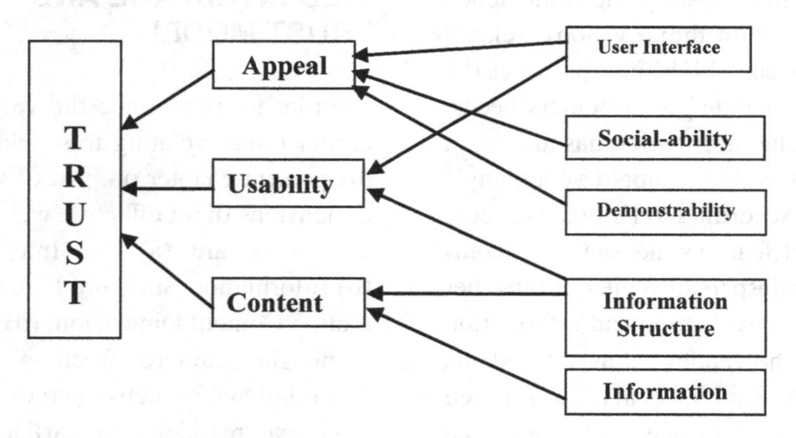

This ease-of- use reflects two characteristics of a trustworthy web site: simplicity and consistency. Buyers appreciate simplicity or a clear design of e-commerce web sites because it reduces the perceived risks of deception, frustration, and wasting time. When the structure and design of the web site are consistent, users feel more confident using the site because they can transfer their learning from one sub-site to the next rather than having to learn everything over again for each new page and trust is build (Neilsen, 1998,). For example, broken links, meaningless images, and similar "hygiene factors" may relate to users dissatisfaction with a web site (Zhang et al., 1999). Key features of structure dimension of a web site may include: (a) Navigation; (b) Accessibility; (c) Functionality; (d) Consistency; and (e) Learnibility.

Information Content Dimension: This dimension refers to the informational components that can be included on the web site, either textual or graphical. A logically structured web site providing comprehensive, correct, and current product information instills trust among the users (Egger, 2001; Neilsen, 1999). If the information regarding the products is precise, factual and contains links of details required, if necessary, then trust is build (Bhattacharya, 2001). Recent market surveys include that some of the companies are using their web sites as a part of integrated communication strategy to create trust and action (Sheehan and Doherty, 2001). In such cases the website interface plays an important role in e-business transactions. The contents should be displayed as being less complex and more users friendly that enriches the visitor's experience and motivates him/her to visit the site again (G. Chakraborty et al 2003). key features of structure dimension of a web site may include: (a) Navigation; (b) Accessibility; (c) Functionality; (d) Consistency; and (e) Learn-ability.

Demonstrability Dimension: Several researchers stress the importance of "demonstrability" in e-commerce, which is to promote the brand reputation of a company online. It is often seen that various features like offline address of the company, details about real people behind etc play the same role of offering certain clues about the credibility of the company as the physical clues in the brick and mortar business transactions. These clues help the potential trading partners to assess the credibility of the company. Especially displaying the seals of approval from various trusted third parties and the accreditations earned help in building trust levels.

Social-ability Dimension: This dimension relates to embedding social cues, such as face-to-face interaction and social presence, into web site interfaces via different communication media, because a lack of the "human touch" or presence may constitute a barrier for at least some consumers to trust online merchants (e.g., Riegelsberger & Sasse, 2002). The effectiveness of a personalization system improves in the long run. Weiner and A. Mehrabian (1968) stated that the choice of language can help create a sense of psychological closeness and warmth.[5] Whereas Nass and Steuer (1993) stressed that the use of natural and informal language can impact perceived social presence.[6] Yoon (2002) also showed that web site trust is influenced by consumer familiarity. Every time a customer interacts with the web site, the personalization mechanism collects new data about the user's preferences, so that a more and more satisfactory service can be offered.

METHODOLOGY AND RESULTS

A list of 100 B2B e-exchanges operating in India was prepared through the use of various search engines like Google, Yahoo, MSN etc. The criterion for the selection of these B2B e-exchanges was random. On visiting these web sites, it was found that some of the web portals were merely a directory of sellers and buyers and not actually carrying out e-commerce transactions. Such web portals were excluded from the sample. Hence, a total of 65 B2B web sites/ e-exchanges were selected for the purpose of the survey. Based on the number of elements present, the web site was to be rated on a 5 point Likert scale for each of the trust inducing dimension. Finally, an overall trust rating of the web portal was also to be obtained for each web interface of the select B2B e-exchanges.

Initially, 25 participants were requested to rate six B2B e-exchanges. The participants were IT savvy in the age range of 30-45 years and came from a variety of background like public sector, business sector, private sector, banking sector and academic sector. The participants were asked to provide their own ratings of each dimension and also the overall rating of the web site based on a questionnaire consisting of 20 questions. Hence, a total of 120 responses were received from each of the 25 participants. Cronbach's Alpha Scale Reliability test was used to test the reliability of the questionnaire and it was found to be 0.8751, which is fairly good degree of reliability.

In order to study the significance of these trust inducing dimensions for the level of trust, linear regression model[7] was used. The model used the level of trust as dependent variable and each of the trust inducing dimensions as independent variable. The purpose was to find out any redundant dimension included in the dimension. The regression analysis was carried out on the data collected from these 25 participants for each of the six B2B e-exchanges. The adjusted R^2 ranged from 86% to 92%. This would imply that more than 85% of the variation in the trust ratings could be collectively explained by the five trust inducing dimensions of an effective web interface. The results of regression analysis for the ratings given for one of the B2B e-exchanges are presented in Table 1.

During the systematic elimination process, all the five trust inducing dimensions were found to be significant. The results showed Demonstrability and Information Content dimension as significant trust inducing dimensions of web interface. This further strengthens the reasons of the present chapter for the inclusion of demonstrability dimension as a trust inducing dimension for web interface to the model proposed by Wang *et al*. The information structure dimension and the User Interface dimension of web interface also were found to be having significant contribution towards building trust in web interface. This could be because of the fact that in India, complete virtual B2B transactions do not occur very often and they are aided with various offline channels like phone, fax, etc. However, the result assures that all these dimensions of an effective web interface as proposed in the model have a potential to enhance trust among the users. Further, socioability dimension, though considered important in B2C e-commerce activities, was not found to

Table 1. Trust ratings of a B2B e-exchange and trust inducing dimensions partial results of regression

Model	Unstandardized Coefficients		Standardized Coefficients		
	B	Std. error	Beta	t	Sig
(Constant)	-0.312	0.460	-	-0.677	0.506
User Interface	0.243	0.115	0.242	2.912	0.003
Info. Structure	0.225	0.116	0.265	2.093	0.009
Info. Content	0.327	0.217	0.327	3.209	0.001
Demonstrability	0.338	0.165	0.264	3.049	0.002
Socio-ability	0. 033	0.097	0.003	1.634	0.020

be contributing much towards trust levels in B2B e-exchanges.

In order to identify any bias in the evaluation of the web interfaces of the six B2B e-exchanges, the author also independently evaluated all the selected B2B e-exchanges. The results of the evaluations were compared with the trust equation obtained of these select B2B e-exchanges. It was observed that the difference between the two results was not significant. This would imply that the evaluations of the B2B e-exchanges done by 25 participants and the author held nominal bias. Therefore, evaluation of the remaining 59 B2B e-exchanges was carried out independently by the author, in the same manner as it was done by the 25 participants.

The regression model used earlier was used on the data so collected from the evaluations of web interfaces of the selected B2B e-exchanges. The results of the linear regression analysis are presented in the Table 2. As may be observed from Table 2, all the five trust inducing web dimensions were found to be contributing towards trust in the web site.

All the dimensions of effective web interface as identified in the model were found to be good predictors of level of trust. The results were fairly comparable with the results obtained from the linear regression analysis carried out on the data collected through 25 participants with respect to six select B2B e-exchanges. Thus, it may be concluded that the five trust inducing web dimensions namely Information Content dimension, Demonstrability dimension, Information Structure dimension, User Interface dimension and the Social-ability dimension (in that order) positively contribute towards trust building processes in e-commerce.

Table 2. Trust ratings of a B2B e-exchange and trust inducing dimensions partial results of regression

Model	Unstandardized Coefficients		Standardized Coefficients		
	B	Std. Error	Beta	t	Sig
(Constant)	-.633	0.232	-	-2.732	0.008
User Interface	0.178	0.114	0.242	2.816	0.008
Info. Structure	0.269	0.102	0.211	2.334	0.011
Info. Content	0.421	0.094	0.382	3.486	0.001
Demonstrability	0.296	0.105	0.261	2.836	0.006
Socio-ability	0.067	0.114	0.060	0.590	0.055

LIMITATIONS AND SCOPE FOR FUTURE RESEARCH

The main limitation of this chapter could be the coverage of only B2B e-exchanges in validating trust in e-commerce. Since, e-commerce involves various other activities, it would be better, if the sample data would contain web interfaces from other segments of e-commerce also. Further, the model proposed in this chapter has tried to include as many trust inducing dimensions as possible. It may, however, be possible that the chapter might have overlooked certain issues that might influence the web interface of the e-commerce sites. However, user experience is a very complex issue related to human-computer interface and may vary from individual to individual. Ideally, different individuals must have rated all the e-exchanges. However, it was not possible for the present scope.

CONCLUSION

E-commerce has the potential to provide a flip to the pace of economic development and provide a unique opportunity to organizations in less developed countries to operate in global markets. This could also have implications on the socio-economic conditions of the less developed countries. Trust has been a major hurdle impeding the growth of e-commerce and the need for enhancing trust cannot be over-emphasized. The chapter, through validation, identifies various trust inducing dimensions for enriching the web interface and there by inducing trust. Focus on these dimensions, which demonstrate 'correct and concise' information, 'relevant structure' of information and 'usability', would help in inducing trust in the faceless environment of e-commerce. This in turn would improve the user experience and the buyer-seller relationship in online trading environment. This supports the earlier findings of Arion et al. (1994) wherein trust in human-computer interactions was stated to be supported by the system. It may, however, be noted that the trust in buyer-seller relationship is also influenced by a number of factors and the enhancement of features of web interface would need to be viewed as an integral part of an overall trust building strategy of an organization. Thus, by effectively incorporating various features of the trust inducing dimensions on the web interface, the merchant is able to provide a trust worthy platform for the customer across global boundaries to transact among each other. This has helped in the development of a virtual society which is driven by trust in technology and enhances the growth of e-commerce. Professional bodies and business organizations need to play an important role in this regard. Development and adherence to globally accepted standards in this regard can go a long way in enriching the 'user experience' in e-commerce.

REFERENCES

ACM Special Interest group on Computer-Human Interaction Group. (1992). *ACMSIGCHI, Technical Report*. New York: ACM.

Arion, M., Numan, J. H., & Pitariu, H. (1994). Placing Trust in Human-Computer Interactions. In*Proceedings of 7th European Cognitive Ergonomics Conference*. (pp. 352-365).

Bart Iakov, Y. Shankar Venkatesh, Sultan Fareena, &. Urban Glen L. (2005). *Are the Drivers and Role of Online Trust the same for all Web Sites and Consumers?*http://ebusiness.mit.edu

Belaal Mohammad Ahmad Ifhan. (2002).*Trust inducing model features for web sites*. http://eprints.uum.edu.

Cao, M., Zhang, Q., & Seydel, J. (2005). B2C e-commerce web site quality: an empirical examination. *Industrial Management & Data Systems*, *105*(5), 645–661. doi:10.1108/02635570510600000

Cassell, J., & Bickmore, T. (2000, Dec.). External Manifestations of Trustworthiness in the Interface. *Communications of the ACM, 43*(12), 50–56. doi:10.1145/355112.355123

Cheskin/Sapient Research and Studio Archetype/Sapient. (1999). *E-Commerce Trust Study*. http://www.sapient.com/cheskin.

Cockburn, A., & McKenzie, B. (2001). What Do Web Users Do? An Empirical Analysis of Web Use. *International Journal of Human-Computer Studies, 54*(6), 903–922. doi:10.1006/ijhc.2001.0459

Connolly R., Bannister F. (Nov. 2007). Consumer Trust in Electronic Commerce: Social & Technical Antecedents. *Proceedings of World Academy of ScienceEngineering and Technology, 25*.

Cyr, D., Kindra, G., & Dash, S. B. (2008). Website Design, Trust, Satisfaction and e-Loyalty: The Indian Experience. *Online Information Review, 32*(6), 773–790. doi:10.1108/14684520810923935

Davis, F. D. (1989). Perceived usefulness, perceived ease of use and user acceptance of information technology. *MIS Quaterly, 13*(3), 319–340. doi:10.2307/249008

Davis, J., Mayer, R., & Shoorman, F. (1995). An integrated model of organizational trust. *Academy of Management Review, 20*(3), 705–734.

Doney, P. M., & Cannon, J. P. (1997, April). An Examination of the Nature of Trust in Buyer-Seller Relationships. *Journal of Marketing, 61*, 35–51. doi:10.2307/1251829

Dowell, J., & Long, J. (1989). Towards a conception for an engineering discipline of human factors. *Ergonomics, 32*, 1513–1535. doi:10.1080/00140138908966921

Egger, F. N. (2001). Affective Design of E-Commerce User Interfaces: How to maximize perceived trustworthiness. *Proceedings of The International Conference on Affective Human Factors Design*. London: Asean Academic Press.

Egger, F. N. (June, 2001). Affective Design of E-Commerce User Interfaces: How to maximise perceived trustworthiness. *Conference on Affective Human Factors Design*.(pp.317-324) Singapore.

Egger, F N. (2003). From interactions to transactions: Designing the Trust Experience for Business-to-Consumer Electronic Commerce.

Fogg, B. J., & Nass, C. (1997). Effects of computers that flatter. *International Journal of Human-Computer Studies, 46*, 551–561. doi:10.1006/ijhc.1996.0104

Fogg B.J, Marshall J., Laraki O., Osipovich A., Varma C., Fang N., Paul J., Rangnekar A, Shon J., Swani P.& Treinen M. (2001). What Makes Web Sites Credible? A Report on a Large Quantitative Study. *ACM sigchi, 3* (1), 61-67.

Fukuyama, F. (1995). *Trust: The social virtues and the creation of prosperity*.New York: The Frees press.

Harridge-March, S. (2006). Can the building of trust overcome consumer perceived risk online? *Marketing Intelligence & Planning, 24*(7), 746–761. doi:10.1108/02634500610711897

Hassanein, K. S., & Head, M. M. (2004). *Building Online Trust through Socially Rich Web Interfaces*. http://dev.hil.unb.ca.

Jarvenpaa, S. L., & Tractinsky, N. (1999). Consumer Trust in an Internet Store: A Cross-Cultural Validation. *Journal of Computer-Mediated Communication, 5*(2).

Keen, P. (1999). *Electronic Commerce Relationships: Trust by design*. Upper Saddle River, NJ: Prentice Hall.

Keen, P. G. W. (1997, April 21). Are you ready for 'Trust' Economy. *Computerworld, 31*(16), 80.

KimJ., MoonJ. Y. (1998). Designing emotional usability in customer interface-trustworthiness of cyber banking system interface, interacting with computers, 10, *1-29.*

King, A. B. (January 2000). *What Makes a Great Web Site?*http://goodmictices.com

Korgaonkar., et al. (2006) as cited in Ganguly B. Dash S. B. Cyr D. Website characteristics, Trust and purchase intention in online stores: - An Empirical study in the Indian context. *Journal of Information Science and Technology, 6*(2), 22-44.

Lee, J., Kim, J., & Moon, J. Y. (April 2000). What Makes Internet Users Visit Cyber Stores Again? Key Design Factors for Customer Loyalty. In *Proceedings of the Computer-Human Interaction Conference on Human Factors in Computing Systems*, The Hague, Netherlands, pp. 305-312.

Lee, M. K. O., & Turban, E. (2001). A Trust Model for Consumer Internet Shopping. *International Journal of Electronic Commerce, 6*(1), 75–91.

Lingaard, G. (1999). *Does emotional appeal determine perceived usability of web sites?*www.cyberg.com

McMahon, K. (2005). *An exploration of the importance of website usability from a business perspective.*http://www.flowtheory.com/KTM-Dissertation.pdf

Mukherjee, A., & Nath, P. (2003). A model of trust in online relationship banking. *International Journal of Bank Marketing, 1*(21), 5–15. doi:10.1108/02652320310457767

Nante, J., & Glaser, E. (2005). *The Impact of Language and Culture on Perceived Website Usability*. http://www.rbcchair.com/chairerbc/fichiers/jetus.pdf

Nass, C., Moon, Y., & Carney, P. (1999). Are respondents polite to computers? Social desirability and direct responses to computers. *Journal of Applied Social Psychology, 29*, 1093–1110. doi:10.1111/j.1559-1816.1999.tb00142.x

Neilsen Normen group. (2000). *Trust: Design guidelines for e-commerce user experience.* Retrieved from www.nngroup.com

Nielsen, J. (1990). Navigation Through Hypertext. *Communications of the ACM, 22*(3), 296–310. doi:10.1145/77481.77483

Nielsen, J. (1993). *Usability Engineering*. New York: Academic Press.

Nielsen, J. (1999). *Designing Web Usability: the Practice of Simplicity*. New Riders.

Nielsen, J. (1999). *Trust or Bust: Communicating Trustworthiness in Web Design. Jacob Nielsen's Alertbox*. http://www.useit.com/alertbox/990307.html.

Nielsen, J. (May 16, 1999). *Who Commits the 'Top Ten Mistakes' of Web Design?* Jacob Nielsen's Alertbox, http://www.useit.com/alertbox/990516.html.

Nielsen, J. (August 19, 2001). *Did Poor Usability Kill E-Commerce*? Jacob Nielsen's Alertbox, http://www.useit.com/alertbox/20010819.html.

Nielsen, J. (May 12, 2002). *Top Ten Guidelines for Homepage Usability*. Jacob Nielsen's Alertbox. http://www.useit.com/alertbox/20020512.html

Ranganathan, C. (2003). Evaluating the options for Business-to-Business E-Exchanges. *Information Systems Management, 20*(3), 22–28. doi:10.1201/1078/43205.20.3.20030601/43070.3

Ranganathan, C., & Ganapathy, S. (2002). Key dimensions of business-to consumer websites. *Information & Management, 39*, 457–465. doi:10.1016/S0378-7206(01)00112-4

Ribbink, D., Allard, C. R., Liljander, V. V., & Treukens, S. (2004). Comfort your online customer: quality, trust and loyalty on the internet. *Managing, 14*(6), 446–456.

Riegelsberger J & Sasse, M.A. (2001). *Trust builders and Trustbusters: The Role of Trust Cues in Interfaces to e-Commerce Applications.*

Rotter, J. B. (1980). Impersonal trust, Trustworthiness and gullibility. *The American Psychologist, 35*(1), 1–7. doi:10.1037/0003-066X.35.1.1

Roy, M. C., Dewit, C., & Auber, B. A. (2001). The *Impact of Interface Quality on Trust in Web Retailers*. http://www.cirano.qc.ca/pdf/publication/2001s-32.pdf

Smith, T. J. Spiers, R. (2008). Perceptions of E-commerce Web sites across two generations. *Informing Science: the International journal of an emerging Transdiscipline,* 12, 159-179

Steel, W. 1999). *Rules of Thumb for Web Designs*. http://www.mcst.edu/webhints.html.

Udo, G. J., & Marquis Gerald, P. (2000). Effective Commercial Web Site Design: An Empirical Study, IEEE Engineering Management Society, 2000. *In Proceedings of the 2000 IEEE*, 313 – 318

Urban, G. L., Amyx, C., & Lorenzone, A. (2009). Online Trust: State of art, New frontiers and research potential. *Journal of Interactive Marketing, 23*, 179–190. doi:10.1016/j.intmar.2009.03.001

Valerie, P. (November 30, 1996). *Good Web Site Design*. Retrieved from www.geocites.com.

Van, D., Landay, J., & Hong, J. (2002). *The Design of Sites: Patterns, Principles, and Processes for Crafting a Customer-Centered Web Experience*. Reading, MA: Addison-Wesley.

Wang, E., & Barrett, S. Caldwell & Gavriel S. (2003). Usability comparison: similarity and differences between e-commerce and world wide web. *Journal of the Chinese Institute of Industrial Engineers, 20*(3), 258–266. doi:10.1080/10170660309509234

Wasserman A. I. (2001). *Principles for the Design of Web Applications*.

Wilson, D. (1999), *Rules of Thumb for Web Designs*. http://www.goodpractices.com

Yen, B., Hu, P., & Wang, M. (2005). Towards Effective Web Site Designs: A Framework for Modeling, Design Evaluation and Enhancement. In *Proceedings. IEEE International Conference on e-Technology, e-Commerce and e-Service*, (pp. 716-721).

ENDNOTES

1 www.techtarget.com

2 In 2000, the Nielsen Norman Group (NN/g) also published a report on trust as part of their E-Commerce User Experience Series. The study confirmed that website design was found to be important, as was content that was out of date, spelling mistakes, long download times and unclear error messages. They asserted that people want to have very detailed information about the company and the products they offer, if possible, with objective reviews.

3 Jeff Johnson

4 Nicolas Virtsonis, 2007

5 M. Weiner and A. Mehrabian. (1968). *Language within language: Immediacy, a channel in verbal communication*. New York, NY: Appleton-Centry-Crofts,

6 C. Nass and J. Steuer. (1993). Voices, boxes, and sources of messages: Computers and social actors," *Human Communication Research*, vol. 19, pp. 504-527.

7 The limitations of linear regression model in this context may be recognized.

Chapter 6
Collaborative Learning: An Effective Tool to Empower Communities

Hakikur Rahman
ICMS, Bangladesh

ABSTRACT

Learning is considered as one of the essential tool to empower a community. Over the past three decades, technology mediated learning has been recognized as an alternate channel strengthening the traditional forms of education. However, the organizational learning at the peripheries and capacity development at the grass roots remain almost unattended, despite recognized global efforts under many bottom-up empowerment sequences. Social components at large within the transitional and developing economies remain outside the enclosure of universal access to information and thus access to knowledge has always been constricted to equitably compete with the global knowledge economy. Despite challenges in designing and implementing collaborative learning techniques and technologies, this article would like to emphasize on introducing collaborative learning at community level and improve the knowledge capacity at the grass roots for their empowerment. The article, further, investigates the relationship of collaborative learning towards improved e-governance.

INTRODUCTION

Learning nowadays is not only comprise of traditional brick-and-mortar classroom sessions, but also encompasses advanced computer assisted collaborative learning and peer learning (Roberts, 2004; Tu, 2004) that support education, knowledge development and research. In recent years, traditional education has shifted towards new methods of teaching and learning through the proliferation

DOI: 10.4018/978-1-60960-497-4.ch006

of information and communication technologies (ICT). At the same time, the continuous advances in technology enable the realization of a more distributed structure of knowledge transfer (Dutton, Kahin, O'Callaghan & Wyckoff, 2005; Iahad, Dafoulas, Milankovic-Atkinson & Murphy, 2005). Furthermore, Internet has allowed the learners and education providers to reach out the sky as the limit in designing, understanding and taking knowledge acquisition processes through various learning techniques. In addition, as classrooms do not remain enclosed to confined peripheries anymore, and the learners do not confined to regularly attended sessions anymore. Anyone, with capability and acceptability can attend a learning session at any time in sequel of his/her career.

Learning is no more a customized pattern of education, but an accumulation of information, content and knowledge to become an accomplished sequence of knowledge acquisition. It is a shift from traditional education to ICT-based personalized, flexible, individualized, self-organized and at the same time collaborative, depending on the demand of a community of learners, teachers, facilitators, experts and researchers (Markus, 2008). Learning has broadened the door of knowledge acquisition processes in multi-disciplinary faculties through multi-dimensional approaches. It is become more dependent on the dynamism of interactive information and content, rather than static information and content that were only available in the form of print. As the society and community that have been based on information, has been more or less, turned into information society and as the economy of information society is mainly based on the creation, dissemination and exploitation of data, information and knowledge, thereby not only learning, but also the whole life system has been inclined towards the dynamism of information and content. In this aspect, Figure 1 illustrates the different component of a computer mediated communication leading to knowledge acquisition and Figure 2 shows the technological evolution in the learning processes based on virtual form of knowledge delivery.

Figure 1. Technology evolution in e-learning (Adopted from Markus, 2008)

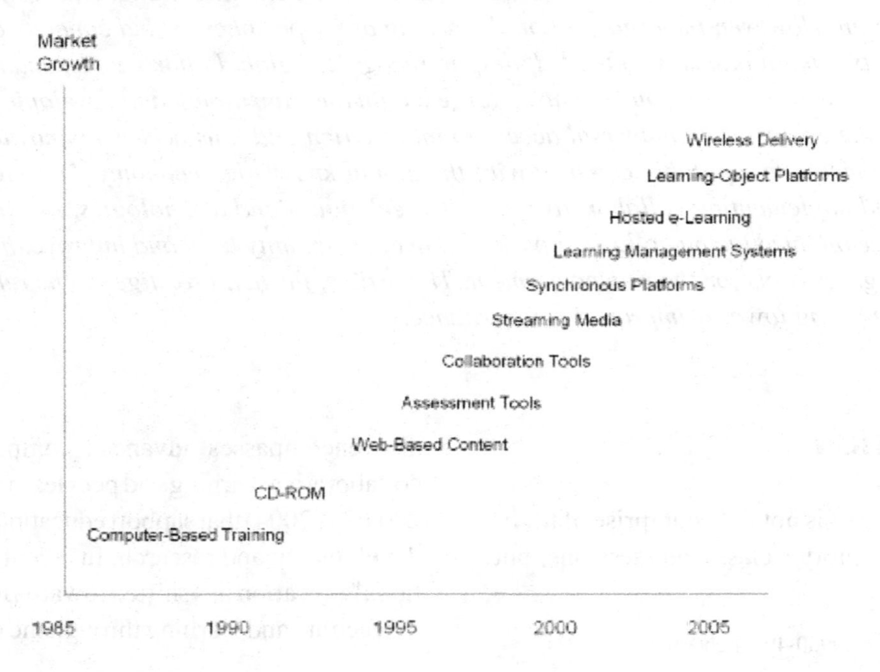

Figure 2. Components of a system of computer mediated communication (Adopted from Silvio, 2001)

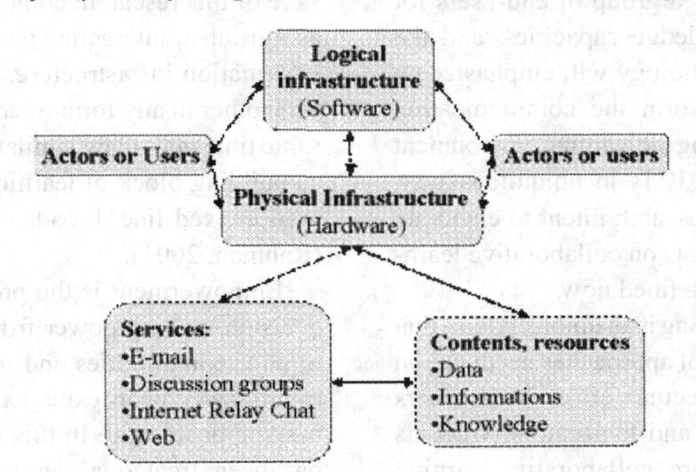

Based on the various forms of virtual knowledge delivery system and depending on technologies available for the dissemination process, this article will concentrate on the need of collaborative learning for the overall societal benefits, including the traditional learning methods. Evident, observations and researches confirm that collaborative learning can strengthen the traditional learning and at the same time, provide the learner a multifaceted window of knowledge acquisition. Incorporating these ideas, the article goes on providing concepts of learning on various society development activities. Along these perspectives, the article will try to relate to the main theme of the special issue, that is e-government and argue that by empowering community people e-governance can be improved. Furthermore, the article put forwards a few cases around the globe, that have been providing capacity development through community empowerment processes and finally it suggests a few research ideas before conclusion.

BACKGROUND

Learning is the acquisition and augmentation of memories and behaviors leading to development of skills, knowledge, understanding, values, and wisdom. In longer term, it is the outcome of experience and the ultimate goal of education[1]. Thereby, learning is the process of acquiring knowledge, attitudes, or skills from study, instruction, or experience (Miller & Findlay, 1996:167).

In the fields of neuropsychology, personal development and education, learning is one of the most important intellectual functions of humans, natural world and artificial cognitive systems. As stated earlier, it relies on the acquisition of different types of knowledge supported by perceptible information. Hence, learning leads to the development of new capacities, skills, values, understanding, and preferences leading to the increasing of individual and group experience (Gadomski, 1993). Numerous literatures are available that are mentioning relevant definitions on learning, but along the social perspective, authentic learning for social development should be the objective of teaching and education.

Furthermore, along the perspective of this research, learning has been thought of a fundamental process of communication (Mantha, 2001), and learning technology has been considered as a means to enhance the communication. However, for sake of understanding, the focus of learning will lead to transform the mental map of the individual end-user or a group of end-users for extending their knowledge capacities, and the focus of learning technology will emphasize on technologies to transform the communication techniques for enhancing the learning environment through utilization of ICTs in ubiquitous way. Furthermore, as this research intent to elaborate on methods and concepts on collaborative learning, the same will be defined now.

Collaborative learning is an umbrella term that incorporates a variety of approaches in education involving joint intellectual effort by learners (students) or learners and facilitators (students and teachers). Therefore, collaborative learning refers to methodologies and environments in which learners engage in a common task where each individual depends on and is accountable to each other. Here, a group of learners work together in searching for understanding, meaning or solutions or in creating an artifact of their learning such as an output, which may include collaborative writing, group projects, and other activities[2]. Collaborative learning has taken on many forms, but this study will emphasize on community based collaborative learning for enhancing the knowledge base of the stakeholders.

In natural terms, a community is a group of interacting elements sharing a common environment. In human communities, intent, belief, resources, preferences, needs, risks, and a number of other conditions may be present and common, affecting the identity of the participants and their degree of partnership. In social format, the concept of community has caused infinite debate, and sociologists are yet to reach agreement on a definition of the term. Traditionally a "community" has been defined as a group of interacting people living in a common location. The word is often used to refer to a group that is organized around common values and social cohesion within a shared geographical location, generally in social units larger than a household. Wider meanings of the word can refer to the national community or global community[3]. In terms of ICTs and for sake of this research, community will be treated as a group of interacting people using same basic information infrastructure, and be connected to each other in any form of connectivity and at the same time may share similar content or resources as building block of learning (intranet, Internet, Wi-Fi, fixed line, broadband, radio, DSL, etc.) (Rahman, 2001).

Empowerment is the process of transferring decision-making power from influential sectors to poor communities and individuals who have traditionally been excluded from the decision-making processes[4]. In this study, empowerment has been treated as an important element of society development, and being the process by which common people take control and action in order to overcome obstacles. Empowerment here particularly refers to the collective action by the oppressed and deprived to overcome the obstacles of structural inequality which have previously put them in a disadvantaged position. Hence, empowerment can only take place if the obstacles preventing it can be identified and removed[5], or means can be found to reduce the gap between empowered and being empowered. Thus, empowerment makes it possible to upgrade the performance of a community through greater delegation of authority to act[6].

Therefore, community empowerment involves individuals or groups acting collectively to gain greater influence and control over the determinants of education, environment, health, living condition and the quality of life in their community[7], and is an important goal in community action for overall development of their livelihood. Hence, community empowerment is the situation of communication that exists when members of a community feel empowered to achieve their self-determined goals, with some measure of significant control over the processes and strategies to attain[8] them by means of self-organization.

Collaborative Learning as an Effective Tool

In simple form, ICT mediated collaborative learning process is mostly governed by interaction with digitally delivered content, network-based services and facilitator's support. Any technologically mediated learning using computers whether from a distance or in face to face classroom setting (computer assisted learning), executed in a group environment forms the core element of collaborative learning. Broadly speaking, as stated earlier too, it is a shift from traditional education or training to ICT-based personalized, flexible, individual, self-organized, group learning based on a community of learners, educators, facilitators, and experts incorporating socio-economic and cultural domains in the learning system. Collaborative learning can act as the foundation of basic education at community level through ICT based learning techniques to raise the knowledge capacity of the individual element of the community. Thus, it is an effective tool to enhance the skill, and knowledge in adapting the dynamically changing scenario of the contemporary interconnected world.

Benefits of collaborative learning facilitates education, research, social cohesion and psychological stability, thus building self-esteem, reducing anxiety, encouraging understanding of diversity, fostering relationships, stimulating critical thinking, increases student retention and encourages group learning. However, in spite of the huge benefits of collaborative learning and being widely known, this learning technique is rarely practices till date (Panitz, 1997; 1998). The instructor-centred paradigm still exists predominantly where resources are limited as in the case of developing countries (Iahad, Dafoulas, Milankovic-Atkinson & Murphy, 2005).

Improvement of E-Governance Through Collaborative Learning

The concept of e-Government has begun to receive increasing attention, adopting new governance models that rely on the extensive usage of ICTs, as well as on re-engineering of business processes. In this context, massive investments have also been made in most developed and developing countries, and extensive researches are also been carried out However, the majority of the research and practical applications in the area of e-Government is mainly focused on carrying out electronic transactions, i.e. on offering citizens and enterprises the capability to perform transactions with the public administration (such as forms, declarations, applications, etc.) via electronic channels (mainly the Internet) at anytime and from anywhere, without having to visit the physical administrative offices (Holmes, 2001; Heeks, 2002; Leitner, 2003; Traunmueller & Wimmer, 2003; Burn & Robins, 2003; Karacapilidis, Loukis & Dimopoulos, 2005). Little efforts have been made towards educating common people through innovative ICTs to increase their knowledge capacity and thereby, by extending e-government services at the grass roots to improve e-governance.

Moreover, these e-Government efforts have been highly influenced by the concepts of economy, rather than knowledge. The usage of ICTs in public administration is not highly innovative, except being trying to provide a one-stop service outlet to their clients, which in many cases could not draw the proper attention. This single access point concept is also dependent on various factors, including infrastructure, content, interconnection among agencies, logistic and physical support. However, it has been constantly emphasized in contemporary researches that it is necessary to exploit the huge innovation potential of ICTs in the public administration to a much larger extent, in order to redesign and support the diversity of e-Government functions, thus heading for a re-engineered e-Government applications to reach greater population (Utsumi, 2005a; 2005b). Bringing them under a broader umbrella of knowledge management through effective collaborative learning, much of these laggings could be eliminated. The most critical functions could incorporate:

- the design, implementation, monitoring and evaluation of knowledge building related to public policy, legislation, development plans and programs;
- empower grass roots communities to take part in the high-level decision-making processes concerning social problems, such as granting of licenses and permissions with high social impact, managing severe environmental problems, etc;
- educate community people at large on basic health, livelihood and environment;
- familiarize mass people on basic intelligence on society development (civic sense, discipline, law and order); and
- familiarize with basic concepts of e-government.

The above mentioned functions are of critical importance for the public administration, because they shape the context of all its lower level activities, which are associated with the production and delivery of public services to the citizens and enterprises. In other words, they are of critical importance for the whole society, having a significant impact on its well-being and development of the society (Celino & Concilio, 2005). By providing capacity development programmes through collaborative learning majority of the community people can be taken under a learning parasol. Along these contexts, this research will try to cover three aspects of collaborative learning (CL); such as the basic elements, methodologies, and technologies need to be adopted for proper implementation. These will be discussed in the main thrust section, next.

However, before going to the next section, some of the e-government issues may be pointed here that could be included in the learning processes (not exhaustive, but a prerequisite and could be dynamically updated):

- Basic education (primary and secondary)
- Higher education (tertiary and lifelong)
- Innovative government services available for the citizen
- Legal, social, financial, and cross-cultural issues
- Methodologies and tools for design and analysis of collaborative practices
- Working with and through collections of heterogeneous technologies
- Innovative technologies and architectures to support group activity, awareness and tele-presence
- Public procurement, bidding, auctions
- Required skills for someone working in a collaborative E-Commerce environment (B2B, B2C, G2C, etc).
- Societal Integration of Business Processes
- Project Management, Groupware, and Workflow
- Information & communication platforms, mobile agents, unified messaging
    - Intra organisational communications
    - E-Health
    - Multilingual access
    - International collaborative programs
    - Mobile devices with PDA/WAP/3G
    - Knowledge discovery & data mining (Banks, 2003); and
    - Social livelihood components, like
    - agriculture
    - environment
    - business process reengineering
    - waste management
    - traffic rules
    - small and medium entrepreneurship
    - responsiveness to basic law and orders
    - food security
    - rural development
    - urban planning, and

- skill development in local industry standards.

MAIN THRUST

In 2004, 1.4 billion students (learners) worldwide spent nearly $2.3 million dollars on education (Beahm, Rogers & Liddle, 2006). These costs are mainly incurred into the traditional system of education. But, through the use of the Internet as a communication medium, education is now shifting towards the learner-centered paradigm (Garrison & Anderson, 2003; Lin & Hsieh, 2001), where the instructor acts as a moderator (educator) primarily responsible for facilitating learning. Based on this paradigm, learners are able to choose how to learn, when to learn, where to learn, and what to learn as far as possible within the resource constraints of any learning environments. This sort of learner-centred paradigm is often supported by e-learning especially through the Internet to facilitate collaborative work in the learning process (Kleimann & Wannemacher, 2004; Iahad, Dafoulas, Milankovic-Atkinson & Murphy, 2005; Hauge & Ask, 2008).

At the same time, millions of people around the world lack the opportunity for higher education due to many reasons, such as high tuition costs, increased competition, and the opportunity costs of leaving employment to study full-time, and these reasons prevent many from learning the skills necessary to enhance their standard of living. Moreover, there are many challenges (skills, knowledge, technology, finance) associated with the use of innovations in technology for extending educational opportunities, especially in developing countries. Based on these, collaborative learning could be an initial start in providing basic education, knowledge building and capacity development. In this context, learning would transform new information into knowledge and collaborative learning would transform new information into knowledge with a group or community. Now, a few elements of collaborative learning will be discussed.

Elements of Collaborative Learning

According to learning theories, different ways of learning exist (Behnken, 2005). In the aspect of collaborative learning, a few of are being described here.

Learning through Innovation

Innovation portrays a new product, process or service (Hauschildt, 2004). Behnken (2005) argued that, collective innovation processes, could then be thought of a process of generating an innovation within a community. If a product, process or service incorporates knowledge then there could be a relationship between innovation and learning, as such;

- Collective learning => produces new knowledge within a group
- Collective innovation => incorporates new knowledge within a group
- Therefore, collective innovation = collective learning (Adapted from Behnken, 2005).

However, collective learning process does not always bring the result that may be termed as innovation. A collective learning process need to be adapted, coordinated and articulated according to the demand of the community.

Blended Learning

In the early days of electronic mediated learning the sole application of technology-oriented concepts was used to be propagated, but majority of current approaches cover hybrid forms of educational methods. The idea of blended learning is to effectively join traditional face-to-face education with technological elements to offer

Figure 3. Different aspects of blended learning (Adopted from Phan, 2007)

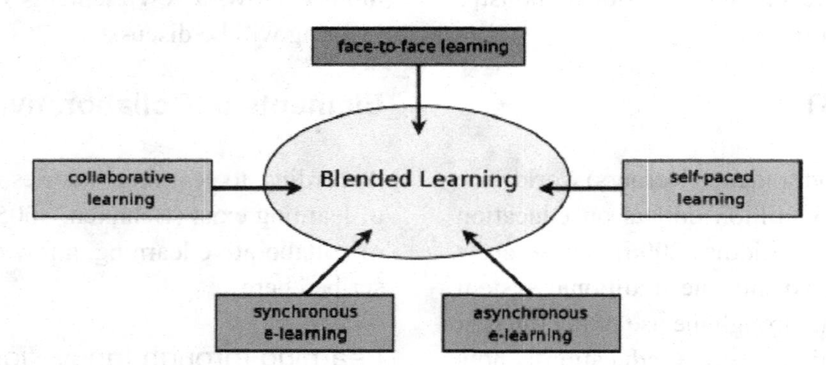

a variety of methods and channels for learning. Blended learning links the effectiveness and flexibility of e-learning with the social aspects of collaborative learning. Blended learning aims at the improvement of learning processes to reach individual learners within the community using all appropriate educational methods. This form of learning integrates virtual components with traditional classroom education. Furthermore, combining face-to-face learning with synchronous and asynchronous forms of e-learning as well as collaborative and self-paced elements aims to compensate drawbacks of the single approaches of learning (Hamburg, Lindecke & ten Thij, 2003; Schmidt, 2005; Hauge & Ask, 2008). Figure 3 and 4 are illustrating two forms of blended learning.

Online Collaborative Learning

Online collaboration involves interaction between learners and educators through any effective means, and it could be intranet, or the Internet. This interaction can occur in one of the following modes:

- Synchronous interaction (on-line interaction with the educator using the Internet)
- Asynchronous interaction (off-line interaction with the educator through other group

Figure 4. A format of blended learning (Adopted from Kumar, 2008)

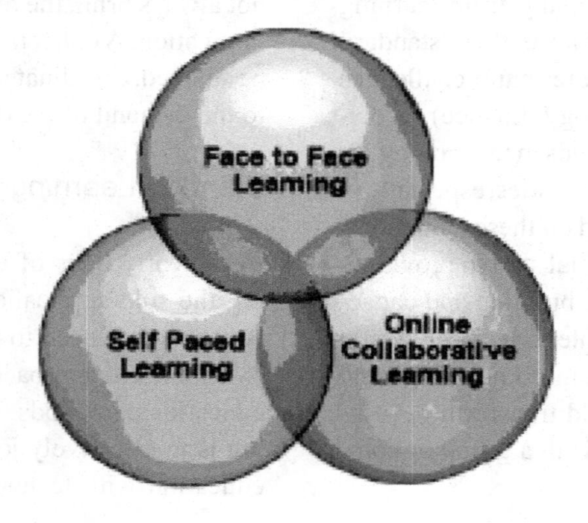

members, or at their own convenience, or by email) (Kumar, 2008).

Online collaborative learning can be used for the development of tools and systems to support group communication, particularly important in building online learning communities. These tools may include:

- Group Web discussion and navigation (In a group they visit HTML based Web pages, explain among themselves and navigate)
- Group Web discussion for streaming contents (In a group they visit audio-video based contents and they discuss among themselves)
- Online Peer Review (A form of task to measure the learning process, and it can be carried out among the groups. A group leader can present, or all can present individually, depending on the task, or the depth of the content) (Chong, 2001).

Demanded Learning with Collective Innovation

As stated earlier, collective innovation generates knowledge within the community, and the performance of a collective innovation depends largely on the extent of synergetic knowledge exchange and on the final quality of the learning capacity. In this aspect, the basic influences are the range of existing knowledge, the depth of the knowledge acquisition processes and the degree of interaction. However, personal influences may produce positive results among the participants to enable individual openness to share knowledge and to learn collectively. This sort of influence grows further through collective learning within vibrant and interactive networks (Behnken, 2005).

Collaborative learning may comprise of demand driven knowledge development, inclusion of indigenous technology providing local level solutions, preservation of environment by providing eco-friendly solutions, techniques to improve livelihood reducing poverty, institutional development to promote grass roots development.

Demanded learning at the grass roots will be effective in solving localized problem locally, acquire knowledge to assist each other in times of routine or emergency life sequences, given the facts that much of the remotely located people merely dependent on their everyday livelihood or may be unaware of the current demand or may be in lack of proper support (skills and infrastructures) during the learning sessions. Furthermore, this sort of learning can create skilled group of personnel ready to support local industry, small and medium scale enterprises, local departmental stores, chain shop outlets, phone-fax stores, or VOIP-based service providing points. This human resources can act as enabler of economic uplift in the nearby societies and in effect, can act as guiding agents for improved governance.

Learning by Interaction

With the help of regular face-to-face interaction community participants under a same network are able to estimate and understand the behavior of their partners. Through positive experiences trust establishes, confidence grows and personal commitment develops. Furthermore, based on regular interactions a common institution may be built up. This sort of common interactions in the form of collective approaches form common framework of action, that can be seen as an enabler of thought and action embedded in the habits of the group. Incorporating mutual expectations of behavior and an allowed regulation, the actor within a group learn gradually how to act and react. Community leaders can also act as informational manipulators by reducing the subjective uncertainty to a level where people are enabled to perform and build the minimum level of trust necessary to be encouraged to share knowledge and become innovative. Their coordinating function is the basis for a capacity to act collectively within the community or network (Elsner, 2000; 2003).

Collaborative Learning Through Social Engagement

Most of the major decision making processes are being taken from the core government system and they take the top-to-bottom route. Whenever community people looks for a related search, they hardly find the proper content or could trace the location of the content. They could be land-records, rule to establish a small trading house, submit tax, or pension policy, apart from complex nature of societal problem like bio-diversity, land degradation, or ecological unbalance. People residing at the remote peripheries of the societal system, or geographical boundaries are always kept outside of these sort of decision making processes, and the result is that they are unaware of the laws/ rules/ gazettes/ proclamations, thus compounding the problem of lawlessness within themselves. If they were been engaged in the policy making processes from the very beginning, then they could draw a boundary between what the law is or at least what would be the consequences of not abiding the law.

Programmes of such nature of immediate social need, including medium and long term implementation could be sought out at community based outlets through collaborative learning processes engaging the community people, and bringing them socially and mentally nearer to the system. This way, they would be easily aware of any upcoming future policies, feel comfortable with the outcome of those policies, and at the same time feel confident to make steps in taking decision at right direction. Multi-purpose village centers, or tele-centers can act as the outlet of this sort of collaborative learning process, and slowly the community people can attain a benchmark from where they can provide economic output.

Collaborative Learning Using Wi-Fi Mesh

With the increasing and unprecedented adoption of wireless technologies in developed and developing countries alike (Keegan, 2003), wireless technologies are revolutionizing the learning processes, thus transforming the traditional ways of learning and teaching into 'anytime' and particularly, 'anyplace' learning. Contemporary studies also reinforce the advantages of using wireless technologies in learning environments, including supporting group work on projects, engaging learners in learning-related activities in diverse physical locations, and even enhancing communication and collaborative learning in the traditional classroom (Barker, Krull & Mallinson, 2005). Educator, being located at a single point can take classes of several classrooms through very low cost wireless intranets.

Moreover, the use of wireless technologies in learning impacts learner motivation, collaboration and mobility, which results in benefits for learners, parents and educators. Those benefits are perceived to outweigh even the difficulties encountered when integrating wireless technologies in learning. The following parameters are the limiting factors in this process: limitations of the connecting devices (wireless devices, servers, routers, switches, hubs, computers), pedagogical issues (skills of the educators handling equipment and the learning process), safety and security concerns (data backup, in-time delivery, proper login), training and support issues (regular and emergency maintenance, training of the trainers), as well as cost considerations (high cost of some connecting devices) (Barker, Krull & Mallinson, 2005). Figure 5 shows a wireless mesh connecting outlets at the village level covering a single or multiple region. The server could be located at a national location, or a district location, depending on the nature of the mesh (depend on geography, population base, etc.) and complexity of the network (depend on distance, number of outlets, etc.).

Figure 5. Wi-Fi Mesh used for collaborative learning (Source: Author)

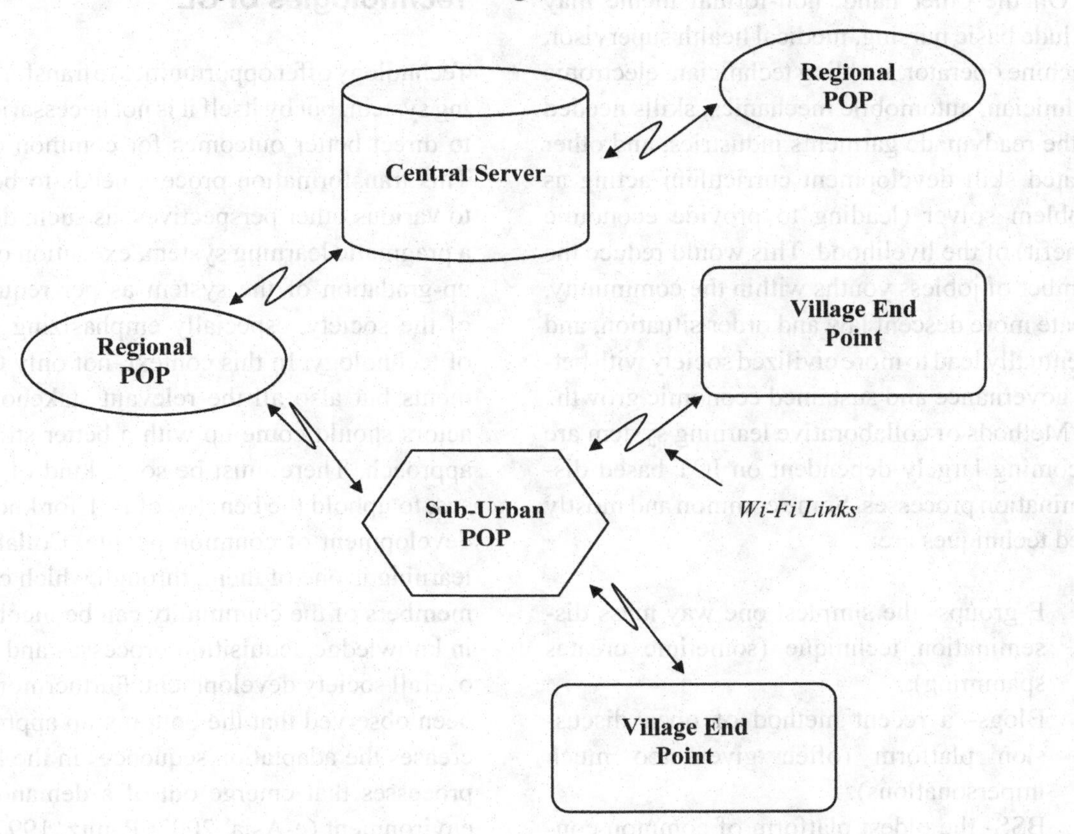

Impact of wireless technology in collaborative learning through its portability, flexibility, availability, collaboration, and motivation would strengthen the grass roots empowerment. Initial invest might seem a bit high in cases, but in the longer term will be economical in terms of their huge benefits (such as, tele-health, e-commerce, consumer commodity, SME activities, and others can easily be incorporated within the network, apart from collaborative learning) and low maintenance costs.

Collaborative learning may take many other forms of implementation, depending on the demand, cost, locality and geography. However, a few of the methodologies of collaborative learning will be discussed next.

Methodologies of CL

At the rudimentary stage, collaborative learning through community driven networks may take two directions, such as the formal education (science, non-science) and non-formal education (skill development programmes). Theme of formal education can accommodate traditional syllabuses of accredited national curriculum, because at the end of the day, this sort of learning need to be recognized by national institutes for future job seekers passed through this process. Without being very ambitious, formal education can take place up to grade eight of national curriculum of a country. However, successful participants from this theme may be given opportunity to higher and studies and research, depending on their mental capabilities and demand of the country.

On the other hand, non-formal theme may include basic nursing, medical health supervisor, machine operator, welding technician, electronic technician, automobile mechanics, skills needed at the readymade garments industries, and other related skill development curriculum acting as problem solver (leading to provide economic benefit) of the livelihood. This would reduce the number of jobless youths within the community, create more descent law and order situation, and eventually lead to more civilized society with better governance and sustained economic growth.

Methods of collaborative learning system are becoming largely dependent on ICT based dissemination processes. Some common and mostly used techniques are:

- E-groups- the simplest one way mass dissemination technique (sometime creates spamming);
- Blogs- a recent method of open discussion platform (often given too much impersonations)
- BSS- the oldest platform of common consequences (not interactive);
- Moderated e-discussions- popularly adopted in recent days to reach at generic consensus utilized in researches to generate reports, create research documents, reach any consensus;
- Virtual seminars- similar to the previous one, but, can be made more interactive through web-techniques and utilization of recently developed utilities (recently becoming very popular in formulating common (Rahman, 2005b). Figure 6 illustrates a simplified format on methodologies of collaborative learning leading to improved e-governance through empowerment of community participants.

Technologies of CL

Technology offer opportunities to transform learning system, but by itself it is not necessarily going to direct better outcomes for common citizens. This transformation process needs to be linked to various other perspectives, as such, design of a pragmatic learning system, execution of it, and up-gradation of the system as per requirement of the society, especially emphasizing the role of technology. In this context, not only Governments but also all the relevant stakeholders or actors should come up with a better strategy or approach. There must be some kind of consensus, to uphold the benefits of ICT for knowledge development of common people. Collaborative learning is one of them, through which common members of the community can be incorporated in knowledge acquisition processes and thereby overall society development. Furthermore, it has been observed that the bottoms-up approach increases the adaptation sequences in the learning processes that emerge out of a demand driven environment (e-Asia, 2007; Panitz, 1997; 1998; Thomas, Howell, Patricia & Angelo, 2001; Venkatesh & Small, 2003; Boud, Cohen & Sampson, 2001; Cohen, 2005).

Following those contexts and consequences, multi-channel delivery systems are always better and should be adopted in the learning processes. In terms of providing higher education through quality digital content has perpetually remained challenge to the academics and researchers, especially when they are being used in open technology platform with interoperability. Moreover, for promoting quality learning through collaborative approach demands formation of appropriate content repositories, geographically inter-connected distributed databases, user friendly access tools, online forums, knowledge banks and interactive but easy access (e-Asia, 2007).

The technologies enabling work-based collaborative learning may include;

Figure 6. Methodologies of collaborative learning (Source: Author)

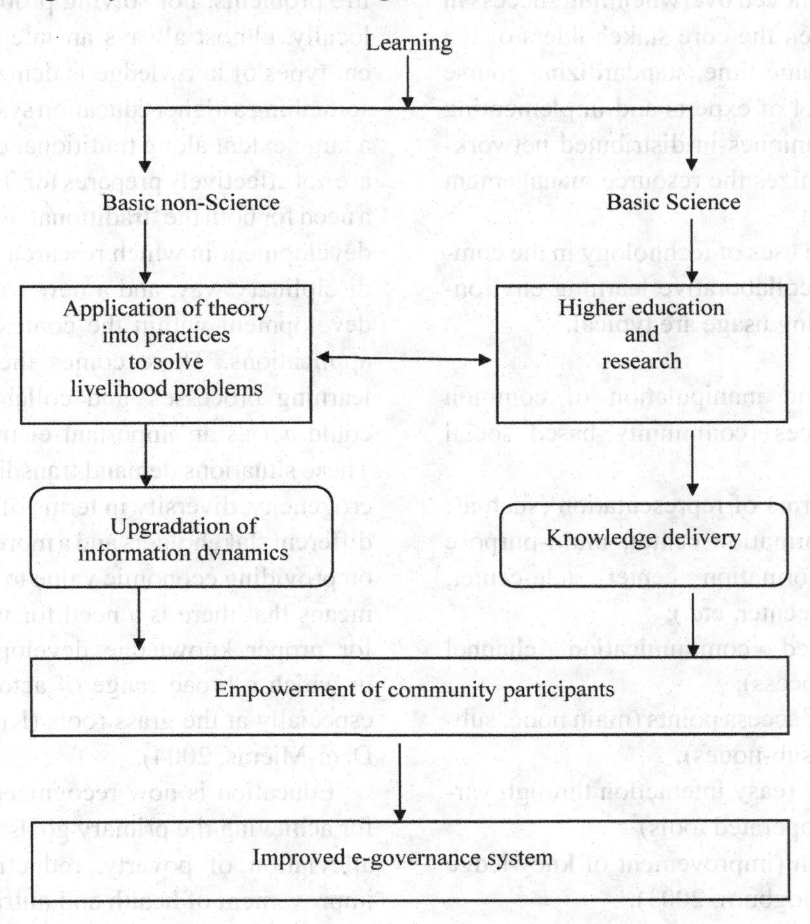

- Computer mediated (personal computers, educational computer conferencing systems, video conferencing, teleconferencing, satellite transmission, etc.)
- On-line interactions (Semantic Webs, Internet, intranet, chatting, messaging, Wiki, etc.)
- Off-line contents (CDROM, television, video-tapes, radio and films, email, blogging, etc.)
- Audio-visual-graphics (audio, video, graphics, multimedia) (Rahman, 2005a; 2007; Bélanger, 2001).

In this aspect, proficient utilization of network resources (TCP/IP, Internet, WWW) in disseminating knowledge-based information in on-line and off-line environment formulates a concrete platform of collaborative learning sequences. There are four technical parameters that are of prime concern in distributed learning platform, namely, Network Infrastructure, File Server, Support Servers (email, egroups, newsgroups, chat/blog/messaging), and Web Server (Rahman, 2005a). Furthermore, Effective utilization of information and communication technology has led the educators to avail means of innovative methodologies to reach out the learners at the outskirts of conventional education arena. Initially intended for dropouts, or residuals of the underdeveloped society, the distant mode of education, at its concurrent diasporas, is no longer lies among them, but, has achieved overwhelming success in consolidating even the core stakeholders of the society. At the same time, standardizing course curricula by panel of experts and implementing the learning techniques in distributed networking aspects optimizes the resource management (Rahman, 2005b).

Regarding the uses of technology in the computer supported collaborative learning environment, the following usage are typical:

- Creation and manipulation of common virtual spaces (community based social networks);
- Multiple forms of representation (such as, village information center, multi-purpose village information center, tele-center, knowledge center, etc.);
- Uninterrupted communication channel (flexible process);
- Diversity of access points (main node, sub-nodes, sub-sub-nodes);
- Interactivity (easy interaction through various easily operated tools)
- Socialization (improvement of knowledge capacity) (Cogburn, 2003).

However, starting from the design stage to the implementation stage, and along the way need assessment, removing impediments from various stages and even at the post implementation, the sustained operation of collaborative learning process may face various challenges, especially that is targeted to empower the community people at the grass roots.

Challenges

Due to the complexity of societal problems, especially in a developing country, attentions were drawn not only for providing more specialized knowledge, which remains necessary as a source of inspiration for innovation, but also for methods to apply knowledge in the right way for solving real life problems. For solving problems that occurs locally, almost always an integration of different types of knowledge is demanded and that is something a higher education system organized to a large extent along traditional disciplinary lines are not effectively prepares for. Therefore, there is a need for both the 'traditional' way of knowledge development in which research is organized in a disciplinary way, and a new way of knowledge development within the context of appropriate applications. There comes the non-traditional learning processes, and collaborative learning could act as an important element of learning. These situations demand transdisciplinarity, heterogeneity, diversity in terms of their input from different stakeholders and a more direct influence on providing economic value to the society. This means that there is a need for whole out efforts for proper knowledge development processes in which a broad range of actors are involved, especially at the grass roots (Kuhlen, 2003; van Dam-Mieras, 2004).

Education is now recognized as the catalyst for achieving the primary goals of development: alleviation of poverty, reduction of inequity, improvement of health and nutrition, population control, social well being, environmental protection, nurturing democracy, and economic growth are among them (UNESCO, 2001). Computer Supported Collaborative Learning (CSCL) can be a tiny solution to those communities, especially where highly educated teachers are not available and technically equipped classes are almost impossible (Bruns & Takahashi-Wetch, 2006; WSIS, 2005). However, empowering marginal and community people through this form of learning will remain a challenge for many years to come, due to a variety of reasons. These could be high initial investments, low cost recovery opportunity, lack of long term subsidy, lesser investment flow from the entrepreneurs due to non-business focus, low acceptability due to financial scarcity, low adaptability due to cultural differences, and many other seen and unforeseen reasons.

Moreover, in terms of using cellular phones and PDAs for collaborative learning, Thornton & Houser (2004) report that some learners had difficulty hearing the audio on both PDAs and cellular phones, and learners suggest headphones would be required when studying in trains, buses and other public places. In addition to these, Thornton & Houser (2004) find few serious technical limitations to the widespread use of wireless technologies in learning. According to the Becta research, possible weaknesses include small screens, not 'rugged' enough for school use, data input (especially free texts) takes time, costs of software and accessories, and the necessity to charge the batteries quite regularly (Perry, 2003; Barker, Krull & Mallinson, 2005).

Before discussing the future research issues in the aspect of collaborative learning for empowering community people, a few cases have been discussed below:

Case-1

A country in South Asia with 141.822 million populations living in 147,570 Sq. Km in a deltaic region faces regular natural calamities like, flood, tsunami, draughts, and tidal weaves. This country, Bangladesh has a literacy rate of 51.6[9] and a lone distance education provider, the Bangladesh Open University (BOU). In spite of potential acceptance, government support and huge demand at the grass roots, BOU could not able to create sufficient scope to facilitate this huge population base. However, there are a few NGOs who are working relentlessly in this sector. BRAC[10] is one of them. This NGO (the largest in the World) is providing primary (pedagogy and non-formal[11]) through over 50,000 village schools and contributing largely to increase the literacy rate. Government has also taken several female student literacy programme[12] since 1993 and included special incentive based programme where female students are given free studentship till grade 12.

In spite of all these, it is a surprising fact that the drop out rate at primary level (till grade five) remains at around 30%, while it increased to over 50% at secondary level (grade nine and ten). Hence, there are immense scopes of conducting mass scale literacy programme up to the primary level leading to secondary level, reaching out the grass roots population. Furthermore, with 450,000 Internet users as of August, 2007[13] (0.3% of the population, according to ITU[14]) Bangladesh is lagging far behind the World class standard in the technology arena.

To uphold the goals of Education For All (EFA)[15] and Millennium Development Goals[16] a programme was initiated in early 2000 with the assistance of the World Computer Exchange, USAID and UNDP. It was a project component of the Sustainable Development Networking Programme (SDNP), a UNDP funded programme through which the project tried to put forward several educational components related to collaborative learning and capacity development. Some others include, establishment of multi-purpose village information centers, establishment of public use cyber centers (pioneer in Bangladesh), establishment of content based web portal/information bank/data bank for common use, establishment of the longest Wi-Fi based radio link connecting several organizations including the largest Agricultural University, establishment of a national Internet exchange (lone in the country), hosting of the F-root server in Dhaka, and various other innovative projects with novel concepts and ideas.

Keeping all these in mind, in a tiny spike, a program has been initiated in a rural corner of Bangladesh to enclave grass roots communities as part of the life long learning processes. The members of the society will be given traditional and non-traditional education, depending on their demands, aspirations and capabilities through a technology mediated educational institution. By clustering them into smaller groups, a micro-credit program will run to empower them economically and socially. Different categories of project (education, health, environment, technology) will run throughout the year for their skill development and knowledge building. Furthermore, adopting appropriate technologies, like introduction of solar power (cooking, water logging, lamps, machineries), wind mills (water logging, electricity) community people will be assisted to carry out their tasks rapidly and easily with limited impact on the local environment and resources. Finally, the surrounding communities will be taken under a Wi-Fi mesh to form a knowledge building network (continued education, self employment, information bank, data mining, improved livelihood, capacity development, market research, food security, basic health promotion, social development, increased governance and sustained advancement) with the intention to propagate knowledge beyond the peripheries.

Established in 1998, the project (SDNP, www.sdnbd.org) started its operation literally from December 1999 and it launched several innovative programmes in Bangladesh. The first component was launching of free cyber centers at public places (July 2000), and with the assistance of the National Press Club the project opened about 15 cyber centers throughout the country (between 2000 and 2004). Establishment of local content based web in local language was another milestone of this project (http://www.mdgbangla.org/). Later on the project started building its own information infrastructure to link several educational institution and build agricultural information based network comprising those institutions and local NGOs, that evolved as multi-purpose village information centers (MVICs, http://www.sdnpbd.org/sdnp/mvic/). Among others, establishment of the first Internet exchange (in 2004) in Bangladesh (http://www.bdix.net/) was another achievement of SDNP and this exchange is emerging as the national data center, as it also hosted the F-Root server. Currently, about 26 largest ISPs are connected to the BDIX.

As the project (SDNP) closed its operation in Bangladesh in December 2006, the project has been transformed into a Foundation, namely the Sustainable Development Networking Foundation (SDNF) in January 2007. SDNF kept a few of those initiative intact and looking forward to continue their operations as long as they can sustain. However, to keep the educational and capacity development activities getting forward, another initiative has been established following the international pattern of SchoolNets, and SchoolNet Foundation Bangladesh has started its operation in Bangladesh since January 2007.

SchoolNet Foundation (www.schoolnetbd.org) would like to carry out the continuation of SDNP School programme (www.sdnbd.org/school_programme/), the Telemedicine activities (www.sdnbd.org/telemedicine/, capacity development activities in terms of providing traditional and non-traditional training at school and college levels for extending knowledge networking at the grass roots (www.icmsbd.org), and a few other activities as relevant to the community development processes. In addition to these, a micro-credit component is in progress, which will only focus on providing financial support for continuing education (primary and secondary). Finally, efforts are there to establish a non-traditional ICT based University in the country, focusing the majority participants of the community.

Case-2

This case is not a project, rather an innovative development of software to enhance collaborative learning at the primary level of education.

In terms of computer use in schools or rural kiosks, it is always observed that the student-to-computer ration is very high. Very often one can see that more than five or six students are using one computer. This is mainly due to insufficient funding. In these cases, one student controls the mouse, while others act as passive onlookers without operational control of the computer. Therefore, learning capacity does not equally distributed over the participants. One student becomes more familiar with the computer, while others are not. Even, if the operation is being carried out on rotation basis, due to increasing number of education, a majority of the learners remain deprived of the learning opportunity. Sometimes, this makes them ignorant of the lessons in the practical sessions.

The accurate technical solution is to provide each child with a mouse and cursor on screen, thus effectively multiplying the amount of interaction per student per computer for the cost of a few extra mice. Despite both the concept and the implementation appear to be unique to date, for the specific application to computers in education in resource-strapped communities, with previous work restricting studies to two mice, or for largely non-educational applications, a recently developed software allows multiple colored cursors to co-exist on the monitor, along with two sample games with some educational content. Initial trials with both single-mouse and multiple-mice scenarios suggest that children are more engaged when in control of a mouse, and that more mice increases overall engagement of the participating students. However, this unique feature of using multiple mouse in a computer deserves attention of researchers in this aspect, especially researching in pedagogy for computers in collaborative learning (Pawar, Pal & Toyama, 2006).

Case-3

The Gravina's Collaborative System (GraviCS) of Gravina, Italy supports collaborative learning to develop process-scenarios in an argumentative, interactive governance environment. It was developed, for the local authority of Gravina, to support decision making in environmental planning as part of a large regional project. That project aims at preparing a preliminary expert analysis for the Apulia territory in creating some Regional Natural Parks. In accordance to the formal procedure, the regional authority organizes public meetings which are not only looked as a consultancy process, but also are searching for political consensus among stakeholders. During such meetings, a preliminary agreement on park infrastructures and management norms is ratified.

In the case of the Gravina's Natural Park (one of the proposed Regional Natural Parks in Gravina), a national financial support permitted the realization of a information system for managing and assisting the weak phase of the public meetings. The environmental planning domain is comprised of multiple, geographically dispersed, participants of diverse professional or private backgrounds, interests, preferences and viewpoints. Therefore, the system required intuitive and easy-to-use interfaces across an Internet based platform. The developed prototype, GraviCS, focuses on distributed and asynchronous collaboration and allows the participants to surpass the requirements of being in the same place and at the same time. Moreover, this system is based on a web platform that provides relatively inexpensive access and it has intuitive interfaces for easier navigation by inexperienced users. The software is divided into two main modules. The first one, represents the web interface of the system, and the second one enables limited access (by self-registration of participants). The system was developed with such intention that it can be used to assist users not only during the process of setting up of the natural park, but also during the subsequent management of the area. The system may support the entire process of decision making by proving a forum (a platform of collaborative learning) in which participants can establish defined protocols through dialogues, and consequently, they can interact in a structured dimension (Celino & Concilio, 2005).

Other Cases

In terms of engaging participants across different countries in collaborative learning, a wide range of programmes and projects on ICTs in education in Africa have activities that involve one or more African countries in varying numbers.

These range from high-level intergovernmental, multi-stakeholder programmes, such as the NEPAD e-Schools initiative, to institutions focused on networking African schools and universities such as the African Virtual University (AVU), to collaborative learning projects that directly involve learners and educators from schools in several African countries (for example, the Global Teenager Project (GTP) and the International Education Resources Network (iEARN).

iEARN (International Education Resources Network (iEARN), www.iearn.org) is one of the largest and oldest global networks of educators and learners that use ICTs in a diverse range of collaborative learning projects. All iEARN projects are designed, initiated, and run by educators and learners. Its network in Africa involves learners and teachers from schools in 29 countries;

The Global Teenager Project (www.globalteenager.org) is an initiative of the Dutch-based International Institute for Communication and Development (IICD). It was launched in 1999, especially to promote the use of ICTs in the classroom. The project focuses on collaborative learning among secondary school students and educators from around the world through a safe, structured virtual environment known as "learning circles." So far, the project involves about 3,000 teachers and students from 200 classes in over 29 countries. Majority of the participants are from Africa, involving learners, teachers, and schools from 12 countries;

SchoolNet Africa (www.schoolnetafrica.net) is an NGO-based in Senegal that promotes education through the use of ICTs in African schools. SchoolNet Africa functions as a network of SchoolNet organisations operating in over 33 countries on the basis of regional programmes on ICT access, training of trainers, and collaborative learning. Till date SchoolNet Africa has produced a range of research reports on the experiences of African countries on the use of ICTs in schools (Farrell, 2008).

FUTURE TRENDS

A collaboratory is not simply formation of a group of learners, not it is simple application of ICTs, it is more than an elaborate collection of concurrent ICTs; it is a new networked organizational form that includes social processes, collaboration techniques, formal and informal communication, and agreement on norms, behaviors, ethics, principles, values and rules within the network. To date, most collaboratories have been applied largely in the field of basic or applied sciences like, physics, mathematics, upper atmospheric research, and astronomy and have been applied recently to additional areas of research such as bio-diversity conservation, ecosystem management or HIV/AIDS. Since the emergence of these collaboratories, a substantial and growing knowledge base has emerged to help communities to understand their development and application of their knowledge in science and industry (Teasley & Wolinski, 2001; Cogburn, 2003).

Furthermore, with transformation of the global economy towards a more knowledge-based, innovation-oriented, ICT-mediated, and geographically-distributed form, it has become increasingly important for higher education and research institutions to equip learners with additional skills. These skills include an interdisciplinary approach like; livelihood problem identification and solving; self organization, self control and motivation; capacity to acquire, manage and disseminate knowledgeable information; increased participation in cross-national and cross-cultural negotiations; adaptability to work under a diversified collaborative environment; and ability to work in geographically distributed virtual scenarios.

Among older theories, Tiffin & Rajasingham (1995) suggest that the balance between human-interaction and computer-interaction is a critical factor in the success of a virtual learning environment. In similar context, Brown & Duguid (2000) suggest that this balance is even more important when the learning environment becomes more complex, vibrant, and geographically distributed. They stated that, these learning teams are further challenged by the adoption of a "stakeholder" perspective in the global-system (i.e., global and multi-national corporations, developed country national governments, developing country national governments, intergovernmental organizations, and non-governmental organizations). They also argued that learning is a social process, and that "peer networks" are an equally important resource to higher education institutions and research institutions. Along this perspective, Hiltz (1990) finds that "collaborative learning" enhances student ratings of virtual courses. Thus further, Cogburn (2003) anticipated that learners engaged in virtual teams (Global Syndicates) that evolve into "learning communities" will have more collective and individual success in their working environments. Therefore, future collaborative learning sequences should be able to synchronize the above mentioned situations and proceed accordingly.

In terms of successfulness of any system, evaluation is a must. Taking into account the input in an established learning environment, the collaborative learning system constructs an illustrative knowledge graph that is composed of the ideas expressed within the system, as well as their supporting arguments. Moreover, through the integrated feedback mechanisms, participants are continuously informed about the status of each item they learn and reflect further on them according to their requirements and interests on the outcome of the learning. In addition, this approach aids group sense-making and mutual understanding through the collaborative identification and evaluation of varied opinions. Such an evaluation can be carried out through either argumentative discussion or e-voting. Figure 7 illustrated an outcome scenario in G2G (Government-to-Government) situation. This has been adopted from Karacapilidis, Loukis & Dimopoulos (2005), but the author argues that in other situations like, G2C, C2G, C2C, or individual-to-individual, the outcome would remain the same. Future research can be carried out to observe any dissimilarity or further improvement when the collaborative learning is being utilized to empower marginal communities.

Furthermore, a system with a shared web-based workspace for storing and retrieving the messages and documents of the participants, using the widely accepted XML document format, exploitation of the web platform renders, among others, low operational cost and easy access to

Figure 7. Outcome of an e-collaboration system (Adopted from Karacapilidis, Loukis & Dimopoulos, 2005)

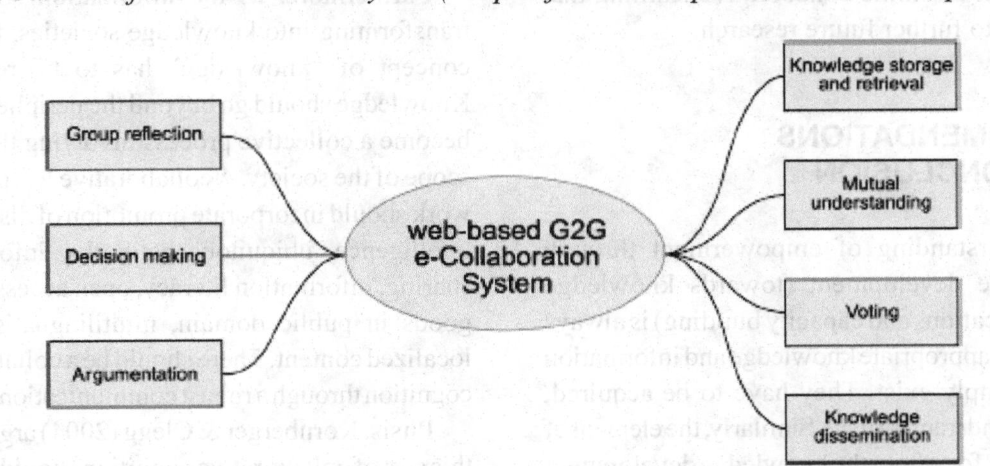

the system. The core of the system maintains all the items (messages and documents), which may be considered as a knowledge base, appropriately processed and transformed, or even re-used in future discussions. Archival of documents and messages being asserted in an ongoing learning takes place in an automatic way that is upon their insertion in the knowledge graph. On the other hand, retrieval of knowledge is performed through appropriate interfaces, which aid users explore the contents of the knowledge base and exploit previously stored or generated knowledge for their current needs. In such a way, this approach builds a "collective memory" of a common community (Karacapilidis, Loukis & Dimopoulos, 2005). However, further research need to be carried out to filter, separate, assemble, and represent in suitable format to different users with diversified patterns and natures.

The assessment of online collaborative learning presents new opportunities and challenges, both in terms of separating the process and product of collaboration, and in the support of skills development. Macdonald (2003) has explored the role of assessment with respect to the process and products of online collaborative learning. He conducted a qualitative case study on two UK Open University courses that have used a variety of models of online collaborative assessment. He also put forwards a number of recommendations for the assessment of online collaborative learning that may lead to further future research.

RECOMMENDATIONS AND CONCLUSION

The understanding of empowerment through knowledge development (towards knowledge communications and capacity building) is always critical, as appropriate knowledge and information do not simply exist. They have to be acquired, nurtured and transformed. Similarly, the element of the society for whom the knowledge development processes are to be devised, their active participation within the processes has also to be ensured. Knowledge and information in all areas are not the same, they are localized and thereby, knowledge development processes should incorporate localized and indigenous inputs. Furthermore, adequate safeguard should be taken for their dissemination, so that they are uniformly and used collaboratively (Kuhlen, 2003).

In this aspect, learning is an important component to not only build a knowledge base among the community participants, but also learning through a collaborative process would strengthen the country's information economy. Moreover, it is essential that the government and private sector ensure appropriate skills are taught through primary, secondary, and tertiary education and at the same time on-the-job training need to be conducted to meet the local industry needs. Additionally, the demand for personnel with IT knowledge (perhaps ICT knowledge), experience and qualifications from the workplace is growing fast, as the workplace is gradually applying computing skills to improve their efficiency in this competitive world (Kelegai & Middleton, 2002). Therefore, learning should not be restricted to individuals and specific perimeters; rather it should take the form of group learning or collaborative learning and should not remain under any comprehensive boundaries.

Furthermore, as the information society is transforming into knowledge societies, the very concept of "knowledge" has to be revisited. Knowledge should go beyond the peripheries and become a collective process involving the entire scope of the society. A collaborative learning network should incorporate promotion of distributed intelligence, ubiquitous networks, information sharing, information literacy, open access, public goods in public domain, multilingualism, and localized content. There should be a collaborative cognition through a robust communication system.

Pitsis, Kornberger & Clegg (2004) argued that this sort of collaborative cognition should include collaborative learning, collaborative knowledge management, knowledge transfer, collaborative memory and collaborative communication. The ultimate purpose is collaborative learning, which can be transformed into the intellectual assets of collaborative knowledge management. Towards this end, participants communicate, making sense of each other, in a collaborative memory. However, (Fagan, Newman, McCusker & Murray, 2006) argued that, they may not necessarily be coming to agreement on a single shared sense of the information, there may still be cultural diversity in understandings, but vibrant communication can reduce the knowledge gap. What it means to the end is a commitment to knowledge transfer and knowledge acquisition in specially designed processes for sharing knowledge, capturing knowledge and empowering them through the acquired knowledge.

Certainly, one of the issues is the use of ICT within the class room or outside the class room, and to innovate traditional educational practices. A interrelated aspect is the importance of human capital. ICT related knowledge jobs require continuous update of knowledge, and at the same time ICT seems to provide technologies for facilitating diversified modes of learning, including the collaborative learning. Literature, however reports many failures especially if the new opportunities are implemented in mere traditional educational situations (van der Meulen, 2003). Successes require incorporation of accurate selection of learning material (based on the demand of the society), appropriate designing (interactive, or non-interactive), indigenous technology (wired, or wireless), and adequate dissemination (synchronous, or asynchronous), as they are vital to the communicative aspects of ICT that are being utilized for collaborative learning.

In recent days, a range of implementation models and approaches are being adopted for the use of ICT in learning. The integration of ICT in learning includes project-based, inquiry-based, individual and collaborative learning models. Moreover, learners studying in the non-formal sector are increasingly required to undertake subjects and courses online or via a blended delivery mode of face-to-face and online learning. Online courses may include audio-visual methods, such as interactive multimedia courseware and online assessments like providing tasks, and taking tests and quizzes. Assessment of the learning may take place through a computerized learning management system, enabling facilitators or educators to track, assess and quantify electronically submitted assignments. Specially, in primary and primary education, learners may undertake project-based learning via Internet-based projects and specially developed online curriculum content (Robbins, 2004; Naidu & Jasen, 2007). As stated earlier, common participants of the society or community may take part in specially design courses of non-formal category, as per their interests and requirements.

Finally, the ability to learn collectively is a significant source of competitive benefit. The benefit goes to the individual, his/her family, his/her community, and his/her nation. According to Boekema, Meeus & Oerlemans (2000), learning is the key for competitive advantage. Despite the financial resources would stay as a decisive booster of innovation, but the synergetic complementation of core competencies (formal education) and knowledge of various backgrounds (non-formal education) are increasingly becoming an essential element of innovation capability (Behnken, 2005), and an important component of an empowered community.

REFERENCES

Banks, D.A. (2003, June). Collaborative Learning as a Vehicle for Learning about Collaboration. *Proceedings of Informing Science InSITE - "Where Parallels Intersect"*, (pp. 895-903), Pori, Finland.

Barker, A., Krull, G., & Mallinson, B. (2005, October 25-28). A Proposed Theoretical Model for M-Learning Adoption in Developing Countries. *mLearn 2005*, Cape Town, South Africa.

Beahm, C. P., Rogers, P. C., & Liddle, S. W. (2006). Opportunities and Challenges of Utilizing Educational Technology in Developing Countries: The eCANDLE Foundation. In Mendez-Vilas, A., Solano Martin, A., Mesa Gonzalez, J. A., & Mesa Gonzalez, J. (Eds.), *Current developments in technology-assisted education* (pp. 1825–1831). Published by Formatex.

Behnken, E. (2005, June 20-22). The Innovation Process as a Collective Learning Process. *ICE 2005: 11th International Conference on Concurrent Enterprising*, Munich, Germany, Springer.

Bélanger, M. (2001). *Work-based distributed learning, Online-document.* Available from http://training.itcilo.org/actrav/library/english/publications/work-based_learning.doc

Boekema, F., Meeus, M., & Oerlemans, L. (2000). Learning, Innovation and Proximity: An Empirical Exploration of Patterns of Learning: a Case Study. In Boekema, F. (Ed.), *Knowledge, Innovation and Economic Growth: the Theory and Practice of Learning Regions* (pp. 137–164).

Boud, D., Cohen, R., & Sampson, J. (2001). *Peer Learning in Higher Education: Learning from & With Each.* Routledge.

Boyd, H., & Cowan, J. (1985). A case for self-assessment based on recent studies of student learning. *Assessment & Evaluation in Higher Education, 10*(3), 225–235.

Brown, J. S., & Duguid, P. (2000). *The Social Life of Information*. Boston: Harvard Business School Press.

Bruns (Jr.), E.L., & Takahashi-Welch, W. (2006). *Implementing Computer Supported Collaborative Learning in Less Developed Countries.* A course paper at the University of Texas at Austin.

Burn, J., & Robins, G. (2003). Moving towards e-government: a case study of organizational change processes. *Logistics Information Management, 16*(1), 25–35. doi:10.1108/09576050310453714

Celino, A., & Concilio, G. (2005, September 13). Open Content Systems for E-Governance: The Case of Environmental Planning. In Z. Irani, T. Elliman, & O.D. Sarikas (Eds.), *Proceedings of the eGovernment Workshop '05 (eGOV05)* (pp. 92-93), Brunel University, West London UB8 3PH, UK.

Chang, C. Y., Sheu, J. P., & Chan, T. W. (2003). Concept and Design of Ad Hoc and Mobile classrooms. *Journal of Computer Assisted Learning, 19*(3), 336–346. doi:10.1046/j.0266-4909.00035.x

Chavan, A. (2004, November). Developing an Open Source Content Management Strategy for E-government. *Proceedings of the 42th Annual Conference on the Urban and Regional Information Systems Association*, Nevada.

Chong Ng, S.T. (2001, May). Taking e-learning education into the future – The global knowledge hall. *Sharing Knowledge and Experience in Implementing ICTs in Universities - Roundtable Papers* IAU/IAUP/EUA, Skagen Roundtable, Skagen, Denmark.

Cogburn, D. L. (2003). Globally-Distributed Collaborative Learning and Human Capacity Development in the Knowledge Economy. In Mulenga, D. (Ed.), *Globalization and Lifelong Education: Critical Perspectives*. New Jersey: Lawrence Erlbaum Associates.

Cohen, E. (2005). (Ed.) *Issues in informing science and information technology, 2*. Informing Science.

Coppola, N. W., Hiltz, S. R., & Rotter, N. G. (2002). Becoming a Virtual Professor: Pedagogical Roles and Asynchronous Learning Networks. *Journal of Management Information Systems, 18*(4), 169–189.

Daniels, S. E., & Walker, G. B. (2001). *Working through environmental conflict—the collaborative learning approach*. Westport, CT: Praeger.

Dede, C. (2004). Enabling Distributed Learning Communities Via Emerging Technologies. *T.H.E. Journal*. Retrieved September 28, 2008 from http://www.thejournal.com/magazine/vault/A4963.cfm

Dringus, L. P., & Terrell, S. (1999). The framework for DIRECTED online learning environments. *The Internet and Higher Education*, *2*(1), 55–67. doi:10.1016/S1096-7516(99)00009-3

Dutton, W. H., Kahin, B., O'Callaghan, R., & Wyckoff, A. W. (Eds.). (2005). *Transforming Enterprise: The Economic and Social Implications of Information Technology*. MIT Press. e-Asia 2007 (2007, February 6-8). Summary report, *e-Asia 2007*. Putrajaya International Convention Center, Malaysia.

Elsner, W. (2000). An Industrial Policy Agenda 2000 and Beyond: Experience, Theory and Policy. In Elsner, W., & Groenewegen, J. (Eds.), *Industrial Policies After 2000, Boston*. London: Dodrecht.

Elsner, W. (2003). Increasing Complexity in the "New" Economy and Co-ordination Requirements Beyond the "Market": Network Governance, Interactive Policy, and Sustainable Action. In Elsner, W., Frigato, P., & Steppacher, R. (Eds.), *Social Costs of the Global "New"*. Economy.

Fagan, G. H., Newman, D. R., McCusker, P., & Murray, M. (2006). *E-consultation: evaluating appropriate technologies and processes for citizens' participation in public policy.* Final Report, e-Consultation Research Project, UK.

Farrell, G. (2008). *Survey of ICT and Education in Africa* (Vol. 1). Association for the Development of Education in Africa (ADEA). International Institute for Educational Planning, Paris, France

Gadomski, A. M. (1993, January/February) (Ed.). Toga: A Methodological and Conceptual Pattern for Modeling of Abstract Intelligent Agent. *Proceedings of the "First International Round-Table On Abstract Intelligent Agent* (pp. 25-27), Rome: Enea.

Garrison, D. R., & Anderson, T. (2003). E-Learning in the 21st Century: A Framework for Research and Practice. London, UK: Routledge Falmer. Hamburg, I., Lindecke, C., & ten Thij, H. (2003, September 25-26). Social aspects of e-learning and blending learning methods. *A Proceedings of the 4th European Conference E-Comm-Line 2003* (pp. 11-15), Bucharest.

Gee, X., Yamashiro, A., & Lee, J. (2000). Pre-class planning to scaffold students for online collaborative learning activities. *Journal of Educational Technology & Society*, *3*(3).

Goodfellow, R. (2001). Credit where it's due. In Murphy, D., Walker, R., & Webb, G. (Eds.), *Online Learning and Teaching with Technology: Case Studies, Experience and Practice* (pp. 73–80). Kogan Page.

Hauge, H., & Ask, B. (2008). Qualifying University Staff in Developing Countries for e-Learning. *iLearning Forum 2008 Proceedings - European Institute for E-Learning (EIfEL)* (pp. 183-188), Paris.

Hauschildt, J. (2004). *Innovationsmanagement*, (3rd ed.), Munich.

Heeks, R. (2002). *Reinventing Government in the Information Age: International Practice in IT-Enabled Public Sector Reform* (2nd ed.). New York: Routledge.

Heinecke, W., Dawson, K., & Willis, J. (2001). Paradigms and Frames for R & D in Distance Education: Toward Collaborative Learning. *International Journal of Educational Telecommunications*, *7*(3), 293–322.

Hiltz, R. (1990). Evaluating the virtual classroom. In Harasim, L. M., & Turnoff, M. (Eds.), *Online education: Perspectives on a new environment* (pp. 133–183). New York: Praeger.

Holmes, D. (2001). *EGov: eBusiness Strategies for Government*. London: Nicholas Brealey Publishing.

Iahad, N., Dafoulas, G. A., Milankovic-Atkinson, M., & Murphy, A. (2005, January 3-6). E-learning in developing countries: suggesting a methodology for enabling computer-aided assessment. In *the proceedings of the 3rd ACS/IEEE International Conference on Computer Systems and Applications (AICCSA-05)* (pp. 847-852), Cairo, Egypt.

Karacapilidis, N., Loukis, E., & Dimopoulos, S. (2005). Computer-supported G2G collaboration for public policy and decision-making. [Emerald Group Publishing Limited.]. *The Journal of Enterprise Information Management, 18*(5), 602–624. doi:10.1108/17410390510624034

Keegan, D. (2003). *The future of learning: From eLearning to mLearning. Hagen*. Germany: Femstudienforchung.

Kelegai, L., & Middleton, M. (2002). Information Technology Education in Papua New Guinea: Cultural, Economic and Political Influences. *Journal of Information Technology Education, 1*(1), 11–23.

Kleimann, B., & Wannemacher, K. (2004). *E-Learning an deutschen Hochschulen*. Hannover: HIS.

Kuhlen, R. (2003, August 3). *Change of Paradigm in Knowledge Management Framework for the Collaborative Production and Exchange of Knowledge*. A paper presented in the Plenary Session, of the World Library and Information Congress: 69th IFLA General Conference and Council, Berlin.

Kumar, R. (2008, July). Convergence of ICT and Education. *Proceedings of World Academy of Science. Engineering & Technology, 30*, 557–569.

Lanzara, G. F., & Morner, M. (2004, 2-3 April). Making and Sharing Knowledge at Electronic Crossroads: the evolutionary ecology of open source. *Proceedings, The Fifth European Conference on Organizational Knowledge, Learning, and Capabilities*, Innskbruck.

Lea, M. (2000). Computer conferencing: new possibilities for writing and learning in Higher Education. In Lea, M., & Stierer, B. (Eds.), *Student writing in Higher Education: new contexts*. UK: Open University Press.

Leitner, C. (2003, July 7-8). E-government in Europe: the state of affairs. *Proceedings of the e-Government 2003 Conference*, Como.

Lin, B., & Hsieh, C. (2001). Web-based Teaching and Learner Control: A Research Review. *Computers & Education, 37*(3-4). doi:10.1016/S0360-1315(01)00060-4

Lin, B., & Hsieh, C. (2001). Web-based Teaching and Learner Control: A Research Review. *Computers & Education, 37*(3-4), 377–386. doi:10.1016/S0360-1315(01)00060-4

Macdonald, J. (2003). Assessing online collaborative learning: process and product. [Elsevier Science.]. *Journal of Computers and Education, 40*(4), 377–391. doi:10.1016/S0360-1315(02)00168-9

Mantha, R. W. (2001, May 4-6). Ulysses: Creating a Ubiquitous Computing Learning Environment. In *Proceedings of the Sharing Knowledge and Experience in Implementing ICTs in Universities EUA / IAU / IAUP Round Table*, Skagen.

Markkula, M. (2006). Creating Favourable Conditions for Knowledge Society through Knowledge Management, e-Governance and e-Learning. A *Proceedings of FIG Workshop* (pp. 30-52), Budapest, Hungary.

Markus, B. (2008, June 11-13). *Thinking about e-Learning, A paper from the Proceedings of Sharing Good Practices: E-learning in Surveying, Geo-information Sciences and Land Administration FIG International Workshop 2008*, Enschede, The Netherlands.

Miller, E., & Findlay, M. (1996). *Australian thesaurus of educational descriptors*. Melbourne: Australian Council for Educational Research.

Naidu, S., & Jasen, C. (2007). *Australia: ICT in Education, UNESCO Meta-survey on the Use of Technologies in Education*. Bangkok: UNESCO.

Oliver, R., & Omari, A. (2001). Exploring Student Responses to Collaborating and Learning in a Web-Based Environment. *Journal of Computer Assisted Learning, 17*(1), 34–47. doi:10.1046/j.1365-2729.2001.00157.x

Panitz, T. (1997). Collaborative Versus Cooperative Learning: Comparing the Two Definitions Helps Understand the nature of Interactive learning. *Cooperative Learning and College Teaching, V8*(2).

Panitz, T. (1998). *Collaborative versus cooperative learning - a comparison of the two concepts that will help us understand the underlying nature of interactive learning*. Available at http://www.capecod. Net/

Pawar, U. S., Pal, J., & Toyama, K. (2006). Multiple mice for computers in education in developing countries. *Proc. IEEE/ACM ICTD* (pp. 64-71).

Perry, D. (2003). *Handheld Computers (PDAs) in Schools*. Coventry, UK: Becta ICT Research, British Educational Communications and Technology Agency.

Phan, M. (2007). *Customisable Checklist to evaluate Learning Management Systems regarding Speci_c Requirements in Viet Nam*. Diploma Thesis for the Department of Computer Science, Albert-Ludwigs-Universität Freiburg, Germany

Pitsis, T.S., Kornberger, M., & Clegg, S. (2004). The Art of Managing Relationships in Interorganizational Collaboration. *M@n@gement, 7*(3), 47-67. Special Issue: Practicing Collaboration.

Rahman, H. (2001, October 16-19). *Utilization of ICT Infrastructure for Collaborative Learning through Community Participation*. A paper presented at the IFUP2001, 4th International Forum on Urban Poverty, Marrakech, Morocco. http://www.ifup.org

Rahman, H. (2005a). Distributed Learning Sequences for the Future Generation. In Howard, (Eds.), *Encyclopedia of Distance Learning* (pp. 669–673). Hershey, PA: Idea Group.

Rahman, H. (2005b). Virtual Networking: An essence of the Future Learners. In Howard, (Eds.), *Encyclopedia of Distance Learning* (pp. 1972–1976). Hershey, PA: Idea Group.

Rahman, H. (2007). Interactive Multimedia technologies for Distance Education Systems. In Tomei, L. (Ed.), *Online and Distance Learning: Concepts, Methodologies, Tools and Applications* (pp. 1157–1164). Information Science Reference.

Robbins, C. (2004). *Educational Multimedia for the South Pacific, Research Report for ICT Capacity Building at USP Project entitled "Maximising the Benefits of ICT/Multimedia in the South Pacific: Cultural Pedagogy and Usability Factors*. Fiji: Prepared for ICT Capacity Building at the University of the South Pacific Media Centre.

Roberts, T. S. (2004). *Online Collaborative Learning: Theory and Practice*. Information Science Publishing.

Robertson, J. (2004). *Successfully deploying a content management system*. Step Two Designs Pty Ltd, www.steptwo.com.au.

Roschelle, J. (2003). Unlocking the learning value of wireless mobile devices. *Journal of Computer Assisted Learning, 19*(3), 260–272. doi:10.1046/j.0266-4909.2003.00028.x

Rovai, A. P. (2000). Building and sustaining community in learning network. *The Internet and Higher Education*, *3*, 285–297. doi:10.1016/S1096-7516(01)00037-9

Salmon, G. (2000). *E-moderating: the key to teaching and learning online*. London: Kogan Page.

Schellens, T., & Valcke, M. (2006). *Collaborative Learning in Synchronous Discussion Groups: What about the Impact of Cognitive Processing. Computers in Human behaviour*. Cape Town, South Africa: University of Cape Town.

Schmidt, I. (2005). *Blended E-Learning*. Saarbrücken: VDM Verlag.

Silvio, J. (2001, May). Building a typology for the comparative analysis of virtual universities worldwide. *Sharing Knowledge and Experience in Implementing ICTs in Universities Roundtable Papers (7), IAU/IAUP/EUA* Skagen Roundtable.

Sorensen, E. K., & Takle, E. S. (2001). Collaborative Knowledge Building in Web-based Learning: Assessing the Quality of Dialogue. *The Proceedings of the 13th World Conference on Educational Multimedia, Hypermedia & Telecommunications (ED-MEDIA 2001)*.

Suthers, D. (2001). Towards a systematic study of representational guidance for collaborative learning discourse. *Journal of Universal Computer Sciences*, *7*, 254–277.

Teasley, S., & Wolinsky, S. (2001). Scientific collaborations at a distance. *Science*, *292*(5525), 2254–2255. doi:10.1126/science.1061619

Thomas, A., Howell, M. C., Patricia, C. K., & Angelo, B. (2001). *Collaborative Learning Techniques*. John Wiley & Sons Inc.

Thornton, P., & Houser, C. (2004, March 23-25). Using Mobile Phones in Education. *In Proceedings of the IEEE International Workshop on Wireless and Mobile Technologies in Education (WMTE)* (p. 3), Taiwan.

Tiffin, J., & Rajasingham, L. (1995). *In Search of the Virtual Class: Education in an Information Society*. New York: Routledge. doi:10.4324/9780203291184

Traunmueller, R., & Wimmer, M. (2003, September 1-5). E-Government at a decisive moment: sketching a roadmap to excellence. *Proceedings of eGov-2003 International Conference: From E-Government to E-Governance, 2739*, Prague.

Tu, C.-H. (2004). *Online Collaborative Learning Communities*. Libraries Unltd Inc.

UNESCO. (2001). *Monitoring Report on Education for All*. UNESCO.

Utsumi, T. (2005a). Global E-Learning for Global Peace with Global University System. In Ruohotie, P. (Ed.), *Communication and Learning in the Multicultural World*. Finland: University of Tampere.

Utsumi, T. (2005b). *Global University System with Globally Collaborative Innovation Network*. Finland: Global University System.

van Dam-Mieras, M. C. E. (Rietje) (2004, May 9-12). Learning In a Global Society. *The Hague Conference on Environment, Security and Sustainable Development*, Peace Palace, The Hague.

van der Meulen, B. (2003, June). Integrating Technological and Societal Aspects of ICT in Foresight Exercises. *Technikfolgenabschätzung*, *2*(12), 66–74.

Venkatesh, M., & Small, R. V. (2003). *Learning-in-Community: Reflections on Practice*. Springer.

Vonderwell, S. (2003). An examination of asynchronous communication experiences and perspectives of students in an online course: a case study. *The Internet and Higher Education*, *6*, 77–90. doi:10.1016/S1096-7516(02)00164-1

Woolgar, S. (Ed.). (2002). *Virtual Society, technology, cyberbole, reality*. Oxford University Press.

WSIS (2005). *Report of the WSIS Education.* Academia and Research Taskforce. Additional ReadingsAlexander, S. (2001). E-learning experiences. Education + Training, 43(4/5), 240-248.

Zurita, G., & Nussbaum, M. (2004). Computer supported collaborative learning using wirelessly interconnected handheld computers. *Computers & Education*, *42*(3), 289–314. doi:10.1016/j.compedu.2003.08.005

Zurita, L., & Bruce, B. C. (2005, June/July). Designing from the users side: reaching over the divide. *Submitted to the Computer Supported Collaborative Learning (CSCL) Conference 2005*, Taipei.

KEY TERMS AND DEFINITIONS

Cooperative Learning: It is an instructional method that allows students to work in small groups within the classroom, often with a division of assignment of specific tasks[24], and it is an instructional strategy in which small, usually heterogeneous groups of students work collaboratively to learn[25]. Cooperative learning was proposed in response to traditional curriculum-driven education[26].

Distance Learning: A form of learning where the instructor and the students are in physically separate locations. The learning process can be either synchronous or asynchronous[21] by which technology is used for mainly continuing education in various ways where the participant does not have to physically be in the place where the educator initiates the learning[22]. It is a type of education where students work on their own at home or at the office and communicate with faculty and other students via e-mail, electronic forums, videoconferencing and other forms of computer-based communication[23].

Distributed Learning: Distributed learning is a term used to describe educational experiences that are distributed across a variety of geographic settings, across time and across various interactive media (Dede, 2004). It is a culture of learning in which everyone is involved in a collective effort of understanding. Distance learning is characterized by four characteristics, such as the diversity of expertise among its members who are valued for their contributions and given support to develop; shared objective of continually advancing the collective knowledge and skills; emphasis on learning how to learn; and incorporate mechanisms for sharing what is learned (Utsumi, 2005a; b).

e-Learning: A form of learning that is enabled by the use of digital tools and content, involving interactivity between the learner and their educator or peers[17] utilizing a network (LAN, WAN or Internet) for delivery, interaction, or facilitation[18]. e-learning can be any technologically mediated learning using computers whether from a distance or in face to face classroom setting (i.e., computer assisted learning)[19]. This form of learning can be used to deliver online courses and/or establish online learning communities; and it supports flexible learning anywhere, anytime for anyone[20].

Experimental Learning: This form of learning is learning by doing[27], or acquired through workplace[28], or based on experience[29]. Experiential Learning is the process of making meaning from direct experience[30]. It is the process of acquiring skills, knowledge and understanding through experience rather than through formal education or training[31]. This process of learning involves the student in his/her learning to a much greater degree than in traditional (pedagogical) learning environments[32]. It addresses the needs and wants of the learner and is seen to be equivalent to personal change and growth[33].

Lifelong Learning: It is the process of acquiring knowledge or skills throughout one's life via education, training, work and general life experiences[34]. Lifelong learning is a form of continuing education, and act as an essential means of accelerating assimilation of new technologies[35]. Usually, these are non-credit instruction of a community service nature other than recreational and leisure time[36]. This form of learning encompasses all learning activity undertaken throughout life, with the aim of improving knowledge, skills and competencies within a personal, civic, social and/or employment-related perspective[37] for professional development with a broad concept where education is flexible, diverse and available at different times and places is pursued throughout life[38].

Online Learning: Online learning is an option for learners who wish to learn in their own environment using technology and/or the Internet[39]. It is a form of learning conducted via a computer network, using the internet and the World Wide Web, a local area network (LAN), or an intranet[40]. Online learning can comprise of any learning experience or environment that relies upon the Internet/WWW as the primary delivery mode of communication and presentation[41].

ENDNOTES

1 en.wikipedia.org/wiki/Learning
2 http://en.wikipedia.org/wiki/Collaborative_learning
3 http://www.geocities.com/Athens/Crete/4060/dictionnaire_etrusque.htm
4 http://www.cpa.ie/povertyinireland/glossary.htm
5 http://www.yhdenvertaisuus.fi/english/what_is_equality/definitions_and_concepts/
6 http://www.novonordisk.com/old/press/socialreports/1998/intro/glossary.html
7 http://www.health.nsw.gov.au/public-health/health-promotion/abouthp/glossary.html
8 http://www.rain.net.au/community_wellbeing/community_wellbeing001.htm
9 Bangladesh Bureau of Statistics data, 2004
10 The Bangladesh Rural Advancement Committee (BRAC) is today one of the largest NGOs working in primary education.
11 Non-Formal Primary Education Programme (NFPE)
12 The Female Secondary School Assistance Project, Female Secondary Education Stipend Project, Higher Secondary Female Stipend Project
13 http://www.internetworldstats.com/asia/bd.htm
14 International Telecommunication Union
15 UNESCO
16 UNDP
17 www.stiltonstudios.net/glossary.htm
18 www.iqat.org/glossary.php
19 www.usd.edu/library/instruction/glossary.shtml
20 www.qmg.com.au/page/glossary
21 www.delmar.edu/distancelearning/student_success/glossary/glossary-d-f.htm
22 http://capso.tamu.edu/glossary.html
23 www.west.asu.edu/achristie/545/webgloss.htm
24 www.nagc.org/index.aspx
25 http://jeffcoweb.jeffco.k12.co.us/isu/gifted/Glossary.doc
26 http://en.wikipedia.org/wiki/Cooperative learning
27 www.unesco.org/education/educprog/lwf/doc/portfolio/definitions.htm
28 www.waveproject.com/glossary/glossary.html/english/e/
29 www.teach-nology.com/glossary/terms/e/
30 http://en.wikipedia.org/wiki/Experiential learning
31 www.warwick.ac.uk/fac/soc/conted/SocratesAPEL/uk/glossuk.htm
32 www.nald.ca/adultlearningcourse/glossary.htm
33 http://hagar.up.ac.za/catts/learner/ameyer/glossaryoflearningtheory.htm
34 www.dest.gov.au/sectors/training_skills/policy_issues_reviews/key_issues/nts/glo/ftol.htm
35 www.et.teiath.gr/tempus/glossary.asp
36 www.pbcc.cc.fl.us/x3922.xml

[37] www.projects.aegee.org/educationunlimited/

[38] www.evaluateit.org/glossary/index.html

[39] www.northislandcollege.ca/students/glossary.htm

[40] www.southbank.edu.au/site/tools/glossary/M-Q.asp

[41] www.usd.edu/library/instruction/glossary.shtml

This work was previously published in Information Communication Technologies and Human Development, Volume 1, Issue 2, edited by Susheel Chhabra and Hakikur Rahman pp. 1-27, copyright 2009 by IGI Publishing (an imprint of IGI Global)

Chapter 7
Enhancing Accessibility to Information Systems by Dynamic User Interfaces

Stefan Richter
Institute for Software Systems in Business, Environment, and Administration, Germany

Norbert Kuhn
Institute for Software Systems in Business, Environment, and Administration, Germany

Stefan Naumann
Institute for Software Systems in Business, Environment, and Administration, Germany

Michael Schmidt
Institute for Software Systems in Business, Environment, and Administration, Germany

ABSTRACT

Many governmental institutions and other organizations have started to provide their customers with access to their documents by electronic means. This alters the way of interaction between authorities and citizens considerably. Hence, it is worthwhile to look at both the chances and the risks that this process of change implies for disabled citizens. Due to different laws or legal directives e.g. governmental authorities have a particular responsibility to consider also the needs of disabled persons. Therefore, they need to apply appropriate techniques for these groups to avoid an "Accessibility Divide". This discussion is built on the observation that governmental and other customer oriented processes are mostly based on the exchange of forms between authorities and citizens. Authors state that such processes can be distinguished into three scenarios, with the use of paper as means of transport on the one end and complete electronic treatment at the other end. For each scenario there exist tools to improve accessibility for people with certain disabilities. These tools include standard technologies like improved Web access by magnifying characters, assistive technologies like document cameras, and more sophisticated approaches like integrated solutions for handling forms and government processes.

DOI: 10.4018/978-1-60960-497-4.ch007

This chapter focuses on approaches that provide access to governmental processes for people with visual impairments, elderly people, illiterates, or immigrants. Additionally, it sees a chance to enable electronic processes in developing countries where the citizens have less experience in handling IT-based processes. The main part of the chapter describes an approach to combine scanned images of paper-based forms containing textual information and text-to-speech synthesis yielding an audio-visual document representation.

It exploits standard document formats based on XML and web service technology to achieve independency from software and hardware platforms. This is also helpful for conventional governmental processes because people within the group of interest stated above often also have problems to access non-digitized information, for instance when they have to read announcements within public administration offices.

INTRODUCTION

In recent years much effort has been spent in Human Computer Interfaces to improve access for disabled persons to computer systems (Muller et al. 1997). To a major extent these activities are enforced by legislative constraints that exist in the US, e.g. the Americans with Disabilities Act (United States of America, 1990) as well as in the European Union (European Commission, 2000), and in its member countries, like in Germany (Bundesrepublik Deutschland, 2006, Bundesrepublik Deutschland, 2002). However, in most countries these efforts have not yet reached their final destination. To a large amount these realizations allow the user only to download particular forms, to print them, and to send it back to the governmental institution after some information has been inserted. While for the web based information systems accessibility aspects are often considered in e-Government platforms, for the procedure of forms filling support for disabled persons is often missing. In many cases it is necessary to process printed documents, yielding a point of media disruption which is difficult to handle for many users with particular impairments.

In this chapter, authors want to evaluate possible scenarios and interim steps while implementing electronic processes in authorities. Thus, more is necessary then supporting electronic forms. They want to take a look at the e-Government sector and the efforts to make it accessible. To clarify the special needs we must have a closer look on the impairments and corresponding assistive tools. Therefore, this chapter discusses in more detail an approach to build interfaces to governmental forms. This exploits different computer science techniques e.g., from the fields of document analysis, language processing, and distributed systems to develop a solution.

Its document representation is based on XML structures and communication is implemented by using web services, which guarantees independence from software and hardware platforms. In most cases when people speak about documents, they have in mind governmental forms, which are used to provide and to maintain information that is necessary to execute governmental processes. This research affirms the stringent necessity of making e-Government processes available for almost all people to lead them to an autonomous and self-determined life.

BACKGROUND

Regarding accessibility to e-Government processes, at first it has to consider the processes that already occur in general. In this aspect, authors explained three main E-Government scenarios in this section showing the diversity of the integration of IT in public authorities. Thereafter, they describe how these scenarios are related to accessibility issues.

E-Government Scenarios

In early days, like many private companies public governments also have started to use IT-systems to accomplish their tasks and substituted paper as means of transport of information by electronic documents. The systems comprise of various software to handle the workload in the offices, ranging from text-processing programs, spreadsheet programs, database systems, up to integrated systems like workflow management and archive systems. Moreover, governments implemented information systems for the contact to their citizens, e.g. geographical information systems, or electronic registers, like the register of residents or the commercial register. For one reason, these efforts are motivated by new public laws, e.g. the Directive of the European Parliament and of the Council on public access to environmental information (Directive 2003/4/EC) which obligates any public authority to inform all citizens about all environmental data they have at hand. This task cannot be achieved economically without the use of IT-systems.

Thus, cost and efficiency considerations are another reason for this process of change. In some areas this development has considerably changed the way of interaction between governmental authorities and citizens. The main focus in this chapter is to evaluate the aspect of accessibility for these newly designed processes.

A first observation is, that the state of progress in the implementation of electronic processes differs from country to country, as well as it appears to be different from city to city within one country. For example, Austria has completely introduced electronic processes. In Germany one may find authorities with a staff of only some decades of persons up to several thousands of members. The size of an authority is a central parameter in cost calculations when asking the question whether an electronic information system should be introduced or not. Thus, the decision to do so is quite different, leading to a rather inhomogeneous situation. Many authorities still have paper-based processes and do not plan to substitute them. Other ones have at hand an electronic version for almost any of their processes. Extracting the core points of these different approaches three major scenarios can be distinguished:

1. An authority relies only on paper-based processes and wants to keep these processes. In this case, for a government-to-citizen process the citizen will receive the forms by mail which he has to complete, print and to return afterwards. This poses several problems for the individuals from the targeted user groups.
2. An authority provides access to digital forms for selected processes. In most cases, these may be accessed via the Internet, where citizens can download these forms, complete them, print them, and return them. A common format for this purpose will be PDF-documents with field information.
3. An authority has converted all its processes to an electronic representation. Here, the citizen has electronic access to these forms. He is e.g. able to complete and to return forms online.

In all scenarios it may be necessary to supply certain help to the citizen to let them know which kind of information is expected while they are completing a particular field within a form. For printed forms there is usually a corresponding guideline document which is sent together with the form.

The subject of integrated help functions must also be considered when moving to an electronic version. All three scenarios can be found in different governmental institutions. Hence, an approach to support users in executing e-Government processes should be able to cope with all these situations.

E-Government and Accessibility

This section elaborates some core points regarding e-Government and accessibility and it discusses some steps to improve accessibility. But, at first, this research has defined the targeted user groups.

User Groups

There exists a wide range of impairments which may prevent people to interact with IT-systems. These impairments may concern either the perception of the systems' output or the input to control the behavior of the system. Also, there exist already many techniques or tools to (partially) overcome these obstacles for general IT-systems which are also applicable to e-Government systems. Some of them will be presented here. This discussion concentrates on users – either citizens or employees within administrations – which have problems to perceive information in its conventional document representation. This group comprises blind people, people with low vision, defective vision, color deficiency or daltonism, as well as illiterates and immigrants. This enumeration involves many different reasons imposing problems for perceiving information. Consequently, a tool to compensate a certain disability has to yield the information in an alternative representation.

For example, illiterates, a lot of the immigrants, or blind people are not able to read text. There are other groups that are different from the blind people but unable for making correct spelling. People with low vision often see information blurred or have problems with the (small) size of the characters. Daltonism allows people to differentiate only between grayscale whilst people with a special color disability have problems to separate some colors and need special color combinations to have the best contrast. In order to provide all of them with access to information one must consider a wide range of possible reasons for their problems.

Regarding e-Government solutions we subsume all these people under the term "users", not distinguishing between disabled citizens and disabled employees.

Core Points of E-Government and Accessibility

When traditional governmental procedures become substituted by digital processes one may encounter in principle the same problems for disabled humans as they exist for the general situation of accessing a computer system. However, as regards accessibility the following main points distinguish e-Government processes from other IT-based private or business processes:

1. Governments are obliged to consider all citizens as clients. Consequence: Also citizens with very special disabilities have to be supported.
2. Employees with disabilities need special support and setups for their working environment. Especially governments are in charge to supply such working environments. Consequence: Appropriate assistive technologies have to be installed.
3. Forms play a vital role in government processes, and therefore, also within e-Government. Often these forms and the related processes have a mandatory character for citizens and the government. Consequently, the processes, the document exchange, and the end devices must ensure high standards regarding security and availability. Consequence: A special assistance for handling forms is necessary.
4. New laws like the German "Behindertengleichstellungsgesetz" (Bundesregierung, 2002a), which is also motivated by the Employment Equality Directive of the European Council (Directive

No. 2000/78/EC), often have at first a special focus on public administrations (Karl 2007). Consequence: e-Government applications have a special responsibility to supply accessible solutions.
5. The usage of devices other than Desktop PCs is increasing; especially mobile clients play an ascending role. Consequence: Also mobile e-Government solutions must be taken into account.
6. The last point is related to the structure of public administrations. Typically, there are several administration levels with different responsibilities and possibilities – from the European Commission to municipal administrations. Consequence: It is not sufficient to provide one method of resolution. Rather, transferable and system-independent concepts with best-practice-solutions are helpful.

Considering these points and their consequences regarding accessibility it is worth to spend a closer look to e-Government processes and their accessibility (West 2003, Peter 2006, Karl 2007).

Steps for Making E-Government Solutions Accessible

From a general point of view one may distinguish three main areas to enable and to improve accessibility to e-Government solutions and processes. They base on the observation, that most e-Government applications are Client/Server applications which use the Internet or an Intranet as means of communication between client and server. For reasons of platform independence and in order to provide the user a well-known interface most of the client applications allow any standard web-browser as user interface.

Thus, an e-Government application appears for the user like an arbitrary web site for which at first accessibility can be improved by exploiting standardized means. In a second step the access can be improved by integrating assistive technologies. A third step of improving accessibility takes into account the important role of forms and processes with a mandatory character and covers all the scenarios of e-Government solutions described above.

In the following, authors describe techniques for these three steps in more detail.

At first, accessibility can be improved by improving Web sites. When standard web technology is used to implement an e-Government application it has to be in line with the rules of construction for web sites. This comprises use of appropriate font sizes, colors, and other guidelines.

Standards for this purpose are described by the Web Accessibility Initiative (2008) and are available from guidelines or laws, like the EU eAccessibility Communication (European Union 2005) or the German regulation for the accessibility of disabled people to Information Technology (Bundesregierung 2002b).

In detail, the Web Accessibility Initiative developed the following "Web Content Accessibility Guidelines" (http://www.w3.org/TR/WCAG10/#Guidelines, last visited 2010-09-24):

1. *Provide equivalent alternatives to auditory and visual content.*
   Provide content that, when presented to the user, conveys essentially the same function or purpose as auditory or visual content.
2. *Don't rely on color alone.*
   Ensure that text and graphics are understandable when viewed without color.
3. *Use markup and style sheets and do so properly.*
   Mark up documents with the proper structural elements. Control presentation with style sheets rather than with presentation elements and attributes.
4. *Clarify natural language usage.*

Use markup that facilitates pronunciation or interpretation of abbreviated or foreign text.

5. *Create tables that transform gracefully.*
   Ensure that tables have necessary markup to be transformed by accessible browsers and other user agents.
6. *Ensure that pages featuring new technologies transform gracefully.*
   Ensure that pages are accessible even when newer technologies are not supported or are turned off.
7. *Ensure user control of time-sensitive content changes.*
   Ensure that moving, blinking, scrolling, or auto-updating objects or pages may be paused or stopped.
8. *Ensure direct accessibility of embedded user interfaces.*
   Ensure that the user interface follows principles of accessible design: device-independent access to functionality, keyboard operability, self-voicing, etc.
9. *Design for device-independence.*
   Use features that enable activation of page elements via a variety of input devices.
10. *Use interim solutions.*
    Use interim accessibility solutions so that assistive technologies and older browsers will operate correctly.
11. *Use W3C technologies and guidelines.*
    Use W3C technologies (according to specification) and follow accessibility guidelines. Where it is not possible to use a W3C technology, or doing so results in material that does not transform gracefully, provide an alternative version of the content that is accessible.
12. *Provide context and orientation information.*
    Provide context and orientation information to help users understand complex pages or elements.
13. *Provide clear navigation mechanisms.*
    Provide clear and consistent navigation mechanisms – orientation information, navigation bars, a site map, etc. – to increase the likelihood that a person will find what they are looking for at a site.
14. *Ensure that documents are clear and simple.*
    Ensure that documents are clear and simple so they may be more easily understood.

Some of these rules concern the syntactical structure of the documents and can be checked automatically. Thus, several tests exist to check if a Website is accessible, e.g. the BITV-Test (DIAS GmbH 2010) or Wave (Temple University Institute on Disabilities 2010). Other criteria of these guidelines have to be checked by human experts.

Table 1. Some selected Web sites of German administrations

URL[1]	Certification Centre	Test Result
http://www.landtag.nrw.de	BIENE Award	"Golden Bee" 2005
http://www.baden-wuerttemberg.de	BIENE Award	"Golden Bee" 2005
http://www.deutschebundesbank.de	BIENE Award	"Silver Bee" 2005
http://www.existenzgruender.de	BIENE Award	"Silver Bee" 2004
http://www.kerken.de	BITV	100 / 100 points (2006)
http://www.arge-kreis-ploen.de	BITV	99 / 100 points (2007)
http://www.bmu.de	BITV	93.25 / 100 points (2006)

In Germany, an award is given for Websites with good accessibility practices, namely the BIENE Award (http://www.biene-award.de, last visited 2010-09-24). It judges Web sites from different fields (including e-Government) and comprises expert interviews and practice tests (Peter 2006).

The following table lists the web sites of some German administrations which have achieved outstanding evaluations as regards the aspect of accessibility.

Other examples for certification centers are the "Accessible Site Certification" (WebAIM 2010) or the German DIN CERTCO "Geprüft barrierefreie Website" (DIN CERTCO 2010).

Beyond standardized software solutions based on Web technologies described in the last section also special assistive technologies are available. In the following tools are being mentioned which are particularly helpful for visually impaired humans. These tools yield better access to different document types (printed or digitalized) and are therefore also suitable to enhance accessibility to e-Government processes.

There is a wide variety of audiovisual aids to provide access for blind or visually impaired people to printed information. This may either be stationary or mobile aids. A classical assistive product is a Closed Circuit Television (CCTV), which is a reading device with an optional XY-sliding-table for easy movement of the reading matter.

Figure 1 shows an example of such a device. It is a combination of an analogue camera and a television monitor which displays the image live stream from the camera. The user has some control elements, e.g. he can adjust the lens for zooming operations or he may choose among different color filters to achieve a suitable color combination for text and background. Often, it has one or two "blinds" which serve as a reading guide pointing out the reading line. A simpler alternative to CCTVs are reading magnifiers. They exist as stationary and mobile variants. They allow magnifying printed information such as newspapers, books or bus schedules.

While a CCTV is helpful for people having poor or defective sight a reading machine is in-

Figure 1. Clear view CCTV from transdanubia

tended for blind people. Often they combine a scanner to capture printed documents and a CPU of a usual computer to process the document image.

The reading machine therefore uses optical character recognition software (OCR) to extract all textual information from an image. Then text-to-speech (TTS) software can convert this plain text into eloquent synthetic speech.

Figure 2 shows an example of such a reading machine, namely the Audiocharta Compact system from SilverCreations. The minimal control elements of such a machine ensure that it is very easy to use. In an ideal case it only needs one button to start the procedure after the user has put the document onto the scanner. A second button allows volume control.

It looms that these tools have three main features. They can magnify printed information, or present the printed information in special binary pseudo colors and they can convert the textual information on printed documents into audible information.

When information is not in printed form but already present in digital form it is easier to change its representation so that it fits to the needs of disabled persons. Therefore, a couple of programs for specific disabilities are available. An example for this is a screen magnifier which adapts the behavior of a traditional magnifier by magnifying the screen content. It can be customized that it uses only a part of the screen and to show a magnified image only of the content which is located near the position of the mouse pointer, but it can also be parameterized to use the whole screen to display a magnified view of the desktop. Often a screen magnifier has a screen reader built-in. A screen-reader uses special tags in the graphical controls to make them audible via synthetic speech. A very familiar and popular screen magnifying tool is Zoomtext from Ai Squared which is shown in Figure 3. Another familiar magnifier is part of the Windows operation systems.

A very special tool for blind people is a Braille display. It enables the user to read the text with his fingertips. The display consists of a couple of 6- or 8-pin Braille segments, each of them being able to display one character in Braille code. The number of segments ranges from very few (4-6) up to 80 segments. Compared to purely electronic devices Braille displays are extremely expensive as they consist of a large number of electro-magnetic switching elements.

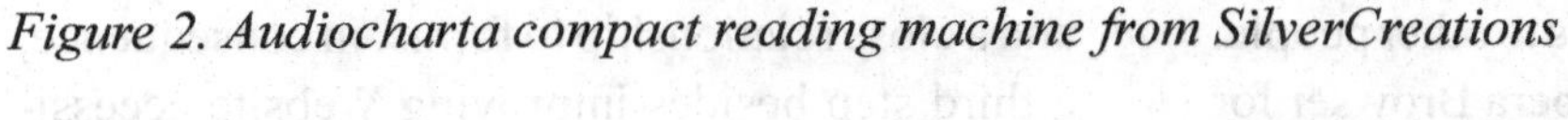

Figure 2. Audiocharta compact reading machine from SilverCreations

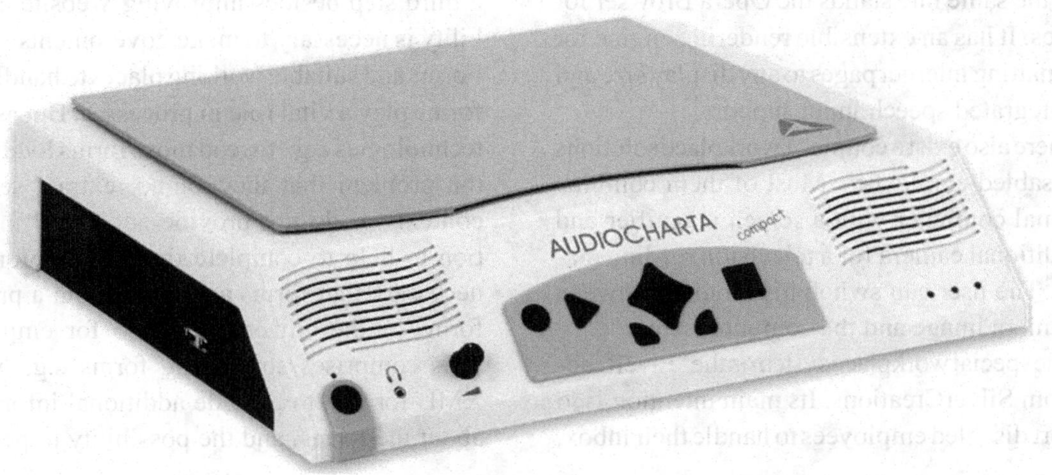

Figure 3. Magnified screen content with Zoomtext

Besides that there exist a lot of small tools that help impaired users in their every-day life. For example the mail-client for blind people (Richter 2003) is a handy tool with a simple user interface and a philosophy to control it with a minimum of keys. This is caused by the projected purpose to enhance an ordinary reading machine with the ability to send scanned documents to other people.

In the same line stands the Opera Browser for devices. It has an extensible rendering engine for reformatting internet pages to any display size and has integrated speech input support.

There also exist a couple of workplace solutions for disabled employees. Most of them combine a normal computer with a screen magnifier and an additional camera for a television reading machine. The user can switch his monitor between the camera image and the computer screen.

One special workplace system is the "LiveReader" from SilverCreations. Its main intention is to support disabled employees to handle their inbox, digitalize it, archive it and send it on demand to another official in charge.

Beyond standard Web technologies and special assistive means the improvement of government processes and their corresponding forms can lead to improved accessibility. As described above e-Government processes have special requirements and a close relationship to forms. In consequence, a third step besides improving Website accessibility is necessary to make governments e-ready. Forms and suitable working places to handle these forms play a vital role in processes. But assistive technologies e.g., to read those forms loudly have the problem that they cannot extract semantic contents or do not provide additional information to help to complete forms. Therefore, it is necessary that forms are provided in a prepared form for the citizens and also for employees. This comprises structuring forms e.g. with an XML-format, to provide additional information about the forms, and the possibility to personalize them. In order to support administrations and citizens two solutions are necessary: one concerns the working environment for employees and the other one concerns access for the citizens. For instance, the GUIDO prototypical system (cf. next section) provides a forms front end as well as a user front end. The former one is installed within a public administration; the latter one can be made available for end users.

Adaptable Interfaces for Governmental Forms

As stated in the introduction this chapter would like to discuss in detail an approach to build interfaces to make governmental forms accessible. So this research will illustrate a possible XML structure for these forms to overcome the limitations of the existing approaches in E-Government. Actual approaches of electronic forms are limited to almost accessible PDF documents or special HTML versions of these documents for online completion (Federal Ministry of Finance Austria, 2008).

As mentioned before in the third step, handling forms play a vital role to make government's e-ready. Therefore, a main target is to develop interfaces which are able to handle governmental forms which may either be available in printed form or electronically. Considering users suffering from reading disabilities documents should adapt their appearance to the needs of an actual user. All these users imply quite different requirements to the user interface design and the interaction with an information system.

For this purpose the GUIDO system (Generating User-specific Interactive Documents) maintains documents that adapt their appearance to the needs of an actual user.

The GUIDO System

In order to be able to economically develop adequate interfaces the system should react almost automatically to the users needs. Figure 4 gives an overview of the system architecture of the GUIDO system.

Figure 4. System architecture

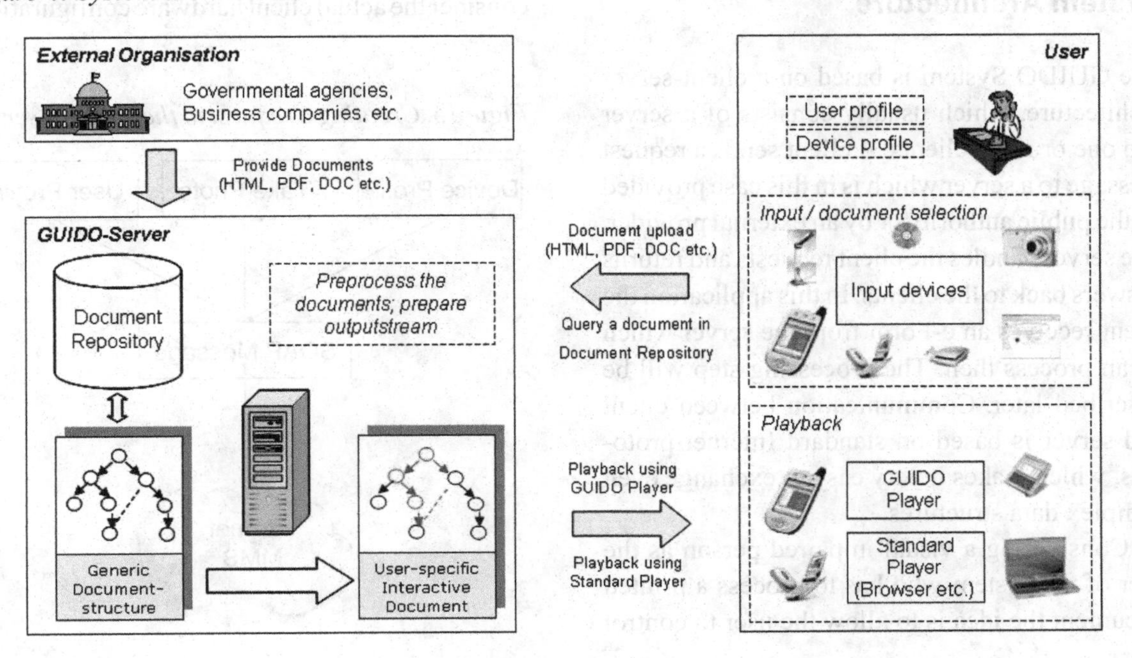

The system consists of different components. The core component is the GUIDO server in the lower left corner of the figure. It is basically a document database where all the documents of a governmental authority may be stored. A database entry of a document consists of different data. This encompasses an image of the document to be able to print or display it on a screen. Furthermore, there should be a semantic description of the document, including information about the fields of a form, their type, and possible relations between them, which ones are already filled with information and which ones have to be completed by the current user.

Moreover, a link to the workflow management in the authority should be present, which could be expressed by a pointer to the description of the workflow where the document is used for. The universe of information associated with a governmental document or form is called the *Generic Document Structure*.

The document repository with all the generic document structures has to be built by the governmental agencies or associated business companies, e.g. the utility company or the water works.

System Architecture

The GUIDO System is based on a client-server architecture, which usually consists of a server and one or more clients. A client sends a request message to a server which is in this case provided by the public authority or by an external provider. The server handles the client requests and returns answers back to the clients. In this application the client receives an e-Form from the server which it can process then. The processing step will be described later. Communication between client and server is based on standard Internet protocols, which makes it very easy to exchange even complex data structures.

Considering a visual impaired person as the user of the system who has to process a printed document the idea is to allow the user to control the client without needing any textual input. Instead, he can use a photo of the document at hand to request a corresponding e-Form. The user can take the photo for example by a scanner, a reading machine, or a digital camera in a desktop environment. In case of a mobile solution this can be done using the built-in camera of a mobile phone. To make the client request more specific in addition to the photo some more information is added to the request, namely the *user profile* and the *device profile*. The tables below show an excerpt from these profiles, while the following figure describes the building process of the client request.

The result obtained from the server should consider the perceptive facilities of the citizen who sent the request. A description of those is given by a user profile contained in the client request. Generally the user profile contains all information that is specific to the user. An example of such an option would be the preferred presentation of the document in contrasting colors. This is to avoid problems with documents that are printed on colored paper which are difficult to read for visually impaired people. Besides of the user specific needs the server should also consider the actual client hardware configuration.

Figure 5. Creating a request to the GUIDO server

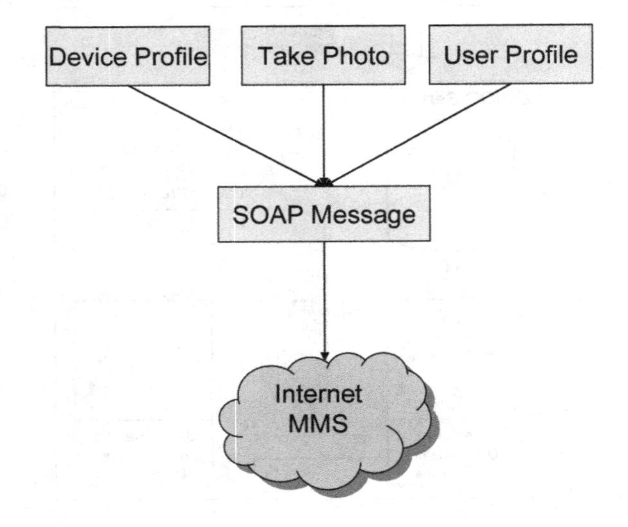

Table 2. Excerpt from profiles and explanation

User profile			Device profile	
Option	Description		Option	Description
Colors	the preferred colors of the user		Display	the display dimensions (for mobile devices)
Audio	user needs audio support		OCR	Can the device run text recognition?
Speed	speed of the audio stream		TTS	Is there a speech synthesis component on the device?

If the citizen uses a cell phone it could be suitable to return only an audio sequence of the document. If the (same) citizen sends his request from his desktop computer the best answer could be a combination of synthesized speech together with a transformed image of the form. Therefore, the device profile lets the server recognize the hardware capabilities of the client and informs it how to transform the document in order to match the needs. The following table mentions some parameters of both profiles.

To retrieve an electronic form of the document from the repository which corresponds to the image contained in the request the server can use both image recognition and text recognition methods. Overall, this leads to a workflow in the GUIDO server which is sketched by the following figure:

The server tries to identify the document description (the template) that fits for the image in the request. If there is no template in the repository the server starts text recognition on the image. If the user profile requests for spoken output the speech synthesis component converts the extracted text into an audio file. After this conversion the server sends the user-specific document to the client.

This document contains the processed image, the optional audio file and additional information to control the further interaction with the user and the client. So every document can be transformed and adapted to the user's needs.

The architecture of the system and the use of a mobile device yield also a mobile document reader. Visually impaired users can get an audible representation of a (traffic) sign, a plate or a bus schedule by just taking a photo and sending it to an appropriate server. Of course, the result is heavily depending on the quality of the photo. If the photo is not focused or if the user has an unsteady hand while taking the photo, the result of the text recognition would be of no use.

To overcome this disadvantage one can try to identify the document using robust image retrieval algorithms which do not need the textual information. With this it is possible to retrieve the document on the photo from the document database. After a document template is found in the database the user specific document will be created using the generic document. The advantage of this approach is that the photo doesn't need to be perfect. Also, for the identification step it could be sufficient to have only a part of the original document. Even if the photo is unfocused the identification process works fine if the document is part of the repository.

Using a client-server solution has some major advantages. A first one is that the service is available for every client device at any time. Furthermore, it results in a platform-independent solution because nearly any device on client side is applicable only if it allows processing the e-Forms. This is because all the necessary work will be done on the server side. With this solution only one server-license for software components is required. If there is an update available then only one update process on server-side is necessary and all clients have access to the improved performance.

Figure 6. Workflow on server side

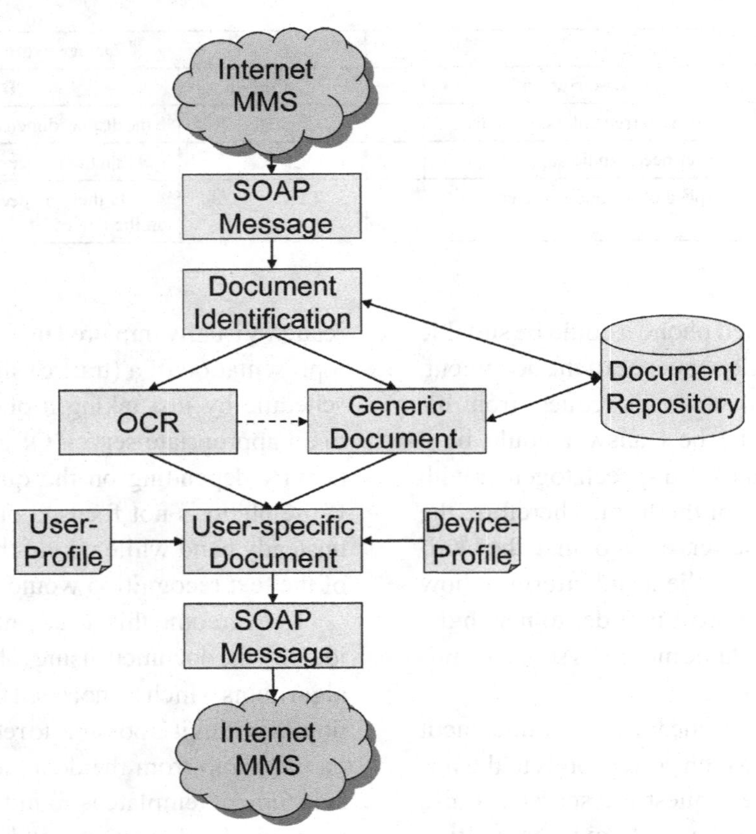

In the next step the idea with these profiles should be extended to a complete model-based interface design for information systems. So the user interface is described by a set of different models, each one covering a different aspect of the design process. Calvary et. al. (Calvary, 2008; Collignon, 2008) described four models, which are arranged in hierarchical order. To consider different input and output devices and varying user needs this approach can be extended with a user-model and a hardware-model which adopts the ideas of the user/hardware profiles discussed before. The user model contains information about the user's physical, perceptual and cognitive abilities, while the hardware model maintains information about the input/output devices. So it will be possible to consider also completely new input devices such as a Remote from Nintendo's Wii (Schlömer, 2008).

Generic Document Structure

The generic document structure is supposed to be the link between paper-based and electronic governmental forms. All information which is needed to process a form is stored in its XML-based structure. Further, this generic document structure contains the combination of the textual and the layout-related information of the original form.

Figure 7 shows on the right-hand side a partially depicted XML-representation of such a governmental form and on the left-hand side the counterparts of some elements in the document.

The input fields have a name, a type, an automatic generated ID, the relative position to an anchor, the value itself, the short help and the guideline. Furthermore it can have a function for semantic checking. The type of the field determines if it is a field of numbers, choice, date, currency, names and so on. Based on this type the User Front End can execute a first syntax check after the user input. It could be proved if a date has the correct format or if a field of numbers contains no letters et cetera.

The field-type could also be used for an input via voice. With special grammars for every type of input field one can achieve satisfying recognition results even for speaker independent speech recognition. The grammars are an instrument to describe the anticipated input and to restrict the possible result from the speech recognition engine. Another advantage achieved by this modular concept is the possibility to reuse these grammar modules for other forms. Because many forms have same standard fields the idea is quite fruitful. The User Front End should use the function which is stored in the <ValidateFunc> tag to make a semantic check of the input.

Figure 7. Original document and (partially) its XML counterpart

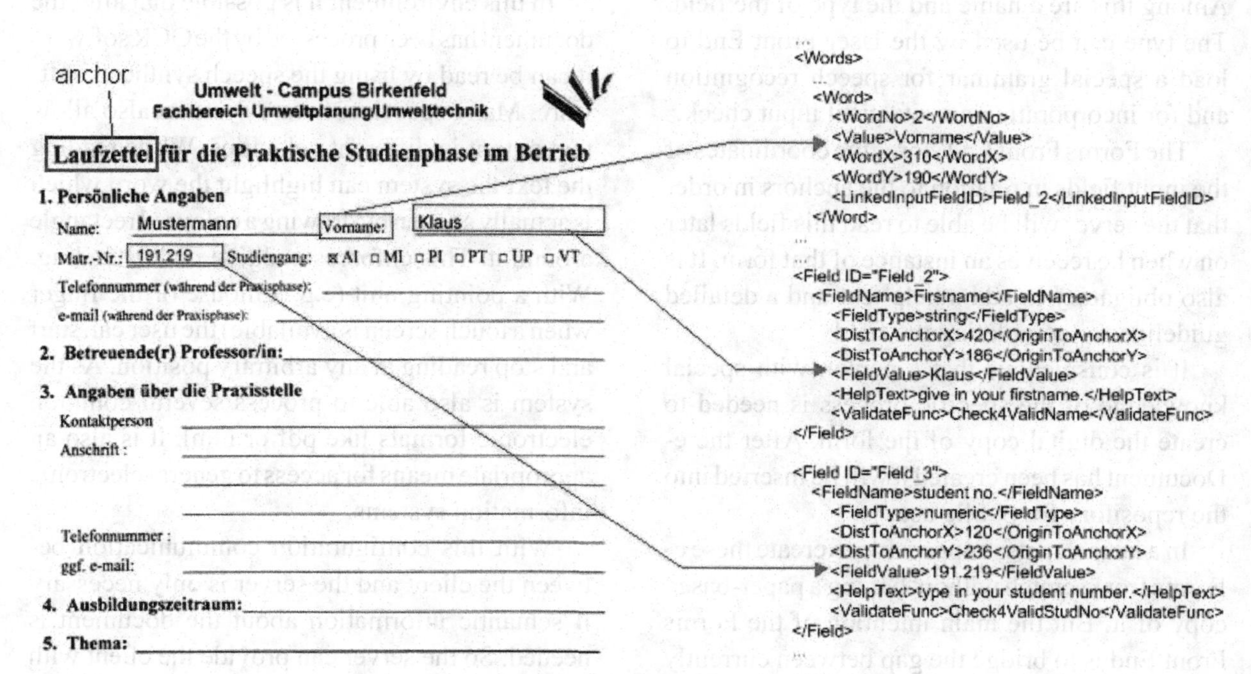

So it is possible to check if a city code matches to the city name or if the name of the person is stored in the database of the authority or if a date is valid, because the date for a building application must be in the future and so on. These functions are known from the use of electronic forms but usually are not available for the handling of printed forms. So this approach is an enhancement which is an improvement not only for disabled persons. The next paragraph shows how to support authorities in creating e-Forms.

Creating E-Forms

The Forms Front End is the component in the GUIDO system which focuses on governmental agencies or external service providers who convert forms on behalf of the authority. It is a tool which assists the agencies creating digital copies of the existing paper-based forms and adds semantic information to them. This unique process of creating the e-Forms is necessary because otherwise the governmental institutions will run into problems when they abruptly reorganize their processes to completely electronic processes. In this essential interim stage the GUIDO system enables the authorities to maintain their current processes in the same manner. This could be desired or necessary due to organizational or financial reasons. The agencies then will be in charge to create additional electronic forms besides the existing paper-based ones and to host these e-Forms in a repository.

For the first step a user of the Forms front end must have available an image and the text of a form. This is either available in the authority or can be created using appropriate tools. For the new document a unique ID will be generated automatically.

Then, the textual information of the form is associated to the layout of the form. In the next step the user has to classify the document. He chooses a name, a document type and an identification number.

Then he defines one or more anchors and denotes the input fields of the form. Therefore, he must border the complete input field on the form. The Forms Front End requests him to input all information which is needed for this field. Among this are a name and the type of the field. The type can be used by the User Front End to load a special grammar for speech recognition and for incorporating a syntactical input check.

The Forms Front End stores the coordinates of the input fields in relation to the anchors in order that the server will be able to read this fields later on when he receives an instance of that form. It is also obligatory to add a help text and a detailed guideline for completing the field.

It is conspicuous that a person with special knowledge of the specific process is needed to create the digital copy of the form. After the e-Document has been created it will be inserted into the repository for public use.

In a later step it is supposable to create these e-Forms from scratch without having a paper-based copy of it. But the main intention of the Forms Front End is to bridge the gap between currently printed forms and the e-Forms and to maintain the interim stage between current paper-based and future purely electronic processes.

User Client

Figure 8 shows a high level configuration of a user client system. It consists of a camera unit shown on the right-hand side to capture printed documents and on the left-hand side the display with built-in speakers. This camera unit enables easy document handling and fast document capture. Alternatively, this could also be a normal scanner.

Such a user terminal could be placed for example in a major administrative department. In this case the speakers would be headphones due to privacy considerations.

The system can vary further according to the software which is installed. For example, there could be an OCR-software, as well as a speech-synthesis and speech recognition software locally available resulting in a highly autonomous system. This setting widely corresponds to the features which is referred to for example in (Kuhn et al. 2007a) and (Kuhn et al. 2007b).

In this environment it is possible that after the document has been processed by the OCR software it can be read by using the speech synthesis software. Many speech software systems also allow text output in different velocities. While reading the text the system can highlight the word which is actually spoken by drawing a coloured rectangle around it. This enforces auditive understanding. With a pointing unit (e.g. a mouse or the finger when a touch screen is available) the user can start and stop reading at any arbitrary position. As the system is also able to process several common electronic formats like pdf or html, it is also an appropriate means for access to general electronic information systems.

With this configuration communication between the client and the server is only necessary if semantic information about the document is needed. So the server can provide the client with

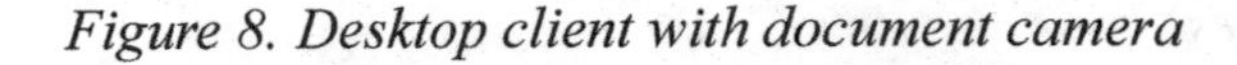

Figure 8. Desktop client with document camera

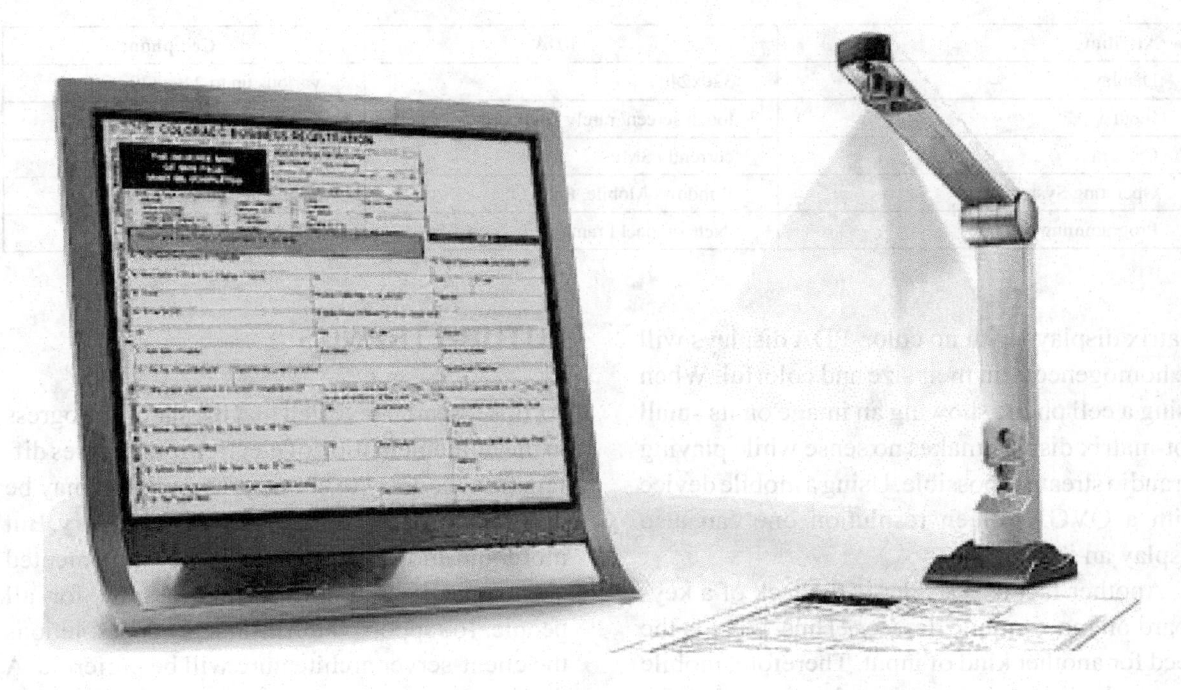

type information about the fields on a form or with information related to a process where the form belongs to. In turn, this information can then be used to verify data from the OCR software which have not yet been recognized with sufficient confidence or to check user input for validity. Another important type of information which the server can deliver are help texts to guide the user through the process of filling the form and explaining to him which information is required in a concrete situation. For the following discussions, the desktop computers are being distinguished from mobile device due to some special features of the latter ones.

In the first projected scenario above the desktop computer is equipped with a scanner. Due to the quality of a scanner or document camera this allows to identify the input fields in the scanned form. This makes sense if the authority sents the citizen a form which has some fields already filled, for instance an identification number for a process, or a name and an address of the citizen which is involved.

By identifying the fields using the relative coordinates in the template the system may recognize the content of these fields. So the user does not need to complete this fields again and if he is connected to the Internet an ID could be used to assign the form directly to the right process in the authority.

A desktop client may vary depending on its configuration. It can be a fat client which is an almost autonomous system for form handling or it can be a thin client. If there is a client where necessary software is missing the server is urged to provide the lacking information. The server can provide this service for every kind of thin clients including mobile devices like PDAs or cell phones. Usually cell phones and PDA's (Personal Digital Assistants) have different configurations. The following list shows a few typical settings for cell phones and PDA's.

The first important difference between cell phones and PDA concerns the display. The size of cell phone displays fairly varies and is often rather small. In many cases they are only dot-

Table 3. Some typical specifications for mobile devices

Attribute	PDA	Cell phone
Display	320x240	various up to 240x320
Input	touch screen/ rarely keyboard	number pad
Camera	currently 3MP	currently 5MP
Operating System	Windows Mobile, Palm OS	Symbian, proprietary
Programming	.NetCompact Framework	often Java

matrix displays with no color. PDA displays will be homogeneous in their size and colorful. When using a cell phone showing an image on its small dot-matrix display makes no sense while playing an audio stream is possible. Using a mobile device with a QVGA screen resolution one can also display an image.

Another fact to consider is the lack of a keyboard on most mobile devices. Thus, there is the need for another kind of input. Therefore, mobile phone devices have a microphone needed for calls. This can be used to record the user's voice and perform speech recognition. The recognition process can be executed on the device itself or on the server that hosts the document repository. However, due to the limitations of actual speech recognition engines the results are often not satisfactory, what makes it tedious to work with it.

Although the main intention is to support governmental processes because of the legal requirements the ideas could be useful for the most processes based on information systems using interfaces to humans. A designer could identify the intent of the user group to specify a user model. This model will reflect the expectations of the users on how a system should react and which result will be expected on a certain input. With such a user model and a corresponding hardware model which models the input/output channels the user interface could be adapted automatically on the client-side, even when the conditions changed on runtime.

FUTURE TRENDS

As this research revealed that the state of progress in the implementation of electronic processes differs from country to country, as well as it may be different from city to city within one country. But more and more processes will be implemented electronically and must be accessible for all people. To support platform-independent solutions the client-server architecture will be preferred. A document transformation server supports the status quo and will be prepared for future progress. Defining user and device profiles to get personalized response from the server just like a browser with cookies is a promising approach. The profiles can grow at the same time as the capabilities of the devices will grow.

As accosted the input via voice is still a topic for further development because of known problems with this technique. Instead, as a workaround it could be reasonable to transfer the spoken input which could not be recognized to the authority. So the official in charge could hear the input as if the citizen would sit in his bureau. The approach to unbuckle the process of document identification and transformation from the client is a little step into the future of computing as stated in (Vanderheiden, 2008). So we can combine both, an individual interface that fits the user's needs and constantly updated software.

The idea of forms which adapt to the needs of a specific user is in general in line with what Vanderheiden calls a pluggable user interface in order to consider ubiquitous accessibility (Vanderheiden, 2008). He says that therefore, "… the ability to "invoke" any assistive technologies or special features that are needed directly from the Net to use on whatever displays are nearby, may be the most effective means of access. …" He also states that combining these personal interface modules with ubiquitous, constantly updated software can allow these special interfaces to work with more and newer types of information as it evolves, increasing the lifetime of the interfaces and thus lowering the cost. Therefore, it will be necessary to change thinking when trying to develop human computer interfaces. It should not be the goal to provide access to an interface element like a particular button but rather to provide access to the underlying function. We believe that this gives need for a semantic description of the function combined with that button which should be given in a standardized description which could be based on XML.

Moreover, such a standard could first be developed for some restricted and formalized domain which can be found in e-Government environments.

Vanderheiden also mentions problems that arise for accessibility from upcoming new devices and human computer interfaces like the iPhone from Apple Corporation or the multi touch display from Microsoft. This results in new interactions between user and device especially in the context of ubiquitous computers. Ladner (Ladner 2008) also addresses the aspect of ubiquitous computing for disabled persons and describes devices for disabled persons. They cover a lot of situations where access is needed even in the web. However, he also mentions the necessity that web designers consider standards and develop web pages that are in accordance with relevant guidelines, e.g. those of the WAI. However, together with spreading Web 2.0 applications there appear a multitude of contributors to web content that are not aware of any standards, not to speak of those dealing with accessibility. But as mentioned before, e-Government applications have to be accessible for all users. So standardization efforts and tools to ensure accordance to standards will be in a main focus in the future.

CONCLUSION

The ongoing process of substituting paper based e-Government processes by electronic processes involves both chances and risks for disabled users. On the one hand there exist more and also more powerful assistive technologies to access electronic information. This gives disabled people a chance for better access to governmental information. However, the new design and execution of the processes implies also some fundamental changes for the interaction between authorities and citizens. Hence, it is necessary to regard how this fits in line with the obligation to give all people access to the information they need. This includes the examination of working environments for disabled employees within administrations.

In the current situation, one can identify different scenarios. In between the range from paper based processes to completely electronic processes any kind of mixture of these forms can appear. The appearance of media disruption imposes problems for all actors involved in a process, but in particular for those people with problems to perceive printed documents.

For them tools which transparently integrate the different media would be very useful. This contribution identified three approaches to handle these difficulties.

At first, authors suggested to apply standard technologies on the Web server and on the client side. Here different standardized solutions exist to magnify fonts, read texts loudly, etc. There also exist several guidelines like the "Web Content Accessibility Guidelines". A second step to improve accessibility is to integrate special assistive technologies like document cameras.

The third approach is more sophisticated and includes the former technologies. It considers the observation that forms play a vital role in e-Government processes. This approach prototypically implemented in the GUIDO system tries to achieve this goal by using information stored in a user profile and a device profile which dispose the server to adapt a governmental form so that is corresponds to the user's needs. This idea is discussed in this chapter for people with visual impairments. However, it can be easily extended to consider users with different needs.

REFERENCES

W3C (2006). Guidelines and resources from the World Wide Web Consortium (W3C). Retrieved from www.w3.org/WAI/, last visited 2010-09-24

Bundesregierung (2002a*). Gesetz zur Gleichstellung behinderter Menschen (Behindertengleichstellungsgesetz – BGG).*http://www.gesetze-im-internet.de/bundesrecht/bgg/gesamt.pdf, last visited 2010-09-24

Bundesregierung (2002a). Verordnung zur Schaffung barrierefreier Informationstechnik nach dem Behindertengleichstellungsgesetz (Barrierefreie Bundesrepublik Deutschland (2002). *Verordnung zur Schaffung barrierefreier Informationstechnik nach dem Behindertengleichstellungsgesetz (Barrierefreie Informationstechnik-Verordnung - BITV) vom 17.* Juli 2002

Calvary, G.; Coutaz J.; Thevenin D.; Limbourg Q.; Bouillon L.; Vanderdonckt J.(n.d.). A unifying reference framework for multi-target user interfaces. *Interacting with Computers, 15*, pp. 289--308

Collignon, Benoît; Vanderdonckt, Jean; Calvary, Gaëlle(n.d.). Model-Driven Engineering of Multi-Target Plastic User Interfaces *Fourth International Conference on Autonomic and Autonomous Systems*, ICAS, 2008

Deutschland, B. (2006). Allgemeines Gleichbehandlungsgesetz der Bundesrepublik Deutschland vom 29.06.2006. http://gesetze-im-internet.de/agg/.

DIAS GmbH. (2010): Barrierefreie Informationstechnik-Verordnung-Test (BITV-Test. http://www.bitvtest.de, last visited 2010-09-24

DIN CERTCO. (2010): Barrierefreie Website. www.dincertco.de/web/media_get.php?mediaid=9080&fileid=13930, last visited 2010-09-24

Edwards, A. D. (1988). The design of auditory interfaces for visually disabled users. In *Proceedings of the SIGCHI Conference on Human Factors in Computing Systems (Washington, D.C., United States, May 15 - 19, 1988)*. (pp.83-88). J. J. O'Hare, Ed. CHI '88. New York: ACM Press.

European Commission (2000). Gleichbehandlung Behinderter in Beruf und Bildung: *Richtlinie des Rates 2000/78/EG vom 27. November 2000*, ABl. L 303 vom 2. Dezember 2000.

European Union. (2005). *eAccessibility. COM(2005)425*. http://eur-lex.europa.eu/LexUriServ/LexUriServ.do?uri=COM:2005:0425:FIN:EN:PDF, last visited 2010-09-24

Federal Ministry of Finance Austria. (2010). *Accessible Documents*. http://formulare.bmf.gv.at/service/formulare/wai_formulare/_start.htm, last visited 2010-09-24

Informationstechnik-Verordnung – BITV. (2010). Retrieved from http://www.gesetze-im-internet.de/bundesrecht/bitv/gesamt.pdf, last visited 2010-09-24

International Adult Literacy Service. (1998). Report for the Organisation for Economic Co-operation and Development (OECD). Retrieved from http://www.statcan.gc.ca/dli-ild/data-donnees/ftp/ials-eiaa-eng.htm, last visited 2010-09-24

Karl, G. (2007). Barrier-free accessibility: Not only talking people can be helped. In Zechner, A. (Ed.), *E-Government Guide Germany. Strategies, solutions, efficiency and impact*. Stuttgart: Fraunhofer IRB Verlag.

Kuhn, N., Richter, S., & Naumann, S. (2007a). Improving Access to EGovernment Processes. In Khosrow-Pour, Mehdi (ed.). *Managing Worldwide Operations and Communications with Information Technology.* Proceedings of the 2007 Information Resources Management Association International Conference (IRMA 2007) Vancouver (British Columbia), Canada. Hershey, PA: IGI Global, pp. 1205-1206

Kuhn, N., Richter, S., & Naumann, S. (2007b). Improving Accessibility to Business Processes for Disabled People by Document Tagging. *Proceedings of the Ninth International Conference on Enterprise Information Systems (ICEIS 2007)*. Funchal (Madeira), Portugal, pp. 286-289

Kurzweil Technologies Inc. (2008). *National Foundation of the Blind: Reader software for a mobile phone.* http://www.knfbreader.com/index.php, last visited 2010-09-24

Ladner, Richard E. (2008). Access and empowerment. *ACM Trans. Access. Comput,. 1 (2),* (October 2008)

Muller, M. J., Wharton, C., McIver, W. J., & Laux, L. (1997). Toward an HCI research and practice agenda based on human needs and social responsibility. *Proceedings of the SIGCHI Conference on Human Factors in Computing Systems (Atlanta, Georgia, United States, March 22 - 27, 1997).* S. Pemberton, Ed. CHI '97, pp. 155-161. New York: ACM Press

Online, S. (2008). *Handy für Blinde liest gedruckte Texte vor*. http://www.spiegel.de/netzwelt/mobil/0,1518,532014,00.html, last visited 2010-09-24

Opera Software ASA. (2004). *New version of Opera Embeds ViaVoice from IBM.* http://www.opera.com/press/releases/2004/03/23/, last visited 2010-09-24

Peter, U. (2006): Accessible E-Government through Universal Design. In Anttiroiko, Ari-Veikko; Mäliä, Matti (2006) *Encyclopedia of Digital Government*, Hershey, PA: Idea Group Inc., pp. 16-19

Richter, S. (2003). Design and Implementation of a communication module for blind and visually impaired humans. *Diploma thesis.* Birkenfeld.

Schlömer, T., Poppinga, B., Henze, N., & Boll, S. Gesture recognition with a Wii controller. Proceedings of the 2nd International Conference on Tangible and Embedded Interaction 2008, Bonn, Germany, February 18-20, 2008. ACM 2008, ISBN 978-1-60558-004-3

Temple University Institute on Disabilities. (2010). *Wave – Web Accessibility Evaluation Tool.* Retrieved from http://wave.webaim.org/, last visited 2010-09-24

United States of America. (1990). Public Law 101-336, 1990. *Text of the Americans with Disabilities Act, Public Law 336 of the 101st Congress,* enacted July 26, 1990.

Vanderheiden, Gregg C. (2008). Ubiquitous accessibility, common technology core, and micro-assistive technology. *ACM Trans. Access. Comput1, (2),* (October 2008)

Web Accessibility Initiative. (2010). *WAI Guidelines and Techniques*. Retrieved from http://www.w3.org/WAI/guid-tech.html, last visited 2010-09-24

WebAIM. (2010). *Accessible Site Certification.* Retrieved from http://webaim.org/services/certification/, last visited 2010-09-24

West, D. M. (2003). Achieving E-Government for All: Highlights from a National Survey. Commissioned by the Benton Foundation and the New York State Forum of the Rockefeller Institute of Government. (2003). Available at http://www.benton.org/archive/publibrary/egov/access2003.html?0#0, last visited 2010-09-24

KEY TERMS AND DEFINITIONS

Accessibility: In this context it means to give all citizens access to governmental information and processes.

Assistive Technology: Standard tools (no matter if hardware or software) offering disabled people the opportunity to compensate some deficiency. In the context of this article: tools to provide access to printed/electronically information.

Visual Impairments: Includes all inherent/not inherent problems preventing people from read and fill out forms.

Standard Protocol: Protocol for general use; usually defined and published by a widely accepted organization. Internet protocols are proposed and defined by the World Wide Web Consortium (W3C).

E-Government Process: Sequence of steps to achieve a certain goal within an authority. The actors involved can either belong to the administrative staff, can be legal units, or citizens. Often, exchange of data is achieved by the exchange of forms.

User Front End: Interface for a citizen or a legal unit to access a service of an authority. This may be a browser plug-in or special software that could run on every capable device unit.

Forms Front End: Interface for the governmental staff for creating the necessary electronically forms and to maintain a document repository.

Document Repository: Storage for the enhanced versions of the governmental forms. Contains for the forms the Generic document structure.

Generic Document Structure: An entry for a form in the document repository. For each form it comprises its digital image, information about the processes where the form belongs to, help texts, links to further information, textual information, and so on.

User Specific Document Structure: Is derived from the Generic document structure. Contains all information a user needs to handle a governmental form.

User Profile: Set of parameters specifying the result the user wants to receive from the server. E.g. it can specify color combinations, or the sampling rate of the audio file.

Device Profile: Set of parameters describing the hard- and software configuration of the device. E.g. informs the server about the size of the display, or whether a speech synthesis program is available.

ENDNOTE

1 All URL last visited 2010-09-24

Chapter 8
Local E-Government Management:
A Wider Window of E-Governance

Hakikur Rahman
ICMS, Bangladesh

ABSTRACT

Despite the immense popularity and potency of electronic government, it is remain uncharted in many countries regarding proper implementation at the local government level. However, technology possess the prospect of improvement in the way government works, and make better interactions with their citizens. National governments are trying to realize this potential by finding ways to implement novel technology in spearheading its utilization to achieve the best services for their citizens. They ranges from awareness raising campaign, knowledge acquisition, social networking to strategic planning, development, and implementation. This article has tried to draw a line of reference by put forwarding the importance of local e-government organizational structure, and their supremacies in terms of utilization of ICT. Along this context, the article has attempted to synthesize a few prospective local e-government scenarios, focus on their adaptation of ICT, and puts forward recommendations to improve local e-government for offering better information services.

INTRODUCTION

Governments throughout the world are in quest of finding ways to deliver public services more efficiently and effectively. Incorporation of electronic governance (e-governance) in the local governments tier is an option often discussed, although the expectations often differ. For example, some expect service delivery costs to be reduced, many hopes for equitable provision of

DOI: 10.4018/978-1-60960-497-4.ch008

public services and others anticipate better planning across a metropolitan area. Various social and political motivations may also be reasons for the change as well. This article is intended to look into various issues of e-government at the local government levels, study the parameters for promotion of e-governance at the grass roots to yield positive economic benefits. In this aspect, there are no straight forward way of improving the e-government system at the local government level, rather many factors control the system, including the structure of the government (tiers of the government system), local demographics (population, size, density), set of responsibilities authorized to local governments and the homogeneity of preferences within the area (Commonwealth, 2004; Fox & Gurley, 2006).

Nowadays, local e-government management includes the extended use of information and communication technologies (ICTs) within government for purposes of improving service delivery to citizens or to enhance back-office operations. The implementation of ICT for overall development and advancement of e-government strategies are likely to have a strong bias towards cities and local towns where most of the citizens reside.

However, it has been observed that, at the national level and in the advantaged localities (central cities, capital cities, and urban areas) ICT's are extensively used to address only key business processes. The national e-government policy does not always apply in devotion to the local government level. Even they are being applied; the policy can not avoid duplication of efforts, problems of interoperability, and inability to leverage economies of scale and security. The key components that drive the local governance and ICT's remain access, content, citizen service, and economic and social development, and for proper implementation of the ICT strategies, the need for these initiatives targeting marginalized areas has also remained not properly identified. Furthermore, in spite of the local governments differ considerably in terms of capacity, content, service delivery, and effectiveness; they have to be dynamic and developmental due to their involvement in local economic development. Local governments need to take the role of the key player in developing integrated rural-based, citizen-centric, information-driven, user-friendly, easily-accessible, and dynamic e-governance system (CPSI, 2005; Samarajiva & Zainudeen, 2008).

Typical services at the local government level incorporates:

- Adults'/children's social care (basic education, pension scheme, retirement plan, primary health, child mortality)
- Economic development (small and medium enterprises, growth centers, consumer commodity, VoIP, call centers, telecenters, multi-purpose information centers)
- Health and Education (nutrition, medicare, continuing education)
- Highways (toll centers, growth centers, village markets)
- Housing and Building control (land use planning, rural and sub-urban planning, zone plans, construction, maintenance)
- Roads and footpaths (mapping, planning, zone-plan, implementation, maintenance)
- Architecture, building control and design (sustainable operational policies and planning, standardization of policies and rules)
- Traffic and transportation (operation and management)
- Art galleries, Leisure, recreation and museums (infotainment, culture)
- Car parking (design, planning, implementation)
- Cemeteries
- Environmental health (awareness campaign and promotional activities on eco-system management)
- Fire service (safety and security)
- Libraries (knowledge building, knowledge promotion, social networking)
- Parks and open spaces

- Police service (law and order)
- Tourism (promotion of local heritage, sustainable eco-tourism)
- Trading standards (law, policy, consumer association)
- Waste collection and disposal (solid waste management), etc. (LGAR, 2006; Wasukira & Naigambi, 2002)

In an underdeveloped country scenario, health and education make up about 75% of all services provided by local governments. Other areas include fiscal decentralization, financial management and accountability, good governance and civic education, infrastructure development, communication and information system, capacity building of local governments, partnership building, institutional strengthening, coordination and integration (Fukao, 1995; Wasukira & Naigambi, 2002).

Governance at the local level matters, and e-governance is a better way of providing government services to the common citizens. However, the method by which governments govern their communities, nationally, regionally and locally, forms an essential element in determining the outcomes which contribute to the quality of life of those communities. In this aspect, good governance, which is governance that allows the collective aspirations of citizens to be fulfilled effectively and efficiently, depends on the way in which public institutions are designed and operate. This includes institutions which balance the ability for citizens to exercise influence with the capacity to allow elected representatives to exercise leadership. The approach in which local government institutions develop and deliver services, and the structure of the local government sector, are matters of concern to all who take an interest in how public sector organizations can achieve effective community outcomes (MDL, 2006; McNabb, 2006).

This article provides a brief overview on local government, discusses about structures of local government in a few countries, tries to synthesize common issues on local government management, looks into parameters related to e-government at the local government level and discusses on issues related to promotion of e-governance services through ICTs at the grass roots, including various aspects of e-government readiness. Furthermore, the article has put forward a few recommendations to improve the local e-government system. The article ends with a conclusion after giving some future research hints in the aspect of strengthening e-government system at the local government level.

BACKGROUND

Local government can be defined as a city, county, parish, township, municipality, borough, ward, board, district, sub-district, or other general purpose political subdivision of a state or a country[1]. In other words, it is a county, municipality, city, town, township, local public authority, school district, special district, intrastate district, council of governments, regional or interstate government entity, or agency or instrumentality of a local government; a tribe or authorized tribal organization, or native village or organization; and a rural community, unincorporated town or village, or other public entity, for which an application for assistance is made by a state or political subdivision of a state[2].

In general term, local government encompasses counties, cities, towns, municipal corporations, and other bodies that govern territorial areas smaller than the state or province or division or other defined geographical boundaries. The authority of these governing bodies is limited to their territorial boundaries. Also, local government authority is limited to subjects of local concern, such as zoning, land management, housing and building codes, and, sometimes, animal control[3].

It is the lowest level of formal state institutions, such as district-level officials or local, publicly accountable decision-making and service-delivery organizations constituted in accordance with national laws (such as in local elections). Local government structures take different forms in different countries and vary in their levels of accountability to local people or to immediate upper-tier of governments[4].

Local governments in different countries are primarily composed of provinces, districts, subdistricts, municipalities, villages and other forms of localities varying by geography, norms, culture, laws, jurisdiction, national integrity, political will and many other parameters. Fox & Gurley (2006:2) supported this by stating, 'Local government size varies dramatically around the world'. The number and size of municipalities differ widely across countries and the differences could have important implications for whether consolidation would be desirable and beneficial. In Sudan, Côte d'Ivoire and the United Kingdom municipalities average more than 125,000 people. Those in many European countries have less than 10,000 people. Countries in South Asia differ considerably from this picture comprising millions of people in a city corporations or urban townships.

As the focus of the article is on local e-government management and local e-governance, efforts have been given here to broadly define these terms in terms of their implications with ICTs.

On one hand, e-government refers to the use of new ICTs by governments as applied to the full range of government functions. Particularly, the networking potential offered by the Internet and related technologies (such as Local Area Networks, Wide Area Networks, Metropolitan Area Networks, Wi-Fi Meshes, and, especially mobile computing) has the potential to transform the structures and operation of the government[5]. It is the process of transforming the governance system, so that the use of the Internet and electronic processes are central to the way that government operates[6], and treated as the umbrella term that refers to the conduct of public sector processes, outputs and services through computer-mediated networks[7]. On the other hand, e-governance can be seen as a network of interconnected organizations (physically or virtually) to include government, business entrepreneurs, nonprofits, and private-sector entities. In e-governance there are no distinct boundaries[8] among these interconnections or integrations.

As an essential building block for e-government, inter-organizational information integration is vital for one-stop shopping for citizens. Simultaneously, integration of local government systems is needed for effective human services, healthcare, public safety, economic development, and homeland security at the grass roots (CTG, 2003). Furthermore, one of the main objectives of establishing local authorities is to provide the public more opportunities to participate in the decision making process regarding the management and development of their respective governing areas. In this aspect, it is assumed that the powers given to the authority of local government in local councils on local authorities have to be exercised in a lawful manner (responsive, transparent, and accountable) in local context. Thus the local commissioner of local government becomes the key official in advising the central authority or immediate higher authority to effect the vested powers of the local government authority (UN-ESCAP, 1999; Socitm, 2003).

Four major themes may act as key to run a local government authority; such as leadership, communication and coordination, immediate risk management, and trust, confidence and transparency (CTG, 2003; Chutimaskul & Chongsuphajaisiddhi, 2004; Anttiroiko, 2004). The local e-government system may comprise of an on-line resource designed to assist electronic access to government delivery intermediaries; provide homogeneous linkage to technology, policy and organizational management; promote inter-organizational integration at the local level to information system development, management and institutional partnership; accommodate subsidies, grants and other facilities to empower local communities with greater autonomies; deliver efficient, citizen-centric and cost-effective contents to accelerate participation and partnership-based e-services; integrate communities, societies, and localities to local, national, regional and global e-government initiatives; produce strategic plan to support efficient delivery of government services; identify level of organizational readiness at the local context to prepare for the effectiveness and efficient service delivery; and lead toward the ultimate goal of transformation to offer better citizen services at the grass roots (CTG, 2002; 2003; Austin City Council, 2008; Hoogwout, 2003; Kolsaker, 2005; Perotti & von Thadden, 2006).

Before going to the next section on study of e-government structures in several countries, if one would like to broaden the mental image on the meaning and application of e-government, it can be found that the meaning has been evolved over time, and application has been shifted over demand. Basically, e-government is about transformation of the government processes through the use of information technology and the Internet, as defined earlier. But, application of e-government is related to transforming the government processes, as how governments work, share information, and delivers services to external and internal clients by harnessing ICT to improve relationships with citizens and business and among various arms of government. It is also related to increasing the operational efficiency and automation. Hence, e-government creates a better opportunity to rethink business processes and reengineer the system to enable to be offered electronically through an on-line resource (UK Government 2002; 2003; Ferguson, 2005; Labelle, 2005; US Government 2006; Austin City Council, 2008; Rahman, 2008).

Countries often consider merging of local governments as a means to lower service delivery costs, improve service quality, enhance accountability, improve equity or expand participation in government (Fox & Gurley, 2006). But, concept of e-government is to make them more independent, provide more autonomy, offer them more power to act within, and formalize their institutional framework by upholding all the benefits of local government through elected representatives. These are supplementary issues that need to clearly justified, attended, and solved by proper authorization, experimentation and validation. Scope of this article will restrain inclusion of similar disputed issues and their justifications. These will demand further extended research on these issues and other concerns that may rise during the investigation.

Furthermore, at the local government level, the system should incorporate more than traditional ICTs, or mere transformation of the government system, but provide insight at the conjuncture of these transformations accompanying diversified nature of localities, cultures, habits, economies, issues, politics, autonomy, trust, accountability, transparency, corruption, and many more. This article has tried to synthesize these facts learning from experiences in contemporary information dynamics across different local government systems and puts forward the realization that e-government is not about straight-forward transformation of the government system; rather it encompasses logical, physical, organizational and managerial aspects of the information dynamics.

MAIN THRUST

From a functional point of view, local e-government may be divided into two main areas; use of ICTs in performing basic administrative, service and democratic tasks, and strategic information system development on policies and related citizen-oriented assessments (Anttiroiko, 2002). In this aspect, ICTs, related telecommunication and other digital networks are considered to be a major driving force of building information societies and economies. These are being increasingly recognized as new factors in improving traditional governance practices. Furthermore, at the local level e-governance appropriate use of ICT can enhance and support economic and social development, particularly in empowering officials and municipal representatives by ensuring linkages and networking, and timely, efficient, transparent and accountable services. Thus, local e-governance means exploiting the power of ICT to assist in transforming the accessibility, quality, closeness, and cost-effectiveness of public services and to facilitate rejuvenating the relationship among clients, citizens, business and the public bodies who work for the benefit of the system (Open Society Institute, 2007).

In addition to being close to citizens and business, local governments constitute the main (or even unique) representation of the government. The relationship of citizens and local authorities tends to be one based on proximity, as the interests of all parties clearly entwined regarding issues related to public services, urban development, school planning, environmental concerns, and even local politics. Moreover, it is at the local level that the impact of ICTs on the relationship between government and citizens can be most effective (Misuraca, 2007). The increased relationship is an important element to enhance the government service delivery mechanisms, and it advances further through appropriate utilization of information technologies. Thus traditional methods are being eventually transformed into electronic means.

While electronic service delivery is the main thrust of e-government policies at all levels, greater community contact is usually seen as more practical and desirable at the local level. In recent times, greater focus has been given to local e-government where a significant amount of 'citizen to government' or vice versa interaction takes place. Therefore, most of the local governments today are under influence to provide efficient and effective e-government information and services as a result of increased accountability and performance management (Shackleton, Fisher & Dawson, 2004). The aim is to intensify consumer's demand and choice, increase local competition, reduce the cost of service delivery at the local level, and better functioning of the government system. However, at the local level, authority, activity and functioning of the local government are imperative.

So far, the thinking of governments on whether to consolidate the local governments differs across countries. In some cases the national government considers whether to mandate consolidation, often across the nation, and in other cases a local decision is made that would only affect governments within a single confined area. The level and authority affect decision-making processes, diverse expected goals and differ political motivations (Fox & Gurley, 2006). Thus division of executive orders, level of the local governments, carrying out of executive powers, and above all role of the executive authorities factors most in the realization, implementation and operation of local e-government systems.

Similarly, interventions taken by various governments to accomplish e-government varies in nature, depending on ground reality, political perseverance, economic strength, cultural acceptability, technological readiness and human adaptability. Henceforth, local e-government relies much on more in-depth intricacy of the governance systems, including basic livelihood infrastructures and government structure at the grass roots.

This article would like to derive recommendations to promote e-governance at the local government level by synthesizing e-government structures and ICT policies of a few countries across the globe. Countries were not selected by any defined criteria, but at random with emphasize on introducing ICT enabled services at the peripheries. At the same time, a few of them are still struggling hard to achieve the ultimate success of e-government. As this is not an analytical study, mathematical modeling has not been adopted; rather empirical study has been conducted. The organizational structure of the government system has been highlighted with their probable activities and at the same time, their preparedness in ICT readiness has also been put forwarded to provide a rational mapping. At the end of this section, a few recommendations were drawn from this study to promote e-government at the grass roots, emphasizing incorporation of information and communication technologies for the government service delivery.

Sri Lanka: The organizational structure of local governance consists of three legal instruments: the Municipal Council Ordinance, the Urban Council Ordinance and the

Pradeshiya Sabhas Act. At present there are 18 Municpal Councils, 37 Urban Councils, and 256 Pradeshiya Sabhas. Local governments used to be divided into wards. Wards have been discontinued under the new proportional system. The composition of a local council is based on the total population of a local authority area and not on a ward basis[9]. *Note: Population is the indicator of local authority area, not geographic boundary.*

The Government of Sri Lanka first recognized the need for the development of ICT through the National Computer Policy of 1983 (COMPOL), and this was the first attempt from the government that was taken by the Natural Resources, Energy and Science Authority of Sri Lanka (NARESA) on the instructions of the then President. A committee appointed by NARESA produced the National Computer Policy Report. The acceptance of the COMPOL report by the Government gave rise to the establishment of CINTEC by Act No. 10 of 1984 as the "Computer and Information Technology Council of Sri Lanka", to function directly under the then President. Later on, the Science and Technology Development Act No. 11 of 1994 changed the name to "Council for Information Technology" but retained the acronym CINTEC. The "e-Sri Lanka", project launched in November 2002 was tasked to develop an ICT Roadmap for Sri Lanka. The e-Sri Lanka roadmap resulted in the implementation of the ICT Act No. 27 of 2003, which resulted in the establishment of the Information and Communication Technology Agency of Sri Lanka, (ICTA), repealing the relevant section of the Science and Technology Act which established CINTEC. The ICTA has been operational since 1st of July, 2003. The mandate of the e-Sri Lanka policy is to build a national information infrastructure, create a framework for the promotion of software and ICT enabled industries, re-engineering the Government and developing ICT-based human resources[10]. *Note: Information infrastructure development, re-engineering of the government, and human resource development are basic ingredients of local e-government development.*

Furthermore, e-Sri Lanka has initiated a multi-prong program targeting to promote peace, equity and growth that has been illustrated in Figure 1.

Republic of Korea: The local government in Korea consists of 248 separate units. The local political system of Korea is broadly distributed into two categories: the general and the special. In Article 117 Paragraph 2 of the Constitution, it is stipulated that the types of local governments in Korea ought to be decided by the law. Based on this provision, the Local Autonomy Act acknowledges the general local governments to comprise of two tiers: the upper-level local governments (i.e., metropolitan cities and provinces) and lower-level local governments (i.e., cities, counties, and districts). There are two levels of local government, Upper-level (provincial) local governments and Lower-level (municipal) local governments[11]. *Note: There are four tiers in the government system.*

The South Korean government adopted the first Master Plan for Informatization Promotion in June 1996, following the enactment of the Framework Act on Informatization Promotion in August 1995. To advance the goals of the first informatization plan, the government also established a national organization for planning and implementation. Later on, in March 1999, the government formulated the second informatization master plan called 'Cyber Korea 21', and in order to overcome the Asian economic crisis and transform the Korean economy into a knowledge-based one

Figure 1. Multi-dimensional approach of the e-Sri Lanka that assisted in growth of local e-government in Sri Lanka (Adopted from Dewapura, 2008)

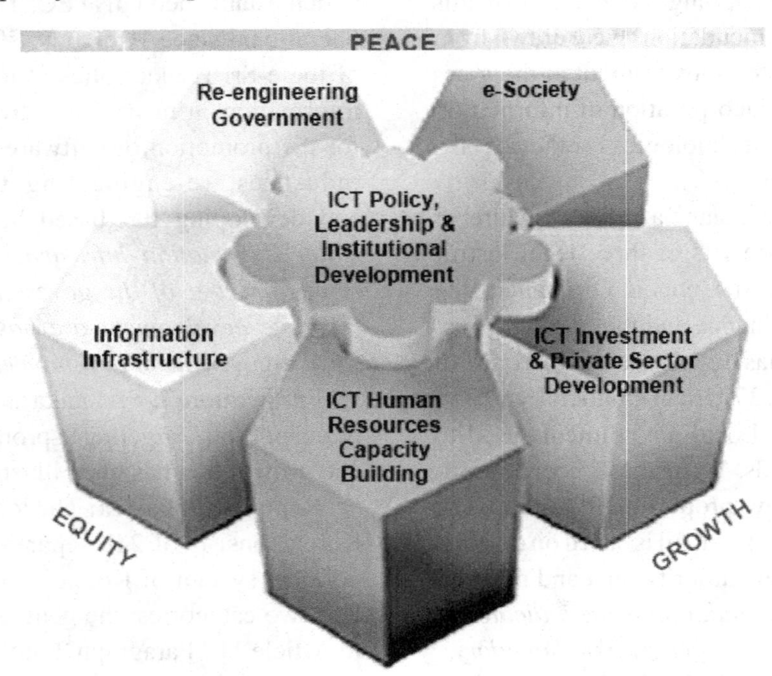

Cyber Korea 21 provides a blueprint for the new information society of the 21st Century. Korea's third informatization master plan, 'e-Korea Vision 2006', embodies the belief that the promotion of informatization in all aspects of the society will lead to an increased effectiveness of all socio-economic activities, higher national performance, and elevated quality of life[12]. Prime objectives of e-Korea Vision 2006 were: to maximize the ability of citizens to actively participate in the information society by utilizing ICT; to strengthen global competitiveness of the economy by promoting informatization in all industries; to realize a smart government structure with high transparency and productivity through informatization efforts; to facilitate continued economic growth by promoting the IT industry and advancing the information infrastructure; and to become a leader in the global information society by playing a major role in international cooperation (Lallana, 2004). *Note: informatization, smart government structure, increased transparency, partnership promotion, and information infrastructure development are ingredients of strengthened local e-governance.*

e-Korea Vision 2006 incorporates various strategic implementation plan to strengthen South Korean e-governance at the grass roots, which is being shown in Figure 2.

Philippines: Local government in the Philippines has its roots in the colonial administration of Spain, which lasted in the Philippines for over 300 years. The establishment of Cebu City in 1565 started the local government system. The legislative initiative promoting local autonomy was strongly supported by academics and public servants, who spearheaded the necessary reforms in changing the government structure and organization of local governments, and included new tasks to enable local governments to address a changing environment. The struggle for decentralization over the past 50 years culminated in the proclamation of the Local Government Code in October

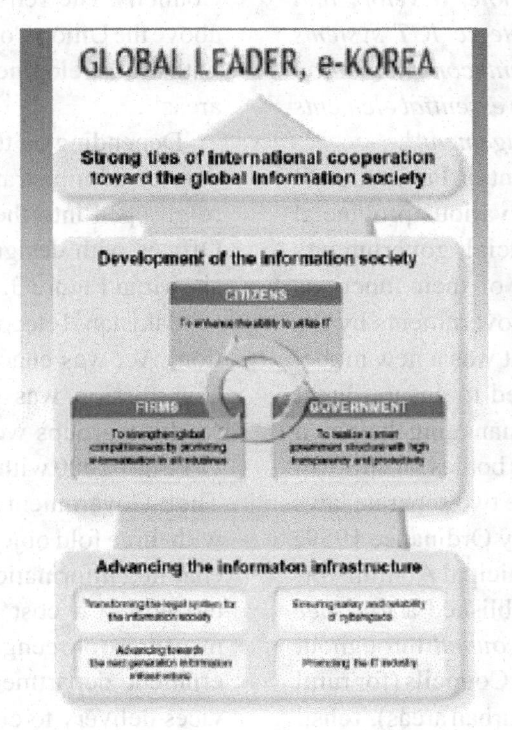

Figure 2. Various steps of e-Korea visioning South Korean local government (Source: http://www.mic.go.kr/eng/res/res_pub_sep_ekv_2002.jsp)

1991. The structures of local governments are both governed by the Local Government Code and by local ordinance passed by the local government concern[13]. *Note: Structure of the local government is essentially needed to be defined for allowing authenticated proclamations to be disseminated at the grass roots.*

The Philippine government launched the National Information Technology Plan for the 21st Century in 1998. In July 2000, a Government Information Systems Plan (GISP) was approved and adopted as a framework and guide for all computerization efforts in the government through the Executive Order #265. The GISP aims to create a system of governance leading to: faster and better delivery of public goods and services; greater transparency in government operations; increased capacities of public sector organizations; and proactive participation of citizens in governance. In 2003, the Information Technology and Electronic Commerce Council (ITECC), the country's ICT policy-making body, envisaged that as an e-enabled society where empowered citizens have access to technologies would provide quality education, efficient government service, greater source of livelihood, and a better way of life. More specifically, the policy adapted to: provide on-line government services to stakeholders; develop an IT-enabled workforce; develop the country as a world-class ICT services provider; create an enabling legal and regulatory environment; and provide affordable Internet access to all segments of the population (Lallana, 2004). In addition to these, the Philippine government created an e-Government fund in 2003. Under the 2003 national budget, the e-Government fund amounted to around 4 billion pesos. Furthermore, the creation of the Commission on Information and Communications Technology (CICT) in 2004 as the government's primary policy, planning, coordinating, implementing, regulating and administrative entity *to promote, develop, and regulate integrated and strategic ICT systems and reliable and cost-efficient communication facilities and services that are essential elements of the local government management*[14].

Pakistan: Local governments in Pakistan exist under the supervision of the various provincial governments, where provincial governments have simply delegated some of their functions and responsibilities to local governments by the promulgation of ordinances. It was a new model of local government pioneered to ensure direct participation of the people in managing their own affairs through representative's bodies set up down to the village level. There were two separate laws for rural (i.e. Basic Democracy Ordinance 1959) and Urban Councils (i.e. Municipal Administration Ordinance 1960). It established a *four tier hierarchical system of local council* throughout the country, namely the Union Councils (for rural areas), Town Committees (for urban areas), Tehsil Councils, and or District Councils and Divisional Councils. The Tehsil Council was the second tier above the Union Council. It was mainly concerned with the development activities in the proclaimed areas[15].

Depending on the specific circumstances, the district administration have been reorganized/re-grouped into the following Offices and Sub-Offices with designated offering of services, as shown in Figure 3.

Pakistan Telecommunications (Re-organization) Act was enacted in 1996, and Pakistan IT Commission was established in 2000. Eleven working groups were created under auspices of IT Policy 2000 with emphasis on e-government[16]. The e-Government Program was launched in 2001 with three fold objectives: to encourage ICT's for enabling information and services delivery to the citizens in a cost effective manner; to initiate measures for reengineering of work flow in government departments enabling electronic services delivery to citizens for bringing efficiency

Figure 3. Services offered by various office and sub-offices under the local government (Source: UN-ESCAP, 2003)

a) **District Co-ordination Office:** Co-ordination, Human Resource Management, and Civil Defense.
b) **Finance and Planning:** Finance & Budget, Planning & Development, and Accounts.
c) **Works and Services:** Housing, Urban & Rural Development, Water Supply and Sanitation, Building & Roads, Energy & Industrial Promotion, and Transport.
d) **Agriculture:** Agriculture (Extension), Livestock, Irrigation & Drainage, Fisheries and Forests.
e) **Health:** Public Health, Environment, Basic and Rural Health Units, Child & Woman Health and Population Welfare. The Medical Superintendent(s) of Hospitals will also function under this Office.
f) **Education:** Boys Schools, Girls Schools, Technical Education, Colleges and Sports.
g) **Literacy:** Literacy Campaigns, Continuing Education and Vocational Education.
h) **Community Development:** Local Government Institutional Development, Community Organization, Labour, Social Welfare & Special Education, Sports & Culture, Registration and Co-operatives.
i) **Information Technology:** Information Technology Development, Information Technology Promotion and Database.
j) **Revenue:** Land Revenue and Estate, Excise and Taxation.
k) **Law:** Litigation, Legal and Legislation.
l) **Magistracy**

in operation; and to bring transparency in government functions and access to information[17]. Promulgation of Electronic Transactions Ordinance in 2002 was another milestone for promoting economic activities at the grass roots strengthening the local government. Furthermore, Telecom Deregulation Policy was approved in 2003.

Nepal: Nepal has a *two-tier system of local governance*; with village development committee (VDC) and Municipality as the lower tier, and district development committee (DDC) as the higher tier. In the case of DDCs and VDCs, there is a provision of classification on the basis of differences in terms of transportation, communication, education and health facilities (including population in the case of VDCs), but such classification has not been completed as yet. Furthermore, the classification does not include the functional bases such as work responsibility, fiscal authority, fiscal attempts and discipline. However, a municipality may be established in any area having access to facilities such as electricity, drinking water supply, roads and transport, education and communication[18]. *Note: Basic human infrastructures are essential elements of the local e-government.*

Nepal's Information Technology Policy was developed in 2000 with the aim to put the country on the global IT map by 2005. Its objectives included: making IT accessible to the general public and increasing employment through this means; creating a knowledge-based society; and establishing knowledge-based industries. Nepal's IT plan also identified 16 activities that constitute its IT policy. These include declaring IT a policy priority; providing Internet access to all of the country's Village Development Committees; computerizing the system in all government offices; developing – with private sector participation – IT parks; promoting e-commerce and e-health; and enacting necessary laws. The action plan for implementing the IT policy includes:

participation of the private sector in infrastructure development; infrastructure development; human resources development; dissemination of IT; promotion of e-commerce, and promotion of e-governance (Lallana, 2003; 2004). *Note: For the promotion of e-governance at the local level, ICT policy should incorporate development activities at the village level.*

China: In 1954, the First National People's Congress made the law The Organizational Rules of the Local People's Congresses and Local Governments. Local people's congresses, local people's administrative committees, local people's courts and local people's procuratorates at various levels were established according to this law. The local people's congress was the local legislative body, exercising legislative power; and the local people's administrative committee was the executive body, exercising administrative power at the grass roots. The people's committee at provincial level consisted of departments such as civil affairs, finance, planning, food, public security, culture, education, health, agriculture and forestry, transport, commerce, industry, supervision, labor, water conservancy, and sports. The people's committee at the autonomous regions and municipalities or cities also had almost the same functional organs similar to those of the provinces. Local governments, according to the Constitution and the related laws, are divided into three main types: local governments at different levels, autonomous governments of nationality regions and governments of special administrative regions. At present, the organizational system of local government is divided into four levels as follows: Provincial level, City level, County level, and Township level[19]. *Note: Local government structures are dynamic and need to re-evaluated and adjusted according to the local demand.*

China's commitment to e-enable the country is *supported at the highest levels and represents a significant national investment in technology*. A clear objective is to make China a major participant in the global economy. China's efforts to connect all major centres with fibre-optic cabling is another clear example of the enthusiasm with which ICTs are being rolled out in that country.

Experimentation in bringing relevant ICT-enabled applications to the people in rural areas is also evidence of the concern that ICTs be relevant to the needs of rural dwellers (Labelle, 2005). China is now experimenting with several models of ICT access at the rural and community levels[20]. In China, access to *information about health* has been demonstrated to be *a priority* in some of the *poorer provinces*.

Bangladesh: The rural/regional local government, as proposed by the latest commission on local government in 1997 has four tiers: Gram (Village) Parishads (being reconstituted in 2003 as Sarkers) (40,392); Union Parishads (4451); Thana/Upazila Parishads (469); and Zila (District) Parishads (64). Urban areas have a separate set of local governments. The Bangladesh Census Commission recognized 522 urban areas in 1991 (with a minimum population of about 5000 or more), but only about 269 of the larger urban areas among these have urban local governments. The six largest cities have a City Corporation status, while the rest are known as Pourashavas or Municipalities, which again are classified according to financial strength[21]. Note: *There are four tiers of local government structure existed in the country.*

In Bangladesh, the focus of capacity-building through ICT is on socio-economic development. Provision of nationwide infrastructure under a national ICT policy, spearheaded by a National ICT task force chaired by the Prime Minister, is designed to facilitate good governance, e-commerce, and as well as e-learning. The human resource development focus of the national ICT policy is to develop ICT professionals and engineers to meet the demand for skilled ICT workers that is growing world-wide, especially for the global software and ICT-enabled services market (Sayo, Chacko & Pradhan, 2004). The National ICT Policy was adopted in October 2002. The Policy aims at building an ICT-driven knowledge based society by the year 2010 (initial target was year 2006). Among fifteen priority areas[22]: agriculture and poverty alleviation, health care, e-government/e-governance, e-commerce, ICT infrastructure, training and human resources development, environment, social welfare, and regional cooperation are found to encourage local government initiatives to be flourished. Furthermore, ICT Act 2006 has been enacted on October 08, 2006; and formulation of rules and regulations are still going on. Very recently, the government has taken an initiative to formulate the national ICT Roadmap for Bangladesh under the Economic Management and Technical Assistance Program (EMTAP), managed by Bangladesh Computer Council with assistance from the World Bank.

In addition to definitive local government structures and enactment of relevant ICT policies to promote grass root governance, national spending on ICT is also an important element for the establishment and growth of strengthened local e-government. Before going to the recommendation section, this study would like to put forward a few insights related to this issue that seems to be pertinent.

ICT Spending

Despite the ICT investment boom in late 1980s and early 1990s, there was a dip period during the new Millennium. However, after quickly recovering, total ICT spending grew in 2003 and continued to grow. Miller (2004) predicted that global economic growth will support continuous ICT spending growth at least through 2007 as shown in Figure 4. He forecasted that from the trough of $US 2.1 trillion U.S. dollars in 2001, total ICT spending will be increasing to over $US 3.2 trillion in 2007.

However, when viewed as a percent of total gross domestic product (GDP), it is clear that the ICT build up during the second half of the 1990s reached its pinnacle in 2000 with total ICT spending accounted for 7.4% of GDP. As ICT spending declined and grew slowly, its share of GDP dropped to 6.8% in 2003. In 2004, ICT's share of GDP fell a little more before increasing to approximately

Figure 4. Global ICT Spending in US$ Billions (Adopted from Miller, 2004)

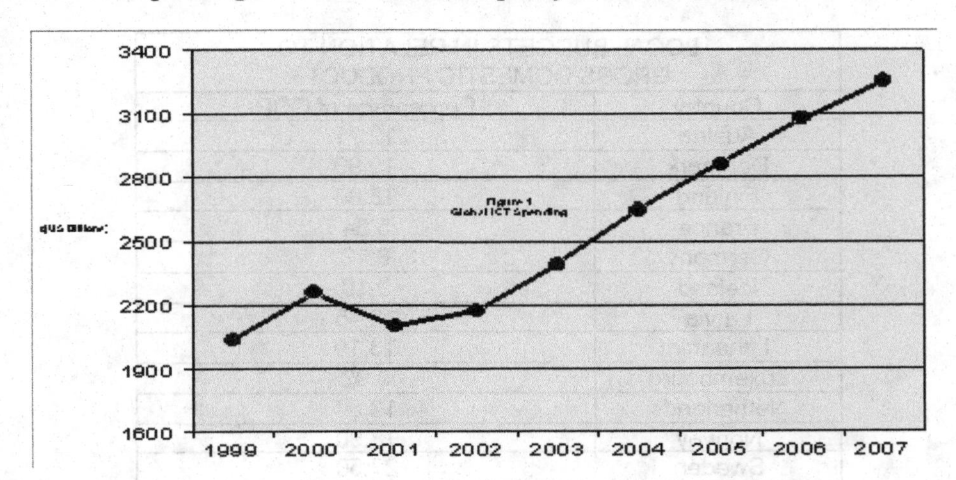

6.9% in the last two years of the forecast interval (Miller, 2004).

According to WITSA's biennial study (Digital Planet: The Global Information Economy, 2004), ICT spending is expected to grow faster than the global economy at approximately 8 percent a year from 2003 through 2007. The global economy at large is projected to grow 7.6 percent annually over that time. The WITSA projection follows a drop-off in global ICT spending from 2000 to 2001, marking a significant rebound in the last two years. However, America is expected to relinquish its lead in per capita spending to second-place after Switzerland in 2007, when the two nations are projected to spend $4,147 and $4,282 per capita, respectively. Asia is projected to be a powerhouse of global growth in ICT spending during that time, growing at a compound annual rate of 9.3 percent

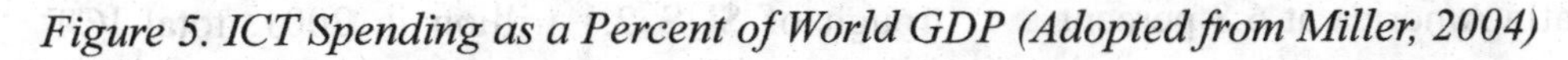

Figure 5. ICT Spending as a Percent of World GDP (Adopted from Miller, 2004)

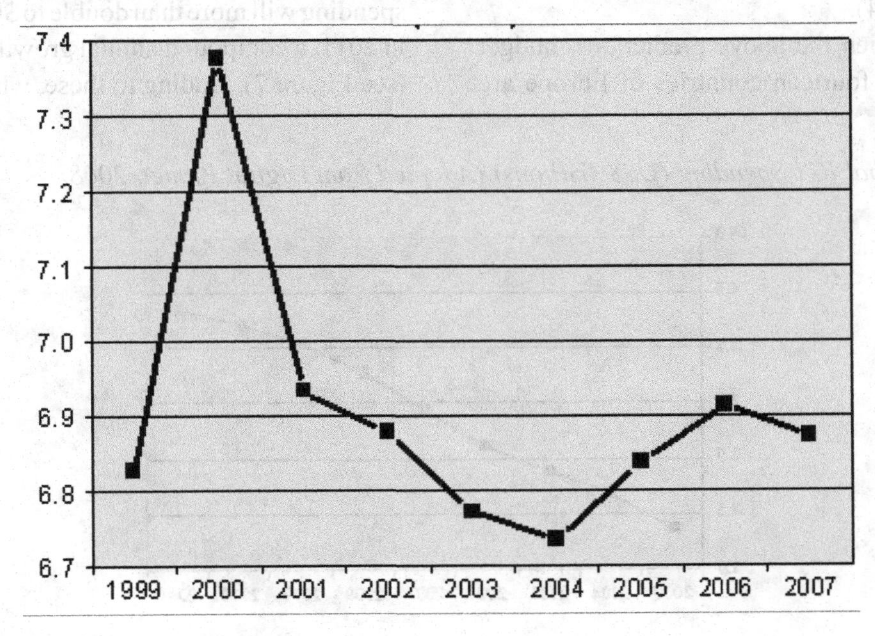

Figure 6. Local budget allocation of European Countries

LOCAL BUDGETS IN RELATION TO GROSS DOMESTIC PRODUCT	
Country	Percentage of GDP
Austria	12.71
Denmark	19.90
Finland	18.00
France	5.54
Germany	8.12
Iceland	9.10
Latvia	12.45
Lithuania	13.10
Luxembourg	9.92
Netherlands	13.30
Norway	18.90
Sweden	27.50
Switzerland	10.80
United Kingdom	11.00

from $568.2 billion in 2003 to $811.1 billion in 2007. Eastern Europe will grow fastest among world regions at a pace of 11.9 percent annually, but will top out at just $68.8 billion in 2007. North America is expected to see the slowest growth rate at 6.7 percent, while Africa ranks third for anticipated pace of growth at 8.8 percent, ahead of Western Europe at 8.7 percent, the Middle East at 8.3 percent and Latin America at 6.8 percent (WITSA, 2004).

To strengthen the above predictions, budget allocations of fourteen countries of Europe are being shown in Figure 6, where it can be seen that most of them are allocating double digit figures for their ICT spending.

Furthermore, Digital Planet (2008) states that moderate global economic growth and increased penetration will support expanding ICT spending growth at least through 2011, albeit it at a more subdued pace than the past few years. From the trough of $US 2.1 trillion in 2001, total ICT spending will more than double to $US 4.4 trillion in 2011, a compound annual growth rate of 7.7% (see Figure 7). Adding to these, when viewed as

Figure 7. Global ICT Spending (US$ Trillions) (Adopted from Digital Planet, 2008)

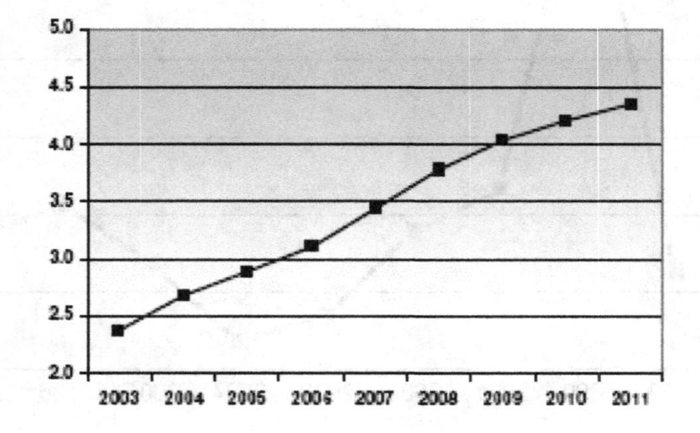

a percent of total GDP, ICT spending remained relatively stable from 2003 to 2006. From 2007 onward, ICT spending as a percent of GDP will trend downwards; reaching 6.3% by 2011 (see Figure 8).

Observations

A three prong approach may simplify delegation of local e-government at the grass roots, such as a merit is made between access policies (aimed at improving access to ICTs for all citizens), content policies (directed to improve the use of ICTs in the city administration and semi-public domains) and infrastructure policies (to improve the provision of broadband infrastructure) (Berg, Meer, Winden & Woets, 2006). In this respect, incorporation of parties actively involved at the grass roots governance processes should be involved. Furthermore, to strengthen local e-government management at the grass roots, three dimensional approach, as suggested by Austin City Council (2008) need to be replaced by a four dimensional approach as shown in Figure 9.

This study observes that a successful e-government system should have the following characteristics:

- Able to meet realistic objectives of the e-government system;
- Able to anticipate emerging technologies and respond accordingly;
- Able to contribute in policy formulation and implementation through grass root stakeholders;
- Able to set up a dependable information system;
- Able to resolve emerging paradox of knowledge superfluity;
- Able to keep updated with most recent information and present them in user-friendly format;
- Able to provide appropriate information at minimum effort and search;
- Able to produce knowledge from the acquired information;
- Able to accommodate open source technology;
- Able to produce confidence among stakeholders enabling atmosphere for attracting investment (local, foreign);
- Able to promote partnership among government, non-government and private entrepreneurs;
- Able to establish stable policies, planning and framework;

Figure 8. Global ICT Spending at a Percent of World GDP (Adopted from Digital Planet, 2008)

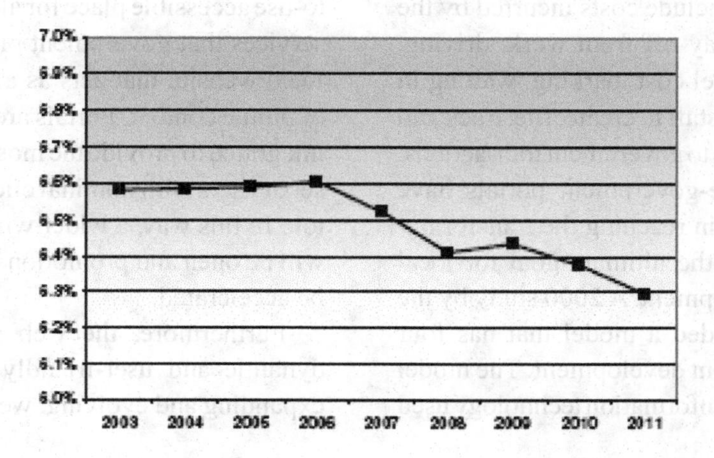

Figure 9. Increased partnership at the local level incorporating all stakeholders (Adopted from CTG, 2002; Author)

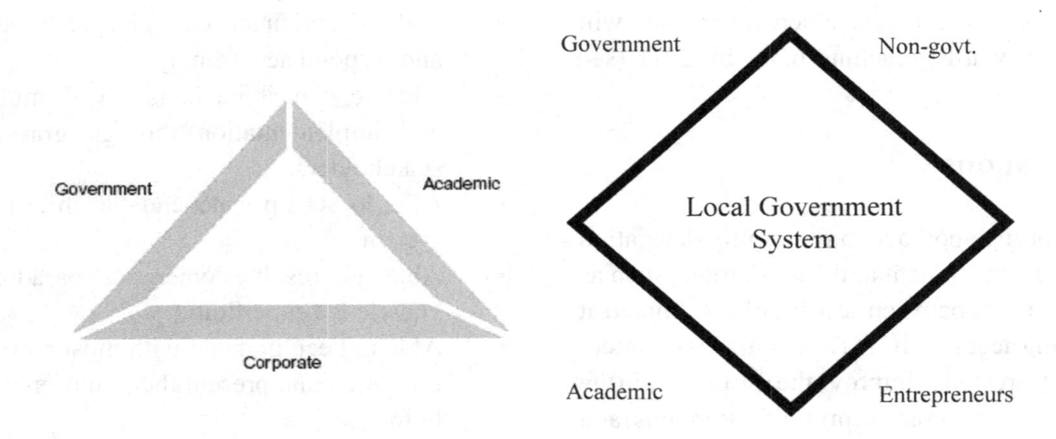

- Able to promote financial transactions leading to e-commerce;
- Able to promote competency among institutions; and
- Able to enhance education system leading to e-learning.

Observations Related to Logical Infrastructure

Recent studies indicate that governments are saving up to 70% on the cost for providing services by moving them online compared to the cost for providing the same services in traditional ways (face to face, telephone, mail, etc.). Moreover, this figure does not include costs incurred by the users for: taking a day off from work, driving, traffic congestion, fuel cost, parking, waiting in queue, hiring special staff to create, file, track and follow-up on business to government transactions.

In this context, e-government portals have evolved to aid cities in reaching the transformation stage, which is the ultimate goal for local e-government development. A 2000 study by the Gartner Group included a model that has four stages of e-government development. The model considers the level of information technology used by the government to relay information online. Moreover, cities or municipalities can progress through the stages of development by expanding upon their service delivery modes (Ridley & Nolting, 2003; Austin City Council, 2008).

An enterprise-wide approach, using standardized policies can aid in implementing cross-organizational solutions. Instead of launching new online services on an organization-by-organization basis, an enterprise-wide approach can be taken at the local level by aggregating services across organizations and accessible through a common web portal. There is now a significant and growing interest in using portal technologies to deliver e-government by bringing together in one easy-to-use accessible place for all the information and services that government provides. A portal is a focal website that acts as a window to an array of online content. Portals are comprehensive and integrated to provide the most usability to a varied set of users with nominal clicks of the mouse button. In this way, a wider window of information will be open and promotion of e-governance will be accelerated.

Furthermore, the web site could be made dynamic and user-friendly. As the Internet is expanding and evolving, web sites are becoming

Figure 10. Use of XML to create user friendly contents (Adopted from CTG, 2000)

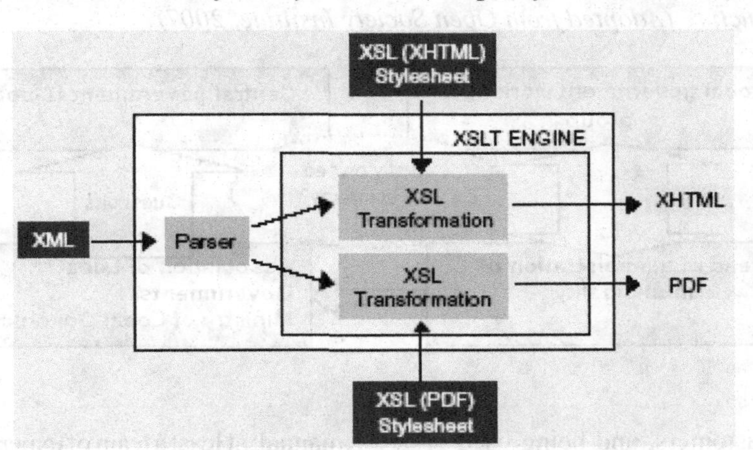

huge, interactive, and complex. With this increased complexity and hugeness, web content management is shifting from static to dynamic techniques. CTG (2000) has been engaged in establishing e-governance technologies for many years and currently bringing the latest thinking in dynamic web technology to a diverse audience of public sector information technology professionals via a presentation series entitled XML (eXtensible Markup Language): From Static to Dynamic Web. The series explains the challenges of cutting-edge Web site management involving content, layout, site map and style using XML. A simplified picture as shown in Figure 10 illustrates how the XML can convert content to more user friendly formats.

Moreover, using electronic communication based social networks; different working groups can be established that can promote better interactions among the grass roots stakeholders and lead to increased e-governance activities at the local government levels. Figure 11 is showing a collaborative approach between the local government and the central government through forming working groups. The more this sort of interactive interactions could be increased, the more the e-governance process at the local level would widen.

Observations Related to Physical Infrastructure

In order to achieve maximum capabilities of ICT tools in E-government practice it is necessary to adjust their organizational structure and usage leading to: actual legal and socioeconomic context in which strategic development planning process is to be performed; pragmatic demands that managing of local development has to be fulfilled; include all participants in local management process; and strengthen institutional capacities and procedures (Lalovic, Djukanovic & Zivkovic, 2004). However, it has been observed that only small percentage of the local governments could meet the criteria for effective e-government management including: leadership, strategic planning, performance measurement and market promotion. In order to be successful, local e-government objectives require strong leadership that champions e-government and works to increase acceptance among stakeholders. In addition to having a local strategy, individual organizations should also include local e-government approach in their strategic plans to ensure each employee is continuously looking for ways to improve processes and service delivery. Regular performance measures are also essential for evaluating whether the particular e-government system (segment of the system) is cost efficient,

Figure 11. Organizational structure of small municipalities for cooperation with central government IT-management structure (Adopted from Open Society Institute, 2007)

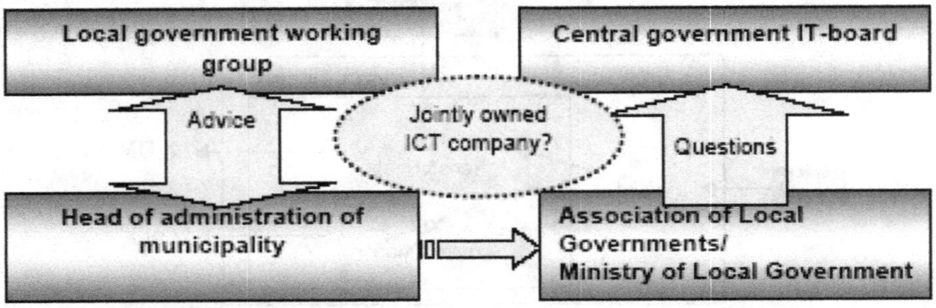

Local government working group

Central government IT-board

Advice

Jointly owned ICT company?

Questions

Head of administration of municipality

Association of Local Governments/ Ministry of Local Government

properly serving customers, and being used effectively (Austin City Council, 2008).

Aiming at a fully functioning information system, the following basic technical preconditions should also be met: establishment of a local network connecting all the computers in the local administration, or at least one computer in each organization or office should be in the network; configuring a central server to host the content, information and supporting software locally; and affordable Internet connection according to the needs of the local government network and its financial capacity. Large municipalities might use a dedicated line (broadband, fiber, ISDN or other available high speed connectivity), providing twenty-four hour connection with the Internet. Smaller local governments may afford to a more limited connection at affordable price. Central government may subsidize the entire operation for a while till the local e-government system becomes self sustainable.

However, to implement a pragmatic information system at the local government level, the following information infrastructure could be thought as a pre-requisite: at least one computer in each department or office (simple PIII or PIV or even clones); one high-end computer (may be a Pentium V) functioning as server; personnel with minimum working knowledge on computer basics and Internet; an user-friendly operating system (perhaps, open source) and a operational manual; at least a team of experts (preferably local) to conduct the basic systemic analysis, install the model, and train the officials (train the trainers).

Furthermore, depending on the governance structure, it has been observed that tiers of the government system may range from three tiers to four tiers. For sake of better governance, a five tier system may be sought of, as depicted in Figure 12.

Observations Related to Financial Infrastructure

It has been observed that the total ICT spending will enter a period of moderating growth through 2011. The slowing economies of developed nations will begin to cool their demand for ICT products. The weak US dollar will fuel ICT exports from the US, especially to emerging economies (WITSA, 2008). Moreover, WITSA (2004) forecasted, in terms of ICT spending: Bangladesh is expected to grow at the highest rate worldwide from 2003-2007, with an annual compound growth rate of 20 percent from 2003-2007; Iran ranks third with a rate of 14.5 percent; China is sixth at 13.9 percent; and India rounds out the top ten at 13.4 percent; China is projected to be the fastest growing top ten ICT spender, with a compound annual growth rate of 13.9 percent from 2003 through 2007; Second place South Korea is the only other top ten nation expected to grow at double digits, with a compound annual growth rate of 11.4 percent, more

Figure 12. A five tier local e-government organizational structure (Source: Author)

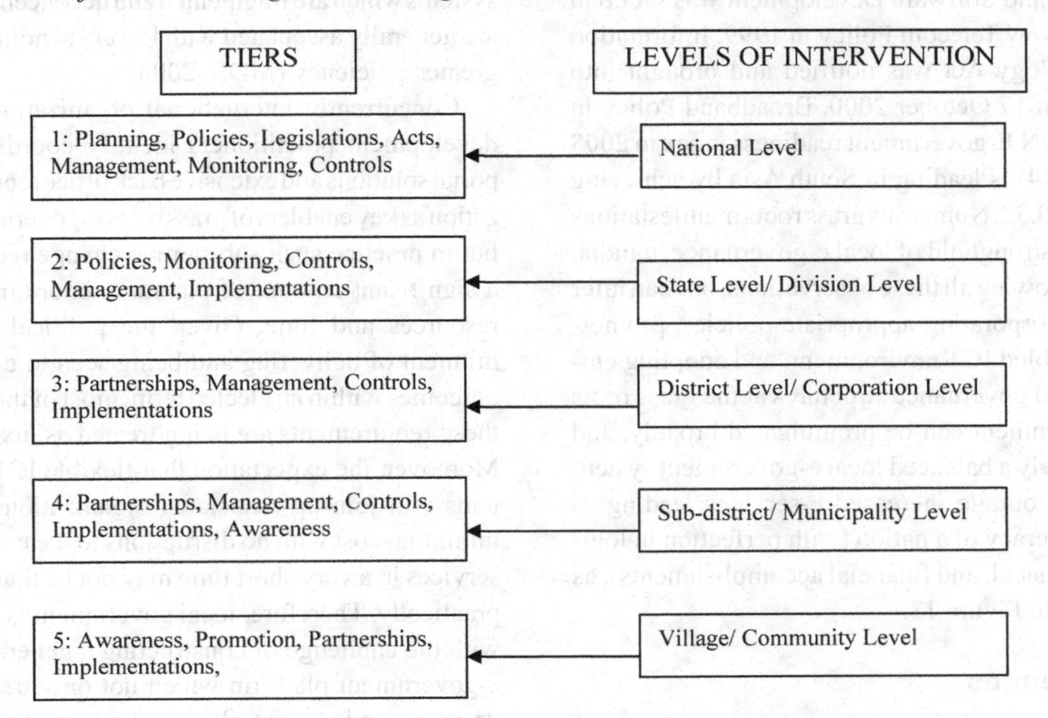

or less the forecast still persists. WITSA (2008) predicts, moderate global economic growth and increased ICT penetration will support expanding ICT spending growth at least through 2011, albeit it at a more subdued pace than the past few years.

Observations Related to E-Government Readiness

It has been observed that, despite Bhutan (established Telecom Act in 1999, Bhutan Information technology Strategy in 1999 and ICT Master Plan in 2001; UN E-government readiness index in 2008 is 0.3074[23]), Nepal (established Telecommunication Regulatory Act in 1997, national Communication Policy in 1992, adopted Telecommunication Policy in 1999, IT Policy in 2000; UN E-government readiness index in 2008 is 0.2725), Bangladesh (established Bangladesh Telecom Regulatory Council in 2001, ICT Policy in 2002; UN E-government readiness index in 2008 is 0.2936), and countries in South Asia and Sub-Saharan region could not implement local e-government in practice. However, Singapore (established National Computerization in 1980, National IT Plan in 1986, Electronic Transaction Act in 1998, IT2000 in 1992, Connected Singapore Master Plan in 2003; UN E-government readiness index in 2008 is 0.7009) has taken lead in implementing local e-government system in Asia.

On the other hand, Pakistan (established Pakistan Telecommunication Re-organization Act in 1996, Electronic Transaction Ordinance Promulgated in 2002; UN E-government readiness index in 2008 is 0.316), and Philippines (established Public Telecommunications Policy Law in 1995, Electronic Commerce Act of the Philippines in 2000; UN E-government readiness index in 2008 is 0.5001) are struggling hard to implement e-government system at the grass roots. Their Digital Access Index (DAI) are

0.24 and 0.43 respectively (ITU, 2003). But, India (established National Telecom Policy in 1994, National Task Force on Information Technology and Software Development was set up in 1998, New Telecom Policy in 1999, Information Technology Act was notified and brought into force on 17 October 2000, Broadband Policy in 2004; UN E-government readiness index in 2008 is 0.3814) is leading in South Asia by achieving DAI of 0.32. Numerous grass roots manifestations are the stronghold of local e-governance in India.

Following all these observations, one can infer that incorporating appropriate policies, promoting enabled ICT environment, and adopting empowered governance structures at the grass roots e-government can be promulgated broadly, and ultimately a balanced local e-government system will encourage increased e-services leading to e-democracy of a nation (with perfection in logical, physical, and financial accomplishments), as shown in Figure 13.

Challenges

The relationship between local government structure and corresponding performance is theoretically and empirically intricate. Local government structure itself is multi-dimensional: fragmentation and deliberation may vary both vertically and horizontally. A structural change on any of the existing dimensions may have a number of theoretical effects and the net outcome may not precisely be predictable at all. However, the empirical evidence suggests that local government systems which are fragmented and deconcentrated are generally associated with lower spending and greater efficiency (MDL, 2006).

Concurrently international organization and development practitioners identify coordinated portal solutions and extensive back office reorganization as key enablers of grass roots e-government, but in practice such substantive change requires a significant amount of planning, coordination, resources and time. Given the political commitment of delivering and being seen to deliver outcomes within an elected term, most of the time these requirements are being treated as luxuries. Moreover, the expectation that flexible ICT systems will join up associated organizations at a minimum cost with no disruptions to the existing services in a very short time may not be that easy practically. Therefore, local government is faced with the challenge of constructing a generalized e-government platform which not only delivers its own services, but also works in partnership with a range of quasi-private agencies to provide services in combined efforts and low costs. It has also been a challenging fact that there may be a hidden motive that closeness to the community may not only encourage greater stakeholder participations, but also provide a useful means of exerting pressure for service improvements upon public sector managers, professionals and front line staff (Shackleton, Fisher & Dawson, 2004; Kolsaker, 2005). This discourages politicians to

Figure 13. Concept of local e-government (Adopted from Open Society Institute, 2007)

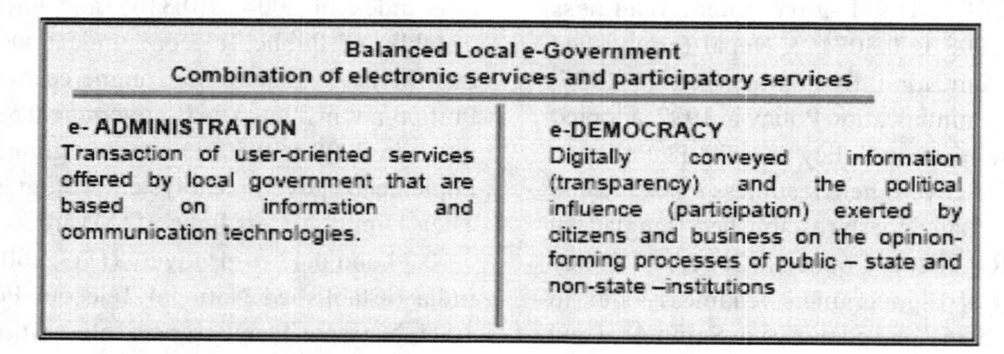

enter very closely into any system that are very near to the grass root participants.

Additionally, a general lack of commitment is found with local e-government strategies (Juana-Espinosa, 2006), where political accountability is a process through which citizens elect their officials to account for their behavior and performance (Aucoin & Heintzman, 2000). Political accountability include ensuring representation of marginalized/vulnerable groups through reserved seats or quota systems; improving the quality of the electoral system with recall, write-in, and independent candidate options; establishing, monitoring, and enforcing transparent rules for financing local elections; reviewing and revising the role of campaign financing rules (that may favor centralized, established parties, and make it more difficult for local candidates or new parties to compete); and securing and strengthening the role of elected body (council) in overseeing local government operations. At the same time, political accountability measures allow for citizen-initiated legislation, referendum, or recall of elected public officials; establish procedures for public petitions to adopt, amend, or repeal an act, legislation, or executive order; empower citizens to demand public hearings on policy decisions and action; and institute citizen ombudsman offices in local governments (World Bank, 2007).

With little or constricted support from the central government, local governments always struggle to provide citizens with adequate social services (USAID, 2006). To rectify this, local government authority should have more autonomy, in terms of administrative and financial aspects. Their ability to make, change, and enforce laws and regulations pertaining to local administrative affairs (i.e., on spatial and sector planning, environmental measures, and local economic development); enjoying a degree of autonomy over the local procurement process; control over local civil service and employment policies (World Bank, 2007) would remain as challenges due to many factors that need extensive study and research.

Contrary to what is often thought, the biggest challenge when developing an information society does not lie in how to acquire the information technology together, but in shaping the organizational, legislative and fiscal framework of the local government to support the development of e-governance at the grass roots. The organizational model may offer the local governments a framework in which to carry out the decision-making processes and project management activities, related to e-government (Open Society Institute, 2007), but designing, adoption and implementation of technology model would remain as challenge in many countries.

One challenge in implementing technology model is bridging the digital divide between urban and rural populations, men and women, and young and old. This requires innovative approaches to increase computer and information literacy, facilitate access to the Internet, and demonstrate the economic value of using the Internet to citizens and businesses. Another challenge is the modernization of public sector infrastructure in terms of increased transparency and accountability, user-focused e-services, and increased efficiency and effectiveness. This challenge includes the development of e-government in local government and increased interactions among various levels of the government, as well as among service delivery agencies. Ensuring appropriate investment in local e-government with priorities remains other challenge from which the system should be evident that benefits are larger than costs (OECD, 2007a;b).

Although the Web has become a major tool for delivering government services, many local governments have yet to obtain benefit from the cost savings and improved citizen communication and e-services due to improper financial and technical support, increased staff demands, and the difficulty of integrating e-government capabilities into existing information technology operations (ICMA, 2005).

In general, public policy and decision-making at local level are relatively challenging and complicated tasks, which are basically characterized by un-structured decision support system and inadequate management, including un-structured data and processes. Moreover, due to the globalization of modern economy, social problems tend to become more and more multidimensional, and often cross the borders of a region or even a country. Therefore, the related issues have to be addressed through close collaboration among relevant public organizations from various administrative layers (e.g. central government organizations, regional administrations, prefectures, municipalities, local development organizations, employment organizations, social service delivery organizations, social security organizations, education organizations, research organizations, environmental organizations, civil society organizations, business associations, professional associations, etc.) and, more often, learning from success stories from other countries (international development agencies, UN agencies, international donors and partners, etc.). In addition to these, enhanced participation of citizens, enterprises and their associations in such tasks is highly desirable (Karacapilidis, Loukis & Dimopoulos, 2005).

FUTURE RESEARCH

Better implementation of local e-government or local e-governance means utilizing the power of ICT to assist in transforming the accessibility, quality and cost-effectiveness of public service and to help revitalize the relationship among customers and citizens and the public bodies who work for their benefits. Planned e-governance and appropriate use of ICT at the local level can enhance and support economic and social development, particularly in empowering officials and municipal representatives by ensuring linkages and networking through timely, efficient, transparent, responsive, and accountable services.

According to European researches (Open Society Institute, 2007) local governments in developed countries are offering up to 77% of public e-services. Often local government portal is the first step to reach even the central government services. Implementation of e-services and broadband strategy is another possibility to overcome problems of different social groups and remote areas. However, well-developed ICT infrastructure with intensive offering of e-services by the local government will remain as challenge for engaging large groups of active citizens into the decision-making processes and supporting development and implementation of proper e-democracy in the locality (Turner, 2004). Moreover, information society development at the grass roots is in large extent remains as an internal issue of local governments in comparison to the central government as local governments are more close to the citizens. Also, after decentralization of local government structures municipalities will be able to offer new, wider variety of services for their citizens. But, it may become a great challenge as a new comer to offer those services through progressive information and communication technologies, rather than the age old traditional way. These issues will need further attention and research in terms of developing new hardware platform, new software platform and new legislation.

To meet the new requirements, an electronic platform needs to be developed, to: allow collection of personal information from the possession of national agencies to be available; allow different degrees of technological maturity among the various services; guarantee high levels of security; deliver high levels of availability; and publish open standards in terms of data format, information exchange, and levels of security. To address these challenges, designs have to be developed by incorporating technological, financial, organizational and institutional surroundings for implementing an integrated electronic services platform. Furthermore, to complement development of the integrated platform, procedures need to be redesigned, standardized, and digitized for a legally compliant process model with a highly calibrated defined structure (Inter-American Development Bank, 2006; Misra, 2008).

Instead of local governments developing e-government applications in a potentially duplicative and isolated manner, central government should assist in promulgating standards and guidelines that encourage local government to collaborate on developing e-services and sharing databases and services across the network (OECD, 2007a;b). Future issues should also incorporate strategies to develop a robust broadband infrastructure with access to all, from all businesses, especially SMEs, to everyone in the community; ensure that the education and the skill base is there to develop and sustain the future workforce; tackle the 'digital divide' and ensure that the information society is open to all; create a business-friendly environment for e-commerce and e-business to develop and reach a critical mass, especially in terms of the digital content (Berg, Meer, Winden & Woets, 2006). In the future local e-government system, local autonomy should persist through participatory budgeting with enhanced participation from women and youth, NGO and civil society.

CONCLUSION

The e-government system at the central should have a clear vision and priorities for local e-government; prepared for the emerging technologies; contain enough political will to lead the e-government endeavors; prioritize selection of e-government projects during implementation stage; competent in planning and managing them; able to overcome resistance from within the government; introduce appropriate monitoring, measurement, feedback and communication paths to follow up the progress of implementation; promote institutional relationship among all implementing partners, especially the private sector implementers; develop adequate human skills to manage the entire operational chain; and able to improve citizen participation in public affairs (Pacific Council, 2002).

E-government is not something that the government can do alone. The private sector, in particular, entrepreneurs from the corporate sectors have a key role to play, from the vision/planning process through implementation, monitoring and evaluation. Citizens need to be treated as the e-government experts, as at the end of the day, e-government is meant to serve citizens. Hence it is critical, especially with projects designed to serve the public directly, to assess their requirements and solicit their input. Most importantly, e-government services should be piloted first with the full participation of local citizens before a government invests in or embarks on a sustained-scale, nationwide version of the project, despite the fact that this sort of project environment vary drastically from locality to locality. Without a form of prior pilot-and-citizen involvement scheme, local e-government project may be risky.

Apart from proper designing, structuring, and launching of an on-line resource, local e-government initiatives involve much more than just a website. A June 2006 report, prepared for the Congressional Research Service by a research team from the University of Texas at Austin's LBJ School of Public Affairs, identified a number of commonly-used factors that contribute to the functioning of state e-government initiatives. They were; appropriate strategies, adequate funding, authentic leadership, adoptive technology, and ample performance measurement (Austin City Council, 2008). Each stage of local e-government development should be built upon the previous one, until government reaches a new consensus for providing further improved e-services to citizens and businesses.

Furthermore, local e-government programs need to be evaluated through public participation. This should be carried out keeping the fact that, access to public services locally is a necessary part of local e-government, but not sufficient.

Rather, facilitating, encouraging, and intensifying openness and increased citizen involvement are fundamentals to local e-government. The entire evaluation process for judging the e-government system's effectiveness or success needs to be carried out through participatory dialogues and interactions. Apart from these, transparent local e-government infrastructure (logical, physical, and financial) is a pre-condition for an effective e-government system at the grass roots. Finally, while talking about utilization of progressive ICTs for improving the local e-government systems, development of adequate human skill is extremely essential.

REFERENCES

Anttiroiko, A. (2002). Strategic Knowledge Management in Local Government. In Grönlund, Å. (Ed.), *Electronic Government: Design, Applications and Management* (pp. 268–298). Hershey, PA: IGI.

Anttiroiko, A. (2004). Towards Citizen-centered Local E-government: The Case of the City of Tampere. In M. Khosrow-Pour (Ed.), Annals of cases on information technology 6, 371-388. Hershey, PA: IGI.

Aucoin, P., & Heintzman, R. (2000). The dialectics of accountability for performance in public management reform. *International Review of Administrative Sciences*, *66*, 45–55. doi:10.1177/0020852300661005

Austin City Council. (2008). *Audit Report 2008: City of Austin's E-government Initiative*. Austin, Texas: Office of the City Auditor.

Chutimaskul, W., & Chongsuphajaisiddhi, V. (2004, May 17-19). A Framework for Developing Local E-Government, In M.A. Wimmer (Ed.), *Proceedings of Knowledge Management in Electronic Government: 5th IFIP International Working Conference, KMGov 2004* (pp. 319-324), Krems, Austria.

Commonwealth (2004). *Commonwealth local government handbook: Modernisation, Council Structures, Finance, E-Government, Local Democracy, Partnerships, Representation,*. Commonwealth Local Government Forum, Commonwealth Secretariat.

CPSI. (2005). *Local Governance and ICTs Research Network for Africa. LOG-IN Africa, COUNTRY REPORT – SOUTH AFRICA, Produced for the LOG-IN Africa Project Planning, Compiled by: Centre for Public Service Innovation (CPSI) and the LINK Centre*. Nairobi, Kenya: University of the Witwatersrand.

CTG. (2000). *Putting information to work: Annual Report 2000*. Center for Technology in Government, University at Albany, State University of New York.

CTG. (2002). *New foundations: Annual Report 2002*. Center for Technology in Government, University at Albany, State University of New York.

CTG. (2003). *Annual Report 2003, Center for Technology in Government*. University at Albany, State University of New York.

de Juana-Espinosa, S. (2006). Empirical Study of the Municipalities' Motivation for Adopting Online Presence. In Al-Hakim, L. (Ed.), *Global E-government: Theory, Applications and Benchmarking* (pp. 261–279). Hershey, PA: IGI.

Dewapura, R. (2008, May). *Enabling Environment for e/m – Government: ICT Infrastructure & Interoperability*. A presentation at the Capacity-building Workshop on Back Office Management for e/m-Government in Asia and the Pacific Region, Shanghai, People's Republic of China. Available at http://unpan1.un.org/intradoc/groups/public/documents/UN/UNPAN030563.pdf

Ferguson, M. (2005). Local E-Government in the United Kingdom. In Drüke, H. (Ed.), *Local Electronic Government: A Comparative Study* (pp. 156–196). Routledge.

Fox, W. F., & Gurley, T. (2006, May). *Will Consolidation Improve Sub-National Governments?* World Bank Policy Research Working Paper WPS3913, World Bank, Washington DC.

Fukao, M. (1995). *Financial Integration, Corporate Governance, and the Performance of Multinational Companies*. Transaction Publishers.

Hanseth, O. (2002). From systems and tools to networks and infrastructures. Toward a theory of ICT solutions and its design methodology implications. Accessed October 25, 2008 from http://heim.ifi.uio.no/~oleha/Publications/ib_ISR_3rd_resubm2.html

Hoogwout, M. (2003, September 1-5). Super Pilots, Subsidizing or Self-Organization: Stimulating E-Government Initiatives in Dutch Local Governments. In R. Traunmüller (Ed.), *Proceedings of Electronic Government: Second International Conference, EGOV 2003* (pp.85-90), Prague, Czech Republic: Springer.

ICMA. (2005). *The Municipality Yearbook 2005. International City/County Management Association*. Washington, D.C.: ICMA.

Inter-American Development Bank. (2006). *Multiphase Program for the Strengthening of Chile's Digital Strategy.* Report # CH-L1001, Document of the Inter-American Development Bank, Washington D.C.

ITU. (2003). *ITU Digital Access Index: World's First Global ICT Ranking*. Geneva: International Telecommunication Union.

Karacapilidis, N., Loukis, E., & Dimopoulos, S. (2005). Computer-supported G2G collaboration for public policy and decision-making. [Emerald Group Publishing Limited.]. *The Journal of Enterprise Information Management, 18*(5), 602–624. doi:10.1108/17410390510624034

Kolsaker, A. (2005. March 2-4). Third Way e-Government: The Case for Local Devolution. In M.H. Böhlen, J. Gamper, W. Polasek, & M.A. Wimmer (Eds.), *Proceedings of E-Government: Towards Electronic Democracy: International Conference, TCGOV 2005* (pp.70-80), Bolzano, Italy: Springer.

Labelle, R. (2005). *ICT Policy Formulation and e-Strategy Development: A Comprehensive Guidebook. United Nations Development Programme-Asia Pacific Development Information Programme (UNDP-APDIP) – 2005*. Reed Elsevier India Private Limited.

Lallana, E.C. (2003, October). Comparative Analysis of ICT Polices and e-Strategies in Asia. *A presentation at the Asian Forum on ICT Policies and e-Strategies*, Kuala Lumpur: UNDP-APDIP.

Lallana, E. C. (2004). *An Overview of ICT Policies and e-Strategies of Select Asian Economies. United Nations Development Programme-Asia Pacific Development Information Programme (UNDP-APDIP) – 2004*. Reed Elsevier India Private Limited.

Lalovic, K., Djukanovic, Z., & Zivkovic, J. (2004). Building the ICT fundament for local E-government in Serbia- Municipality of Loznica example. In M. Schrenk (Ed.), *Proceedings of the 9th International Symposium on Planning and IT*, Vienna, Austia.

LGAR. (2006). *Workforce Map of Local Government*. UK: Local Government Analysis and Research.

McNabb, D. E. (2006). *Knowledge Management in the Public Sector: A Blueprint for Innovation in Government*. M.E. Sharpe.

MDL. (2006). *Local Government Structure and Efficiency. A report prepared for Local Government New Zealand*. New Zealand McKinlay Douglas Limited.

Miller, H. N. (2004). *Presentation by World Information Technology and Services Alliance (WITSA)*. WITSA President Harris N. Miller at the September 27, 2004 WITSA Steering Committee meeting in Bakubung, South Africa.

Misra, D. C. (2008). Emerging E-government Challenges: Past Imperfect, Present Tense but Future Promising. *A presentation at the 4th International Conference on E-government,* Indian Institute of Technology, Delhi.

Misuraca, G. (2007). *E-Governance in Africa, from Theory to Action: A Handbook on ICTs for Local Governance*. IDRC.

OECD. (2007a). *OECD E-government Studies. Organisation for Economic Co-operation and Development*. Turkey: OECD Publishing.

OECD. (2007b). *OECD E-government Studies. Organisation for Economic Co-operation and Development.* Hungary: OECD Publishing.

Open Society Institute. (2007). *ICT for Local Government Handbook.* Budapest: Local Government and Public Service Reform Initiative. Open Society Institute.

Pacific Council. (2002). *Roadmap for E-government in the Developing World: 10 Questions E-government leaders Should Ask Themselves. The Working Group report on E-government in the Developing World.* CA: Pacific Council on International Policy.

Perotti, E.C., & von Thadden, E-L. (2006, March). Corporate Governance and the Distribution of Wealth: A Political-Economy Perspective. *Journal of Institutional and Theoretical Economics JITE, 162*(1), 204-217(14). Mohr Siebeck.

Pironti, J.P. (2006, May). Key Elements of a Threat and Vulnerability Management Program - Information Systems Audit and Control Association Member Journal, ISACA, 6.

Planet, D. (2004). *Digital Planet 2004: The Global Information Economy, a biannual report of the World Information Technology and Services Alliance*. VA, USA: WITSA.

Planet, D. (2008). *Digital Planet 2008: The Global Information Economy, Executive Summary of the report of the World Information Technology and Services Alliance*. VA, USA: WITSA.

Rahman, H. (2008). An overview on Strategic ICT Implementations Toward Developing Knowledge Societies. In Rahman, H. (Ed.), *Developing Successful ICT Strategies: Competitive Advantages in a Global Knowledge-Driven Society* (pp. 1–39). Hershey, PA: IGI.

Ridley, C. E., & Nolting, O. F. (2003). *The Municipal Year Book. International City Managers' Association*. International City Management Association.

Samarajiva, R., & Zainudeen, A. (2008). (Eds.) *ICT Infrastructure in Emerging Asia: Policy and Regulatory Roadblocks.* SAGE Publications & IDRC.

Sayo, P., Chacko, J. G., & Pradhan, G. (2004). (Eds.) *ICT Policies and e-Strategies in the Asia-Pacific: A critical assessment of the way forward.* United Nations Development Programme-Asia Pacific Development Information Programme (UNDP-APDIP) – 2004. Reed Elsevier India Private Limited

Shackleton, P., Fisher, J., & Dawson, L. (2004). E-Government Services: One Local Government's Approach. In Linger, H. (Eds.), *Constructing the Infrastructure for the Knowledge Economy: Methods and Tools, Theory and Practice*. Springer.

Socitm (2003). *Managing e-government – a discussion paper.* A Socitm insight publication. Socitm. www.socitm.gov.uk

Turner, T. J. (2004). *Local Government E-Disclosure & Comparisons: Equipping Deliberative Democracy for the 21st Century*. University Press of America.

UK Government. (2002). Www.localegov.gov.uk: The National Strategy for Local E-government, Office of the Deputy Prime Minister, Great Britain, Local Government Association (England and Wales), UK Online for Business, Office of the Deputy Prime Minister.

UK Government. (2003). Www.localegov.gov.uk One Year on: The National Strategy for Local E-government, Great Britain Office of the Deputy Prime Minister, Office of the Deputy Prime Minister, Local Government Association (England and Wales)., Great Britain, UK Online for Business, Office of the Deputy Prime Minister, 2003

UNESCAP. (1999). (Ed.) Local Government in Asia and the Pacific: A comparative Study of Fifteen Countries. Bangkok, (UN Economic and Social Commission for Asia and the Pacific), Thailand: UNESCAP.

UNESCAP. (2003). *Country Reports on Local government Systems*. Pakistan: United Nations Economic and Social Commission for Asia and Pacific.

US Government. (2006). FY 2005 Report to Congress on Implementation of the E-government Act of 2002, Office of Management and Budget, Executive Office of the President, United States.

USAID. (2006). *The Global Development Alliance: Public-Private Alliances for Transformational Development, U.* USA: S. Agency for International Development.

van den Berg, L., van der Meer, A., van Winden, W., & Woets, P. (2006). *E-governance in European and South African Cities: The Cases of Barcelona, Cape Town, Eindhoven, Johannesburg, Manchester, Tampere, The Hague*. Venice: Ashgate Publishing, Ltd.

Wasukira, E., & Naigambi, W. (2002). *Report on the Usage of ICT in Local Governments in Uganda*. Canada: IICD.

WITSA. (2004). *WITSA Study: World IT Spending Rebounds Thanks Largely to Developing World*. World Information Technology and Services Alliance.

WITSA (2008, May 28). *Global ICT Spending Tops $3.5 Trillion: Industry Experiences Subdued Spending Growth*, Press release, World Information Technology and Service Alliances (WITSA), Kuala Lumpur, Malaysia

World Bank. (2007). Local Government Discretion and Accountability: A Local Governance Framework, Social Development Department in collaboration with the Finance, Economics and Urban Department (FEU) and the Social Protection Team (HDNSP), Report No. 40153, The World Bank, Washington DC.

ADDITIONAL READING

Anttiroiko, A.-V., & Savolainen, R. (1999). The role of local government in promoting IS development in Finland. *Finnish Local Government Studies*, *27*(3), 410–430.

Barber, B. (1984). *Strong democracy. Participatory politics for a new age*. Berkeley: University of California Press.

Becker, T., & Slaton, C. D. (2000). *The Future of Teledemocracy*. Westport, CT: Praeger.

Blank, J. L. T., & Lovell, K. C.A. (2000). Performance Assessment in the Public Sector. In J. L.T. Blank (Ed.), *Public Provision and Performance: Contributions from Efficiency and Productivity Measurement* (pp. 3-21), Amsterdam: Elsevier.

Desai, N. (2005). Internet Governance: Asia-Pacific perspectives. In Butt, D. (Ed.), *UNDP-APDIP, Elsevier*.

Gronlund, A. (2000). *Managing electronic services. A public sector perspective*. London: Springer.

Gronlund, A. (Ed.). (2002). *Electronic government: Design, applications & management*. Hershey, PA: IGI Global.

Gross, T. (2002). E-democracy and community networks: Political visions, technological opportunities and social reality. In Gronlund, A. (Ed.), *Electronic government: Design, applications & management* (pp. 249–266). Hershey, PA: IGI Global.

Harris, R., & Rajora, R. (2006). *Empowering the Poor: Information and Communications Technology for Governance and Poverty Reduction - A Study of Rural Development Projects in India, UNDP-APDIP*. Elsevier.

Heeks, R. (2001). Reinventing government in the Information Age. *International practice in IT-enabled public sector reform*. London/New York: Routledge. Available online at: http://idpm.man.ac.uk/idpm/rgia.htm.

Lovell, C. A. K. (2000). Measuring Efficiency in the Public Sector. In Blank, J. L. T. (Ed.), *Public Provision and Performance: Contributions from Efficiency and Productivity Measurement* (pp. 23–53). Amsterdam: Elsevier.

Macintosh, A., Davenport, E., Malina, A., & Whyte, A. (2002). Technology to support participatory democracy. In Gronlund, A. (Ed.), *Electronic government: Design, applications & management* (pp. 226–248). Hershey, PA: IGI Global.

Norris, P. (2001). *Digital divide? Civic engagement, information poverty & the Internet in democratic societies.* Consulted in May 28, 2003. Available online at: http://ksghome.harvard.edu/~.pnorris.shorenstein.ksg/book1.htm.

Obi, T. (2007, August). (Ed.) *E-Governance A Global Perspective on a New Paradigm Volume 1 Global E-Governancees*. IOS Press

Stephens, G. R., & Wikstrom, N. (2000). *Metropolitan Government and Governance: Theoretical Perspectives, Empirical Analysis, and the Future*. New York: Oxford University Press.

KEY TERMS AND DEFINITIONS

e-Governance: E-Governance is a network of organizations to include government, nonprofit, and private-sector entities

e-Government: Refers to the use of new information and communication technologies (ICTs) by governments as applied to the full range of government functions.

e-Government Management: Management of e-government activities for improved socio-economic environment, ideology, public affairs, social life, and promote innovation in sustainable e-government system for the government. In this aspect, information technology may effectively support the government's innovation activities to establish e-government in innovative ways.

e-Government Structure: Different layers of the e-government services for the development of an electronic infrastructure to support e-protocol, e-applications / e-petitions and internal organizational function of the public organization.

Information Infrastructure: An information infrastructure is defined by (Hanseth, 2002) as "a shared, evolving, open, standardized, and heterogeneous installed base" and by (Pironti, 2006) as all of the people, processes, procedures, tools, facilities, and technology which supports the creation, use, transport, storage, and destruction of information.

Local e-Government: Local e-government could be seen as the improvement, effectiveness and efficiency of local government in leading and delivering services to all communities through the use of ICTs with a vision of central and local government working in partnership to deliver better outcomes for people and places, including real challenges for local government in terms of political and managerial leadership, thus enhancing citizen engagement and participation for better service delivery

Local Government: The lowest level of formal state institutions, such as district-level officials or local, publicly accountable decision-making and service-delivery organizations constituted in accordance with national laws (such as in local elections). It is a city, county, parish, township, municipality, borough, board, district, or other general purpose political subdivision of a state.

Local Government Institutions: Local government institutions may comprise of municipal governments, boards of education, district health boards and a variety of other special purpose institutions. These institutions provide services needed by their residents, such as land use planning, roads, utilities, public transit, economic development promotion, education, health services, and infotainment.

ENDNOTES

1. Google definition
2. http://www.fema.gov/oer/reference/glossary.shtm
3. http://from-feral2domestic.com/legal-terms.html
4. http://www.idrc.ca/en/ev-105151-201-1-DO_TOPIC.html
5. http://web.worldbank.org/wbsite/external/topics/extinformationandcommunicationandtechnologies/0,,contentMDK:21035032~pagePK:210058~piPK:210062~theSitePK:282823,00.html
6. www.agimo.gov.au/archive/publications_noie/2001/11/ar00-01/glossary
7. www.nrw.qld.gov.au/about/policy/documents/33/definitions.html
8. http://en.wikipedia.org/wiki/E-Governance
9. http://www.unescap.org/huset/lgstudy/newcountrypaper/SriLanka/SriLanka.pdf
10. http://www.ifip.or.at/minutes/GA2005/Rep_SriLanka1.pdf
11. http://www.unescap.org/huset/lgstudy/newcountrypaper/RoK/RoK.pdf
12. http://www.mic.go.kr/index.jsp
13. http://www.unescap.org/huset/lgstudy/newcountrypaper/Philippines/Philippines.pdf
14. Executive Order No. 269, 'Creating the Commission on Information and Communications Technology', available at http://www.ops.gov.ph/records/eo_no269.htm
15. http://www.unescap.org/huset/lgstudy/newcountrypaper/Pakistan/Pakistan.pdf
16. http://www.apdip.net/projects/dig-rev/info/pk/
17. http://www.apdip.net/projects/dig-rev/info/pk/
18. http://www.unescap.org/huset/lgstudy/newcountrypaper/Nepal/Nepal.pdf
19. http://www.unescap.org/huset/lgstudy/newcountrypaper/China/China.pdf
20. Wang, D.D. 'Note to the UNDP SURF-IT discussion list under the heading: ICTD practice note on rural poverty', First draft, October 8, 2003.
21. http://www.unescap.org/huset/lgstudy/newcountrypaper/Bangladesh/Bangladesh.pdf
22. http://www.sdnbd.org/sdi/issues/IT-computer/itpolicy-bd-2002.htm
23. http://www.itu.int/newsroom/press_releases/2003/30.html

This work was previously published in Information Communication Technologies and Human Development, Volume 1, Issue 2, edited by Susheel Chhabra and Hakikur Rahman, pp. 48-76, copyright 2009 by IGI Publishing (an imprint of IGI Global)

Chapter 9
National Culture and E-Government Readiness

Zlatko J. Kovačić
The Open Polytechnic of New Zealand, New Zealand

ABSTRACT

Diffusion of information and communication technologies is a global phenomenon. In spite of rapid globalization there are considerable differences between nations in terms of the adoption and usage of new technologies. Several studies exploring causal factors including national cultures of information and communication technology adoption have been carried out. The focus of this article is slightly different from other studies in this area. Rather than concentrating on the individual information technology an overall e-Government readiness is the focus. This research conducted an analysis of the impact national culture has on e-Government readiness and its components for 62 countries. E-Government readiness assessment used in this study is based on the UN E-Government Survey 2008, while the national cultural dimensions were identified using Hofstede's model of cultural differences. The research model and hypotheses were formed and tested using correlation and regression analysis. The findings indicate that worldwide e-Government readiness and its components are related to culture. The result has theoretical and practical implications.

INTRODUCTION

Analysis of electronic government readiness worldwide is difficult for conceptual and methodological reasons. Furthermore, little quantitative assessment of the factors that might cause a country to become ready has been conducted or completed so far. Therefore, it is too early to make any comparative or even meta-analysis of various research efforts. This article aims to contribute to empirical research literature in the area of electronic government, focusing on national

DOI: 10.4018/978-1-60960-497-4.ch009

culture that might have an impact on the country's readiness for e-Government. Before setting up a theoretical framework for the analysis, we begin by defining the core concepts and identifying the main issues.

The concepts of electronic governance (hereafter labeled e-Governance), electronic government (e-Government), and electronic democracy (e-Democracy) have not been uniquely defined and used in literature. The term e-Government (also called digital or virtual government) is sometimes confused with e-Governance and the two terms are often used interchangeably. For example, Fountain (2004) defining e-Government says it refers to governance affected by Internet use and other information technologies and also includes e-Democracy (see also Fountain, 2001, for an alternative definition). However, e-Governance is a broader concept, which includes the use of information and communication technology (ICT) by government and civil society to promote greater participation of citizens in the governance of political institutions. According to Fang (2002) e-Government can be defined as a way for governments to use the most innovative information and communication technologies, particularly web-based Internet applications, to provide citizens and businesses with more convenient access to government information and services, to improve the quality of the services and to provide greater opportunities to participate in democratic institutions and processes.

Though most of the e-Government definitions focus more on use of technology, management and delivery of public services (for example Edmiston, 2003), Pardo (2000) stated that e-Government is about transforming the fundamental relationship between government and the public. In other words, eGovernment initiatives are complex efforts to change intended to use new and emerging technologies to support a transformation in the operation and effectiveness of government. Grönlund (2003) pointed to the strategic aspect of e-Government initiative by stating that "electronic government refers certainly to more use of information technology (IT), but more importantly to attempts to achieve more strategic use of IT in the public sector" (p. 55). This strategic aspect of e-Government opens discussion of some societal and technical topics and the interactions between the two, as was noted by DiMaggio, Hargittai, Neuman, Robinson & John (2001). On the societal level, they suggested that the adaptation of government and civic engagement to increasingly computerized environments raises political, organizational, and social questions concerning use, context, reciprocal adaptation mechanisms, learning, the design of government work, the design of political and civic communities of interest, and the design of nation states in addition to international governance bodies.

In this article, we have accepted the definitions and classification provided by Rogers Okot-Uma. As a starting point he uses the Good Governance concept to clearly explain the relationships between e-Governance, e-Government and e-Democracy Okot-Uma (2004) defines Good Governance as processes and structures that guide political and socio-economic relationships, with particular reference to "commitment to democratic values, norms & practices; trusted services; and to just and honest business". E-Governance includes all processes and structures by means of which the new ICTs can be used by government to enable:

- Administration of government and delivery services to the public; this constitutes e-Government;
- All forms of electronic communications between government and citizen with the aim of informing, representing, encouraging to vote, consulting and involving the citizen. This constitutes e-Democracy;
- Transact business with its partners, clients and the markets. This constitutes government electronic business.

In the last decade we have witnessed a rapid rate of Internet penetration worldwide. Although this Internet diffusion happened on a global scale there are significant differences between countries in terms of how far they went and how fast they have adopted new information and communication technology (hereafter labeled ICT) as was shown by Maitland & Bauer (2001). Since the adoption of a new technology varies between countries it is important to construct a composite measure of the country's overall readiness to adopt and use a new technology and also to measure factors that contribute to the adoption of ICT. Various factors influencing Internet adoption have been considered in several studies. It was confirmed that telecommunication infrastructure (Hargittai, 1999), socio-economic factors (Robinson & Crenshaw, 1999) and cultural values (Maitland & Bauer, 2001) have a significant influence on ICT adoption among countries.

A country's overall readiness to adopt, use and benefit from using ICT is called country's e-Readiness. Knowledge of the factors which make a significant contribution to e-Readiness and the country's position on the e-Readiness scale would help the country's leaders to identify the strengths and weaknesses of the country's current position and to concentrate on the areas where improvement and further integration of ICT could be made (Bridges.org, 2001). An important component of the country's overall e-Readiness is its government readiness to operate and benefit from the new environment.

The label 'e-Government readiness' is used to describe government readiness to adopt, use and benefit from ICT, and it also forms one of the main focuses of analysis. The concept of e-Government readiness is important because of the opportunities it creates for each country in terms of benefiting from e-Commerce activities, openness to globalization, potential to strengthen democracy and make governments more responsive to the needs of their citizens, increasing citizen wellbeing, etc.

The second focus in analysis is on the role that culture has in the adoption of ICT. Cultural differences between countries in general and particularly in relation to information technology adoption is a highly researched subject. The concept of culture adopted and used in this article is based on works of Dutch anthropologist Geert Hofstede who defines culture as "a system of collectively held values". The following authors identified cultural values as one of influential factors on adoption of ICT: Bagchi, Cerveny, Hart & Peterson (2003), Johns, Smith & Strand (2003), Maitland & Bauer (2001) and Sørnes, Stephens, Sætre, & Browning (2004). Others also recognize the role culture could have in adopting ICT; for example, Bridges.org (2001) suggests that: "... unique cultural and historical environment of a region must be taken into account as part of a national ICT policy to truly gauge the country's e-Readiness for the future." In other words, each country should find its own way to the optimal e-Government readiness which is consistent with the national culture.

The main objective of this research is to investigate the relationship between national culture and e-Government readiness. More specifically the purpose of this research is to provide a theoretical framework for the impact of national culture on e-Government readiness and to test whether the national cultural dimensions have significant impact on the e-Government readiness. While most of other papers in this area are focused on an individual indicator of a country's e-Readiness (for example, the number of Internet hosts or the number of PCs per 100 citizens) this article is the first to use a synthetic indicator to measure e-Government readiness. In addition, the data set for this article includes the largest number of countries in comparison to data sets in other papers.

In the next two sections we review e-Government readiness frameworks and the relationship between national culture and ICTs, providing the theoretical foundation for our empirical analysis. Based on deduction from theory and previous empirical work the third section will provide the answer to the question, how does culture influence e-Government readiness? The following two sections report data, method of analysis, results and a discussion of the results. In the final section some implications of this research will be presented.

E-GOVERNMENT READINESS FRAMEWORKS

In this article, e-Government readiness is defined as the aptitude of a government to use ICTs to move its services and activities into the new environment (a similar definition was given in UN, 2008). The reason why e-Government readiness is important to monitor and assess is explained in Bridges.org (2001), for country's e-Readiness: "It is increasingly clear that for a country to put ICT to effective use, it must be 'e-Ready' in terms of infrastructure, the accessibility of ICT to the population at large, and the effect of the legal and regulatory framework on ICT use. If the digital divide is going to be narrowed, all of these issues must be addressed in a coherent, achievable strategy that is tailored to meet the local needs of particular countries."

There is a proliferation of e-readiness assessment tools, methods, guidelines, and results. Based on its study of assessments, Bridges.org (2001) finds that at least eighty-four countries have been assessed using one of the instruments, sixteen countries have been assessed by five different organizations, and many countries have not had any e-readiness assessments. While the assessment indicators vary, most tend to measure ICT connectivity, ICT use and integration, training, human capacity, government policies and regulations, infrastructure, security, and economy. Bridges.org (2001) lists a few studies in which a list of indicators were expanded to explore historical background and socio-cultural and political variables such as ethnic homogeneity, population density, political openness, political structure and culture, and key players.

The most complete assessment of e-Government readiness including 190 countries was initially undertaken by the United Nations in 2001. The United Nations Division for Public Economics and Public Administration, together with the American Society for Public Administration, started a project in 2001 analyzing the e-Government environment of 190 UN Member States. Their conceptual frame postulates that the state of e-Government readiness is a function of the combined level of a country's state of readiness, economic, technological development and human resource development. A final product of their analysis was the construction of a synthetic indicator named the e-Government Index. Two years later in 2003, the UN Department of Economic and Social Affairs and the Civic Resource Group presented a second survey slightly changing the definition of the e-Government index and naming it the e-Government Readiness Index. Since then the same framework were used in 2004, 2005 and 2008 to assess e-Government readiness of 192 UN member states. Three important indices contribute to this index: the web measure index, the telecommunication infrastructure index and the human capital index as described in Table 2. An alternative measure of e-Government readiness was provided by West (2008a). He focuses on the features national government web sites are offering. It may be said that this index tries to capture the same phenomenon as the UN web measure index.

The adoption of an e-Government initiative or UN framework for assessment of an e-Government readiness is basically a supply-side approach to e-Government analysis using data about the national government web sites, telecommunication infrastructure and human capital rather than a demand-side approach, which is based on the real use of the e-Government web sites by the citizens, businesses and government or their perceptions about quality of the online services delivery. In this article measurement and assessment of e-Government readiness was based on the United Nations (2008) framework. There are two reasons for adopting the UN e-Government readiness framework: their framework includes more countries than any other study and the data is gathered in a consistent manner covering key areas of any e-Readiness assessment study, i.e. technological infrastructure and human capital component. Unfortunately this framework does not provide data to apply a demand-side approach while the other studies which took both, a supply and demand-side approach (Altman, 2002; Graafland-Essers & Ettedgui, 2003) are limited to Latin American and European data only.

There were also other assessments of e-Government readiness worldwide (West, 2008a) or at the different levels of federal, state or local governments (West, 2008b; Holden, Norris & Fletcher, 2003) or regions of the world (Altman, 2002). West (2008b) assessed federal and state e-Governments in US. He claimed that "Although considerable progress has been made over the past decade, e-government has fallen short of its potential to transform public-sector operations" (p. 1). Altman (2002) assessed e-Government in Latin America. Surprisingly he didn't find a direct proportional relation between those countries with high potentiality (readiness) and those with actual broad use of e-Government. His research is of particular interest because it brings together the supply-side and demand-side approaches to e-Government analysis. Graafland-Essers & Ettedgui (2003) assessed e-Governments in Europe also taking both supply-side and demand-side approaches. Bridges.org (2001) provides a very detailed list and comparison of e-Readiness assessment models which were developed until 2001.

Choucri, Maugis, Madnick, & Siegel (2003) critically considered these, what they called 'first generation' e-Readiness models and setup a theoretical framework for the 'next generation' of e-Readiness models. Defining e-Readiness "as the ability to pursue value creation opportunities facilitated by the use of the Internet" (p. 4), they derived a key element of their framework from the answer to the question: e-Readiness for what? According to them, an e-Readiness indicator should measure the degree of ability and the capacity to pursue, but emphasis in the framework should be put on value creation opportunities. Another framework of national e-Readiness was given in Bui, Sankaran & Sebastian (2003). Their framework is based on eight factors: digital infrastructure, macro economy, ability to invest, knowledgeable citizens, competitiveness, access to a skilled workforce, culture, and the cost of living and pricing. A total of 52 indicators were used to quantify these 8 factors. A detailed analysis of each individual country's e-Readiness with all components included would probably require the use of the 'next generation' of e-Readiness models, to make a decision about the optimal approach to implementation of ICT. However, to identify the impact various factors could have on the global e-Government readiness, derivation of a synthetic indicator is sufficient. The main reason for not using one of proposed 'next generation'

Table 1. Countries with highest and lowest cultural dimensions values

Power Distance		Individualism		Masculinity		Uncertainty Avoidance	
Max	Min	Max	Min	Max	Min	Max	Min
Slovakia 107	Austria 11	US 91	Guatemala 6	Slovakia 110	Sweden 5	Greece 112	Singapore 8
Malaysia 104	Israel 13	Australia 90	Ecuador 8	Japan 95	Norway 8	Portugal 104	Jamaica 13
Iraq 95	Denmark 18	UK 89	Panama 11	Hungary 88	Iceland 10	Guatemala 101	Denmark 23

Table 2. Description of variables, acronyms and data sources

Acronym	Description
eGOV	*e-Government Readiness Index* is a composite index based on the Web Measure Index, the Telecommunication Infrastructure Index and the Human Capital Index. [Source: United Nations (2008)]
WMI	*Web Measure Index* is a quantitative index which measures the generic aptitude of governments to employ e-Government as a tool to inform, interact, transact and network. [Source: United Nations (2008)]
TII	*Telecommunication Infrastructure Index* is a composite, weighted average index of six primary indices, based on basic infrastructural indicators that define a country's ICT infrastructure capacity. These six indices are: PCs/1,000 persons; Internet users/1,000 persons; Telephone lines/1,000; On-line population/1,000 persons; Mobile phones/1,000 persons; and TVs/1,000 persons. [Source: United Nations (2008)]
HCI	*Human Capital Index* is based on the United Nations Development Programme "education index". This is a composite of the adult literacy rate and the combined primary, secondary and tertiary gross enrolment ratio, with two thirds of weight given to adult literacy and one third to the gross enrolment ratio. [Source: United Nations (2008)]
PDI	*Power Distance Index* is a cultural construct developed by Geert Hofstede and interpreted as the degree of equality, or inequality, between people in a country's society. [Source: Hofstede (2004)]
IDV	*Individualism* is a cultural construct developed by Geert Hofstede and interpreted as the degree a society reinforces individual or collective achievement and interpersonal relationships. [Source: Hofstede (2004)]
MAS	*Masculinity* is a cultural construct developed by Geert Hofstede and interpreted as the degree a society reinforces, or does not reinforce, the traditional masculine work role model of male achievement, control, and power. [Source: Hofstede (2004)]
UAI	*Uncertainty Avoidance Index* is a cultural construct developed by Geert Hofstede and interpreted as the extent to which the members of a culture feel threatened by uncertain or unknown situations, i.e. unstructured situations. [Source: Hofstede (2004)]
GDP	*Gross Domestic Product* per capita in 2007 - purchasing power parity. [Source: CIA's World Factbook, CIA (2008)]

of e-Readiness models is a lack of the full set of data for most of the countries. Usually complete and reliable data is available only for developed countries.

All these measures of e-Readiness (United Nations, 2001, 2008; West, 2008a, 2004; Kirkman, Osorio & Sachs, 2002), including the 'next generation' of e-Government readiness measures (Bui, Sankaran & Sebastian, 2003; Choucri, Maugis, Madnick, & Siegel, 2003), provide what is known as a supply-side approach to e-Government research. This means they studied the features that are available on national government web sites and in the e-Government environment (telecommunication infrastructure and human capital) rather than the real usage of e-Government websites by the citizens, businesses and government, or their perceptions of online services delivery. The rationale for a supply-side approach to the analysis of e-Government in this article stems from the expectation that people will act rationally and will use the newly opened channel for communication with government agencies and for participation in public affairs. However, according to Graafland-Essers & Ettedgui (2003, pp. 35-36) there are differences in consumer preferences towards use of a particular online e-Government service even when highly sophisticated online services are available. For example, for services such as tax declaration only a few people prefer performing this operation online (demand-side), though the level of sophistication of this service (supply-side) is the highest when compared to other online services. The most preferred online services on the e-Government websites are library book search and job search.

Research studies which considered the various factors having an impact on the ICT adoption confirmed that telecommunication infrastructure (Hargittai, 1999), socio-economic factors (Robinson & Crenshaw, 1999) and cultural values (Maitland & Bauer, 2001) contributed to the explanation of differences in Internet diffusion between countries. We would also expect that in a democratic political system the government will foster the design and development of various channels for providing their services to the citizens. Indeed, research has examined the impact of democracy, corruption and globalization on e-Government readiness and found that more democratic countries are higher ranked on the e-Government readiness list than the less democratic countries (Kovačić, 2005). He found significant positive correlations between e-Government readiness and democracy (Freedom House index) and between e-Government readiness and globalization. Of course the degree of e-Government service adoption does not depend only on the level of democracy in the country but also on the cost of implementation, the perceived political benefits for the government from implementing an e-Government initiative and other factors. As Bretschneider, Gant & Ahn (2003) suggested, the degree of e-Government service adoption could be explained in terms of the perceived administrative benefit from adopting e-Government services, the political nature of online applications, the government's organizational capacity in adopting new information technology, and the diffusion effect of e-Government service technology.

NATIONAL CULTURE AND ICTs

The concept of culture is not uniquely defined in literature. As Sørnes, Stephens, Sætre & Browning (2004) pointed out over 400 definitions of culture have been identified. Fortunately, in most of these definitions a commonly held view is that the cultural environment influences and shapes the values shared by the members of the society. Hofstede (1981), whose four-dimensional cultural model was used in this article, wrote that "... culture is the collective programming of the human mind that distinguishes the members of one human group from those of another. Culture in this sense, is a system of collectively held values" (p. 24). He emphasized that "in the center is a system of societal norms, consisting of the value systems (the mental programs) shared by most of the population" (p.24). According to him, culture is an "interactive aggregate of common characteristics", "a collective phenomenon" which "is learned, not inherited" (p. 24).

Though the Hofstede model of culture is the most well-known classification of culture it is not the only one used in literature. Chanchani & Theivanathampillai (2002) investigate and discusse alternative classification of culture to Hofstede's classification based on the works of Triandis, Trompenaars and Fiske. They have set up a framework for comparing alternative classifications, evaluating the sufficiency and adequacy of these classifications. One of their suggestions is to use a classification of culture based upon the research objective. The Hofstede model is recommended in the following case "... if the researcher wishes to use an instrument or has collated data then correlation with Hofstede's data may be considered" (p. 15). McSweeney (2002) also criticized Hofstede's model of national cultural differences. He focused his critique on the Hofstede research methodology arguing that the quality of evidence in the Hofstede model of national culture is poor and the set of assumptions are not justified. However, in spite of criticisms the Hofstede model of culture has been widely used in the literature in the last two decades. There have been also numerous studies on the relationship between national culture and the use and adoption of ICTs. The following authors: Bagchi, Cerveny, Hart & Peterson (2003), Johns, Smith & Strand (2003), Maitland & Bauer (2001), Robinson & Crenshaw (1999) and Veiga, Floyd & Dechant (2001) concluded that the significant variation in Internet diffusion, IT implementation and acceptance between countries could be attributed to national culture as described by Hofstede's cultural model. Sørnes, Stephens, Sætre & Browning (2004) provided an excellent overview of the literature and a list of relevant studies on how ICTs impact culture and how culture impacts on ICT practices.

Based on 116,000 questionnaires Hofstede (1980, 1983) collected data from 50 countries and 3 regions about the work-related value patterns of employees in IBM, a large multinational firm. By using data from one firm only Hofstede controlled for a number of industry and company variables so that he could focus on cultural differences. Using correlation and factor analysis he revealed four largely independent dimensions of differences between national value systems: (1) power distance (large vs. small), (2) individualism vs. collectivism, (3) masculinity vs. femininity, and (4) uncertainty avoidance (strong vs. weak). Later Hofstede identified a fifth dimension, dealing with long versus short-term orientation, replying to those who criticized his cultural model to be biased toward Western culture.

The Power Distance dimension reflects the perception that members of society have about unequal distribution of power in institutions and organizations and the extent to which it is accepted in a society. People in countries where power distance is large accept a hierarchical order in which everybody has a place that needs no further justification. Countries with small power distance allow upward social mobility of its citizens and their participation in the process of decision making. One of the conditions for such citizen's participation would be the implementation of various communication technologies which would support and help this participation happen. Therefore it could be argued that a country with a larger power distance would have a negative attitude toward implementing and using ICTs.

The Individualism/Collectivism dimension describes the relationship between individuals and the group in a society. For the countries with low individualism, i.e. high collectivism, people consider the group as the main source of their identity. On the other hand, an individualistic culture would pay more attention to the performance of the individual. Time management would be important and any technology that could help individuals to perform more efficiently would be highly regarded and quickly accepted. Therefore it could be argued that the country with a strong individualistic culture would have a positive attitude toward implementing and using ICTs.

The Masculinity/Femininity dimension describes the achievement orientation in a society. When the preferences in society are for achievement, assertiveness, and material success then the country is ranked high on masculinity. On the other side, cultures that rank low on masculinity, i.e. high on femininity, prefer relationships, caring for the weak, and the quality of life. A high masculinity index indicates a culture that emphasizes masculine values and has very separate and rigid gender roles and expectations. Some authors, such as Bagchi, Cerveny, Hart & Peterson (2003) argued that "ITs promote more cooperation at work, better quality of life and these values are espoused in nations with low MF index" (p. 960). However, it could be argued equally well that in a country with high masculinity there would also be a positive attitude toward implementing ICTs if these technologies improve performance, increase the chance of success and support competition, which are all key factors of a masculine culture. In other words the masculinity/femininity dimension could have at least at the conceptual level a mixed impact on the ICTs.

The Uncertainty Avoidance dimension describes the degree to which members of a society feel uncomfortable with uncertainty and ambiguity, preferring structured over unstructured situations. Members of societies with strong uncertainty avoidance would tend to avoid or reduce the risk induced by the unknown, i.e. unstructured situation, while people from countries with weak uncertainty avoidance could be described as 'risk takers'. It could be expected that countries with strong uncertainty avoidance would be slow in the adoption and use of new ICTs, while the countries on the opposite end of this scale would be leaders in implementing new ICTs and willing to take the risk of failure. Therefore it could be argued that the country with a strong uncertainty avoidance culture would have a negative attitude toward implementing and using ICTs.

All four dimensions of the Hofstede cultural model were included in the later empirical analysis. As statisticians say 'let the data speak for itself'. However, in the literature not all four dimensions were considered to be relevant for research on the impact of national culture on the ICTs adoption. For example in Maitland & Bauer (2001) only uncertainty avoidance dimension from the Hofstede model has been included. However, they have added two other variables which might be considered as cultural variables: gender equality and English language. Also, Johns, Smith & Strand (2003) included the individualism/collectivism and uncertainty avoidance dimensions only. They felt that achievement orientation (masculinity/femininity dimension) has a mixed impact on the use of technology. The same conclusion was drawn for power distance dimension and its impact on the use of technology.

To illustrate the four Hofstede cultural dimension values, three countries were selected from the list of all countries, those with extreme values (maximum and minimum) on each dimension and their scores were presented in Table 1. For example, Slovakia scores 110 on masculinity and Sweden 5 reflecting the fact that Slovakia is a 'masculine' society where men are tough and concerned with material success, whereas women are more tender and interested in quality of life. On the other side of the masculinity/femininity scale Sweden is a 'feminine' society where both men and women are equally concerned with quality of life.

HOW DOES CULTURE INFLUENCE E-GOVERNMENT READINESS?

Figure 1 describes the model of influence that national culture has on e-Government readiness. The arrow in the cultural environment block illustrates the assumption that national culture affects society's basic values. People of the country are using these basic values as a foundation to build and shape the whole legal environment and a legal system with its three constitutive components: legislature, executive and judiciary. Then the legal environment and the legal system influences whether and how the government will use the new ICTs to support its internal and external activities. External to this model are socio-economic, technological and other factors which may influence e-Government readiness.

Though in his conceptualization Hofstede treated national culture as systematically causal, we can argue along the same line with Sørnes, Stephens, Sætre & Browning (2004) that "the

Figure 1. Model of the impact of national culture on the e-Government readiness

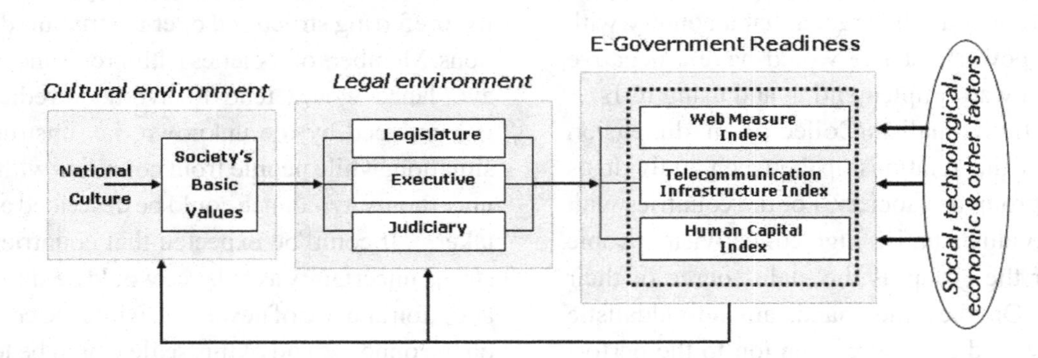

relationship between organizational cultures and ICTs is not simply causal. Either one can cause changes in the other, because technology is part of culture and vice versa." In other words, there is a reflexive and dynamic relationship between national culture and ICTs rather than causal. Therefore, arrows, i.e. feedback links from e-Government block to legal and cultural environment blocks in Figure 1 have been added to take into account the impact that e-Government may have on the national culture and legal system. However these feedback links were not analyzed further for the methodological reasons explained later.

Based on the model in Figure 1, the above discussion of Hofstede's four cultural dimensions and the attitude that the country and its government might have toward using ICTs the following research hypotheses are offered:

- ***Hypothesis H1:*** The government of a country with a larger power distance would have a negative attitude toward increasing the level of e-Government readiness
- ***Hypothesis H2:*** The government of a country with a strong individualistic culture would have a positive attitude toward increasing the level of e-Government readiness
- ***Hypothesis H3:*** The government of a country with a high/low masculine culture would have a positive attitude toward increasing the level of e-Government readiness
- ***Hypothesis H4:*** The government of a country with a strong uncertainty avoidance culture would have a negative attitude toward increasing the level of e-Government readiness

DATA AND METHODOLOGY

Data for this article was collected from three different sources and was available for 62 countries. While the data for e-Government readiness and GDP per capita were available for 192 countries, the major constraint came from a database containing cultural dimensions scores (Hofstede, 2004), i.e. data for only 62 countries was available. Generally, one of the main difficulties in assessing worldwide e-Readiness including e-Government readiness and the effect that national culture might have on ICTs adoption and their use, is a lack of data which would cover most of the countries around the world and would be available for all indicators to be included in analysis. Table 2 describes the definition of variables in detail, their acronyms and data sources used.

The reason for including GDP per capita in an analysis is explained by Hofstede (1980). He suggested including economic variables such as GDP per capita when examining the effect of national culture. When the effect of others hard variables (economic variables, for example) are significant, then the cultural variables are redundant. If the cultural variables are still significant in spite of included economic variables, then the effect of culture on observed phenomenon, i.e. e-Government readiness and its components could be confirmed.

Methods of correlation and regression analysis were applied to the data. To estimate e-Government readiness regression models ordinary least squares estimation method was used. For all calculations in this article the SPSS for Windows version 13 was used.

RESULTS

Is there a relationship between cultural dimensions and e-Government readiness and its components? Pearson's correlation coefficients were calculated to test the hypotheses that the e-Government readiness and its components correlate with the Hofstede's national cultural constructs. Results are presented in Table 3.

Three results emerge from the correlation matrix presented in Table 3. Firstly, e-Government readiness and all its components are highly

Table 3. Pearson's correlation matrix

	eGOV							
WMI	.885**	**WMI**						
TII	.945**	.736**	**TII**					
HCI	.802**	.559**	.702**	**HCI**				
PDI	-.651**	-.519**	-.705**	-.433**	**PDI**			
IDV	.696**	.528**	.753**	.507**	-.648**	**IDV**		
MAS	-.132	-.168	-.107	-.060	.038	.053	**MAS**	
UAI	-.196	-.235	-.231	.035	.262*	-.253*	.000	**UAI**
GDP	.584**	.628**	.861**	.584**	-.591**	.662**	-.040	-.159

negatively correlated with Power Distance Index (correlation coefficients are in range from -0.433 to -0.705) and positively with Individualism (correlation coefficients are in range from 0.507 to 0.753). High individualism (countries where individual rights are paramount) accompanied with smaller power distance (those countries which allow upward social mobility of its citizens) characterized a society in which e-Government readiness is at the higher level than in the collectivistic countries with larger power distance. There are no statistically significant correlations between e-Government readiness and other two cultural dimensions, Masculinity and Uncertainty Avoidance Index. Contrary to other studies where Uncertainty Avoidance Index was argued to be one of the most relevant cultural construct which explain ICTs adoption, in this analysis Uncertainty Avoidance Index was not statistically related to any other variables included, beside two cultural constructs, i.e. Power Distance Index (correlation coefficient 0.262, significant at 5% level) and Individualism (-0.253, significant also at 5% level).

Secondly, within a subset of national cultural components there is a highly significant negative correlation between Power Distance Index and Individualism, while all the other correlation coefficients are insignificant at the usual 5% level. This result, i.e. corr(IDV, PDI) = -0.648 confirms Hofstede's proposition that a collectivist country is also likely to be a high power distance country. However, from methodological point this result could cause a multicollinearity problem when it comes to the estimation and interpretation of regression models for e-Government readiness and will be address later.

Thirdly, the absolute value of the correlation coefficient between GDP per capita and all e-Government readiness indices is higher than the correlation coefficients of e-Government readiness indices with any cultural dimension. This result might suggest that economic factors, as measured by GDP per capita, are more important than any other cultural construct, or even the only one which explain variation in the level of e-Government readiness. This will be tested using a regression model which includes both GDP and cultural variables.

Furthermore, significant correlation coefficient corr(GDP, IDV) = 0.662 confirmed another proposition from Hofstede's work. He indicated a strong relationship between a country's national wealth and the degree of individualism in its culture. Richer countries tend to be more individualistic, whereas poorer countries are more collectivist. As a poor country grows richer it tends to move away from a collective pattern to an individualistic one. Also, positive correlation coefficients between GDP and e-Government readiness indices mean that developed countries are better prepared for implementation of e-Government initiative.

After the correlation analysis a regression analysis was used to get further insight into the relationship between the variables considered in this article. However, at this stage it cannot be assumed that the national culture is truly exogenous, in other words, that there is one-way causation between national culture and e-Government readiness (i.e. national culture → e-Government readiness). The same was pointed out by Slack & Wise (2002) who argued that there is a reflexive relationship between cultures and ICTs, i.e. the relationship between culture and ICTs is not simple causal. Therefore we could argue equally well that the e-Government readiness could have an impact on national culture. So, if we have a two-way causation in a function such as e-Government readiness (i.e. national culture ↔ e-Government readiness), this implies that the e-Government readiness function cannot be treated in isolation as a single equation model, but belongs to a wider system of equations that describe the relationships between the relevant variables. This system of equations, known in econometrics as a simultaneous equation system, would be more appropriate to use for a full description of such complex social phenomena as a national culture, e-Government and its components. However, at this stage we have estimated a single regression equation for each of the key variables (e-Government readiness indices) in spite of the fact that the estimation method used (ordinary least squares) will produce a biased estimate of the effects that national culture has on e-Government readiness. This result is due to a violation of the assumptions of the estimation method used, which creates what is known as simultaneous equations bias.

Regression analysis was carried out in two steps. In the first step e-Government readiness indices were regressed on all four cultural indicators. A summary of the regression results is presented in Table 4.

Based on the results of the correlation analysis, i.e. significant correlation between cultural indices, we checked to see if there is any problem with multicollinearity. Multicollinearity simply means a high correlation between the independent variables, i.e. cultural indices. One effect of "too much" collinearities between independent variables is that the standard error of ordinary least squares estimates tends to be inflated. This also means that we get a less efficient estimate of the regression coefficients. To detect degree of multicollinearity a collinearity diagnostic tool known as variance-inflation factor (VIF) was used. An arbitrary, but common cut-off criterion for deciding when a given independent variable displays "too much" multicollinearity is VIF value of 4. Since VIF was well below 2 in all regression models reported in Table 4, the multicollinearity as such was not a problem in these regression models.

The first column in Table 4 lists dependent variables, the second column shows which cultural index is significant at the 5% level and the value of its standardized β coefficient. Standardized β coefficients are used to make statements about the relative importance of the independent variables in a regression model. A higher β value means

Table 4. Summary of regression results (models with cultural variables only)

Dependent variable	Variable with significant t-test at 5% level (β-coeff.)	R^2
Web Measure Index	PDI(-0.27), IDV(+0.34)	0.37
Telecommunication Infrastructure Index	PDI(-0.36), IDV(+0.53)	0.66
Human Capital Index	IDV(+0.43)	0.32
e-Government Readiness Index	PDI(-0.33), IDV(+0.49)	0.57

that the particular variable is more important that the others. The coefficient of determination R^2 measures the proportion of the variation in the dependent variable "explained" by the regression model (last column in Table 4). The results reported in Table 4 suggest that the cultural variables explained between 32% and 66% variation in the e-Government readiness indices when treated as the only explanatory variables in regression models. Observed jointly, cultural variables made a significant impact on e-Government readiness. However, when observed individually, only two of the cultural variables (Individualism and Power Distance Index) are significant at level 5%. The sign of each coefficient matches our expectation: IDV has a positive sign, while PDI has a negative impact on e-Government readiness. Furthermore, standardized β coefficients suggest that IDV is a relatively more important cultural construct in predicting e-Government readiness than PDI. Interestingly MAS was not significant in any regression model, suggesting there are both masculine and feminine countries with a strong attitude toward implementing e-Government initiative.

PDI coefficient has a negative sign in the regression model for TII. This result is consistent with the findings of Veiga, Floyd & Dechants (2001). According to them, attitude toward ICT use will be enhanced by a decision and implementation process that increases users' sense of participation in the choice of new ICTs. We should therefore expect to find evidence of resistance in situations where new ICTs, and policy pertaining to their use, are implemented without the participation of members below top management.

UAI was used in many studies as a predictor of the likelihood of ICT adoption. Surprisingly, UAI appears not to be significant in regression models presented in Table 4.

In summary, based on regression models with cultural variables only, the first two hypotheses, i.e. H1 and H2 were confirmed.

For the reason explained in the data and methodology section GDP per capita was added to the list of explanatory variables in each regression model of e-Government readiness. A summary of all regression models is presented in Table 5.

Similarly to regression models in Table 4 the VIFs were calculated for models in Table 5. The values of VIF increased slightly, but they were still well below 4 in all regression models reported in Table 5. The largest value of VIF was in case of IDV variable, where VIF was about 2.24. This result would suggest that the multicollinearity was not a serious problem in the regression models where GDP and cultural indices were independent variables and the conclusion could be made that among cultural variables in regression models which include GDP only IDV and PDI seem to be significant in some models.

Based on regression model WMI was found not to be related to any cultural dimension, which means that all governments accepted that they have to implement ICTs as a tool to inform, interact, transact and network. Controlling a level of economic development (measured by GDP) in the regression model for WMI (general aptitude

Table 5. Summary of regression results (models with cultural variables and GDP)

Dependent variable	Variable with significant-test at 5% level	R^2	Part of R^2 attributed to culture
Web Measure Index	GDP(+)	0.47	0.032
Telecommunication Infrastructure Index	PDI(-), IDV(+), GDP(+)	0.83	0.072
Human Capital Index	GDP(+)	0.40	0.062
e-Government Readiness Index	IDV(-), GDP(+)	0.73	0.052

of government to employ e-Government) it appears that general globalization trend in this area, i.e. wide acceptance of e-Government initiative was not influenced by any cultural dimension. Similar result was obtained for HCI, i.e. based on its regression model HCI was found not to be related to any cultural dimension.

Results of the regression model for TII are comparable with the results from Bagchi, Cerveny, Hart & Peterson (2003) and confirm their findings. They have tested the impact of national culture on adoption of six information technologies. Since the TTI is a synthetic index composed from six primary indices similar to their six information technologies, it would be expected to get similar results for the synthetic indicator as they got for each individual indicator. In their case all cultural variables were significant with the same sign, though not the same cultural variables in each regression model for an individual IT. Also the coefficient of determination was in the same range as in their study.

For the overall e-Government readiness, measured by eGOV the regression model suggests that IDV and GDP are significant variables. Contributions of cultural dimensions to variation in e-Government readiness regression models are from 3.2% to 7.2% (last column in Table 5).

Returning back to four postulated hypotheses it could be said that the hypothesis H1 was weakly supported, i.e. the government of a country with the larger power distance do have a negative attitude toward increasing the level of e-Government readiness. Hypothesis H2 was moderately supported, i.e. the government of a country with a strong individualistic culture does have a positive attitude toward increasing the level of e-Government readiness. However, hypothesis H3 was not supported, i.e. masculinity of the country has nothing to do with the attitude toward increasing the level of e-Government readiness. The same conclusion was reached in case of hypothesis H4, i.e. the uncertainty avoidance culture has nothing to do with the attitude toward increasing the level of e-Government readiness.

CONCLUDING REMARKS AND LIMITATIONS

This study is designed to examine whether differences in worldwide e-Government readiness levels are explained by cultural variables. Our results give some support to this statement. Based on results from Table 3 - Table 5 it was found that national cultural indicators have a moderate impact on the e-Government readiness worldwide. Among four cultural dimensions Individualism and Power Distance are the only significant variables that could be used to explain differences in level of e-Government readiness.

This study has implications both for practice and for theory. It shows that cultural variables are relevant to the worldwide e-Government readiness. Indeed, the empirical analysis found that the model with both economic and cultural variables explains between 40% and 83% of the variability in e-Government readiness indices (cultural variables alone contributed from 3.2% to 7.2%). Among cultural variables in regression models which include GDP only IDV and PDI seem to be significant in some models. From regression models for WMI and HCI it follows that general aptitude of government to employ e-Government via governmental websites and that general aptitude of government to increase value of the human capital were not influenced by any cultural dimension. Results of the regression model for TII confirm that cultural variables such as PDI and IDV were significant.

In addition to this empirical finding the study also has implication for diffusion theory, or adoption of a new technology theory. Empirical results justify the inclusion of cultural variables and demonstrate the need to broaden the adoption of a new technology theory in the area of the influence of social norms.

Bridges.org (2001) suggests: "… the unique cultural and historical environment of a region must be taken into account as part of a national ICT policy to truly gauge the country's e-readiness for the future."

This study is subject to certain reservations. National culture constructs were derived from the Hofstede cultural model. Since there are other cultural models it would be necessary to check whether cultural constructs based on an alternative theory of culture to Hofstede's theory confirm the impact that national culture has on the e-Government readiness. Also, alternative definitions and indicators of e-Government readiness should be used to see how robust the results in this study are, where the e-Government readiness is based on the United Nations definition. Finally, other factors such as culture of government and ruling system should be considered in the regression model as they might have impact on the e-Government readiness.

ACKNOWLEDGMENT

This article is a revised version of the paper "The impact of national culture on worldwide e-government readiness" published in the *Informing Science Journal, 8*, 143-158. I take this opportunity to thank John Green, Senior Lecturer in the School of Information And Social Sciences at the Open Polytechnic of New Zealand for valuable comments on the first draft of this paper. However, the author should be held responsible for any remaining errors.

REFERENCES

Altman, D. (2002). Prospects for e-government in Latin America: Satisfaction with democracy, social accountability, and direct democracy. *International Review of Public Administration*, 7(2), 5–20.

Bagchi, K., Cerveny, R., Hart, P., & Peterson, M. (2003). The influence of national culture in information technology product adoption. In *Proceedings of the Ninth Americas Conference on Information Systems* (pp. 957-965).

Bierens, H. J. (2004). *EasyReg International*. University Park, PA: Department of Economics, Pennsylvania State University.

Bretschneider, S., Gant, J., & Ahn, M. (2003, October 9-11). A general model of e-government service adoption: Empirical exploration. *Public Management Research Conference*, Georgetown Public Policy Institute Washington, D.C. Retrieved May 10, 2004, from http://www.pmranet.org/conferences/georgetownpapers/Bretschneider.pdf

Bridges.org. (2001). *Comparison of e-Readiness assessment models*. Retrieved April 1, 2004 from http://www.bridges.org/e-Readiness/report.html

Bui, T. X., Sankaran, S., & Sebastian, I. M. (2003). A framework for measuring national e-readiness. *International Journal of Electronic Business*, *1*(1), 3–22. doi:10.1504/IJEB.2003.002162

Chanchani, S., & Theivanathampillai, P. (2002). Typologies of culture. University of Otago, *Department of Accountancy and Business Law Working Papers Series, 04_10/02*. Dunedin: University of Otago.

Choucri, N., Maugis, V., Madnick, S., & Siegel, M. (2003). Global e-Readiness – for what? *MIT Sloan School of Management Research Paper 177*.

CIA. (2003). *The World Fact book*. Retrieved April 1, 2004, from http://www.cia.gov/cia/publications/factbook

DiMaggio, P., Hargittai, P., Neuman, E., Robinson, W. R., & John, P. (2001). Social implications of the Internet. *Annual Review of Sociology*, *72*, 307–336. doi:10.1146/annurev.soc.27.1.307

Edmiston, K. D. (2003). State and local e-Government: Prospects and challenges. *American Review of Public Administration, 33*(1), 20–45. doi:10.1177/0275074002250255

Fang, Z. (2002). E-Government in digital era: Concept, practice, and development. *International Journal of the Computer, the Internet and Management, 10*(2), 1-22.

Fountain, J. E. (2001). The virtual state: Transforming American government? *National Civic Review, 90*(3), 241–251. doi:10.1002/ncr.90305

Fountain, J. E. (2004). Digital government and public health. *Preventing chronic disease – Public Health Research. Practice, and Policy, 1*(4), 1–5.

Graafland-Essers, I., & Ettedgui, E. (2003). Benchmarking e-government in Europe and the US. *RAND, MR-1733-EC, 2003*. Retrieved May 10, 2004, from http://www.rand.org/publications/MR/MR1733/MR1733.pdf

Grönlund, Å. (2003). Emerging electronic infrastructures: Exploring democratic components. *Social Science Computer Review, 21*(1), 55–72. doi:10.1177/0894439302238971

Hargittai, E. (1999). Weaving the Western web: Explaining differences in Internet connectivity among OECD countries. *Telecommunications Policy, 23*(10/11).

Hofstede, G. (1980). *Culture's consequences: International differences in work-related values*. Beverly Hills, California: Sage Publications.

Hofstede, G. (1981). Culture and organizations. *International Studies of Management and Organization, 10*(4), 15–41.

Hofstede, G. (1983). National cultures in four dimensions – A research-based theory of cultural differences among nations. *International Studies of Management and Organization, 13*(1-2), 46–74.

Hofstede, G. (2004). *Geert Hofstede cultural dimensions*. Retrieved November 19, 2004, from http://www.geert-hofstede.com/hofstede_dimensions.php

Holden, S. H., Norris, D. F., & Fletcher, P. D. (2003). Electronic government at the grass roots: Contemporary evidence and future trends. In *Proceedings of the 36th Hawaii International Conference on System Sciences*. Big Island, Hawaii, January 06 - 09, 2003. Retrieved May 10, 2004, from http://csdl.computer.org/comp/proceedings/hicss/2003/1874/05/187450134c.pdf

Johns, S. K., Smith, M., & Strand, C. A. (2003). How culture affects the use of information technology. *Accounting Forum, 27*(1), 84–109. doi:10.1111/1467-6303.00097

Kaufmann, D., Kraay, A., & Mastruzzi, M. (2008). *Governance matters VII: Governance indicators for 1996-2007*. World Bank Policy Research June 2008

Kim, K.-J., & Bonk, C. J. (2002). Cross-cultural comparisons of online collaboration. *Journal of Computer-Mediated Communication, 8*(1). Retrieved April 1, 2004, from http://www.ascusc.org/jcmc/vol8/issue1/kimandbonk.html

Kirkman, G. S., Osorio, C. A., & Sachs, J. D. (2002). The network readiness index: Measuring the preparedness of nations for the networked world. In Dutta, S., Lanvin, B., & Paua, F. (Eds.), *The global information technology report 2001 – 2002: Readiness for the networked world* (pp. 10–29). New York: Oxford University Press.

Kovačić, Z. (2005). A brave new eWorld? An exploratory analysis of worldwide e-Government readiness, level of democracy, corruption and globalization. *International Journal of Electronic Government Research, 1*(3), 15–32. doi:10.4018/jegr.2005070102

Maitland, C. F., & Bauer, J. M. (2001). National level culture and global diffusion: the case of the Internet. In Ess, C. (Ed.), *Culture, technology, communication: towards an intercultural global village* (pp. 87–128). Albany, NY: State University of New York Press.

McSweeney, B. (2002). Hofstede's model of national cultural differences and their consequences: A triumph of faith – a failure of analysis. *Human Relations*, *55*(1), 89–118.

Okot-Uma, W'O R. (2004). *Building cyberlaw capacity for eGovernance: Technology perspectives*. The Commonwealth Centre for e-Governance, London, United Kingdom.

Pardo, T. A. (2000, October). Realizing the promise of digital government: It's more than building a web site. *IMP Magazine*.

Robinson, K. K., & Crenshaw, E. M. (1999). *Cyber-space and post-industrial transformations: A cross-national analysis of Internet development*. Retrieved May 10, 2004, from http://www.soc.sbs.ohio-state.edu/emc/RobisonCrenshawCyber1a.pdf

Slack, J. D., & Wise, J. M. (2002). Cultural studies and technology. In Livingstone, S., & Lievrouw, L. (Eds.), *Handbook of new media* (pp. 221–235). London: Sage.

Sørnes, J.-O., Stephens, K. K., Sætre, A. S., & Browning, L. D. (2004). The reflexivity between ICTs and business culture: Applying Hofstede's theory to compare Norway and the United States. *Informing Science Journal, 7*. Retrieved May 10, 2004, from http://inform.nu/Articles/Vol7/v7p001-030-211.pdf

United Nations. (2008). *UN e-government survey 2008: From e-government to connected governance*. Retrieved August 25, 2008, from http://unpan1.un.org/intradoc/groups/public/documents/un/unpan028607.pdf

United Nations, Division for Public Economics and Public Administration and American Society for Public Administration. (2001). *Benchmarking e-government: A global perspective --- Assessing the progress of the UN member states*. Retrieved May 10, 2004, from http://www.unpan.org/e-government/Benchmarking%20E-gov%202001.pdf

Veiga, J. F., Floyd, S., & Dechant, K. (2001). Towards modeling the effects of national culture on IT implementation and acceptance. *Journal of Information Technology*, *16*(3), 145–158. doi:10.1080/02683960110063654

West, D. M. (2008a). *Improving technology utilization in electronic government arount the world, 2008*. Governance Studies at Brookings. Retrieved August 24, 2008, from http://www.brookings.edu/~/media/Files/rc/reports/2008/0817_egovernment_west/0817_egovernment_west.pdf

West, D. M. (2008b). *State and federal electronic government in the United States, 2008*. Governance Studies at Brookings. Retrieved August 24, 2008, from http://www.brookings.edu/~/media/Files/rc/reports/2008/0826_egovernment_west/0826_egovernment_west.pdf.

This work was previously published in Information Communication Technologies and Human Development, Volume 1, Issue 2, edited by S. Chhabra and H. Rahman, pp. 77-93, copyright 2009 by IGI Publishing (an imprint of IGI Global)

Chapter 10
Socio-Cultural Context of E-Government Readiness

Lech W. Zacher
Kozminski University, Poland

ABSTRACT

E-government is not only an innovative idea but, more and more in a growing number of countries, a practical activity of high priority. It reflects the emergence and development of information societies (IS). Socio-cultural context is a framework of e-government strategies and practices. The context will determine the effects of such efforts. It is important to consider and understand the socio-cultural characteristics and functioning of society while its e-government undertakings are planned and introduced. From this point of view, the presently emerging worldwide information societies can be grouped in classes. It may help to analyze the classes' needs and possibilities and to formulate proper e-government agenda to be implemented. The real specificities and diversities among classes make the IS development multi-trajectory. In our diversified world, the effects will vary greatly.

INTRODUCTION

E-government readiness has several dimensions. Readiness on the side of government depends on government's *propensity* to use information-communication technologies (ICTs) to exchange information and provide services to citizens and business. Such propensity, which is psychological, political, social and cultural in character, varies from society to society. Psychological, political, social and cultural factors and conditions can act in positive or negative directions with regard to e-readiness. Either way, it can be assumed that governments play the role of enlightened leaders and are under the influence of the external world (that is, the networked world). However, governments may strive for democratic governance or prefer tough rules based on control of people,

DOI: 10.4018/978-1-60960-497-4.ch010

surveillance, and manipulation. Moreover, they can be effective in both cases. ICTs can be used in both ways.

So the historical heritage, norms and values, social and religious customs and attitudes, orientations (e.g., proactive, future-oriented, openness), social aspirations and national ambitions, social structure, level and span of education, competences in administration and business, technological culture, political and legal systems, media status, advancement of civil society, relations with other countries etc. are *non-technological determinants* of e-government readiness of all its stakeholders – public administration, business and civil society. These factors constitute the socio-cultural context of e-readiness of both government and society. It is worth noting that business has an economic motivation and arguably a more "natural" technology-driven propensity to use ICTs.

To make government and social e-readiness work properly and effectively, all factors and their feedbacks should be considered by *all stakeholders*, including central and local public authorities, business, and NGOs, while the stakeholders develop strategies, plans, and policies and while they react to various pressures and challenges imposed by ICTs and globalization. Socio-cultural context ought to be treated dynamically—as changing and as creating potential for change. Unfortunately, quite often this context is overlooked or underestimated what diminishes possible advantages and positive effects for all sides involved in ICTs applications and diffusion in all spheres of social activities and life.

The fundamental component of e-readiness is *technology*—computers, telecommunications infrastructure, and ICT organization and management. Tools and techniques of e-government are connected with the use of some hardware, software and orgware. The latter has evident political and cultural dimensions. Security and privacy standards also have legal dimensions.

E-government is nowadays necessary in order to take advantage of ICTs for better public administration, for more effective business performance, and for citizen activism. However, it is costly. After the initial period, e-government functioning should be measured and evaluated from the point of view of all stakeholders. Moreover, failures, incompatibilities, negative side effects, and exclusions should be also identified and recognized, and strategies and policies toward diminishing or liquidating the unintended effects elaborated and implemented.

Media and *educational institutions* (being in fact the components of socio-cultural context) can play an important role in ICTs assimilation and use in the public and private sectors.

BACKGROUND

Selected Approaches and Cultural Challenges

The discourse on the socio-cultural context of e-readiness is a rather new area of research. From the point of view of the history of science, it can be located close to such themes as technology and culture, technology and society, STS studies, or social assessment of technology. Without the technological component e-government would not exist. However, more types of innovation are necessary to make it work. It is worthwhile to note that technology also emerges in certain cultural, social, economic, and international settings. This is the first socio-cultural condition of technology. The second lies in technology's social use (i.e., for education, training, strategies, policies, building infrastructure, legal framework). The third condition is connected with the social response (will, propensity, attitudes, expectations, competencies).

The analysis of social contextual aspects of e-government should not be ignored by policymakers and developers. Social aspects of technology have been investigated from the point of view of already classic sociological theories like structuration theory, actor-network theory, social construction of technology, strategic choice approach and so on, not to mention more philosophically oriented approaches concerning man – technology relationship. However for the practical understanding of e-readiness issues and for the pragmatic goals (like developing strategies and undertaking deliberated actions) the best approach should be policy-oriented approach, not a general theoretical perspective. The transdisciplinary experience of STS studies can be helpful here, especially if linked to governance, public management and politics (see Dunleavy et al., 2006).

The socio-cultural conditions combine to create strong positive feedbacks. But the processes involved here are not totally planned or totally spontaneous, they are mixed and may be chaotic. Moreover, external factors and pressures (global competition, international trade, technology transfer, FDIs, migrations, brain drain) are influential as well. External influences—consumption patterns, attitudes, worldviews, behaviors presented by the global media—have distinctive cultural and socio-political characteristics. They can be disruptive for some aspects of local or national cultures, but may at the same time create or contribute to a *potential for change* and a more effective *cultural ability* to transform and reform. So the imitation or adoption of external influences, when reasonably implemented, may be instrumental for e-readiness undertakings.

To overcome the technocratic temptation, it is vital to underline that "culture matters" (Harrison & Huntington 2000), that cultural values powerfully shape social, political and economic performance, and therefore also shape e-government readiness. Cultural values determine individuals', governments', and business' thinking, decision making and actions. Culture shapes perceptions of technological change, risks, opportunities, and positive and detrimental effects. It forms as well the principles around which activity is organized and strategies and actions performed. Culture plays a significant role in shaping attitudes about work, trust, responsibility, authority, democracy, and visioning of the future. Religious beliefs can be considered part of culture.

There are progress-resistant cultures and progress-prone cultures (Lindsay in Harrison & Huntington, 2000, p. 284, see also Gascó, 2007). For better understanding of culture's role and its mechanism of influence, it is useful to refer to *mental models* (or mindsets). Mental models are the underlying beliefs that influence the way people behave. While culture is a broader macro-level variable, a mental model is a micro-level variable. Mental models apply to individuals and groups of individuals, and are identifiable and changeable. Culture reflects the aggregation of individuals' mental models. The two are linked in a perpetually evolving system (Lindsay, *ibidem*, p. 284). Mental models are differentiated in communities, companies, societies, and regions. Globalization is producing global-oriented mindsets. There is a difference between mental models in advanced countries and in LDCs. Important for e-readiness are the mental models of political leaders, government officials, business leaders, and NGO leaders.

Changing mental models, by the introduction of ICTs and their applications, can generate changes in the culture of a society. Those mental models which drive the strategic choices that are being made should become the focus for change efforts in the e-government area. There are many challenges for e-government utilizing ICTs and connected with cultural dimensions. For example, rather massive migration of people from LDCs (e.g., to Europe) requires a special tailoring of e-government information and services exchange. Not only do migrant inhabitants but also tourists in their masses need special ICT treatment. On the other hand, the globalization processes have cultural, or rather multicultural, dimensions. Culture is being changed globally, nationally, locally and on the level of individual mindsets (on cultural aspects of globalization see Appadurai, 1996, Rajaee, 2000).

Much research has been done on themes connected with culture, technology and organization interactions. For example, cultural politics of technology (Sørensen, 2004) versus normative politics of technology (Garrety & Badham, 2004) has been discussed. The former type of politics combines critical and constructive interventions; the latter is based on user-centered design. These approaches can be applied to ICTs in the context of e-government. Since ICTs are in fact science-based innovation, they can be an objective of innovation policy. Their creation, transfer and diffusion may capitalize on the idea of a triple helix of university, industry, and government relations (Etzkowitz, 2006).

Cultural values are present where the new centers of electronic and information technology were created. Interestingly, the communities involved in such processes (e.g., in Silicon Valley – Lécouyer, 2005) generated the dynamic pattern of knowledge sharing. So communities of learning, practice and collaboration developed the *culture of knowledge sharing*. Needless to say, such centers were and are at present an important source of technological and cultural innovations that can be helpful in e-readiness undertakings.

E-readiness in the networked environment, which is the growing density of relations between government, business and the public, needs a kind of intermediary. Information brokers can play such a role. The brokers link sources and users of new e-government ideas, concepts, actions and services. In fact, we can think in terms of information, knowledge and technology brokers. They are *ex definitione* innovation brokers (on innovation brokers as actors in social networks, see Winch & Courtney, 2007).

ICTs have both great dynamic and transformational powers so the process of their introduction into e-government activities should be smoother and more effective if some principles of so-called *constructive technology assessment* are adopted. However, the two-track approach (promotion and control of technologies separated) does not resolve problems of resistance or risks (*nota bene* the present societies are called *risk societies*).

It is important to understand that linear models are no longer relevant. The e-government model should not be "producer-centered" (sender sending coded messages to receiver who decodes them). Transfer of information and knowledge is not simple because coding and decoding is culturally embedded; moreover, both parties should be interested in the process and users should not be just reactive but interactive. It cannot be assumed that there is *a priori* consensus on values and goals. So the challenge is how to reshape, transform, put to use, and translate into a domestic setting this information, knowledge and ICTs' possibilities and programs. The concept of *domestication* can be applied (e.g., domestication of computers, the Internet, e-government information and services exchange). It is a kind of *cultural appropriation* (Sørensen, Aune & Hatling, 2000).

E-government activities are more successful in the *information and innovation intense environments*. Such special environments have been internationally known since the 1970s as industrial clusters, information or knowledge cities, technopolises, and science parks (Roberts, 2005). In other words, these factors constitute an *intelligent ambiance* which is very advantageous for stimulation of e-readiness. However, the networks of e-government have various densities, junctions, and operating speeds in different countries and regions.

E-readiness of all parties (government, business, citizens) depends greatly on their *understanding* and *trust* in science and technology, in their positive effects and in the possibility of avoiding serious detrimental impacts. Social assessment of e-government can be a part of social learning. A *reflexive version of social learning* (see Wynne, 1995) involving the systematic investigation and debate of ICT's role, impact, risks and transformational strength imposes adequate re-organization of society and relations between public administration, business and citizens. Under democracy, *social paternalism* (deciding from above what is good for society) cannot be accepted. The same can be said about the *technological expert* approach neglecting non-technological consequences.

E-readiness may be analyzed from various points of view reflecting *different rationalities* (sets of criteria of choice): political, technological, business, social. Technological commitments reflect social values, cultural identities, interests and incipient social orders. So reflexivity, social criticism and negotiatory discourse can be helpful to promote e-readiness and to make socially wise use of it. The users' involvement in ICT systems design and functioning can generate important knowledge but it should be rewarding for the users by offering a feeling of being in control and having certain advantages. Thus a more *participatory policy* approach is needed. A proper institutional framework for e-governance should be built on the basis of *communicative partnership* of all actors involved (i.e., scientific and technological, government, business, civil society – Burgess & Chilvers, 2006). In common opinion, e-governance is connected primarily with *participatory* or deliberative (or discursive) forms of *democracy*, which is an important value constituting a new socio-cultural context for decisions, activities and life.

The transformation from the traditional government structures and mechanisms to e-government is a complex process and very difficult from the beginning (especially for less advanced countries). It requires some policy and *transition management* (the term adopted from Elzen, Geels & Green, 2005) in which long-term targets, visions and transition agendas will be present. For example, in the European Union, there were general recommendations and directives in the Bangemann Report in 1994 and then in the Lisbon Strategy in 2000 (recently updated).

From the point of view of *social learning* strategies, it is interesting to use the concept of *bounded socio-technical experiments* (Elzen et al., 2005). To avoid risks, it is reasonable to introduce ICTs and services and social arrangements connected with them on a small scale. The results should diffuse on a larger scale what can lead to further experimentation, to a new applications, and single-loop learning (among the experiment actors and their immediate professional networks) and higher-order social learning (including society at large) (*ibidem*).

Aside from top-down policies, bottom-up initiatives are important for stimulating e-government readiness. Bottom-up initiatives express public interest in finding advantageous solutions. But all actors in the process should show good will to achieve consensus. The process is in fact of a civilizational character; that is, it epitomizes the co-evolution of technological and societal developments. Technological innovations and market-type instruments alone are not sufficient for the transition. Social, cultural and human capitals play no less significant roles.

Cross-cultural studies, investigations and comparisons can be insightful and practically helpful. Mobile phones, which have become the primary info-communication tool, are a good example. They are already *embedded* in the lives and activities of more than a billion people, not only young, but people of all professions and social strata. The mobiles are used similarly in different cultures (e.g., in Japan and Europe; see Ito, Okabe & Matsuda, 2005). However, there are indications that cultural heritage (e.g., in Africa) may influence mobile phone users' behavior. Still, the new technologies are actually focused on the developed world. It can be postulated then to design technologies, including mobile phones, important for e-readiness that satisfy culturally specific needs. This resembles somewhat the old idea of *appropriate technology* (originated in E. F. Schumacher's works in the 1960s. and 1970s). There are, however, opinions that we experience growing cultural divides and the need for *indigenous technology* appears again (especially in big, diversified countries like India) (see van Dijk, 2005).

Digital divides—on the local, national and global levels—constitute a serious problem when we debate e-government readiness. These divides mean the information exclusion of groups and individuals, countries and regions. There are national and international efforts to bridge these divides, but low-cost laptops and free software may be not enough. The learning capacity and the cultural ability to use the new opportunities can be lacking. Moreover, some studies show that the access to ICTs does not eliminate inequalities, injustice and poverty (see Eubanks, 2007). They are not only the distributional problems.

The development of the techno-landscape and electronic gadgetry can positively influence people's understanding of technology and facilitate its use. However, the immense diversification of the world and of societies requires that a kind of special approach be taken for less advanced countries and regions (see Ali & Bailur, 2007). More attention should be given to the multifaceted impacts of ICTs on humans (Bradley, 2006). It may stimulate e-readiness too.

E-exclusion and the digital divides can also be caused by the growing differences in skills access and usage access. Users with only basic computer literacy and basic, not high-speed, Internet connections will be left far behind; the same concerns apply for elderly and disabled persons. This disparity may greatly affect e-readiness and e-government effectiveness. Can the new generations ("computer children," network generation) overtake the growing share of ageing and disabled persons?

The worst case for e-government readiness would be the significant resistance to ICTs because of lack of trust, computer crimes, overtechnization of life, or computerphobia because of widespread electronic surveillance (see special issue of International Sociology, 2004). Some Neo-Luddism is still possible (Jones, 2006).

CONSIDERING SOCIO-CULTURAL CONTEXT IN E-GOVERNMENT DEVELOPMENT

The Timely Discourse

Any formulation of ICT strategies for developing efficient and effective government systems, for building appropriate institutions and for the empowerment of communities requires a solid recognition of the socio-cultural situation and trends. The proper reference point is the broader framework, i.e., information society (or e-society).

Assumptions and Issues

For better understanding of the problematique, it is necessary to make some general explanations.

The e-government idea and its practical applications are often viewed in a rather abstract way. Therefore, some silently accepted assumptions should be disclosed and considered in this discourse:

- E-government is an integral part (a set of institutions and activities) of an information society (IS);
- The more developed the IS, the more chances for e-government's broad and effective use;
- E-readiness cannot be separated from a given (investigated) IS or its class and its stage of development;
- An important characteristic of an IS is always its socio-cultural context. This context should always be considered both in analyzing e-readiness of a particular IS and in recommending formulation of ICT strategies and policies;

E-readiness has both internal (domestic) and external dimensions; the latter is connected with the globalization processes (making the networked world) and sometimes with regional integration processes (e.g., in the European Union).

Moreover, e-government as a concept has several definitions and interpretations (Heeks, 2005, Anttiroiko, Mälkiä, 2006, Zacher, 2007). It is connected with other terms such as *e-administration, e-democracy, e-governance, e-economy, e-business, e-society, e-world* and more. It is worthwhile to note that when prefacing a word with "e" we underline the technological aspect (electronic technology), while "i" (info) is often used as its substitute and signifies the more general aspect of information in its socio-cultural setting.

Some immediate provisional conclusions can be formulated:

- In information societies, there are complex and interactive *networks* of power (political and economic) and, connected with them, activities (like e-government, e-governance), so a particular IS should be investigated from the point of view of existing networks. (Such approaches will be helpful in this regard, such as the theory of networks or web theory; Wellman, 1999, 2001; Burnett & Marshall, 2003.);
- Some characteristics of IS development are especially important: general education, technical education, computer literacy, e-inclusion, societal prospective orientation, innovativeness, drive for new knowledge, legal and policy framework, strategies for ICTs development and use, empowerment of communities, potentials of technology, knowledge and skills transfer, international cooperation and exchange. There are measurable indicators for analyzing and forecasting these characteristics;
- Span and depth of e-government applications depend strongly on these indicators—precisely, on their values and levels;
- E-readiness is determined to a great extent by the stage of advancement of the IS and strategies of government, businesses and people (social, cultural and human capitals).

IS Classes: Diversities and Specificities, Some Possible Recommendations

The real information societies (in plural) are just emerging in recent decades, mostly in highly developed countries. These countries make up the highest class of ISs—they are *pioneers* in ICT production, diffusion and application. They are sometimes called *high-tech economies/societies*. They are technologically, economically, educationally, and culturally advanced. They are often rich and expansionistic, exporting not only technology and goods but also patterns of development (via media, tourism, migrations). But even these countries are still far from info- or e-maturity. The question emerges to what extent their patterns of ICT use are universal and applicable in effective ways in other countries. Opinions vary since the important factor here is the socio-cultural context, which is very different in the networked world. Even in the highest class of ISs, some regions, communities, and individuals (e.g., poor, unemployed, disabled) can be excluded from informational development and its advantages. That is why e-inclusion policies are needed. The good example is the EU initiative called European e-Inclusion Initiative (Lisbon, 3 Dec. 2007 – see http://ec.europa.eu/informationsociety/einclusion, also www.epractice.eu).

The second class of information societies is in fact somewhat divergent. Arbitrarily, one can separate the second class into one subclass—the *emerging powers* (like China and India)—and a second subclass—the *transitional countries* (so called "post-Communist" countries, mostly in Eastern Europe, recently introducing and stimulating the market economy and joining the EU). These two subclasses differ greatly, yet their similarity lies in some of their problems. The first subclass has started in fact from the position of LDCs, having all of the problems and barriers typical for LDCs like poverty, illiteracy, big traditional agriculture, low technology, unemployment, cultural backwardness. In other words, their *social* and *cultural capitals* were rare resources. However, due to economic dynamics and technological progress (from imitation to their own original creations, e.g., computer software, techno-service, highly skilled personnel, science and technology parks), they are becoming more and more world players. Nevertheless, they still experience substandard living conditions, economic dualism (only enclaves of high tech), informational exclusion of significant segments of their (immense, in the cases of China and India) societies, and brain drain.

The problem of the emerging powers is how to harness ICTs into processes of *development of masses of people* and how to introduce e-government on such scale. E-readiness may not be difficult at the level of regions, communities, and individuals, thanks to infrastructure, access to the Internet and education (all stimulated by both state efforts and market forces), but the *immense scale* of these undertakings can make it difficult and generate specific problems not known in smaller economies and societies. What strategies and policies should be adopted in ICT production, diffusion and use, and—no less important—how can those strategies and policies be made really participative for the masses of people in order to diminish dualism, exclusion, and brain drain? It is necessary to consider the socio-cultural and political contexts and their transformations. Positive adaptation, creativity and synergy are badly needed. E-government, e-democracy, e-governance can be *goals* as well as *instruments* of change. However, when comparing the emerging powers with highly advanced countries, the difference is that the countries in question, which constitute a big part of the world and humankind, will have still serious problems with poverty, medical care, pension systems, overpopulation, unemployment, excessive armaments, and migration. Could the emerging powers imitate the trajectory of the pioneers or perhaps learn from the transition economies, or should they do it their own way (not easy in the age of globalization, TNCs activities, hegemonic policies of the old powers)?

Transitional countries have their specific transitory problems but they are definitely not LDCs. They usually have a fairly good education level and are culturally developed, with more modern attitudes and some technological culture in society. They have quite good general infrastructure and do not occupy as much land mass as the emerging powers. Their political and administrative institutions function well. What they lack is sufficient expenditures for research and development (R+D) and proper R+D strategies. This is, however, partly substituted by R+D Framework Programs of the EU. Moreover, they have problems with the transfer of innovation from the R+D sphere to industry and to social life.

Transitional countries are subject to brain drain and excessive migration (e.g., from Poland). However, they capitalize on European integration (cooperation programs, aid funds). Their businesses and bureaucracies are not fully ready for complex electronization, computerization, informatization. Barriers are plentiful: insufficient technical infrastructure, high cost of access to the Internet, traditional organization (hierarchical and bureaucratic) of decision-making processes, often poor management, and psychological obstacles in the case of less educated, poor, provincial people. Intellectual and cultural elites are good, political elites not; the rest of society is in the difficult process of transition to a market economy and democratic system. E-government is perceived more as a tool of politics than an instrument enhancing dialogue between authorities and citizens, so stimulating people's participation and activism is rather difficult (NGOs are relatively new and have poor financing). Moreover, many legal regulations are counter-effective to e-government or e-governance (e-democracy included). The vast sector of shadow (informal) economy and corrupt networks are not conducive to e-undertakings in administration, politics, business, and finances.

Transition economies/societies have some problems that are *similar* in nature to the emerging powers and LDCs and that contribute negatively to e-readiness; they are, however, of a much smaller scale. Nevertheless, these countries can take advantage of integration and of participation to a growing extent in international networks of research, innovation, exchange of experts and the like. Their socio-cultural setting is rather conducive to e-change (besides the fact that some societal segments are excluded). These countries try to orient (or re-orient, rather) on strategies and patterns of more developed countries. However, the difference and distance is significant so the simple imitation does not seem possible or feasible. Such countries ought to improve their technical infrastructure, make access to the Internet easier and cheaper, stimulate info-building capacities, re-orient legal regulations and attitudes of authorities on all levels (from central to local), make the people and their NGOs more active and participative. All this requires some *social innovations*—new procedures, new organizational and institutional structures, some regulatory efforts, more responsiveness and friendly attitudes of public authorities, more e-inclusive policies more social networks (computer-aided) and commitment, more e-trust and e-democracy (Zacher, 2001), not only electoral but also participatory.

The next class of information societies are countries which are *less advanced* (e.g., many in Latin America and Africa) and not developing sufficiently in economic, technologic, education, social, or cultural spheres. Do they have chances for a really "indigenous" development of an IS? Globalization—technology transfer, international trade, FDIs, technological cooperation and aid, impacts of global networks, potentials of Internet use—pushes toward *e-development*, at best in enclaves. Such countries should not be passive observers of the globalization and informatization of the world. They should have *proactive* policies and should elaborate comprehensive strategies to participate in the contemporary civilizational development. Societal activism is necessary for e-government readiness in such countries (see e.g. Rahman, 2007, Mutula & Wamukoya, 2007).

What seems to be the most important and necessary is the *comprehensive approach* to ICT development, to economic and socio-cultural dimensions of ICT diffusion and their massive use. Internal conditions and factors ought to be identified, forecasted and evaluated and a *model of change* should be elaborated for policy needs. Each society's specificities, negative and positive, should be analyzed in depth. External factors are also to be considered. Experiences of other countries (especially on similar stage) may help greatly. The countries may also capitalize on international institutions. The question for overall strategy (of government, business and citizens and their organizations) is how to *connect* in a synergetic way some domestic ICT enclaves with external factors and nets and to stimulate socio-cultural change which is at the same time *conditio sine qua non* of success. So it is not a problem of ICTs *per se* but of connectivity and of making relevant the socio-cultural context. Market mechanisms ought to be supported by a variety of government policies, by a learning propensity, and by properly profiled activities of citizens (education, consumption patterns, learning from media, NGOs). Both supply and demand sides are important. Simplistic imitation of the pioneering countries is not recommended.

To take advantage of ICTs as an engine of human development is difficult in less advanced countries because there is a mutual conditioning of technology use and of cultural progress. Persons are both the agents and beneficiaries of development, provided they are empowered, able to participate, and feel included. Unfortunately, in many cultures the model of family life is often male-dominated and authoritarian, excluding women and children from the modern pattern of progress. For many religious traditionalists, such model is mandated by religion. A new *mindset* is needed as well as new public family *support systems* to change the unjust tradition and overcome practically the barriers in education, work and public activities of women, children and youth.

It is, thus, possible to talk about *cultural access* to ICT. On a societal level, such access requires the acceptance of other traditions and cross-cultural harmony, also in many cases ethnic peace-building. In societies of ethnic, religious, racial and social tensions and conflicts, e-government readiness is difficult to promote and stimulate. Moreover, e-literacy can be used for fights and terrorism. Positive values should not be forgotten in education, public policy and media.

It is often pointed out that building modern ICT infrastructure is not enough (e.g., in a society of castes or ruled by religious fundamentalists). "Physical" use of ITCs should be accompanied by cultural, psychological, even spiritual transformations. Moreover, technological appliances should be properly tailored and appropriately used (this is reminiscent of the E. F. Schumacher concept of intermediate or appropriate technology which can be creatively used in the present). A good example, elaborated in India by Cisco, is the so-called information kiosk, a wheeled box with computer and wireless Internet access. It is designed to be used even by illiterate persons. Another promising "technological fix" is a mobile phone (about 140 millions in India) that is radically changing communication possibilities. "A $100 computer" is one of the Global Millennium Goals initiated by N. Negroponte from MIT. To make the computer common and readily available, the Indian government finances research on a $10 laptop for children (there are similar initiatives by private companies in the hardware and software areas).

The supply side of ICTs is necessary to stimulate social and cultural change. However, *techno-solutions* will not automatically achieve success without costs and negative side effects. But is there a better way? It is not the philosophical recognition of technological determinism but the acceptance and practical utilization of *technologically driven dynamic forces of change*. Moreover, it seems that technological systems and human (social) systems develop asymmetrically throughout history. From the middle of the last century (this time point coincidences with J.D. Bernal's concept of the beginning of the scientific and technological revolution), the leading force has been technology and its systems. So the technologization of politics, government, and democracy is the present civilizational trend. Of course we can postulate that socio-cultural contexts should make it more humane. Moreover, we can strive for this in order to come closer to an equilibrium between power of technology and power of culture and people. Such equilibrium will be beneficial for e-government readiness.

CONCLUSION

The socio-cultural context of development determines to a great extent its directions, dimensions and effects. The same refers to e-government development, of which a crucial part is the e-readiness of all parties involved in the process, including public administration, business, citizens and their organizations. All conditions of development, including material and technological, constitute a potential for change. The important component of this potential is culture in a broad sense, including for example gender ideal, aesthetic patterns, understanding of justice, beneficence, privacy, security, autonomy, corruption, liberty, duty, sacrifice, loyalty, sanctity, beliefs, trust, cooperation, competition, innovation, and future.

To make culture conducive to e-development, some *cultural ability* is needed (i.e., innovativeness, future orientation, entrepreneurship, proactivity, efficacy). Both the potential for change and the cultural ability to capitalize on ICT development and applications vary in communities, societies, countries and regions.

The process of e-development, and e-government development in particular, is very complex. It is determined by the aforementioned conditions (potentials, capacities) but it can be to some extent modified, changed, or improved by deliberate strategies, policies, actions and behaviors of all stakeholders of the process. Positive feedback and learning contribute to final success. Undoubtedly, mental models of individuals and groups and the capacity to learn define the future changes. Social learning capacities are based on social, cultural and human capitals. In the emerging information societies (e-societies), in spite of diversities there are some common propensities and aspirations to capitalize on the development occasions resulting from present technological and economic trends (e.g., globalization).

The transformational potential of ICTs is enormous. The same may be said of the emerging and possible effects and impacts, also detrimental. Because of the diversity of societies and countries, the situation is multi-optional and the future is open-ended.

In addition, the agents of change—actors, subjects, stakeholders—are diverse. They use various types of rationalities: technological, economic, environmental, political, social or utilitarian vs. moral. Information societies are now driven more by technology than by culture and social choice. Hence, there are quite distinct tendencies toward technocracy, meritocracy, netocracy as forms of control and governance.

To summarize:

- E-government is an important dimension (and factor) of IS;
- ISs emerging at present are developing according to multi-trajectory patterns;
- ISs are widely differentiated as to scale, span and effectiveness and impacts of ICTs used;
- A factor (and set of conditions at the same time) greatly determining IS "shape" is socio-cultural context (often overlooked or underestimated);
- To change this context is difficult since it cannot be just bought or imported; moreover, it has some potential both to create and adapt to new technologies and also to restrain or even refuse them;
- A list of quantitative and qualitative indicators can be set up to identify potentials and gaps of a society trying to capitalize on the use of ICTs; so no one, universal, "good for all" strategy and policy can be recommended. Specificity of the country ought to be seriously considered. (This is sometimes difficult for foreign experts, and also for domestic experts not sufficiently experienced, often educated abroad and too ambitious and optimistic.) Some exemplary indicators are: social and psychological propensity to novelty and change, to risk and challenge; level of education; level of technological culture and technical skills; willingness to imitate and adopt patterns of development of advanced countries; social customs and habits stimulating (or not) entrepreneurship and technology use; institutional structures of government (all levels); public activism and potentials of NGOs; business openness to consider all central and local stakeholders (also social). It is worth noting that positive feedback exists between the introduction of ICTs (and e-government undertakings) and change in socio-cultural context;
- At least two kinds of general strategies should be elaborated to stimulate e-government readiness: (1) strategy directed toward technology *per se* (in connection with financing, management, organizational infrastructure), and (2) strategy directed toward improving socio-cultural context and to make it advantageous to ICT diffusion, effective use, absorption;

- Not only is the introduction of e-government technologies and procedures important, but so too is making them work properly (which is not merely technical matter); impact assessment methods (e.g., anticipatory social impact assessment) should be used to make it possible. In government, strong subsidiary attitudes are necessary (not easy at all), some de-bureaucratization and transparency in decision-making processes, as well as openness to citizens' demands and requests. Conversely, citizens should be apt to use ICTs and be publicly more active and participative (a condition very difficult in many countries having weak democratic tradition and authoritarian rulers). The next step, in the not-too-distant future, will be to take care of improving economic effectiveness of e-government;
- Also external factors (integration, globalization, international trade, cooperation, joint actions, global media) will impose introduction and acceleration of e-government structures, mechanisms and procedures. More and more, the national e-governments will be a part of international and global networks of power, politics, information, dialogue with citizens (netizens), and public activism. Of course, in some cases in some regions other trends are possible as well. For example, a kind of Orwellian scenario would mean e-government better invigilating, controlling and censoring citizens and their activities. Much will depend on generations not yet born.

FUTURE RESEARCH DIRECTIONS

There are already well-established research agendas concerning the information civilization, information society, e-society, knowledge-based society, and of course e-government problematique (see e.g., IGI Encyclopedia of Digital Government and other series of this publisher).

The most important task is to make e-government activities a kind of *learning system*. E-government initiatives should thus be taken not only by the authorities, but also by citizens, the business sector, and the R&D sphere. They all may contribute to the e-government agenda which ought to be all increasing one. Moreover, e-government's practices, methods, opportunities and barriers should be the subject of multidisciplinary research evaluating the intersection of theory, method and empirical findings.

E-government agendas in general are well known since they are often the subjects of conferences and public and political debates. The following list exemplifies issues considered in e-government agendas:

- Strategies for development, implementation, maintenance and best practices;
- ICT investment planning and decision-making, including models and financing;
- Transforming inter- and intragovernmental information systems, integration of systems, reengineering administrative processes;
- Assessing risks, possible failures, costs and benefits, and barriers limiting e-government initiatives and undertakings;
- Models, frameworks and implementation guidelines;
- E-government business models (economic effectiveness);
- Public procurement and e-government;
- Redesigning linkages and cooperation within and between government agencies;
- E-democracy and e-governance (citizen participation);
- E-government and social exclusion, especially in regard to unemployed or disabled persons, women, and minorities;

- E-government discourse and implementation, and the role of the media (public in particular);
- Other social, political and cultural issues in e-government.

The list clearly shows that e-government initiatives and their implementations are *de facto* transforming government and transforming the people, processes and policies involved. The complexity of issues and problems must be recognized by all actors. This recognition is especially important in less advanced countries, though their map of problems will be unique. It is worthwhile to focus research on one, usually underestimated factor of change and development. It is in its nature *generational*. The subsequent generations of people are better and better educated especially in the areas of computers, the Internet, and other technological objects and processes. Symbolically, one can talk about a "books, press and films generation," then "TV (media) generations," and presently "computer (or Internet) generations." These terms illustrate civilizational way of people from a "Gutenberg Galaxy" to "Internet Galaxy". So to imagine now what the future behavior of the next generations will be is difficult and risky. It is also conceivable that a kind of rule of *generational acceleration* may emerge. If so, there is the chance for faster development but also a danger of a *new digital divide* ("worlds of two speeds"). However, the chance can mean that further progress is achieved as a result of the higher level of computer literacy and use, ICT infrastructure, and the experience of living from birth in an emerging IS. G. Small supposes that the *digital natives* (already born and raised in the digital environment) will have new abilities to live easily and creatively in such setting (perhaps this can be even genetically imprinted – Small, 2008). This means a new possible level of e-readiness, but also – new digital divides.

REFERENCES

Ali, M., & Bailur, S. (2007). The Challenge of Sustainability in ICT4D – Is Bricolage the Answer? *Proceedings of the 9th International Conference on Social Implications of Computers in Developing Countries*, Sao Paulo.

Anttiroiko, A.-V., & Mälkiä, M. (2007). *Encyclopedia of Digital Government*. Hershey, PA: Idea Group Inc.

Appadurai, A. (1996). *Modernity at Large. Cultural Dimensions of Globalization*. Minneapolis: University of Minnesota.

Bradley, G. (2006). *Social and Community Informatics: Humans on the Net*. London: Routledge.

Burgess, J., & Chilvers, J. (2006). Upping the ante: a conceptual framework for designing and evaluating participatory technology assessment. *Science & Public Policy*, *33*(10), 713–728. doi:10.3152/147154306781778551

Burnett, R., & Marshall, P. D. (2003). *Web Theory – An Introduction*. London, New York: Routledge.

Dijk, van J.A.G.M. (2005). *The Deepening Divide. Inequality in the Information Society*. London: Sage.

Dunleavy, P. (2006). *Digital era governance: IT corporations, the state and e-government*. Oxford: Oxford University Press.

Elzen, B., Geels, F. W., & Green, K. (Eds.). (2004). *System Innovation and the Transition to Sustainability – Theory, Evidence and Policy*. Aldershot: Edward Elgar.

Etzkowitz, H. (2006). The new visible hand: an assisted linear model of science and innovation policy. *Science & Public Policy*, *33*(5), 310–320. doi:10.3152/147154306781778911

Eubanks, V. (2007). Popular technology: exploring inequality in the information economy. *Science & Public Policy*, *34*(2), 127–138. doi:10.3152/030234207X193592

Garrety, K., & Badham, R. (2004). User-Centered Design and the Normative Politics of Technology. *Science, Technology & Human Values*, *29*(2), 191–212. doi:10.1177/0162243903261946

Gascó, M. (2007). *Civil Servants' Resistance toward E-Government Development*. In: Encyclopedia of Digital Government. Hershey, PA: Idea Group Inc.

Heeks, R. (2005). *Implementing and managing e-government*. London: Sage.

International Perspectives on Surveillance: Technology and Management of Risk, 2004, *International Sociology*, vol. 19, No. 2, June, 131-254.

Ito, M., Okabe, D., & Matsuda, M. (Eds.). (2005). *Personal, Portable and Pedestrian: Mobile Phones in Japanese Life*. Cambridge, Ma: MIT Press.

Jones, S. E. (2006). *Against Technology: From the Luddites to Neo-Luddism*. New York: Routledge.

Lécouyer, Ch. (2005). *Making Silicon Valley: Innovation and the Growth of High Tech, 1930-1970*. Cambridge, MA: MIT Press.

Mutula, S. M., & Wamukoya, J. (2007). *E-Government Readiness in East and Southern Africa*. In: Encyclopedia of Digital Government. Hershey, PA: Idea Group Inc.

Rahman, H. (2007). *Community-Based Information Networking in Developing Countries*. In: Encyclopedia of Digital Government. Hershey, PA: Idea Group Inc.

Rajaee, F. (2000). *Globalization on Trial. The Human Condition and the Information Civilization*. Ottawa: International Development Center.

Roberts, R. (2005). Issues in Modeling Innovation Intense Environments: The Importance of the Historical and Cultural Context. *Technology Analysis and Strategic Management*, *17*(4), 477–495. doi:10.1080/09537320500357384

Small, G. (2008). *iBrain: Surveying the Technological Alteration of the Modern Mind*. New York: HarperCollins.

Sørensen, K. H. (2004). Cultural Politics of Technology: Combining Critical and Constructive Interventions. *Science, Technology & Human Values*, *29*(2), 184–190. doi:10.1177/0162243903261944

Sørensen, K. H., Aune, M., & Hatling, M. (2000). *Against Linearity – On the Cultural Appropriation of Science and Technology*. In: M. Dierkes, C. von Grote (Eds.) (2000). Between Understanding and Trust – The Public, Science and Technology. Amsterdam: Harwood.

Taylor, P. A., & Harris, J. L. (2005). *Digital Matters – Theory and culture of the matrix*. London, New York: Routledge.

Wellman, B. (1999). *Networks in the global village*. Boulder, Col.: Westview Press.

Wellman, B. (2001). Physical place and cyberplace: the rise of networked individualism. *International Journal of Urban and Regional Research*, No. 1.

Winch, G. M., & Courtney, R. (2007). The Organization of Innovation Brokers: An International review. *Technology Analysis and Strategic Management*, *19*(6), doi:10.1080/09537320701711223

Wynne, B. (1995). Technology Assessment and Reflexive Social Learning: Observations from the Risk Field. In Rip, A., Misa, T. J., & Schot, J. (Eds.), *Managing Technology in Society – The Approach of Constructive Technology Assessment*. London, New York: Pinter.

Zacher, L. W. (2001). *Between Risk and Trust – Values, Rules and Behaviour in the E-Society*. In: Innovations for an e-Society – Challenges for Technology Assessment, Berlin (conf. proc.), ITAS – VDI.

Zacher, L. W. (2007). *E-Government in the Information Society*. In: Encyclopedia of Digital Government, vol. II. Hershey – London – Melbourne – Singapore: Idea Group Inc.

ADDITIONAL READING

Abbot, J. P. (2001). Democracy@internet.asia? The challenges go the emancipatory potential of the net: lessons from China and Malaysia. *Third World Quarterly*, *22*(2).

Browning, G. (1996). *Electronic Democracy: Using the Internet to Influence American Politics*. Wilton, CT: Pemberton Press.

Bugliarello, G. (1997). Telecommunities: The Next Civilization. *The Futurist*, 31.

Castells, M. (2000). *The Rise of the Network Society, 1* (2nd ed.). Oxford: Blackwell.

Castells, M. (2001). *The Internet Galaxy: Reflections on the Internet, Business and Society*. Oxford: Oxford University Press.

Currie, W. (2000). *The Global Information Society*. Chichester, New York: Wiley.

Davis, J. (Eds.). (1997). *Cutting Edge – Technology, Information Capitalism and Social Revolution*. London, New York: Verso.

Ester, P., & Vinken, H. (2003). Debating Civil Society: On the Fear for Civic Decline and Hope for the Internet Alternative. *International Sociology*, *18*(4). doi:10.1177/0268580903184002

Everard, J. (2000). *Virtual States – The Internet and the Boundaries of the Nation – State*. London, New York: Routledge.

Fisher, D., & Wright, L. (2001). On Utopias and Dystopias: Towards an Understanding of the Discourse Surrounding the Internet. *Journal of Computer-Mediated Communication*, *6*(2).

Florida, R. (2002). *The Rise of Creative Class*. New York: Basic Books.

Gassler, R. S. (2001). Globalization and the Information Economy. *Global Society*, *15*(1). doi:10.1080/13600820123743

Gray, Ch. H. (2001). *Cyborg Citizen*. New York: Routledge.

Heeks, R. B. (2001). *Reinventing Government in the Information Age*. London: Routledge.

James, J. (2001). Bridging the digital divide with low-cost information technologies. *Journal of Information Science*, *27*(4), 211–217. doi:10.1177/016555150102700403

Kamarck, E. C., & Nye, J. S. (Eds.). (1999). *democracy.com? Governance in a Networked World*, Hollis, NH: Hollis Publishing.

Kapucu, N. (2007). Ethics of Digital Government. In *Encyclopedia of Digital Government*. Hershey, PA: IGI Global.

Katz, J. (1997). The Digital Citizen. *Wired, 12*.

Kerckhove de. D. (1997). *Connected Intelligence – The Arrival of the Web Society*. Toronto: Somerville House.

Levinson, P. (2004). *Cellphone. The Story of the World's Most Mobile Medium and How It Has Transformed Everything*. New York: Palgrave Macmillan.

May, Ch. (2002). *The Information Society – a skeptical view*. Cambridge: Polity.

Negroponte, N. (1996). *Being Digital*. New York: Vintage Books.

Rheingold, H. (2002). *Smart Mobs – The Next Social Revolution – Transforming Culture and Communities in the Age of Instant Access*. Cambridge, MA: Basic Books.

Rifkin, J. (2000). *The Age of Access – The New Culture of Hypercapitalism Where All of Life is a Paid – for Experience*. New York: Jeremy P Tarcher/ Putnam.

Schiller, D. (1999). *Digital Capitalism*. Cambridge, Ma – London: The MIT Press.

Van de Donk, W. (2000). *Infocracy or Infopolis? Transparency, Autonomy and Democracy in an Information Age*. In J. Hoff, J. Horrocks. & P. Tops (Eds.), *Democratic Governments and New Technology*, New York: Routledge.

Wellman, B., & Haythornthwaite, K. (Eds.). (2002). *The Internet in Everyday Life*. Malden, Ma: Blackwell.

Zacher, L. W. (2000). The Way Towards a Knowledge Society – Some Barriers not only for Countries in Transition. In Banse, G. (Eds.), *Towards the Information Society*. Berlin, Heidelberg: Springer.

Zacher, L. W. (2007). E-transformations of Societies. In *Encyclopedia of digital government* (*Vol. 2*). Hershey, PA: Idea Group.

Zacher, L. W. (2009). Information Society Discourse. In *Encyclopedia of Information Science and Technology* (2nd ed.). Hershey, PA: IGI Global.

KEY TERMS AND DEFINITIONS

Cultural ability (capacity) to capitalize on ICT development and applications: refers to all IS stakeholders, to organizations, social groups, individuals; their innovativeness, entrepreneurship, level of proactivity, efficacy, future-orientation, aspirations, technological culture, educational advancement, managerial skills etc. constitute this ability.

E-inclusion: a set of policies and activities leading to "e-inclusive society" where every person has equal opportunities to participate, including those people who are physically, mentally, socially or economically disadvantaged.

E-Government: institutionalized practices and activities using ICTs to provide information and services by public administration to the society and to interact with various stakeholders (like business, NGOs), also within government itself.

E-government readiness: can be understood in two ways: as readiness of a government to use ICTs to exchange information and provide services to business and citizens; this readiness depends on pro-modern attitude, strive for efficiency, understanding of world trends, political will and also on availability of ICTs (including technical infrastructure, proper skills and administrative structures etc.);as a societal readiness – with the idea, procedures and mechanism of e-government broadly understood – to use ICTs in all spheres of social activities; this depends on an existing socio-cultural context (i.e. historical heritage, norms and values, social and religious customs, psychological attitudes and propensities, social aspiration and ambitious social structure, education, competences in public administration and business, technological culture, political and legal systems, media status, advancement of civil society, relations with other countries and so on).

E-readiness stakeholders: all subjects involved in the development of e-government, namely central and local public authorities, businesses, NGOs – together with their plans, strategies, policies, also attitudes, reactions and behavior.

E-Transformations of Societies: all social changes driven by various electronic devices and systems (mostly ICTs) used in all areas of human activities and life (including e-economy, e-banking, e-trade, e-media, e-government, e-democracy, e-health, e-learning etc.). E-transformations stimulate e-government readiness.

Information Society (IS): a society, which predominantly deals with production and applications of information in all fields of economy and social activities and human life. Many particular indicators are elaborated. It is assumed that the mass info-activities in IS are based on sufficient technical infrastructure, access, computer literacy, cultural capacity, efficiency etc. There are closely related terms as: information-rich society, cyber-society, e-society, network (or networked) society, virtual society, digital society, information society based on knowledge.

IS classes: the real world's societies are very diversified in terms of the advancement of IS characteristics and indicators. The most advanced countries create ICTs and use them widely and effectively. They are *pioneers* in ICTs production, diffusion and applications (they are often called *high-tech economies/societies*). The second class comprises two subclasses: the *emerging powers* and the *transitional countries*. The third class are countries which are *less advanced*, not participating sufficiently in e-development, if so – only in enclaves. They are subject both of digital exclusion and divide.

Multitrajectory IS development: the great diversity of determinants and conditions, also of strategies and policies of various information societies makes their development multi-optional and *de facto* multitrajectory; this diversity determines their differentiated progress on the way towards a "mature info-society".

This work was previously published in Information Communication Technologies and Human Development, Volume 1, Issue 2, edited by Susheel. Chhabra and Hakikur Rahman, pp. 94-109, copyright 2009 by IGI Publishing (an imprint of IGI Global)

Chapter 11
Developing Peer-to-Peer Supported Reflection as a Life-Long Learning Skill:
An Example from the Translation Classroom

Eva Lindgren
Umeå University, Sweden

Kirk P H Sullivan
Umeå University, Sweden

Huahui Zhao
Umeå University, Sweden

Mats Deutschmann
Umeå University, Sweden

Anders Steinvall
Umeå University, Sweden

ABSTRACT

Life-long learning skills have moved from being a side-affect of a formal education to skills that are explicitly trained during a university degree. In a case study a University class undertook a translation from Swedish to English in a keystroke logging environment and then replayed their translations in pairs while discussing their thought processes when undertaking the translations, and why they made particular choices and changes to their translations. Computer keystroke logging coupled with Peer-based intervention assisted the students in discussing how they worked with their translations, enabled them to see how their ideas relating to the translation developed as they worked with the text, develop reflection skills and learn from their peers. The process showed that Computer Keystroke logging coupled with Peer-based intervention has to potential to (1) support student reflection and discussion around their translation tasks, (2) enhance student motivation and enthusiasm for translation and (3) develop peer-to-peer supported reflection as a life-long learning skill.

DOI: 10.4018/978-1-60960-497-4.ch011

IINTRODUCTION

Lifelong learning and the importance of generic skills have gained a clearer position in higher education in Europe thanks to the Bologna Declaration (1999) in general and the Bergen Communiqué (2005) in particular as it "explicitly mentions the chance to further implement lifelong learning in higher education through qualification frameworks" (Jakobi & Rusconi, 2009: 52), and policy development in the European Union (see Dehmel, 2006 for a good overview of this policy development). What is worth mentioning here are the subtle changes in the definition of lifelong learning between 2000 and 2001 from "all purposeful learning activity, undertaken on an ongoing basis with the aim of improving knowledge, skills, and competence" (CEC, 2000, p. 3) to "all learning activity undertaken throughout life, with the aim of improving knowledge, skills and competences within a personal, civic and/or employment-related perspective" (CEC, 2001, p. 9). Dehmel highlights the removal of *purposeful* as informal learning with no specific purpose is a core element of lifelong learning, and the change from on *an ongoing basis* to *throughout life* that stresses continuous learning from the cradle to the end of life. Dehmel also points out a shift in the understanding of life-long learning from the 1970s to today; a shift from humanistic ideals of Bildung to "primarily utilitarian, economic objectives" (p. 52), even if these have been nuanced recently to combine the social and cultural with the economic.

One generic skill of central importance for life-long learning is being able to reflect upon what you are working with in order to learn. It is important that formal learning experiences at university support the development of reflective learning skills, that is the ability to "demonstrate self-awareness and motivation, awareness of the process of learning and independence" (Pickering, 2005). Micelli (2006) wrote that without reflective learning "learners cannot accept responsibility for their own learning" (p. 1), and Boud, Keogh & Walker (1985) view learner reflection as a "a term for those intellectual and affective activities in which individuals engage to explore their experiences in order to lead to a new understanding and appreciation" (p. 3) and one that involves mentally revisiting the experience or event. Some, e.g. Posner (1996), argue that more learning can be derived from retrospective reflection on an event's process of learning than from the experience itself (Posner, 1996).

This chapter examines peer-to-peer learning training in the context of a university level translation class. Lindgren, Sullivan, Deutschmann and Steinvall (2009), of which this chapter is an expanded version containing much textual overlap, focussed on how peer-to-peer learning could be used to improve student learning in the translation class and to overcome some of the criticism directed at the use of the Grammar-Translation Method of language teaching over the past 20 years as this method focuses on form as rather than on language for communication (see, for example, Levefere & Bassnett, 1998). Peer-to-peer supported reflection skills are defined here as both giving support and being sufficiently reflective to respond to peer learning support. The development of peer-to-peer supported reflection as a life-long learning skill is important; in working life much learning occurs in non-formal settings and is supported by peers who are often work colleagues, and in personal life much learning occurs in hobby and home settings where the supporter can be a family member peer, including your children, in sporting and social clubs and other daily activities.

This chapter begins with a look at peer-to-peer learning before turning to the computer tool we used in the classroom to train both peer-to-peer learning and the students' Swedish-English translation skills. The computational technique used in this paper to enhance language learning and support the development of peer-to-peer reflective learning is Computer Keystroke Logging. In an earlier study Lindgren, Stevenson and Sullivan

(2008) demonstrated how Computer Keystroke Logging together with Peer-Based Intervention could be used to enhance language acquisition. In this chapter we show that this is a potential approach to develop peer-to-peer supported reflection skills that can become a life-long learning skill, and that the approach we use can be generalised to other areas of the university curriculum.

REFLECTIVE LEARNING VIA PEER-TO-PEER LEARNING

The values of peer-to peer learning are encapsulated in the post-to-peer learning interview found in Zhao (2008). Her informant Shu said, about creating opportunities to read and then learn from peers' drafts, the following:

I feel it will help me because I may learn some words from them. If I meet words I don't know, I will look them up, finding out whether it is right or not. If it is wrong, I will pay attention to it next time when I write. In addition, I can learn some styles and use them in my own writing later. When I provide peer feedback, I am learning. (Shu/prei/P49)

Shu's view can also be seen in the what was said by Mo:

I can learn from their writing by reading it, another resource for learning. Two students will write in two different styles. In addition, I can learn some new words, sentence structures and organisation from theirs. (Mo/posti/P31)

Zhao (2008) also found that students saw that reading peers' work also encouraged them to be more reflective about their own writing:

Their mistakes can also reflect some of mine. I can also see my shortcomings via reading their errors. (Zhang/posti/P5)

Similarly, Ping also perceived peer-to-peer learning as a resource to reflect on her own writing:

For example, in poems, I wrote some words she was not familiar, so she commented that it was beautiful but I used some too flowery words. So I have to think about my purpose of writing a poem, to make it understood or to make it beautiful. So I have to weight it up and integrate it into final versions. (Ping/posti/P41)

Zhao suggested that these learner statements confirmed the value of reading peers' drafts and taking part in peer-to-peer learning, and that it may play a metacognitive role in prompting learners to reflect further on their own writing.

Zhao's (2008) result aligns with the existing studies that have reported the value of peer-to-peer learning in terms of reading peers' writing. Leki (1990), for instance, found that 16 of 17 respondents made positive comments about the value of reading peers' writing and 15 of them reacted positively to the helpfulness of peer feedback, through asking students to write in response to the following two questions:

1. How useful was it to you to read other students' papers?
2. How useful was it to you to read/hear other students' comments on your papers? (pg.6 - 7)

Similar findings were reported by Mangelsdorf (1992). She asked 5 ESL teachers to take 20 minutes of class time towards the end of a semester of using peer assessment to let students answer these questions in writing:

1. Do you find it useful to have you classmates read your papers and give suggestions for revisions?
2. What kinds of suggestions do you often receive from your classmates?

3. What kinds of suggestions are most helpful to you?
4. In general, do you find the peer-review process valuable? (pg.275 - 276)

Based on the 40 English as a Second Language (ESL) college freshmen's responses, she found that 22 students held positive views of peer assessment overall and 12 students showed mixed perceptions. Among these positive views, 68% were about the value of peer feedback in helping them with clarifying, developing, generating, and comparing ideas. In accord with this, Mendonca and Johnson (1994) ascertained through interviewing 12 ESL college students that peer assessment was valued by all students in helping them to spot problematic points that they themselves could not find, and in learning from peers by reading their writing. Positive attitudes towards peer feedback were also reported by the Chinese ESL college learners in Hu's (2005) study, including (1) learning about writing by reading peers' drafts and by providing peers with feedback, (2) motivating them by having a peer read their writing, (3) making them understand the importance of audience, and (4) peers helping them to locate problems they themselves would not have noticed.

Teacher Intervention in Peer-to-Peer Learning

Learners' trust in teachers over peers is a natural phenomenon in educational contexts, given that researchers in various research contexts have also observed the same phenomenon (e.g. Villamil and Guerrero 1996 in the Puerto Rico context; Falchikov 1998 in the British context; Ferris 2003 in the American context; Min 2005 in the Taiwan context). Teacher monitoring of the process of peer assessment has been explicitly articulated by the learners in interviews and diaries in this study and also raised by other empirical studies on peer assessment.

The approach which was most frequently suggested by the students in Zhoa's (2008) post-assessment questionnaire and interviews was to provide training in peer-to-peer learning, in particular in regard to the influence of the entrenched teacher-guided learning in China, as Zhu argued:

We Chinese students have been used to teachers' guidance. The tutor at lease tells us what to do next when we are asked to comment and discuss our peers' writing. It can make peer assessment more efficiently within such a limited class time. (Zhu/posti/P21)

A similar assertion was made by Shu, who claimed this:

Sometime he [referred to the writing tutor] provided guidelines on how to provide feedback so that we could focus on those aspects he suggested. This helps me to be more organised when participating in peer assessment. (Shu/posti/P37)

The effectiveness of the tutor's guidance on peer assessment was exemplified by Li's diary data, before and after she received the tutor's guidance on how to comment the organisation of writing. Before receiving directions, she wrote this in her diary:

I suppose that the situation will be better if the teacher can tell us some standards of commenting on a poem before the peer assessment period. We'd better be told that which kind of poem is good and which kind is bad before writing poems on own or assessing others' poems. Then everyone can focus on several aspects of the poem while reading the partner's poem. Otherwise, after writing poems according to our criteria, we still don't know how to write and judge a good poem. (Li/D/2803/P46)

After the tutor, Art, provided the instruction in how to tell good organisation, she wrote this:

In the writing class last week, Art first made a demonstration focusing on the organisation of an essay. Then we were asked to do peer assessments as usual and paid special attention to the organisation of the essay. The assessment on organisation between my partner and I was mainly based on Art's demonstration. I think his demonstration is helpful for us to comment on each other's organisation. (Li/D/2303/P20)

Other studies have articulated how teachers should interact in peer-to-peer learning settings. Mindham (1998), for example wrote:

Tutors can become involved in the interaction which takes place during peer assessment, they may direct, or at least monitor, the learning processes rather than retiring to a quiet and isolated hermitage to read and mark essays and examination scripts. (p.102)

and, Flachikov (1998) identified three factors for peer assessment (i.e. teachers, students and the institution) and described teacher factors as follows:

Teacher factors seem to involve traditional conceptions of student and teacher roles, in which teachers 'run the show' and students receive the benefits of teacher experience rather than of their own. Involving students in important process such as assessment requires a change in the traditional teacher (and student) role. Moreover, any change is stressful and upsetting (pg.18).

Zhao's (2008) study suggested Teachers need to acknowledge and address students' concerns about peer assessment from the outset in using peer assessment. In this respect, Ferris (2003) suggested the following steps:

- Ask the students to think of some reasons why having classmates read their papers and give suggestions might be a good idea;
- Ask the students to think of problems or concerns they might have about peer feedback;
- Mention research that demonstrates that peer feedback can benefit student writing and that other L2 learners enjoyed peer assessment;
- Point out that even very accomplished writers often receive and even solicit peer response (in writers' groups, peer-reviewed journals or books, etc.) and that getting feedback from classmates and co-workers may be a regular part of their future academic and professional lives. (pg. 168 - 189)

Although it might be time-consuming at the very beginning to raise these concerns with students in respect of the limited class time and teachers' overloaded teaching program, it is valuable to mitigate students' resistance to peer-to-peer learning and stimulate their enthusiasm for peer-to-peer assessment and learning.

A subsequent question following the necessity of training is how to provide such training. An analysis of this issue revolves around three main considerations:

First, what steps could be followed to "hold students responsible for taking peer feedback opportunities seriously (Ferris 2003:165)"? Its desirability has been suggested by the negative impact of peer collaborator' inattention to peer assessment on peer assessment and peer-to-peer learning. One way to make students responsible for peer assessment suggested by Min (2005) is to grade peer commentary. A follow-up question would be: how much should the grade for commentary account for a student's final writing mark? Ritter (1998) attached a 10% mark for peer commentary when introducing peer assessment on writing in her history class. However, she also noted that such a way made a distinct grade student hard to find a peer who was willing to evaluate his essay. In this case, grading peer feedback to encourage students to be responsible for their feedback only works when each student is guaranteed to be paired with a peer collaborator. Furthermore, how to give the grade is another essential issue that requires empirical investigation.

Second, how to set up peer assessment and peer-to-peer learning criteria? The learners in this research seemed to prefer to get the criteria from their writing tutor, as suggested by other researchers (e.g. Mangelsdorf 1992; McGroarty and Zhu 1997; Berg 1999; Patri 2002; Min 2005). However, there is also the possibility of getting the criteria from students (e.g. Lockhart and Ng 1995; Brown 1998; Race 1998; Hyland 2000), i.e. the use of indigenous writing criteria.

Thirdly, should training be an ongoing process? The writing tutor in this case study research, while maintaining the necessity of training, believed that he should foster learning autonomy (see 6.6). This resonates with the two criteria presented by Aljaafreh and Lantolf (1994, pg. 468) for efficient scaffolding in the zone of proximal development (Vygotsky, 1978), namely, graduated and contingent. This, however, runs counter to Rollinson's (2005) and Liu and Sadler's (2003) remarks that training should be an ongoing process.

Based on the above unresolved positions, it might be more reasonable to say that the specific classroom and cultural context represent the key factors in the development of appropriate training schedules. The writing tutor in this research recommended one way in our casual talk about how to train students. That is, ask students to write a short essay with the title: if I were the writing tutor, how would I train my students in conducting peer assessment? In this way, students' perceptions of training in peer assessment are captured, which can then be integrated into the tutor's own ideas of training techniques based on related empirical studies. By doing this, the specific steps of training would be more likely to cater to learners' and tutors' needs and thus, possibly be more effective. Further, this training schedule can be tested out and revised on the basis of these pilot results.

Figure 1 identifies how peer assessment may be influenced by five principal factors, namely: the existing culture of learning, institution effects, task effects, student effects, and teacher effects.

1. the role of examinations and teachers in the culture of learning in which the peer-to-peer learning is taking place;
2. institution effects: design of syllabus, teaching loads and support for peer assessment;
3. task effects: time allocated for the peer-to-peer task, setting in which peer-to-peer task is conducted (i.e. in class or outside class), organisation of students (e.g. in pairs or in groups, and with whom), and the genre of peers' writing;
4. student effects: language proficiency, the gender of the peer collaborators, familiarity with the peer collaborators, peer collaborators' commitment to peer-to-peer learning, and students' affective factors (e.g. perceptions of peer-to-peer learning);
5. teacher effects: how teacher and peer assessment via peer-to-peer learning are integrated; roles of teachers in the peer-to-peer task (i.e. modelling, intervention in peer assessment such as agreement with peer feedback, and training for peer-to-peer learning), and teachers' affective factors (e.g. perceptions of learning and peer-to-peer learning).

Having overviewed peer-to-peer learning and the factors that influence its uptake, we turn to the computer tool we used in the classroom to train both the students' peer-to-peer supported reflection skills as a life-long learning skill and the students' Swedish-English translation skills.

COMPUTER KEYSTROKE LOGGING

A computer keystroke logging program is a one that records the keystrokes, cursor and mouse movements made by the user of the logging

Figure 1. Factors influencing peer assessment (Zhao, 2008)

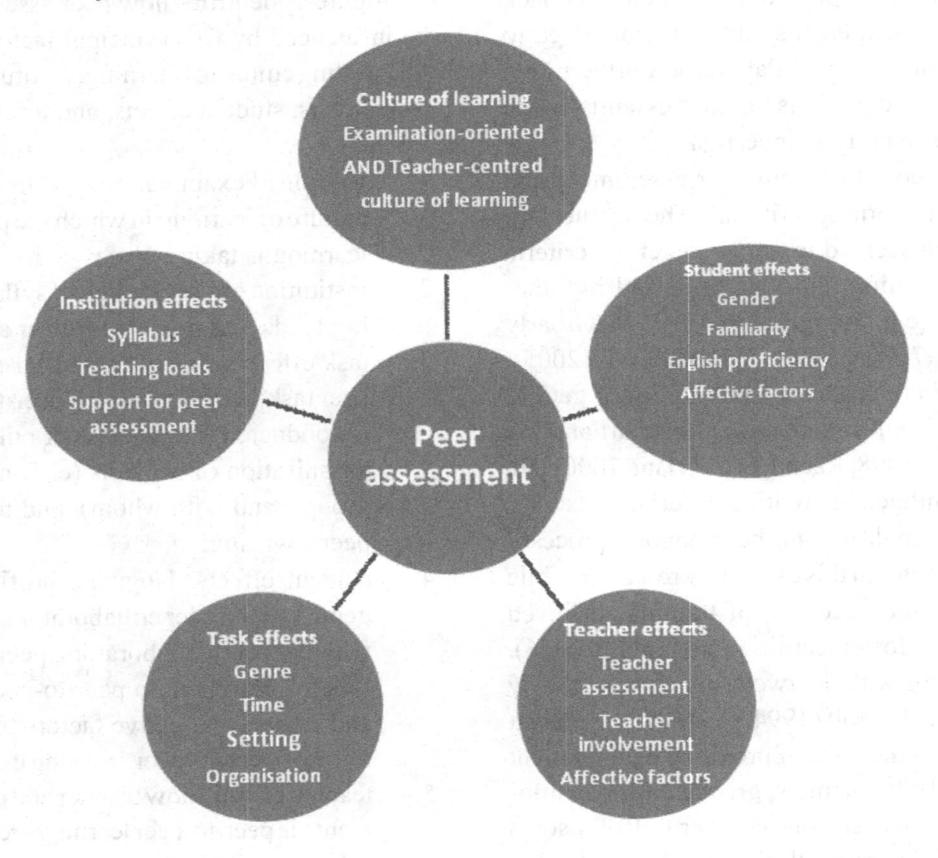

program's editing window. Cut and paste actions are also recorded. After the translator has finished their translation the entire translation process as recorded in the editing window can be replayed in real-time as the keystroke logging program also recording the time details of each recorded event. Figure 2 shows how the replay screen looks in the computer keystroke logging program, JEdit (Cederlund & Severinson Eklundh, n.d.). All the data is recorded to a log file which permits details automatic and manual analysis of the composition process, the pauses and the revisions. The log file distinguishes computer keystroke logging from programs such as Camtasia that also facilitate recording and play-back of activities on the computer screen. The study presented in this paper could also have been conducted using Camtasia.

Computer keystroke logging affords the teacher and the learner the possibility of analysing the translation process in more detail than is possible with programs such as Camtasia, and may as Ransdell (1990) wrote in relation to the essay class "provide insight for the composition instructors and students" (p. 143). There are a number of keystroke logging programs available: four computer keystroke logging programs are overviewed in Sullivan and Lindgren (2006).

As Lindgren (2005) pointed out, as long as the writer [translator in this study] is familiar with the computer keyboard, the writer [translator] is able to work undisturbed and as they would normally. The logging can be complemented with think aloud protocols where the translators voice how they are thinking while translating, or with retrospective protocols where the translators

Figure 2. The replay function in JEdit

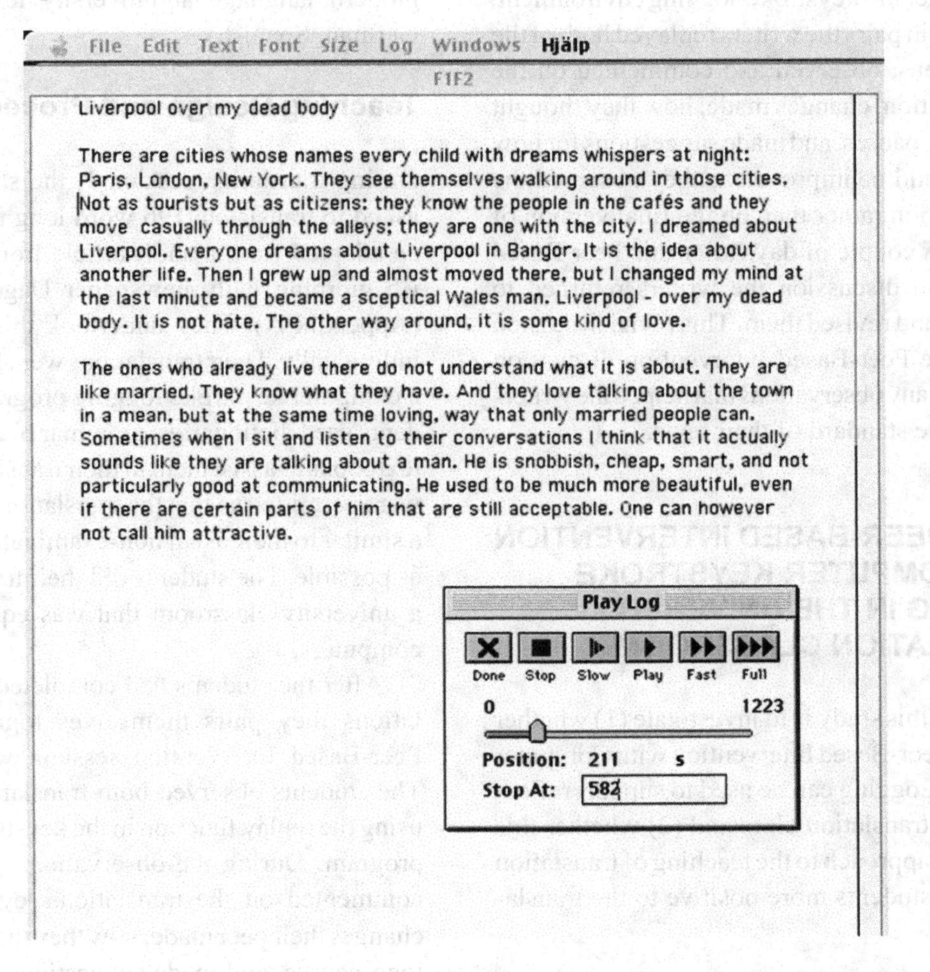

talk their way through how they worked when observing the replay of their translation session. Retrospective reflection is a key element of the study presented in this chapter. Translation studies using computer keystroke logging include Englund Dimitrova (2005, 2006), Hansen (2005), Jakbosen (2003), Jenssen (2001), Lorenzo (1999, 2001, 2002), Lundquist (2002) and Rothe-Neves (2003). To our knowledge no one has considered how computer keystroke logging could be used for teaching and learning in the translation class, nor how it can be used to support the development of peer-to-peer reflective learning skills for life-long learning.

COUPLING PEER-BASED INTERVENTION WITH KEYSTROKE LOGGING TO SUPPORT REFLECTION

Lindgren, Stevenson and Sullivan (2008) demonstrated that the reflective language learner could be supported logging in the language classroom by coupling Peer-Based Intervention with Computer Keystroke. In their study that built upon the ideas of Braaksma, Van den Bergh, Rijlaarsdam and Couzijn (2001) and Braaksma, Van den Bergh, and Rijlaarsdam (2002), Lindgren and Sullivan (2003) and Sullivan and Lindgren (2002), the writers independently undertook composition tasks on the computer in a keystroke-logging environment. Thereafter in pairs the writers replayed both of the compositions, observed and commented on the texts evolution, changes made, how they thought during long pauses, and made suggestions for how the text could be improved as they observed the text's creation rather than on the final version of the texts. A couple of days after this Peer Based Intervention discussion the writers returned to their texts and revised them. This revision session showed the Peer-Based Intervention discussion provided many observations that helped the writers improve the standard of their texts.

USING PEER-BASED INTERVENTION WITH COMPUTER KEYSTROKE LOGGING IN THE UNIVERSITY TRANSLATION CLASSROOM

The aim of this study is to investigate (1) whether coupling Peer-Based Intervention with Computer keystroke Logging can be used to support reflection in the translation class and (2) whether this innovative approach to the teaching of translation makes the students more positive to the translation class.

METHOD

The Participants

The participants were ten students in the final semester BA English translation class, six female and four male. All the students had good computer keyboard skills and were used to word-processing their assignments. Eight of the students had Swedish as their first language and two had English as their mother tongue. Six of the class considered themselves to be English-Swedish bilinguals. Seven of the students had studied other subjects at university of which four had studies another modern language at university level (French, German, Spanish).

Teaching Design and Procedure

During translation session 1, the students were asked to translate a 196 word long text that was an adapted from a short article from the Swedish morning daily newspaper Dagens Nyheter (Appendix A). They undertook this translation individually. Their translations were logged using a computer keystroke logging program. The students used dictionaries, grammar books and their textbooks to assist them in their translation. These were permitted so that the translation session was a similar to their usual non-examination situation as possible. The students did their translations in a university classroom that was equipped with computers.

After the students had completed their translations they pairs themselves together and a Peer-Based Intervention session was initiated. The students observed both translation sessions using the replay function in the keystroke logging program. During the observation sessions they commented on the translation's evolution, the changes their peer made, how they thought during long pauses, and made suggestions for how the translations could be improved. This phase was informally observed.

During the next translation class, translation session 2, the student pairs were asked to select one their translations and improve it based on their Peer-Based Intervention discussions of their two translations during the previous translation class.

After the students had revised and agreed upon a translation the students completed a short evaluation questionnaire that included demographic questions and question about how the students usually worked with a translation (See Appendix B).

Figure 3. Two students' individual translations after Translation Session 1 (top and middle boxes) and their joint translation after Translation Session 2 (lower box). The underlined segments in the top and middle boxes indicate differences between the two students' translations after Translation Session 1. In the bottom box the underlined segments indicate they come only from Writer 1 (1), only from Writer 2 (2), or are new (3)

Writer 1 – translation session 1

Liverpool over my dead body

There are cities which names all children with dreams whisper at night: Paris, London, New York. They see themselves walk around in these cities. Not as tourists but as residents: they know people at the coffee shops and move comfortly through the allies: they are one with the city. I dreamed about Liverpool. In Bangor everyone dreams about Liverpool. That is the idea about a different life. Then I became an adult and almost moved there, but I changed my mind at last minute and became a sceptical Walesman. Liverpool-over my dead body. It is not hatered. The opposite, it is some kind of love.

The ones who already live there of course does not know what it is all about. They are like married. They know what they have got. And they love to talk about the city in a mean but at the same time a loving way that only married can. Sometimes when I sit and listen to their conversations my thought is that it actually sounds as if they were talking about a man. He is snobbish, cheap, smart, and not very good at communication. He used to be more beautiful, even if some parts of him still are okay. Attractive however one can not call him.

Writer 2 – translation session 1

Liverpool over my dead body

There are cities whose names all children with dreams whisper about at night: Paris, London, New York. They see themselves walking about in those cities. Not as tourists but as residents: they know people in the cafés and move confidently though the alleys. I dreamt of Liverpool. In Bangor everyone dreams of Liverpool. It constitutes the idea of a different life. Then I grew up, and almost moved there, but I changed my mind and became a sceptical Welshman. Liverpool - over my dead body. It is not hatred. On the contrary, it is a kind of love.

Those who already live there don't understand what it is about. It is as if they were married. They know what they've got. And they love to talk about the city in the mean but at the same time loving way that is reserved for the married. Sometimes when I sit and listen to their conversations I think that it actually sounds as if they were talking about a man. He is posh, cheap, clever, and not very good at communicating. He used to be much more handsome, even if there are certain parts of him that still are acceptable. However, one cannot say that he is attractive.

Writer 1 and 2 – translation session 2

Liverpool over my dead body

There are cities which (1) names all children with dreams whisper at night: Paris, London, New York. They imagine themselves walking around (1) in these cities. Not as tourists but as residents: they know people at (1) the coffee shops (1) and move comfortly (1) through the allies (1) : they are one with the city (1). I dreamed about (1) Liverpool. In Bangor everyone dreams about (1) Liverpool. It is (3) the idea of (2) a different life. Then I became an adult (1) and almost moved there, but I changed my mind at the last minute (1) and became a sceptical Welshman. Liverpool-over my dead body. It is not hatered (1). On the contrary (2), it is some (1) kind of love.

Those who (2) already live there cannot (3) understand what it is all (1) about. They are as if they were (3) married. They know what they (3) got. And they love to talk about the city in a (1) mean but at the same time a (1) loving way that only married couples can (3). Sometimes when I sit and listen to their conversations my thought is that it (1) actually sounds as if they were talking about a man. He is snobbish (1), cheap, smart (1), and not very good at communication (1). He used to be more beautiful (1), even if some parts of him still are okay (1). Attractive, however, one cannot call him (1, 3).

FINDINGS AND DISCUSSION

Collaborative Revision: An example

Before moving on to the questionnaire findings, we present an illustrative example of how one pair of students worked with their texts (see Figure 3). The figure includes two writers' texts from translation Session 1 (texts 1 and 2) and their joint revised translation, which was the product of translation Session 2 (text 3). During this second session they developed one of their translation based on the content of their Peer-Based Intervention session. There were 32 differences between their session 1 translations, i.e. texts 1 and 2. Text 3 shows how the writers dealt with the differences between their original translations and the input from the Peer-Based Intervention session. In the majority of cases they kept Writer 1's version; In three cases the student pair exchanged Writer 1's text for Writer 2's text and in seven cases they jointly created a new version based on their Peer-Based Intervention triggered reflection and discussion. Arguably, this methodology stimulated reflection, discussion and development.

During the task there was minimal teacher intervention: the students decided between themselves which texts to start from and which revisions were to be made. The teachers did not interfere with this process since the purpose of the exercise was to stimulate student reflection and discussion rather than to produce "correct answers". Thus, the students were empowered to work without fear of embarrassment and at their particular level of readiness. However, it is a teaching challenge to appropriately engage with this empowerment to optimize individual student learning.

An examination of Figure 3 reveals a number of problematic didactic challenges. It is clear that text 2 would achieve a higher grade than text 1. In spite of this, or perhaps because of this, text 1 was chosen as the base for the joint revised translation. The students may have thought they chose the better text, may have actively chosen the poorer text as it would be more of a challenge (although little support for this hypothesis can be found in text 3's quality), or due to factors such power (e.g. gender and class), character traits (e.g. level of self-efficacy and self-confidence) and cultural expectations. The internal dynamics of a pair defines the level of the pair's readiness. A dynamically imbalanced pair, for example one in which the dominant peer is less proficient in the second language, is less likely to be able to take full-advantage of their collective knowledge and gain from this approach — the pair's level of readiness will thus be the level of readiness of one of the peers and not of both. This we suspect is true of the pair whose texts are presented in Figure 3. In contrast a dynamically balanced pair would be able to negotiate an appropriate level of readiness that would be advantageous to both peers. It is, thus, central for the success of this approach that how pairs are constructed is given due consideration.

Questionnaire

The questionnaire (see Appendix B) was constructed to collect demographic information, information about the students' approach to the translation class and their perception and experience of coupling Peer-Based Intervention with Computer keystroke Logging in the translation class. Here we report and discuss those questions that relate to supporting the development of peer-to-peer reflective learning. The interested reader is directed to Lindgren, Sullivan, Deutschmann and Steinvall (2009) for the discussion of the questions relating to the students' approaches to translation.

The demographic questions revealed that the 10 participants in the class we observed were typical for this level of English class at the university. This typicality means that the findings of this case study have validity even if as with case studies they cannot be generalised.

Four of the ten students had undertaken translations for others for which they had been paid.

Figure 4. The average number of actual revisions made by the individual student during translation session 1, and by the peer pairs during translation session 2

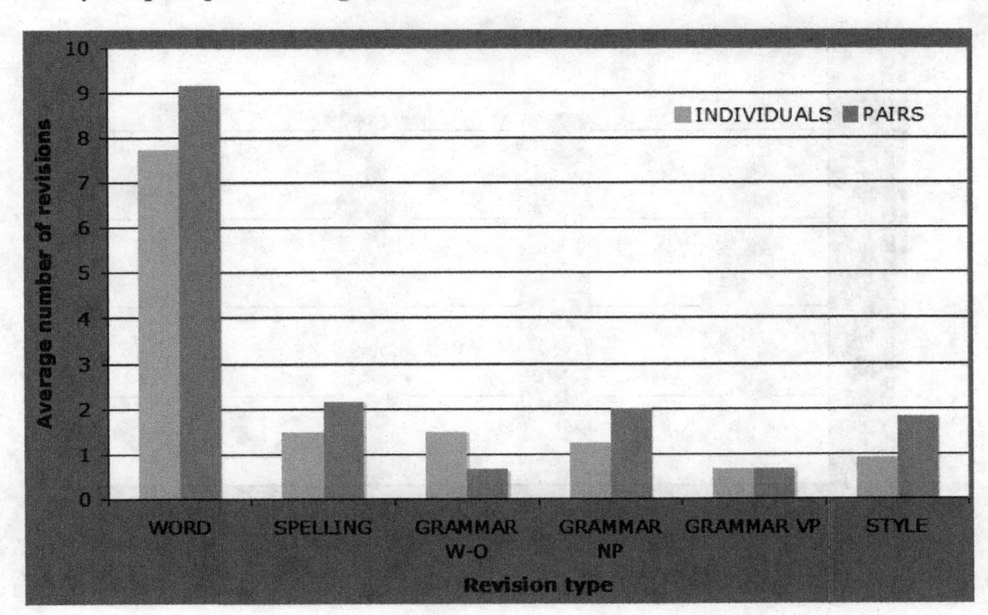

Two had undertaken Swedish-English translation and two English-Swedish translations. Half of the students usually undertook their translations using pen and paper and half directly in a word processor: one of the students wrote that she used pen and paper as it her ability to spell sat in her fingers. The students also had different approaches to the beginning of a translation: half of the students usually read the entire text before beginning, the others only read a sentence or less before starting to translate.

At the end of the translation after completing the translation of the final sentence one student usually felt this was the end of the task, three students read through their translations and the remaining students both read through their translations and made changes to their texts. Thus, just over half of the students were used to revising their texts and this suggested that the students would be open to the observed approach and to developing peer-to-peer supported reflection as a life-long learning skill.

One way in which the teaching design used in this case study differs from the students' normal way of working with a translation was that all students usually undertake their translations at home on their own rather than partly in pairs in a classroom setting. Working in class rather than at home on the translation also affected the students' ability to prepare for the class; they were not used to not coming to the class without have seen the translation in advance. This difference made the task more similar to many real life peer-to-peer reflective learning situations.

The students reported that they neither prepared in detail nor at all for their translations classes. This suggests that the students did not make sure that the translation they presented in class was of the absolute best quality. This finding has implications for the case study; the presentation of one's translation process to a peer may increase how careful the students are with their translations. Unfortunately the questions relating to how the student worked under and experienced the coupling Peer-Based Intervention with Computer keystroke Logging

Figure 5. The students' reporting on how instructive, meaningful, fun and stimulating they had found the coupling of peer-based intervention with computer keystroke logging in the translation classroom

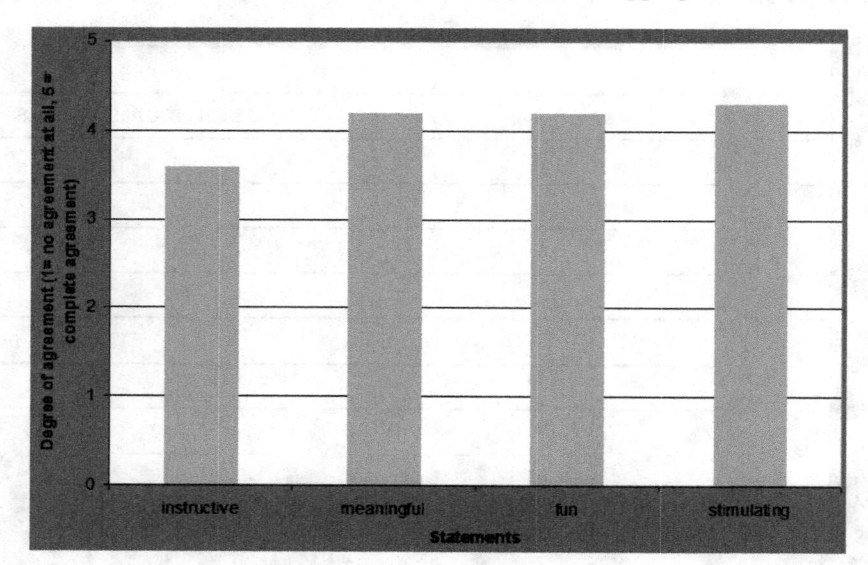

failed to pick up this particular possible impact of the teaching method; this was not expected considering the importance of the teacher's role in peer-to-peer learning as reported earlier in this chapter. An examination of this possible impact should form part of any future study.

Figure 4 reveals that the students have little perception of how they work with their translations. The most frequent revision is at the word level. This figure also gives an indication of how the teaching approach impacts upon the changes made. Even if more style revisions are made during the peer translation session 2, the focus remains at the word level. The focus suggests that when using peer-based intervention in conjunction with computer keystroke logging that detailed and focused instructions need to be used to lift the discussion and focus during both the replay and discussion session and translation session 2.

Although one student thought that the teacher should have been more active:

Maybe the teacher could have come with some tips and ideas so that one could get better. As it was this time, it felt like everything we did was good.

This comment reveals an underlying problem with the teaching approach as presented in this study and that is that it is the student themselves who define the standard. As research in self and peer assessment has shown, the students need to be trained in how to do apply the criteria for assessment (e.g. Broadfoot, James, McMeeking, Nuttal, & Stierer, 1988). Future application of the technique used in this study should be framed by a pre-activity training class in peer assessment, peer feedback, and Peer-Based Intervention. Further, Peer-Based Intervention sessions should have a clearer teacher-defined task focus. These findings reflect the factors of importance of peer-to-peer learning from Zhao (2008) and presented in Figure 1.

Another possible explanation for the focus on vocabulary is that although the students found the replay and discussion instructive they also found it embarrassing and awkward. The focus on form could indicate an active choice to avoid discussion of difficult and subjective aspects of the translation. Here again training in self- and peer-assessment and the giving of feedback may result in a difficult student focus during the discussions and the revisions made during translation session 2; peer internal dynamics play a central role as to whether subjective aspect can be discussed or not.

A majority of the students reported that during translation session 1 that they gained an increased awareness of their translation processes and that the replay and discussion helped them see new things in their own translation that they would have liked to change. A majority (7:3) also reported that the replay and discussion around their peer's translation helped them see new details that could be used to improve their own translations. A slightly larger majority (8:2) reported that they increased their awareness of their own translation process and strategy during Translation session 2.

The students positive reaction to the reflective teaching approach and their belief in it prompting greater awareness are seen in the following four free text comments:

It was the small changes that I myself had thought about when I wrote the translated text. Then when I was able to talk about them with my peer it helped a lot to discuss what the different alternatives that I had thought about could mean

I naturally saw things that I could have done better, partly because I saw things that I hadn't seen earlier and partly become I saw that my peer had done something better.

I don't change things once I had written them. I usually write only when I am certain of what I want to write. I also noticed that I don't read very far in the text before I start to translate. Maximally I read one sentence at a time. It was really useful to get a better understanding of how I work with a translation.

I notice how much I change my texts, and that it is correct to do so as my texts most often were better after the changes.

Figure 5 shows how positive the students were to this approach to teaching translation on a Likert scale, where 5 = complete agreement. That the students were positive to this teaching approach and as the majority found they could work in their usual way and were not disturbed by knowing their work was being logged suggests that this approach to the teaching of translation has the possibility of increasing student participation and of using pairs to individualize feedback at an appropriate level in a way that is usually not possible in the translation classroom. Further, if the approach makes use of the changes suggested in this chapter, the quality of the impact of this approach on both translation skills and the life-long learning skill of peer-to-peer learning.

CONCLUSION

From our informal observations when using this method and from the questionnaire responses we conclude that Peer-Based Intervention coupled with computer keystroke logging has the potential (1) to support student reflection and discussion around their translation tasks, (2) to enhance student motivation and enthusiasm for translation, and (3) support the development of reflective peer-to-peer learning skills that support life-long learning. This parallels the findings of Lindgren, Stevenson and Sullivan's (2008) study of the same methodology in the second language-writing classroom.

Larger scale studies coupled with student and teacher pre-activity training related to this method are needed to confirm the potential of this approach in the translation and other classrooms as a way to develop peer-to-peer supported reflection as a life-long learning skill. This study has focused on the use of computer keystroke logging in combination with peer-based intervention as a way of supporting the development of reflective peer-to-peer learning in the translation classroom. The approach can easily be transferred to other parts of the formal curriculum, where it is possible to record the process, either as here using keystroke logging or using other approach such as video recording, that can form a basis for peer discussion, intervention, reflection and learning, before pairs of students, or groups of students, proceed to work, reflect and support each other in producing a final product on the basis of peer-to-peer reflection and learning. The development of this life-long learning skill cannot occur in a single class or course, but is one that required scaffolding through formal education so that once in a work situation the seed of a the skill that was sown in school, college and university can develop throughout the individual's working and post-working life.

REFERENCES

Aljaafreh, A., & Lantolf, J. P. (1994). Negative feedback as regulation and second language learning in the zone of proximal development. *Modern Language Journal*, *78*(4), 465–483. doi:10.2307/328585

Berg, E. C. (1999). The effects of trained peer response on ESL students' revision types and writing quality. *Journal of Second Language Writing*, *8*(3), 215–241. doi:10.1016/S1060-3743(99)80115-5

Bergen Communiqué. (2005). *The European Higher Education Area – Achieving the goals*. Bergen, May 19–20.

Bologna Declaration. (1999). *Joint declaration of the European Ministers of Higher Education*. Bologna, June 19.

Boud, D., Keogh, R., & Walker, D. (1985). *Reflection: Turning Experience into Learning*. London: Kogan Page.

Braaksma, M. A. H., Rijlaarsdam, G., & Van den Bergh, H. (2002). Observational Learning and the Effects of Model-Observer Similarity. *Journal of Educational Psychology*, *94*(2), 405–415. doi:10.1037/0022-0663.94.2.405

Braaksma, M. A. H., Van den Bergh, H., Rijlaarsdam, G., & Couzijn, M. (2001). Effective learning activities in observation tasks when learning to write and read argumentative texts. *European Journal of Psychology of Education*, *1*, 33–48. doi:10.1007/BF03172993

Broadfoot, P., James, M., McMeeking, S., Nuttal, D., & Stierer, S. (1988). *Records of achievement: report of the national evaluation of pilot schemes*. London: HMSO.

Brown, S. (1998). *Peer assessment in practice*. Birmingham: SEDA Administrator.

Carless, D. (2005). Prospects for the implementation of assessment for learning. *Assessment in Education*, *12*(1), 39–54. doi:10.1080/0969594042000333904

Cederlund, J., & Severinson Eklundh, K. (n.d). *JEdit: The logging text editor for Macintosh*. Stockholm, Sweden: IPLab, Department of Numerical Analysis and Computing Science, Royal Institute of Technology (KTH).

Commission of the European Communities (CEC). (2001). *Communication from the Commission: making a European area of lifelong learning a reality (COM (2001) 678, final of 21.11.01)*. Luxembourg: Office for Official Publications of the European Communities.

Dehmel, A. (2006). Making a European area of lifelong learning a reality? Some reflections on the European Union's lifelong learning policies. *Comparative Education*, *42*(1), 49–62. doi:10.1080/03050060500515744

Englund Dimitrova, B. (2005). *Expertise and Explicitation in the Translation Process. (Benjamins Translation Library 64)*. Amsterdam: John Benjamins Publishing Company.

Englund Dimitrova, B. (2006). Segmentation of the writing process in translation: experts versus novices. In *Sullivan, K. P. H., & Lindgren, E. (2006). Computer keystroke logging: Methods and Applications* (pp. 189–201). Oxford, England: Elsevier.

Falchikov, N. (1998). Involving students in feedback and assessment. In Brown, S. (Ed.), *Peer assessment in practice* (pp. 9–23). Birmingham: SEDA Administrator.

Ferris, D. R. (2003). *Response to Student Writing: Implications for Second Language Students*. Mahwah, NJ: Lawrence Erlbaum Associates.

Hansen, G. (2005). *Störquellen in Übersetzungsprozessen. Eine empirische Untersuchung von Zusammenhängen zwischen Profilen, Prozessen und Produkten*. Doctoral dissertation. Copenhagen: Copenhagen Business School.

Hu, G. (2005). Using peer review with Chinese ESL student writers. *Language Teaching Research*, *9*(3), 321–342. doi:10.1191/1362168805lr169oa

Hyland, F. (2000). ESL writers and feedback: giving more autonomy to students. *Language Teaching Research*, *4*(1), 33–54.

Jakobi, A. P. & Rusconi, A. (2009). Lifelong learning in the Bologna process: European developments in higher education. *Compare: A Journal of Comparative and International Education*, 39 (1), 51 — 65

Jakobsen, A. L. (2003). Effects of Think Aloud on Translation Speed, Revision and Segmentation. In Alves, F. (Ed.), *Triangulating Translation* (pp. 69–95). Amsterdam, Philadelphia: John Benjamins Publishing Company.

Jensen, A. (2001). *The Effects of Time on Cognitive Processes and Strategies in Translation. PhD dissertation.* Copenhagen Working Papers in LSP 2. Copenhagen: Copenhagen Business School.

Lefevere, A., & Bassnett, S. (1998). Where are we in Translation Studies? In Bassnett, S. & A. Lefevere: *Constructing Cultures. Essays on Literary Translation*. Clevedon: Multilingual Matters.

Leki, I. (1990). Coaching from the margins: Issues in written response. In Kroll, B. (Ed.), *Second Language Writing* (pp. 57–68). Cambridge, UK: Cambridge University Press.

Lindgren, E. (2005). *Writing and Revising: Didactic and Methodological Implications of Keystroke Logging. (Skrifter från moderna språk, No. 18)*. Umeå, Sweden: Umeå University, Department of Modern Languages.

Lindgren, E., Stevenson, M., & Sullivan, K. P. H. (2008). Supporting the reflective language learner with computer keystroke logging. In Barber, B., & Zhang, F. (Eds.), *Handbook of Research on Computer Enhanced Language Acquisition and Learning* (pp. 189–204). Hershey, PA: IGI Global Inc.

Lindgren, E., & Sullivan, K. P. H. (2003). Stimulated recall as a trigger for increasing noticing and language awareness in the L2 writing classroom: A case study of two young female writers. *Language Awareness*, *12*, 172–186. doi:10.1080/09658410308667075

Lindgren, E., Sullivan, K. P. H., Deutschmann, M., & Steinvall, A., (2009). Supporting learner reflection in the language translation class. *International journal of information technologies and human development*, 1(3),26-48.

Lockhart, C., & Ng, P. (1995). Analysing talk in ESL peer response groups: stances, functions and content. *Language Learning, 45*(4), 605–655. doi:10.1111/j.1467-1770.1995.tb00456.x

Lorenzo, M. (1999). Apuntes para una discusion sobre metodos de estudio del proceso de traduccion. In Gyde Hansen (Ed.), *Copenhagen Studies in Language, vol. 24: Probing the Process in Translation. Methods and Results* (pp. 21–42). Copenhagen: Samfundslitteratur.

Lorenzo, M. (2001). Combinación y contraste de métodeos de recogida y análisis de datos en el estudio del proceso de la traducción - Proyecto del grupo TRAP. Quaderns. *Revista de traducció, Barcelona, 6*, 33–38.

Lorenzo, M. (2002). ¿Es posible la traducción inversa? - Resultados de un experimento sobre traducción profesional a una lengua extranjera. In G. Hansen (Ed.), *Copenhagen Studies in Language, vol. 27: Empirical Translation Studies. Process and Product* (pp. 85–124). Copenhagen: Samfundslitteratur.

Lundquist, L. (2002*). L'anaphore associative: Etude contrastive et expérimentale de la traduction de l'anaphore associative du français en danois*. Romansk Forum XV, No. 16.

Mangelsdorf, K. (1992). Peer response in the ESL classroom: What do the students think? *ELT Journal, 46*(3), 274–293. doi:10.1093/elt/46.3.274

Mendonça, C. O., & Johnson, K. E. (1994). Peer review negotiations: Revision activities in ESL writing instruction. *TESOL Quarterly, 28*(4), 745–769. doi:10.2307/3587558

Miceli, T. (2006, December). Foreign language students' perceptions of a reflective approach to text correction. *Flinders University Languages Group Online Review, 3*(1). Retrieved October 3 2008 from http://ehlt.flinders.edu.au/deptlang/fulgor/volume3i1/papers/Miceli_v3i1.pdf

Min, H.-T. (2005). Training students to become successful peer reviewers. *System, 33*(2), 293–308. doi:10.1016/j.system.2004.11.003

Mindham, C. (1998). Peer assessment: report of a project involving group presentations and assessment by peers. In Brown, S. (Ed.), *Peer assessment in practice* (pp. 45–66). Birmingham: SEDA Administrator.

Patri, M. (2002). The influence of peer feedback on self- and peer-assessment of oral skills. *Language Testing, 19*(2), 109–133. doi:10.1191/0265532202lt224oa

Pickering, A. (2005). *Facilitating reflective learning: an example of practice in TESOL teacher education.* Retrieved May 29 2008 from http://www.llas.ac.uk/resources/goodpractice.aspx?resourceid=2395

Posner, G. J. (1996). *Field Experience: A Guide to Reflective Teaching*. White Plains, NY: Longman.

Race, P. (1998). Practical pointers on peer assessment. In Brown, S. (Ed.), *Peer assessment in practice* (pp. 108–113). Birmingham: SEDA Administrator.

Ransdell, S. E. (1990). Using real-time replay of students' word processing to understand and promote better writing. *Behavior Research Methods, Instruments, & Computers, 22*(2), 142–144.

Ritter, L. (1998). Peer assessment: lessons and pitfalls. In Brown, S. (Ed.), *Peer assessment in practice* (pp. 79–86). Birmingham: SEDA Administrator.

Rothe-Neves, R. (2003). The influence of working memory features on some formal aspects of translation performance. In Alves, F. (Ed.), *Triangulating Translation. Perspectives in Process Oriented Research* (pp. 97–119). Amsterdam: Benjamins.

Sullivan, K. P. H., & Lindgren, E. (2002). Self-assessment in autonomous computer-aided second language writing. *ELT Journal*, *56*(3), 258–265. doi:10.1093/elt/56.3.258

Sullivan, K. P. H., & Lindgren, E. (2006). *Computer keystroke logging: Methods and Applications*. Oxford, England: Elsevier.

Villamil, O. S., & de Guerrero, M. C. M. (1996). Peer revision in the L2 classroom: social-cognitive activities, mediating strategies, and aspects of social behavior. *Journal of Second Language Writing*, *5*, 51–75. doi:10.1016/S1060-3743(96)90015-6

Vygotsky, L. S. (1978). *Mind in society: The development of higher psychological processes*. Cambridge, MA: Harvard University Press.

Zhao, H. (2008). *Who takes the floor: peer assessment or teacher assessment?: a longitudinal comparative study of peer- and teacher-assessment in a Chinese university EFL writing class*. Unpublished PhD thesis, University of Bristol, England.

APPENDIX A: THE SWEDISH TEXT THAT WAS TRANSLATED IN ENGLISH

Liverpool Över Mina Döda Kropp

Det finns städer vars namn alla barn med drömmar viskar om natten: Paris, London, New York. De ser sig själva gå omkring i de där städerna. Inte som turister utan som fast bosatta: de känner folk på kaféerna och rör sig vant genom gränderna; de är ett med staden. Jag drömde om Liverpool. I Bangor drömmer alla om Liverpool. Det är idén om ett annat liv. Sen blev jag vuxen och flyttade nästan dit, men jag ändrade mig i sista stund och blev en skeptisk walesare. Liverpool – över min döda kropp. Det är inte hat. Tvärtom, det är någon sorts kärlek.

De som redan bor där förstår inte vad det handlar om. De är liksom gifta. De vet vad de har. Och de älskar att prata om stan på ett elakt men samtidigt kärleksfullt sätt som bara gifta kan. Ibland när jag sitter och lyssnar på deras samtal tänker jag att det faktiskt låter som om de pratade om en man. Han är snobbig, snål, smart, och inte särskilt bra på kommunikation. Han var mycket vackrare förr, även om det finns vissa delar av honom som fortfarande går an. Attraktiv kan man emellertid inte kalla honom.

APPENDIX B: THE QUESTIONNAIRE

The questionnaire was run online. The questionnaire thus looked different to the students. The possibility to comment on an answer with free text was permitted after many of the questions. The questions are grouped here after topic and not after there order of presentation. In some places clarification to the reader is given in italics; these clarifications did not form part of the presented questionnaire.

Demographic and Background Questions

How many university credits did you have at the beginning of this semester?
(One Swedish university credit at the time this data was collected was the equivalent of one week's full-time study, and the academic year was 40 weeks long.)

1. 0-20
2. 21-40
3. 41-60
4. 61-80
5. 81-100
6. 101-120
7. 121-

I am

1. Female
2. Male

As well as the languages I have studied at university, I am able to communicate well in speech and writing in:

1. No other language
2. German
3. French
4. Spanish
5. Russian
6. Finnish
7. Other

I am

1. 18-25 years old
2. 26-30 years old
3. 31-35 years old
4. 36-40 years old
5. 41- years old

I would describe myself as bilingual

1. Yes
2. No

My mother language is

1. Swedish
2. English
3. German
4. Spanish
5. French
6. Other

I have studied the following languages at university

1. I have not studied another language
2. Swedish
3. German
4. French
5. Spanish
6. Finnish
7. Russian
8. Other

Approach and Experience of Translation

I have experience of translating outside of university. My experience includes:

1. I have no experience at all
2. Swedish to English translation
3. English to Swedish translation
4. Other

Do you usually translate using

1. A computer
2. Pen and paper

How much of the text do you usually read through before you begin to translate?

1. the whole text
2. a paragraph
3. a sentence
4. less than a sentence

How much do you usually prepare for a translation class?

1. very thoroughly
2. thoroughly
3. not particularly thoroughly
4. Not at all

Where do you usually prepare your translation?

1. At home
2. In the library
3. in a café
4. Somewhere else

Which of the following do you usually use when working with a translation?

1. Swedish-English/ English-Swedish dictionary
2. English Dictionary (e.g. Longman)
3. Thesaurus
4. Grammar book
5. Text corpus
6. Internet
7. Nothing

What do you after you have translated the sentence in the text?

1. Read through the text
2. Revise the text
3. Both 1 and 2
4. Neither 1 nor 2

When translating it is sometimes necessary to revise the translation. How often do you think you revise on average when doing a translation.

Within the sentence

Within the paragraph

After having the translation

1 = Very infrequently or never; 5 = very often

What do you find difficult with translation?

word

spelling

grammar: word order

grammar: NP

grammar: VP

Style

1 = very easy; 5= Very difficult

How do you usually work when preparing a translation?

1. Alone
2. Together with another student
3. Together with several other people

Experience of the Teaching Approach

Were you affected by the environment (room) in which the translation sessions took place

1. Yes
2. No

During the replay and discussion about your PARTNER's translation did you notice details that could led to improvements in your own translation

1. Yes
2. No

How much were you disturbed knowing that your work was being logged?

1 = not at all; 5 = very disturbed

During the replay and discussion did you notice something new about your person translation process?

1. Yes
2. No

Were you about to work as usual with your translation?

1. Yes
2. No

Do you think that Session 1 gave you an increased awareness of your translation strategy and your language?

1. Yes
2. No

During the replay and discussion about your own text did you detect new things that you would like to have done differently?

1. Yes
2. No

How much were you affected by there being other students in the room?

1= not at all; 5= Lots

How did you react to the replay of and discussion around your own translation? How instructive did you find it?

1 = very little

5 = very much

How embarrassing or awkward did you find it?

1 = very little

5 = very much

Do you think that Session 2 gave you an increased awareness of your translation strategy and your language?

1. Yes
2. No

Who text did you work with during session 2?

1. My text
2. My peer's text

During translation session 2 your task was to improve the translation. Estimate how many changes of the following types you made on the 1-5 scale for the following levels:

word
expressions and idioms
spelling
grammar: word order
grammar:NP
grammar: VP
style
1: none; 5: lots
How did you enjoy this way of working with translation?
1= not at all
5= Lots

Suggestions for Improvements to this class; E.g. What was Missing?

The teacher played a passive role in this class. Would you have preferred it if the teacher has been more active, explained more and made corrections during the reply and discussion?

1. Yes
2. No

The importance of the discussion
(1 = none; 5 = lots)
Discussion 1
Discussion 2
How did you find the teaching approach
Instructive
Meaningful
Fun
Stimulating
1= not at all; 5 = very

Chapter 12
New Trends in ICTs and Computer Assisted Language Learning (CALL)

Bolanle A. Olaniran
Texas Tech University, USA

ABSTRACT

This chapter explores information communication technologies (ICTs) (e.g., computer-mediated communication) and the implications for use in language learning and second language learning (L2). Further, the chapter presents culture and new trends in ICTs for L2 learning. Specific modality, challenges, and issues for future considerations in L2 learning are discussed. The chapter argues for the need to understand culture and contextual appropriateness of L2 learning in ICT environments. Finally the chapter contends that ICTs are best relegated as a supplemental role or tools, rather than as an outright substitute for traditional L2 learning and curricula.

INTRODUCTION

The increasing role of technology in fabrics of our lives is noteworthy. Technology, the likes of computer-mediated communication (CMC), is changing the way and where learning occurs. E-learning is one direct result of the technology revolution that is continuously taking place. One of the appealing aspects of communication technologies in learning is the ability to use CMC to offer courses in both asynchronous and synchronous environments (Olaniran, 2009). Specifically, CMC is considered an important tool for learning because it facilitates interaction and active learning (Driscoll, 2000; Kanuka & Garrison, 2004; Olaniran, 2004).

One area of learning that is gaining increased traction with communication technology is lan-

DOI: 10.4018/978-1-60960-497-4.ch012

guage learning or second language learning (i.e., L2). With communication technology students or L2 learners are able to interact with speakers of target languages at convenience (Egorov, Jantassova, & Churchill, 2007). Technologies allow people to communicate either one-on-one or one-to-many while allowing teachers to use technology to disseminate curricula or assist instruction. In an earlier work the author addressed the question of whether language, especially new language learning, can be facilitated by CMC especially when accounting for culture? After all, it is not sufficient to learn just the rudiments of a language; instead, the contextual appropriateness of a language is what determines the competency at which a second language (L2) learner will be judged and evaluated. Hence this chapter revisits this questions and attempts to present new trends in L2 learning with technologies.

It has been argued that the contextual appropriateness of L2 reinforces the importance of culture in language learning (Olaniran, 2009). To this end, Garrett (2009a) identifies two factors that are crucial to contextual appropriateness of technology in language learning. First that *language context* "be correct, authentic and appropriate" (p. 709). Second, that the technology or program run as it should without problems or crashing (Garrett, 2009a). Both of these are what Garrett (2009a) called absolute in language learning and technology. Garrett (2009a) argued that a technology package or platform that is presumed inadequate by one teacher may be deemed appropriate by another. She went on to claim that no reviewer of a particular technology package for language learning can establish a rating that is equally valid for all users regardless of their technological expertise. In essence, Garrett (2009a) indirectly acknowledges the role of other variables, especially culture and contexts for L2 learning in technology environments. Thus, culture plays an important role in language learning. However, what is not certain is whether anything has changed and if so to what ends? It is the goal of this chapter to identify newer trends in L2 learning to answer the proposed question. First however is a general review of relevant literature.

CULTURE AND LANGUAGE IN ONLINE ENVIRONMENTS & CMC

The challenge of culture in e-learning environments has been identified (Olaniran, 2006a). Culture introduces certain complexities to learning as a whole and specifically to second language (L2) comprehension with communication technologies. To this end, a number of scholars have called for the importance of considering culture and language when designing curriculum for international students in computer environments (Morse, 2003, Olaniran, 2007, 2007; Osman & Herring, 2007; Patsula, 2002; Usun, 2004). Addressing L2 learning or culture will be incomplete without addressing the dimensions of cultural variability, which offers a way to identify how culture influences human communication interaction and in particular L2 learning. Furthermore, dimensions of cultural variability helps to draw implications for L2 learning in technology environments. Following is a discussion of the dimensions of cultural variability proposed by Geert Hoftede and is often used to explain culture.

Dimensions of Cultural Variability

Originally, there are four dimensions of cultural variability consisting of: uncertainty avoidance, individualism, power distance, and masculinity (Hofstede, 1980, 2001; see also Olaniran, 2007). These four categories result from data collected from fifty countries and three world regions (Hofstede, 1980). Past research has used these four dimensions to evaluate cultural differences and their effects on uncertainty reduction in intercultural communication encounters (Gudykunst, Chua & Gray, 1987, Olaniran, 1996, 2004; Roach & Olaniran, 2001).

Uncertainty avoidance describes "the extent to which people feel threatened by ambiguous situations and have created beliefs and institutions that try to avoid these" (Hofstede & Bond, 1984, p. 419). Individualism-collectivism acknowledges the fact that in individualistic cultures, "people are supposed to look after themselves and their family only," while in collectivistic cultures, "people belong to in-groups or collectivities which are supposed to look after them in exchange for loyalty" (Hofstede & Bond, 1984, p. 419). Power distance is explained as "the extent to which the less powerful members of institutions and organizations accept that power is distributed unequally" (Hofstede & Bond, 1984, p. 418). Masculinity refers to cultures "in which dominant values in society are success, money and things," while femininity refers to cultures "in which dominant values are caring for others and quality of life" (Hofstede & Bond, 1984, p. 419-420). While there has been extensive critique of Hofstede's dimensions, as too simplistic to handle modern cultures and complex intercultural interaction (see Ess & Sudweeks, 2006; Fougere & Moulettes, 2007; Martin & Cheong, 2007; McFaydyen, 2008), his contribution to cultural understanding has withstood the test of time (Olaniran, 2007). For Instance, Macfadyen, (2008) points to the challenge of viewing culture as national only. In spite of the critique, Hofstede's dimensions offer a general and useful framework especially when one takes into account other variables such as socioeconomic development, relationships, and others.

Hofstede's dimensions of individualism-collectivism and power distance are particularly useful when exploring communication interactions and cultural impacts in e-learning such as L2. One reason is that the dimensions of individualism and power distance directly influences how one communicates or interacts with people in technology environments (Olaniran 2004; 2007). The other reason is that both dimensions focus on relationships and relational development, which are believed to be less subject to change within culture (Smith, 2002). For instance, in spite of globalization leading to economic changes that may be moving people, including those in collectivistic cultures, towards individualistic ideals of wealth acquisition, those individuals still maintain their identity and hold on to norms that guide different relationships (i.e., co-workers, families, etc.) and communication rules (Smith, 2002).

Looking at the dimensions of cultural variability, Olaniran (2007) addressed extensively, the implication for e-learning. However, specific implications for L2, or language learning in general, over ICT deserve more attention. For instance, differences in culture influence expectations that students and teachers hold about learning. One feature that differs in a power distant culture is that students look up to instructors as the primary source of knowledge and attempt to bring about constructivist types of learning which involves putting students in participative control of their learning, which is bound to be met with resistance. Other scholars addressed this problem when they concluded that students hesitate to challenge or question teachers in cultures where teachers are the main source of knowledge (Bates, 1999; Bodycott & Walker, 2000, Usun, 2004). Collectivist cultures, on the other hand, tend to accept traditional roles, where adults are reluctant to accept subordinate or student roles—specifically, students usually spoke in the classroom only when instructors called on them (Gunawardena, 1996; Olaniran, 2007; Osman & Herring, 2007). In a study of Azerbaijan adult learners and U.S. facilitators using synchronous chat, it was found that the constructivist idea and increased interaction patterns were not present, to which the authors attributed cultural differences among learners, designers, and facilitators (Osman & Herring, 2007). The authors concluded that because the education system in Azerbaijan is a teacher-centered system (i.e., power distant), the learners expect to be listening rather than actively contributing to the discussion. At the same time, the study brought the issue of learners' language competence to the forefront given that the learners had limited English knowledge; thus, contribution to the synchronous chat was limited at best. The question remains how does culture influence L2 acquisition, where individual participation tends to be more critical than group interactions? To answer the question one needs to explore the characteristics of L2 or language learning, which is addressed in the next section.

The collectivistic dimension of culture was also specific in how people apply ICTs. For instance, a study of minority ethnic groups in the Netherlands found that individuals with closer ties to their home countries tend to sympathize with people in the home countries more and have greater contact via ICTs with people in their primary cultures. Specifically, Moroccan young girls used the Internet for maintaining as well as widening social contacts and more so than their counterparts from the host culture (D'Haenens, Koeman, & Saeys, 2007; Linders & Goosens (2004). In essence, ICT use represents a way for ethnic minority youths to navigate and establish their cultural identity between two cultures.

CMC and Language Learning

The Internet and other computer programs have added to the text-based dimension of CMC such that hypermedia has made technology more versatile and interactive via graphics, audio, and video embeddedness. Multimedia is believed to add certain advantages for e-learning and in particular L2, where users are provided opportunities to write and listen as in real world or traditional settings (Egorov, et al., 2007; Garrett, 2009b). Thus, language skills are integrated into technology media in a manner that makes it *natural* to combine different language skills, as well as giving students greater control and freedom in how they learn the language at their convenience (Egorov, et al. 2007, Olaniran, 2009). Specifically Egorov et al. (2007) contended that students are able to move at their own pace, where they can select their own paths, *going forward and backwards* to different parts of the curricula at their own pace. Further, learners can focus on content, without sacrificing focus on language form. Alternately, computers can be programmed to simulate certain contexts in which L2 learners can use to practice their skills. For example, *Dustin* was developed at the Institute for Learning Sciences at Northwestern University (Schank & Cleary, 1995*). Dustin simulation offers students a scenario of arriving at a U.S. airport where students must navigate customs and seek transportation to the city. The students interact by responding (e.g., typing) to series of video clips. If students provide correct responses they move to another task, whereas, if an incorrect response is provided they are referred to a remedial solution. Students/learners are able to control their participation by asking what to do, what to say, and when, while listening to what is being said or they may request translation (Egorov, et al., 2007).

Language learning, especially L2 mastery, is characterized by individuals' motivation to learn a language or through job and career obligations. For the most part global and multinational corporations (MNCs) operate using English or in some cases, the home language of the parent organization. However, with English referred to as the Internet or commerce language (Kayman, 2004), it is gaining more wide appeal; thus, increased numbers of individuals are increasingly learning English in addition to their indigenous language. The majority of the learning is taken place via World-Wide Web (WWW) and the Internet, along with structured ICTs dedicated toward such language learning.

As people migrate from one place to another, there is increasing need to learn the language and the culture of the host societies to facilitate the adaptation process (D'Haenens, et al., 2007; Putnam, 2000; Thorne, Black, & Sykes, 2009). Notwithstanding, the increased availability of computer-enhanced or enabled language learning has brought on a few challenges. One problem includes the ability to offer a certain group from the home country information in the dominant language to assist members in establishing and maintaining relations. For example, D'Haenens et al. (2007) offer the example of how Turkish and Moroccan websites allow the ethnic groups in the Netherlands to deal with their religion and with the feature of anonymity, they offer a forum to discuss social issues such as discrimination freely. Culture influences religious practices in a way that parents encourage their children to strive for linguistic mastery such that children will be able to read, speak, and write Arabic in an attempt to read Koran for Moroccan youths (D'Haenens, et al., 2007). Similarly, parents of ethnic minority groups tend to attribute greater importance to ICT use in an attempt to help their children adapt to the new culture and future careers.

Furthermore, different languages and backgrounds create different degrees of difficulty for learners (Olaniran, 2009). Brown and Iwashita (1996) found students who were native speakers of English and Chinese languages experience different levels of difficulty in computer-adaptive Japanese grammar tests. To such an end, it is offered that students learning a language would find the target language easier provided that the target language is similar to the primary language. Three factors are important from the findings. First is the degree of congruence between the original language and the target language. Second, is the zero contrast that affects acquisition of articles and inflection morphology (tone) in the target language. For example, Japanese speakers who have no articles in their language were found to take longer in learning definiteness and indefiniteness in English language when compared to Germans whose language consists of articles (Zobl, 1983). Third, is the constraint in linguistic markedness, which consists of the complexity or infrequent use of certain features of a language (Larsen-Freeman & Long, 1991). That is when a certain feature is more pronounced in the target language than the primary language, learners would experience increased difficulty. For instance, Chinese speakers who do not distinguish between plural and singular will take longer to learn English where such a distinguishing feature is present (Brown & Iwashita, 1996). These are a few examples in which culture affects language learning, given that a key component of language is culturally rooted.

One of the greatest challenges to L2 learning via ICTs is the cultural appropriateness. It is not enough to learn phrases and words in a particular language; rather, it is pressing to understand the contextual use of the language phrases and terminologies to facilitate cross-cultural communication appropriateness in order to be effective. Different CMC modality can be applied to language learning. There are two major categories of technological structure—asynchronous and synchronous. There is a clear distinction between synchronous and asynchronous CMC and consequently implications for classrooms and subsequent learning (Olaniran, 2009; Smith, Alvarez-Torres, Zhao, 2003). Asynchronous CMC structure includes mediated interactions where participants are not constrained by time. On the other hand, synchronous CMC structure provides an environment that requires simultaneous interactions among participants and users, and is sometimes constrained by time (e.g. chat, computer conferencing, video-conferencing) (Olaniran, 2006). Interactivity facilitates greater levels of immediacy in the synchronous CMC environment. Therefore, a synchronous CMC allows learners to comment, ask questions, or interact with instructors at the same time (Olaniran, 2006; 2009). The interactivity allows instructor to manipulate how students learn the materials (McAlister, Ravenscroft, & Scanlon, 2004; Olaniran, 2004; 2009). While synchronous CMC has the advantage of speedy feedback (e.g., immediacy), which is essential in negotiation of meaning and synthesizing ideas (Chou, 2001; Lobel, Neubauer, & Swedburg, 2005), it is not without its drawbacks. One drawback includes overwhelming interactions that can make chat sessions difficult to manage and avoid chaos (Olaniran, Stalcup, & Jenson, 2000, see also Table 1 in Olaniran 2006 for comparison of synchronous and asynchronous CMC advantages and disadvantages). Aside from chaotic interaction, chat makes it difficult to see the relationship between different messages, which becomes more pronounced when multiple conversations are going on (Bober & Dennen, 2001; Olaniran, 2009; Teng & Taveras, 2004-2005). As for asynchronous CMC applications in language learning, the lack of instant feedback appears to render it useful, where learners can post messages or questions and answers in order to get additional help in learning basic concepts. An asynchronous environment may also be suitable for composition of messages in the primary language while using software applications (e.g., Google) to translate the message to the intended or targeted language (Olaniran, 2009).

All in all, ICT appears to be best used in language learning as a supplemental tool. Language learning may be best in a face-to-face (FtF) environment (Garrett, 2009b; del Puerto & Gamboa, 2009; Olaniran, 2009). This is not to say that CMC can not play an important role in language learning environments. For instance, synchronous CMC can offer an additional dimension to the learning taken place in FtF settings. The added dimension includes the ability to facilitate students' ability to think about how they learn, gather information, compose arguments, and learn about the contextual implications of language phrases (Davie & Wells, 1991; Olaniran, 2009; Walker, 2004). For instance, Zapata and Sagarra (2007) illustrate that Spanish language learners overwhelmingly conclude that in-class FtF activities lead to their use of ICTs, which help them in listening, reading, and pronunciation. More importantly, the students suggest that ICTs provide them with redundancy that helped in their L2 lexical knowledge. Furthermore, both synchronous and asynchronous CMC offers students the opportunity to seek help from non-instructors and peers in language and culture learning, which is central to the principles of deep learning and transformation learning that facilitate active and reflective thinking to learners while increasing their cognitive skills (Clawson & Choate, 1999; Olaniran, 2006; 2009; Olaniran, Savage, & Sorenson, 1996; Osman & Herring, 2007).

The supplemental role of ICT tools (e.g., CMC) in language learning is also echoed in Reinders and Lazaro's (2007) study, which evaluated self-access technology centers and their use of language learning. They found that all the centers investigated provided very limited support materials for language learning. Specifically, the authors found that rarely were synchronous chat and asynchronous discussion boards used. Instead, less interactive technologies such as language learning software, e-mail, and Internet resources were used and sometimes for administrative purposes such as advising, direct access to resources, and evaluation processes. Del Puerto and Gamboa (2009) also found that L2 instructors seldom used computers, although they believed they could be helpful to learners. When learning support services are offered, they focused on language learning materials, advice, needs analysis, process monitoring, planning the learning process, and assessment. Thus, self-access centers by their nature tend to be more conducive to autonomous training rather than group or curriculum instruction geared toward learners' primary language needs. It is also difficult to match learners' use of resources at the self-access centers directly to performance, and training materials (Lazaro & Reinders, 2006; Reinders & Lazaro, 2007).

NEW TRENDS IN ICTs AND L2 LEARNING

It has been argued that technological breakthroughs in CMC have made possible the integration of meaningful and authentic communication into all aspects of the language learning curriculum (Egorov, et al., 2007; Thorne, et al. 2009). For instance, Thorn and associates argued the need for extended periods of language socialization and the salience of creative expression in L2. To this end, they discussed new media tools that offered some of the necessary socialization processes including: 'fan communities' and 'virtual gaming spaces,' each of which are discussed below.

Fan Communities

The notion that L2 learning is no longer confined to traditional classroom environments but rather to social network spaces is illustrated by the fan community. Fan community is centered on the hybrid and participatory nature of language practices. Egorov et al. (2009) use the example of Japanese *anime* (animation) and comics to illustrate how the Asian popular culture allows the youths learning English to use hybrid linguistic practices such as coded switching and writing in multiple languages to get their ideas across. As a result, the fan community is believed to offer non-proficient speakers opportunities to practice their skills without the added anxiety of formal criticism (see Lam, 2000). Furthermore, Black (2006, 2008) describes how English learners (i.e., ESL) compose and publicly post their fictional narratives on *Fanfiction.net*. Fan fiction involves what is known as remix composition, where participants begin with preexisting media, such as a book, movie, or video game and rework it or add new materials such as music or images to extend the original work. To this end, Thorne et al. (2009), conclude that remixing in fan fiction demonstrates ways by which already available media provides resources for utilizing language and other semiotic resources to create symbolic works. Furthermore, because users were able to use multiple languages and their knowledge of Asian culture, they were able to create highly popular fan fiction stories and display expertise as writers.

At the same time, their expert status served as a way to overcome their struggles with composing in English or L2 (Black, 2005; Thorne, et al. 2009). Black (2008) illustrates with a participant who received more than 7000 comments or feedback over a three-year period. While some of the feedback was supportive of the participant, other reviews point to the participant's struggle with English (e.g., grammatical error and rhetorical choices) (Black, 2008; Thorne, et al., 2009). However, the dialogue between the participant and reviewers was also viewed as collaborative and participatory writing given that the participant used the reviews to edit and repost the revised text (Black, 2008; Thorne, et al., 2009)

Virtual Environments and Online Gaming

Virtual environments (VE) consist of the Multiuser virtual environments such as Second Life, 3-D gaming spaces designed to facilitate fantasy world settings (e.g., World of War Craft) and online environments targeted for education goals (e.g., Quest Atlantis, Second Life) (Thorne et al., 2009). Specific L2 research focusing on VE is very limited. Nonetheless, Thorne et al. (2009) contend that, "participation in virtual environments constitutes a set of international cultural practices that have contributed to an overall shift in the perception and construction of reality, including the political, economic, educational, and social choices that people make in the real world" (p. 807). World of War craft is considered the most popular online game, hosting more than 14 million players world-wide and supporting the following languages: Chinese, English, French, German, Korean, Russian, and Spanish (Thorne et al., 2009).

The communication mode in VE is synchronous text-based with digital avatars that serve as on-screen identity representations of players or participants. VE provides multiple synchronous text channels to allow communication with virtual co-participants and players. It also offers a *whisper* channel for one-to-one interaction anywhere within the virtual world, an asynchronous mail-style tool, along with increasing voice capabilities. A study explored how VE, in particular Second Life, was helpful in L2 (i.e., Zheng, Li, & Zhao (2008). Zheng et al. (2008) explored Second Life for Chinese language and culture, where preliminary findings showed that Second Life offered opportunities for negotiation of meaning while developing high levels of engagement with participants. Notwithstanding, Thorne et al. (2009) called this study into question because it did not extend findings on previous negotiations, such studies that focused on text-based synchronous CMC chat communities (e.g., Abrams, 2006, Smith, 2004). Thorne et al. (2009) suggested the need for CALL in VE to move beyond text-based CALL paradigms and specifically addressed avatar-rendered identity in VE. Specifically, they argued as to whether communication, behavior, and self-perception differ when human agents assume graphical avatar? Other studies explored whether one's avatar impacts ones behavior and found that the attractiveness of one's avatar influenced self-disclosure and perception of friendliness of the user, along with the fact that users with taller avatars demonstrated more confident verbal behaviors (Yee & Bailenson, 2007; Thorne, et al., 2009). However, Yee and Bailenson (2007) also found support for the *proteus effect*, where digital self-presentations are different of other's perception. Consequently, possible experimentation and different use of avatars in VE is believed to offer useful opportunities for L2 learning (Lee & Hoadley, 2007; Thorne, 2009). For instance, VE through several identity switching is believed to aid L2 learners in skills acquisition for analyzing socio-cultural contexts affecting L2 (Thorne et al., 2009). Notwithstanding, it is yet to be demonstrated the actual impacts of VE on L2 learning outside of these speculations of possible potentials. Therefore, whether the potential translates to actual L2 learning is not clear. As a matter of fact, Thorne, et al., (2009) acknowledged similar limitation with online gamers that their specialized communication may have limited applicability or transfer to other contexts.

Thorne (2008c) explored specific multilingual interaction in World of War Craft between two players (a Russian and an American) interacting in their own language, which created a curiosity and eventual desire to learn each other's languages. For instance, it was reported that during the 30 minutes interaction, the American player would periodically IM friends raised in Ukraine to ask for phrases in Russia in order to facilitate communication with the Russian player. Unfortunately, it was reported that some of these phrases were ridiculously vulgar, though the Russian player was good-natured about the interaction. Thorne et al. (2009) contended that the transcript offers positive ground for L2 learning in terms of providing a venue for *natural* and *unscripted* interaction, reciprocal alterations, and explicit self correction at levels of linguistic form, while offering an emotional bond and motivation for learning others' language. However, given the concern for pedagogical use of technology, demonstrated mastery of L2 is questionable, although the technology provided opportunities for L2 learners to gain good exposure to co-participants' cultures (Olaniran, 2009). In essence, exposure to surface levels, rather than deep levels, of linguistic form is the norm.

Implications

There appears to be a disconnection about ICTs and L2 learning. The literature often assumes that simply because technologies can aid L2 learning, it is a good substitute. However, this is not the case. Consequently, Garrett (2009) argues that language teachers need to access the goal of ICTs in L2 learning regarding the level of language learning and the type of language learners. Answering these questions would help determine the degree to which technology can be incorporated into L2 learning along with how such usage can be measured and evaluated to determine contextual appropriateness and effectiveness in L2.

Furthermore, the need to determine the contextual appropriateness reinforces the impact of culture and technology affordability and access. For instance, Computers and VE technologies are taken for granted in particular in the U.S. and there is an assumption that everyone has access to these technologies when that is not the case elsewhere. The issues of digital divide and technology access and adoption have been addressed in the literature. However, these issues play a significant role on if and how one uses technology. The disparity between economically developed countries and less economically developed ones often determines access to technologies in e-learning and in L2 in particular. It is difficult to convince administrators to commit their scarce resources to acquisition of CALL technologies especially in the less economically developed region of the world (Garrett, 2009b; Olaniran, 2007, 2009). Even in economically vibrant societies, where access to the WWW and other resources abound, appropriate use of technologies in L2 courses is questionable. For example, simply providing Web links to L2 material does not constitute adequate or appropriate CALL (Garrett, 2009). Garrett contends that "the real challenge is, as it always has been, developing the activities that will integrate the content of authentic materials into the language learning process and engage students" (p. 723). Notwithstanding teaching L2 skills with technology requires teaching in-depth non superficial trans-cultural concepts that require adequate presentation in CALL technologies or in traditional environment. According to Garrett (2009b), not every teacher is adept in this mode of instruction which she called the *top-down processing* that requires more class time than can be devoted.

The role in L2 learning may be secondary rather than primary. Nothing in the new trends or advancement in technology for L2 language has changed that. If the goal of L2 learning is to develop surface rather than in-depth comprehension of a language, ICTs may be able to stand alone or function as the primary mode of learning. If on the other hand comprehension in terms of grammar and cultural or contextual accuracy is the goal, then a more traditional supplemented by CALL appears to be better equipped to address the situation (Garrett, 2009b, Olaniran, 2009; Thorne et al., 2009). Garrett (2009) discussed how to become trans-lingual and trans-cultural in saying "To be successful historians, environmental engineers, social psychologists or economist in the 21st century they must function comfortably in cultures other than their own" (p. 724). The economic globalization made this even more so today than in the past (Olaniran, 2001, 2004, 2007, 2009). ICTs can provide innovative applications of technologies to language learning curriculum. For instance, the applications of ICTs (synchronous or asynchronous) can add increased flexibility to teachers and language learners in how the course is taught, while helping to adapt students to innovative ways of learning language. For instance, with the aid of ICTs, students can become the focus of attention in learning. Language teachers can bring into the courses ICTs and accompanying resources to help student grasp the course contents while allowing students to access the resources at their own pace and adapt the technologies in ways that accommodate different learning needs and styles and perhaps at their own pace (Olaniran, 2009). The use of new technologies, in essence, can hold the promise of flexibility and give greater accessibility to secondary resources for students but not as stand-alone knowledge dissemination (Olaniran, 2004; Reinders & Lazaro, 2007; Zapata & Sagarra, 2007).

However, in order to realize the promise of ICTs in language learning, it is imperative that some degree of structure and guidance needs to be in place for learners (e.g., Garrett, 2009b; Salmon, 2004; Olaniran, 2009). First of all, it is apparent that basic or first time language learners need not be left alone to fend for themselves in language learning (Olaniran, 2009). Thus, a Web assisted course structure in which there are skilled instructors that understand how to adapt ICTs into traditional course content delivery are needed to make this happen. From within the classroom, instructors can then apply ICTs, such as interactive Web boards, videoconferencing and others to create interactive environments that facilitate student-to-student interaction, as well as student-to-teacher interaction in a learning environment. This approach would allow students to not only understand language phrases, but also to practice cultural appropriateness of those phrases with their classmates and others (i.e., language native speakers) on and offline (Olaniran, 2009).

FUTURE TRENDS

If ICTs are to take hold or benefit L2 learning, administrators would need to play a bigger role. Currently, It appears that administrators do not realize that L2 technology needs is different from the general IT needs. Technology use for L2 learning is very different from general IT needs of other instructors. However, Garrett (2009b) argues that on many campuses, language media and resource centers have been taken over by IT services that turned those centers into general multipurpose computer labs. This is a tremendous oversight and disservice to L2 curricula that needs to change. Similarly, with increasing globalization, there is a huge need to figure out a way to address or teach less commonly taught languages (LCTLs). Perhaps this is one area where ICTs may actually hold promise. Not all institutions can afford to teach some of these languages, thus, application of technology to handle most of the knowledge distribution will be appropriate where the final certification will be done by teachers or examiners from other institutions that offer them (See Garrett, 2009). With most L2 technologies used to assist practices of form and conversational or oral communication skills, there needs to be a renewed commitment and emphasis on grammar skills. Garrett (2009b) argues that if L2 learners are to move from basic levels to advanced ones, they will need a foundation in *grammar*, *reading* and *writing* to justify participation in advanced level courses. To this end, she suggests a rethinking of the way grammar is taught in advance L2 courses in terms of following non-linear and non-hierarchical while addressing the complex relationships between grammar and vocabulary.

Social networking is the current rave in ICTs, however its use for L2 learning is still in its infancy. Cases and scenarios presented in this chapter about their use are still anecdotal and inconclusive. The degree to which they can help address some of the challenges with L2 learning needs to be explored further. For instance, can they be used to address the complex relationship between grammar and vocabulary? To what degree do these technologies help deep learning and does the learning in the forum translates to actual language mastery or comprehension? Can Artificial Intelligence (AI) help in L2 learning? These are few questions that needs to be addressed as one moved forward about L2 learning.

CONCLUSION

This chapter explored the role of ICTs in language learning, while alluding to the impact of culture in learning. The chapter presented the dimensions of cultural variability to address implications for language acquisition, while exploring synchronous and asynchronous ICTs along with their position in language learning. New trends in technology and L2 were explored as well as identifying the latest developments in CALL and L2 learning. From the argument and literature review, the author concludes that the role of ICTs in L2 learning has not changed and that ICTs are better at supplementing, rather than acting as stand alone in L2 learning. The chapter concludes with some issues that need to be addressed as this area of research moves forward.

REFERENCES

Abrams, Z. (2006). From Theory to practice: Intracultural CMC in the L2 classroom. In Duncate, L., & Arnold, N. (Eds.), *Calling to CALL* (pp. 181–209). San Marcos, TX: Computer Assisted Language Consortium.

Bates, T. (1999, September). *Cultural and ethical issues in international distance education.* Paper presented at the *UBC/CREAD* conference, Vancouver, Canada. Available at: http:ilbate.o.cstudies.ubc.ca/pdfiCR~"AD.pdt.

Black, R. W. (2005). Access and affiliation: The literacy and composition practices of English language learners in an online fan fiction community. *Journal of Adolescent & Adult Literacy, 49*(2), 118–128. doi:10.1598/JAAL.49.2.4

Black, R. W. (2006). Language, culture, and identity in online fan fiction. *E-learning, 3*(2), 170–184. doi:10.2304/elea.2006.3.2.170

Black, R. W. (2008). *Adolescent and online fan fiction.* New York: Peter Lang.

Bober, M. J., & Dennen, V. P. (2001). Intersubjectivity: Facilitating knowledge construction in online environments. *Education Media International, 38*(4), 241-250.

Bodycott, P., & Walker, A. (2000). Teaching abroad: Lessons learned about inter-cultural understanding for teachers in higher education. *Teaching in Higher Education*, 5(1), 79–94. doi:10.1080/135625100114975

Brown, A., & Iwashita, N. (1996). Language background and item difficulty: The development of a computer-adaptive test of Japanese. *System, 24*(2), 199–206. doi:10.1016/0346-251X(96)00004-8

Chou, C. (2001). A model of learner-centered computer-mediated interaction for collaborative distance learning. In M. R. Simonson (Ed), *Annual Proceedings of Selected Research and Development [and] Practice Papers Presented at the National Convention of the Association for Educational Communications and Technology* (AECT) (pp. 74-80), Atlanta, Georgia.

Clawson, R. A., & Choate, J. (1999). Explaining participation on a class newsgroup. *Social Science Computer Review, 17*, 455–459. doi:10.1177/089443939901700406

D'Haenens, L., Koeman, J., & Saeys, F. (2007). Digital citizenship among ethnic minority youths in the Netherlands and Flanders. *New Media & Society, 19*(2), 278–299. doi:10.1177/1461444807075013

Davie, L., & Wells, R. (1991). Empowering the learner through computer-mediated communication. *American Journal of Distance Education, 5*, 15–23. doi:10.1080/08923649109526728

del Puerto, F., & Gamboa, E. (2010). The evaluation of computer-mediated technology by second language teachers: collaboration and interaction in CALL. *Educational Media International, 46*(2), 137–152. doi:10.1080/09523980902933268

Driscoll, M. P. (2000). *Psychology of learning instruction* (2nd ed.). Boston: Allyn and Bacon.

Egorov, V., Jantassova, D., & Churchill, N. (2007). Developing pre-service English teachers' competencies for integration of technology in language classrooms in Kazakhstan. *Educational Media International, 44*(3), 255–265. doi:10.1080/09523980701491732

Ess, C., & Sudweeks, F. (2005). Culture and computer-mediated communication: Toward new understandings. *Journal of Computer-Mediated Communication, 11*(1), Retrieved on July 17, 2006 from, Http://jcmc.indiana.edu/voll11/issue1/ess.html.

Fougere, M., & Moullettes, A. (2007). The Construction of the Modern West and the Backward Rest: Studying the Discourse of Hofstede's Culture's Consequences. *Journal of Multicultural Discourses*, *2*(1), 1–19. doi:10.2167/md051.0

Garrett, N. (2009a). Technology in the service of language learning: Trends and new Issues. *Modern Language Journal*, *93*, 697–718. doi:10.1111/j.1540-4781.2009.00968.x

Garrett, N. (2009b). Computer-Assisted Language Learning Trends and Issues Revisited: Integrating Innovation. *Modern Language Journal*, *93*, 719–741. doi:10.1111/j.1540-4781.2009.00969.x

Gudykunst, W. B., Chua, E., & Gray, A. J. (1987). Cultural dissimilarities and uncertainty reduction processes. In McLaughlin, M. (Ed.), *Communication Yearbook* (*Vol. 10*, pp. 457–469). Beverly Hills, CA: Sage.

Gunawardena, C. N., Lowe, C. A., & Anderson, T. (1997). Analysis of a global online debate and the development of an interaction analysis model for examining social construction of knowledge in computer conferencing. *Journal of Educational Computing Research*, *17*(4), 397–431. doi:10.2190/7MQV-X9UJ-C7Q3-NRAG

Hofstede, G. (1980). *Culture's consequences*. Beverly Hills, Ca: Sage.

Hofstede, G., & Bond, M. (1984). Hofstede's culture dimensions: An independent validation using Rokeach's value survey. *Journal of Cross-Cultural Psychology*, *15*, 417–433. doi:10.1177/0022002184015004003

Hofstede, G. H. (2001). *Culture's consequences: Comparing values, behaviors, institutions, and organizations across nations*. Thousand Oaks, CA: Sage.

Kanuka, H., & Garrison, D. R. (2004). Cognitive presence in online learning. *Journal of Computing in Higher Education*, *15*(2), 30–49. doi:10.1007/BF02940928

Kayman, M. (2004). The state of English as a global language: Communicating culture. *Textual Practice*, *18*(1), 1–22. doi:10.1080/0950236032000140131

Lam, W. S. E. (2000). Literacy and the design of self: A case study of a teenager writing on the Internet. *TESOL Quarterly*, *34*, 457–482. doi:10.2307/3587739

Larsen-Freeman, D., & Long, M. (1991). *An introduction to second language acquisition research*. Harlow, Essex: Longman.

Lazaro, N., & Reinders, H. (2006). Technology in self-access: an evaluative framework. *PacCALL Journal*, *1*(2), 21–30.

Linders, L., & Goosens, (2004). Bruggen bouwen met virtuele middelen [Building bridges with virtual tools] In J. de Haan and O. Klumper (eds.) *Jaarboek ICT en samenleving: beleid in praktijk [ICT Yearbook and Society: Policy and Practice]* pp. 121-139. Amsterdam: Boom Publishers.

Lobel, M., Neubauer, M., & Swedburg, R. (2005). Comparing how students collaborate to learn about the self and relationships in a real-time non- turn-taking online and turn-taking face-to-face environment. *Journal of Computer-Mediated Communication*, *10*(4), 18. Available at http://jcmc.indiana.edu/vol10issue4/lobel.html.

MacFadyen, L. (2008). The perils of parsimony: "National culture" as red herring? *Proceedings of the Sixth International Conference on Cultural Attitudes towards Technology and Communication. Nîmes, France, June 2008*

Martin, J., & Cheong, P. (2008). Cultural Considerations of Online Pedagogy. In St.Amant, K., & Kelsey, S. (Eds.), *Computer-Mediated Communication across Cultures: International Interaction in Online Environments*. Hershey, PA: IGI Global.

McAlister, S., Ravenscroft, A., & Scanlon, E. (2004). Combining interaction and context design to support collaborative argumentation using a tool for synchronous CMC. *Journal of Computer Assisted Learning, 20*, 194–204. doi:10.1111/j.1365-2729.2004.00086.x

Morse, K. (2003). Does one size fit all? Exploring asynchronous learning in a multicultural environment. *Journal of Asynchronous Learning Networks, 7*(1), 37–55.

Olaniran, B. (2006). Challenges to implementing e-learning and lesser developed countries. In Edmundson, A. L. (Ed.), *Globalized e-learning cultural challenges* (pp. 18–34). Hershey, PA: Idea Group, Inc.

Olaniran, B. A. (1994). Group performance and computer-mediated communication. *Management Communication Quarterly, 7*, 256–281. doi:10.1177/0893318994007003002

Olaniran, B. A. (1996). Social Skills Acquisition: A closer look at foreign students and factors influencing their level of social difficulty. *Communication Studies, 47*, 72–88.

Olaniran, B. A. (2001). The effects of computer-mediated communication on transculturalism. In Milhouse, V., Asante, M., & Nwosu, P. (Eds.), *Transcultural Realities* (pp. 83–105). Thousand Oaks, CA: Sage.

Olaniran, B. A. (2004). Computer-mediated communication as an instructional Learning tool: Course Evaluation with communication students. In P. Comeaux (Ed.), *Assessing Online Teaching & Learning*, 144-158.

Olaniran, B. A. (2006). Applying synchronous computer-mediated communication into course design: Some considerations and practical guides. *Campus-Wide Information Systems. The International Journal of Information & Learning Technology, 23*(3), 210–220.

Olaniran, B. A. (2007). Culture and communication challenges in virtual workspaces. In K. St-Amant (ed.), *Linguistic and cultural online communication issues in the global age* (pp. 79-92*)*. Hershey, PA: Information science reference (IGI Global).

Olaniran, B. A. (2009). Culture and Language Learning in Computer-Enhanced or Assisted Language Learning. *International Journal of Communication Technologies and Human Development, 1*(3), 49–67. doi:10.4018/jicthd.2009070103

Olaniran, B. A., Savage, G. T., & Sorenson, R. L. (1996). Experiential and experimental approaches to face-to-face and computer mediated communication in group discussion. *Communication Education, 45*, 244–259. doi:10.1080/03634529609379053

Olaniran, B. A., Stalcup, K., & Jensen, K. (2000). Incorporating Computer-mediated technology to strategically serve pedagogy. *Communication Teacher, 15*, 1–4.

Osman, G., & Herring, S. (2007). Interaction, facilitation, and deep learning in cross-cultural chat: A case study. *The Internet and Higher Education, 10*(2), 125–141. doi:10.1016/j.iheduc.2007.03.004

Patsula, P. J. (2002). Practical guidelines for selecting media: An international perspective. *Usableword Monitor* (February 1). Available at: http:/uweb.txstate.edu/~-db15/edtc5335/docs/mediaselection_criteria.htm

Putnam, R. (2000). *Bowling alone, the collapse of and revival of civic America*. New York: Simon & Schuster.

Reinders, H., & Lazaro, N. (2007, April). Innovation in language support: The provision of technology in self access. *Computer Assisted Language Learning*, *20*(2), 117–130. doi:10.1080/09588220701331428

Roach, K. D., & Olaniran, B. A. (2001). Intercultural willingness to communicate and communication anxiety in International Teaching Assistants. *Communication Research Reports*, *18*, 26–35.

Salmon, G. (2004). *E-tivities. Der Schlussel zu aktivizen online-lernen*. Zurich: Orell Fussili Verlag AG.

Schank, R. C., & Cleary, C. (1995). *Engines for education*. Hillsdale, NJ: Lawrence Erlbaum Associates.

Smith, B. (2004). Computer-mediated negotiated interaction and lexical acquisition. *Studies in Second Language Acquisition*, *26*, 365–398. doi:10.1017/S027226310426301X

Smith, B., Alvarez-Torres, M., & Zhao, Y. (2003). Features of CMC technologies and their impact on language learners' online interaction. *Computers in Human Behavior*, *19*, 703–729. doi:10.1016/S0747-5632(03)00011-6

Smith, P. B. (2002). Culture's consequences: Something old and something new. *Human Relations*, *55*(1), 119–135.

Teng, T. L., & Taveras, M. (2004-2005). Combining live video and audio broadcasting, synchronous chat, and asynchronous open forum discussions in distance education. *Journal of Educational Technology Systems*, *33*(2), 121–129. doi:10.2190/XNPJ-5MQ6-WETU-D18D

Thorne, S. (2008). Transcultural communication in open internet environments and massively multiplayer online games. In Magnan, S. (Ed.), *Mediating Discourse online* (pp. 305–327). Amsterdam: Benjamins.

Thorne, S., Black, R., & Sykes, J. (2009). Second Language Use, Socialization, and Learning in Internet Interest Communities and Online Gaming. *Modern Language Journal*, *93*, 802–821. doi:10.1111/j.1540-4781.2009.00974.x

Usun, S. (2004). Factors affecting the application of information and communication technologies (ICT) in distance education. *Turkish Online Journal of Distance Education,* 5(1). Available at: http://tojde.arladolu.edu.tr/tojde13/articles/us\m.html

Walker, S. A. (2004). Socratic strategies and devil's advocacy in synchronous CMC debate. *Journal of Computer Learning*, *20*, 172–182. doi:10.1111/j.1365-2729.2004.00082.x

Yee, N., & Bailenson, J. (2007). The proteus effect: The effect of transformed self-representation on behavior. *Human Communication Research*, *33*, 271–290. doi:10.1111/j.1468-2958.2007.00299.x

Zapata, G., & Sagarra, N. (2007). CALL on hold: The delayed benefits of an online workbook on L2 vocabulary learning. *Computer Assisted Language Learning*, *20*(2), 153–171. doi:10.1080/09588220701331352

Zheng, D., Li, N., & Zhao, Y. (2008). *Learning Chinese in Second Life Chinese language School*. Paper presented at CALICO Annual Conference. San Francisco, CA.

Zobl, H. (1983). Markedness and the projection problem. *Language Learning*, *33*(3), 293–313. doi:10.1111/j.1467-1770.1983.tb00543.x

Chapter 13
Enhancing the Educational Experience of Calabrian Cultural Heritage: A Technology-Based Approach

Eleonora Pantano
University of Calabria, Italy

Assunta Tavernise
University of Calabria, Italy

ABSTRACT

The aim of this chapter is to illustrate a technology-based approach for promoting and diffusing Calabrian cultural heritage of the ancient Magna Graecia period (VIII cent. B.D.- I cent. A.D.) in a global perspective.

To achieve this goal, the chapter focuses on the use of 3D technologies, on virtual and augmented reality, with emphasis on the stereoscopic Virtual Theatre. These innovative tools support the creation of a global vision of the fragmentary archaeological Calabrian heritage, as well as the possibility to play with the virtual findings as in a videogame, by choosing what to explore and the contents to access. Moreover, these technologies exploit the entertaining components of the systems in order to provide personalized and interactive educational contents.

INTRODUCTION

In recent years, a great deal of research in the development of efficient Information and Communication Technologies (ICTs) has driven to an increasing effort in the realization of virtual cultural goods, in order to support and promote knowledge transfer related to Cultural Heritage (Knipfer et al., 2009). As a consequence, the new

DOI: 10.4018/978-1-60960-497-4.ch013

concept of "Virtual Heritage" has been introduced, referring to the use of three-dimensional computer modelling in order to virtually reconstruct monuments, buildings, and finds (Roussou, 2002; Drettakis et al., 2005). Hence, an increasing amount of both digitalized museum materials (Styliadis et al., 2009) and cultural heritage educational contents have become available on the web, stimulating interest and curiosity especially in young people (Parry, 2005; Cutri et al., 2008; Tonta, 2008). These materials offer a new kind of experience that is formative and amusing at the same time (Petric et al., 2003; Mason & McCarthy, 2006; Owston et al., 2009; Adamo et al., 2010): this new kind of learning has been called "Edutainment", thanks to the mixture of the two terms "education" and "entertainment" (Bilotta et al., 2009; Pantano & Tavernise, 2009).

Knowledge transfer and acquisition in a both powerful and stimulating way enriches the experience, by focusing on a technology approach capable of supporting a deeper understanding of the artistic heritage in all of its components. Moreover, an archaeological site acquires visibility at a global level through the emphasis on its distinctive aspects (Bertacchini et al., 2007). In this view, Calabria region in Southern Italy represents a meaningful example of a very important but almost unknown archaeological site. In fact, it is a surely underestimated territory that offers a wide cultural and artistic patrimony as heritage of the Greek colonial expansionism in the Mediterranean area, which dates back from VIII cent. B.D. to II cent. A.D. (for this reason, the zone is also known as "Magna Graecia").

The present work illustrates how a technology-based approach linked to Virtual Heritage, based on virtual tours, navigation system, virtual theatre, and virtual museum have been applied to Calabrian cultural patrimony in order to enhance users' learning experience, as well as to promote the territory.

BACKGROUND

Emerging technologies such as virtual worlds have been heralded as powerful tools capable of radically transforming learning and teaching. Moreover, mobile technologies and internet are daily used by learners for both accessing to rich digital media contents and communicating with others in order to reach new didactic experience beyond the traditional ones in classroom (Bertacchini et al., 2008; Kuznik, 2009). For instance, educational involvement in the 3D online world "Second Life" has become fashionable (Herold 2010), as well as engaging web-based communication platforms have been realized in order to allow students to easily access different learning tools, such as program information, course contents, teaching assistance, discussion boards, document sharing systems, and learning resources (Chen, 2009). Many claims have been made about the "added value" which can be gained from interacting with these kinds of virtual representations, such as easier learning, better understanding and training, more engagement and pleasure (Scaife & Rogers, 2001), social skill development, high motivation (Hamalainen, 2008). For this reason, some advanced tools have been used as powerful device for training people with certain disabilities (Parsons et al., 2006), whereas user interfaces have become more intuitive by both following the requirements of the individual learner and reinforcing the drive towards more personalized didactic and greater educational autonomy (de Freitas & Neumann, 2009). Therefore, the aim of an edutainment virtual tool is to provide students with challenges related to the learning task (Kiili, 2005), by exploiting one of the main characteristic of cultural heritage virtual environments related to the possibility to provide new experiences, by enabling users to interact with objects and navigate in 3D space in ways usually possible in the physical world (e.g. ruined or fragmentary objects, disappeared locations). In fact, anything that has been present in ancient daily life is virtually realized on a computer through the interpretation of the remaining tangible evidences: cities, specific buildings and squares, houses and the objects in them. Hence, "intangible heritage" is displayed, according to the UNESCO (2005) perspective which see this kind of heritage as "the practices, representations, expressions, knowledge, skills - as well as the instruments, objects, artefacts and cultural spaces associated therewith - that communities, groups and, in some cases, individuals recognize as part of their cultural heritage". Therefore, a great number of associated information (i.e. archaeological data, aerial photos, texts) are available and can be fruited through a mouse click.

Regarding the immersion in the virtual reconstructions, simulated environments enable learners to assume roles in particular contexts and have meaningful, authentic experiences (Slator et al., 1999).

Like Alice walking through the looking glass, learners immerse themselves in distributed and synthetic environments, though "avatars", who represent virtual *alterego* and support users in the learning-by-doing process using virtual artifacts to construct knowledge (Walker, 1990). Hence, this new kind of learning overcome the experiences provided by static 2D images, which usually consist of photographs of cultural ruins, buildings, drawings, and fragmentary information referring to specific finds as well as to wide scenarios. In fact, users may exploit the navigation in virtual reconstructed worlds in order to build different paths for an interactive fruition of 3D objects. In this way, the learning process becomes more attractive and exciting, as well as particularly successful due to its visual characteristics, caused by the almost immediacy of images in relation to the sequential nature of texts and sounds (Pantano & Tavernise, 2009). Moreover, although current Virtual Navigation Systems (VNS) have been developed as a desktop application to stimulate visits for a wide range of environment, the capabilities of VNS have also been combined with the high performance of mobile devices, providing exciting features such as allowing the user to have a combined real and virtual tour depending on his/her location (Cutrì et al., 2008). The most used devices are tablet pc, pocket pc, smart-phone, Personal Digital Assistant (PDA) (Cutrì et al., 2008; Alfano & Pantano, 2010), iPhone, iPod and iPad (Pantano & Servidio, in press). For instance, many museums, art galleries and cultural parks have developed podcasts (files collections for iPod, iPhone, iPad) free downloadable by their web site, in order to promote collections and/or attract more visitors. In this way, these devices support users during their experience, by providing several detailed and interactive contents (videos, photos, texts). Therefore, since the rapid diffusion of iPad due to their high quality of display and computational efficiency, in the field of cultural heritage knowledge transfer has become more enjoyable and attractive. Figure 1 shows a possible visualization of archaeological park of Lokroi on an iPad.

Hence, users can access to the information by both a "remote connection", which requires a previous download of the favourite contents, and a real time one, by using an internet connection. Furthermore, the integration with GPS (Global Positioning System) improves the VNS functionalities, by providing more exciting features such as allowing the user to have an effective combination of real and virtual information depending on him/her location. Since GPS recognizes user's geographical position, the system supports him/her to easily identify the most convenient access path to the findings (Worboys, 2004; Mohino, 2005; Tait, 2005; Taylor, 2006), and to manage various types of data, traditional contents such as images, textual documents, photos, videos or innovative ones such as three-dimensional models, virtual reconstructions), through an innovative approach devoted to a geographical access to information. In addition, GPS allows the management of thematic maps consisting of different superimposed layers (each layer is a spatial dataset containing a common feature type represent-

Figure 1. Example of visualization of virtual reconstructions though iPad

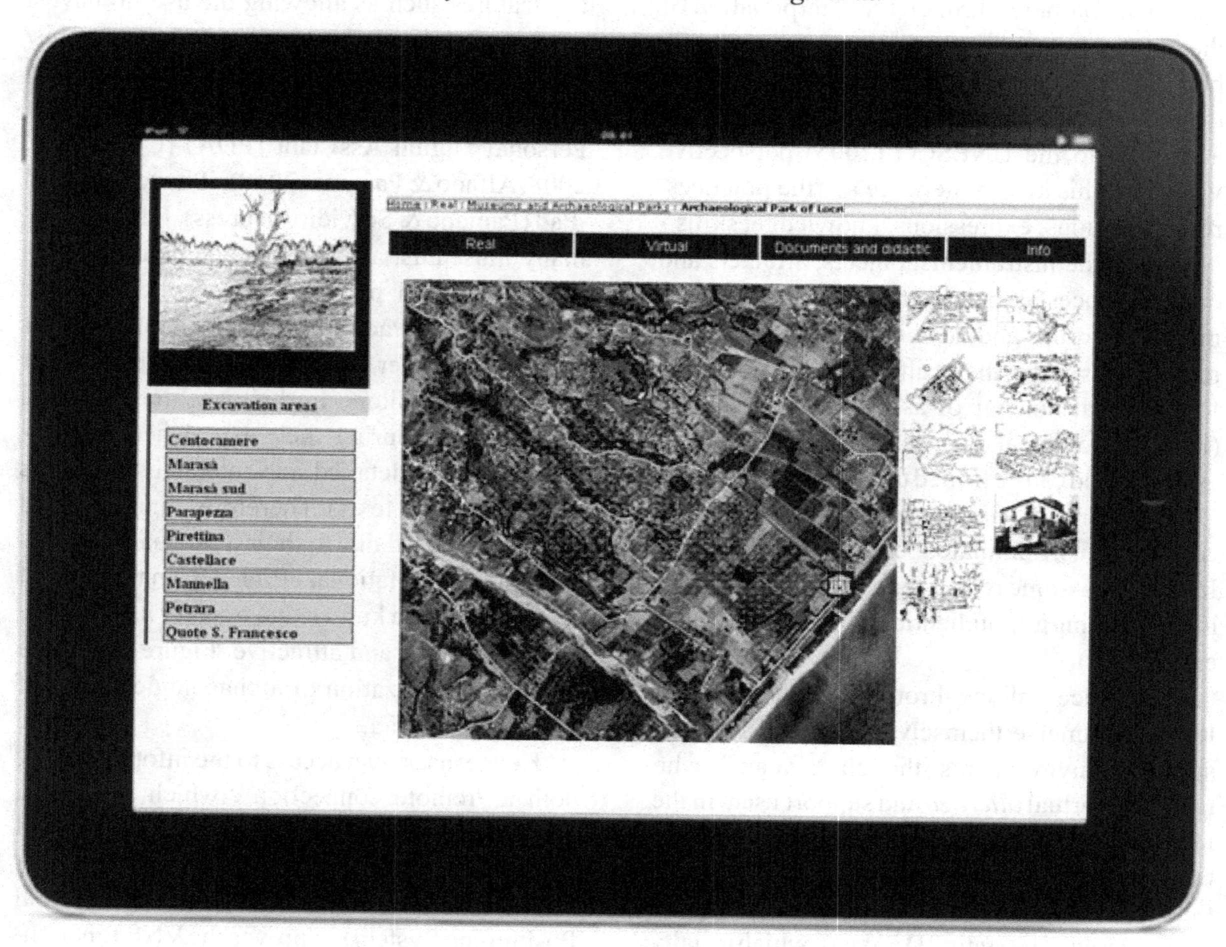

ing objects, events and phenomena from real world) and extracting interesting aspects of a territory (Pantano & Servidio, in press). Therefore, user is able to virtually travel across the area, increasing or reducing the scale level of the representation and choosing the features more interesting to visualize. Hence, these systems are expected to be capable of attracting a wider audience.

MAIN FOCUS OF THE CHAPTER

In this chapter a technology-based approach linked to Calabrian Cultural Heritage for enhancing users' learning experience is introduced. In particular, virtual tours, a virtual theatre, and a virtual museum, for the exploitation of the Magna Graecia cultural patrimony are presented. Regarding virtual tours, the "Connecting European Culture through New Technology - NetConnect" project (http://www.netconnect-project.eu), has been carried out with the purpose of investigate the current use of technology to make heritage more visible and accessible, as well as to share and highlight common cultural patrimonies of European significance. Hence, it has concerned the definition of interconnections among three archaeological sites from an historical point of view, as well as the 3D reconstruction of three

Figure 2. The multimedia section in the Virtual Tour: video (links) and text (right)

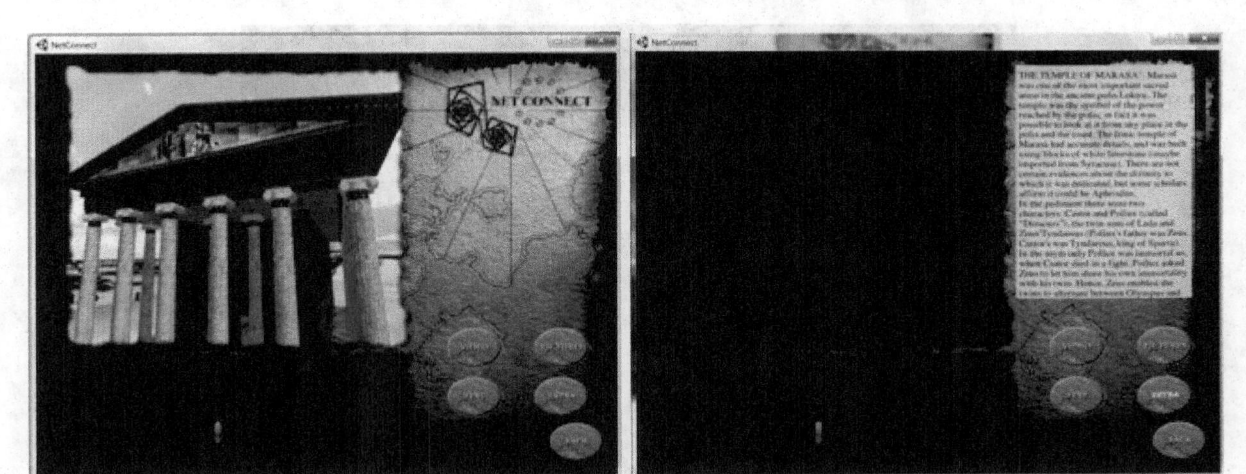

virtual sites incorporating information about the daily life of the age of reference: Lokroi in Italy, Glauberg in Germany, and Biskupin in Poland. In these virtual worlds "Virtual Tours" (VTs) are allowed, representing the chance of a convincing immersive experience in computer environments, emerging from the model of cities/landscapes which existed in the past and are lost nowadays (available on the portal of NetConnect project, through the section "3D Reconstructions").

Each environment follows an approach similar to videogames, which provides users innovative, attractive and stimulating experiences. In fact, user can utilize the standard mouse, keyboard or joystick. Furthermore, the navigation can be manipulated through a choice in a menu of the different points of interest; a map present on the right side of the screen shows the user's position in the virtual environment. Therefore, users have the opportunity of exploring the different elements of the cultural patrimony, carrying on unique and personalized experiences through the placement of activities and tasks that can support learners' own interests. Moreover, in these virtual worlds 3D imagery is used to create immersive experiences, making possible innovative ways of approaching learning by integrating a range of different tools via a single user interface (de Freitas, 2006) such as videos, texts, pictures, and animations (Figure 2).

Furthermore, the superimposed and interchangeable view of the virtual object and the real one supplements and enhances learning solutions. Since the value of manipulation of virtual objects, according to learning principles of constructivism, user can interact with all objects present in the scenario and manipulate them in order to improve the experience (Figure 3).

Humans are virtually represented by avatars, who show ancient clothes, whereas animals cross the streets like in the past (Figure 4). The characters animating the VT have been realized on the basis of vascular paintings and other documentation from bibliographical researches realized on purpose. In fact, the mere presence of characters has a positive impact on students' perception of learning experience, called the "persona effect" (Lester et al., 1999). On the contrary, the learner acts as himself/herself, there is not the mask of an "avatar", a surrogate persona in the virtual world, but there is an "I-vision" that enhances his/her active role in the virtual world.

Moreover, the scenario presents also an "ancient" theatre, where it is possible to attend a performance carried out by virtual actors, which are animated 3D Talking Heads displaying facial

Figure 3. An example of objects that is possible to manipulate in the virtual house of Lokroi

Figure 4. Characters and animals in Lokroi virtual world

Figure 5. The three interfaces of Face3D software

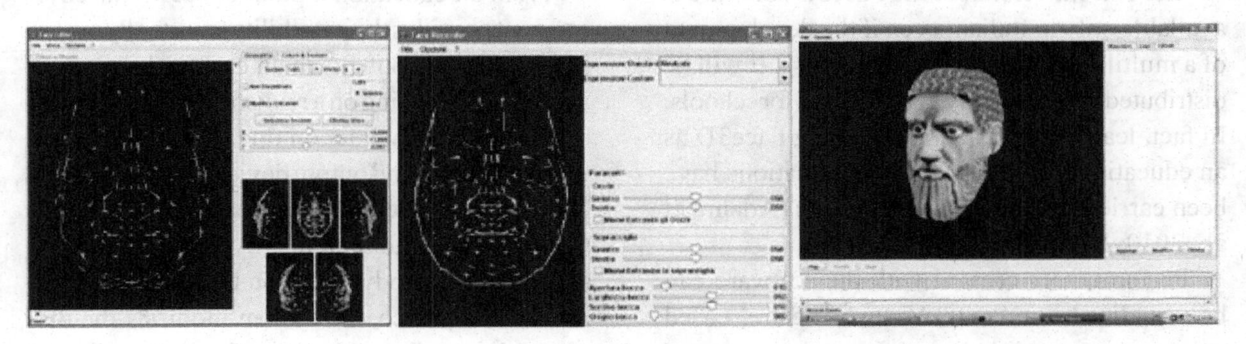

expressions. For their modelling and animation the software Face3D has been implemented. (Bertacchini et al., 2007). It consists of three Graphical User Interfaces: Face3DEditor for modelling the virtual heads, Face3DRecorder for animating them, and Virtual Theatre for realizing video performance. In particular, in Face3DEditor a parameterized head model constitutes the starting point for the generation of each mask; its vertices (n. 131) and triangles (n. 113) can be moved in order to easy create a new face. Facial expressions are synchronized with pre-recorded files of speech in Face3DRecorder: eight emotions (neutral, anger, surprise, sadness, fear, joy, disgust, attention) can be used, but also small alterations in the facial expression can be created.

In Virtual Theatre it is possible to import the animations of the Talking Heads created in Face3DRecorder, as well as to create and to manage the performance of different virtual agents. In Figure 5 the three interfaces of Face3D software are shown.

Figure 6. The Virtual Theatre

The current version of the Face3D software is available only in Italian, even if the development of a multilingual version is in progress. It will be distributed to accompany a publication for schools. In fact, learners' satisfaction in using Face3D as an educational tool in classroom situations have been carried out with positive results (Adamo et al., 2010; Bilotta et al., 2010).

Performances created with the software can be visualized in a "real" Virtual Theatre, based on stereoscopic tools in order to enhance users' experience. The technological infrastructure of this theatre consists of two special video-projectors, a wide screen, a tracking system, and a 3D workstation. It represents an effective pervasive environment, which supports the 3D exploration of the virtual reconstruction of the findings of Calabrian Magna Graecia period (Hansmann et al., 2003; Steventon and Wright, 2006). Figure 6 shows how users can visualize the 3D objects, by wearing glasses with polarized lens.

From an educational point of view, this environment provides the possibility to virtually access to educational contents in an easy and entertaining way, by focusing on a realistic and comfortable user-computer interaction due to the integration of several input and output devices (Oviatt, 2008). In fact, the interactive system can recognize user' position and movements by identifying the shadow or the body direction through a camera. As a consequence, the system modifies the displayed images by giving an effective "feeling of presence". Furthermore, several researches have been focused on its application for enhancing tourism attractiveness of destinations (Pantano & Servidio, in press).

The Virtual Theatre has also been realized in a virtual version (Figure 7), in order to have access to 3D contents also from a distance.

In fact, web-based tools seem to be very efficient for allowing a broad range of users to access knowledge related collections and educa-

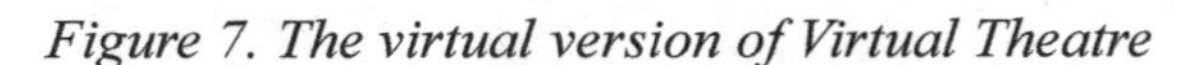
Figure 7. The virtual version of Virtual Theatre

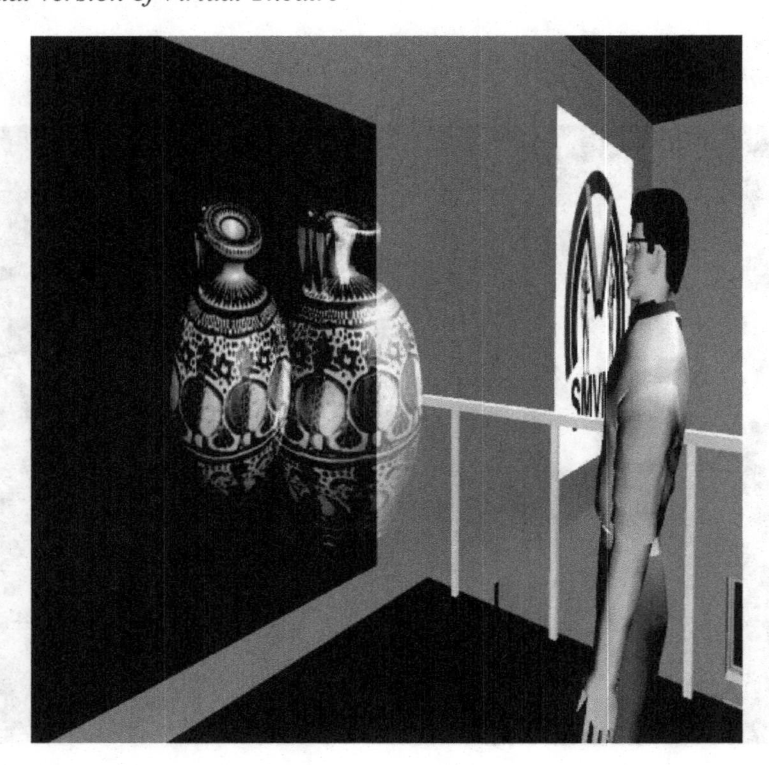

tional contents, which can be virtually experienced in an easy and interactive way (Ghiani et al., 2009; Chen, 2009; Kotler and Kotler, 2009). In this view, the Virtual Museum Net of Magna Graecia (http://www.virtualmg.net) offers a global vision of Calabrian archaeological heritage by exploiting the 3D graphics techniques (Bertacchini et al., 2006, 2007). It provides several educational contents (texts, virtual exploration of archaeological findings, videos, etc.) which users can free access to improve their knowledge. The system offers different personalized paths of fruition, thus different targets can focus on the favorite displayed modality and on the most interesting information according to their own needs.

Regarding portals, a website for NetConnect Scientific Community have been realized: its name is "NetConnect International Network on new Technologies in Europe for Cultural Heritage - NET-IN-TECH" (http://netconnect-project.eu/index.aspx). It has been designed and developed in order to provide visibility on NetConnect activities and to facilitate relations among institutions of the sector in order to obtain an efficient network of co-operating experts and a creative know-how exchange (Pantano & Tavernise, 2009).

FUTURE RESEARCH DIRECTIONS

Future studies might focus on the comparison of the use of the proposed technology for different educational contents, in order to deeply understand the most efficient for each field of application. In fact, only few researches have focused on the effective users' intention of using the most advanced technologies such as adaptive museum guides (Pianesi et al., 2009).

With this purpose, researchers might exploit the use of Technology Acceptance Model (TAM), developed by Davis (1989), which focuses on four main variables: perceived ease of use, perceived usefulness, attitude and behavioral intention. The model aims at investigating users' attitude towards a new technology, as well as their acceptance in terms of perceived ease of use (the degree to which users believe that the use of the technology requires no effort), perceived usefulness (the degree to which users believe that the use of the technology enhances their performance), intention of use and affective use (Hossian & de Silva, 2009; Hsu & Lin, 2008; Kwon & Wen, 2010). Different quantitative analysis might be based on sample of different ages, in order to understand which target is most influenced by one specific technologies.

In addition, users' consumption of 3D educational contents might become a social experience, thus they could prefer creating and consuming their favourite contents in a virtual scenario with respect to a traditional one. In this way, teachers could identify how exploit the use of these technologies *in loci*, understanding how learners make use of these systems and which factors affect their participation to the virtual scenarios. In fact, the use of advanced technologies in didactics might support the birth of innovative methods of teaching, strictly linked to Edutainment, and with a strong impact on education. Some researches have already been carried out, but further improvements are needed.

Finally, NetConnect avatars could become guides in virtual environments or be used for educational purposes.

CONCLUSION

In this chapter, an overview of the recent advances in technologies for promoting cultural Calabrian patrimony in Southern Italy as heritage of the ancient Magna Graecia (VIII cent. B.D.- I cent. A.D.) has been presented. In fact, a global vision of the fragmentary archaeological Calabrian heritage and the superimposed and interchangeable view of the real find and its reconstruction allow a promotion of the territory, providing visibility through internet divulgation as well as an emphasis on its distinctive aspects. Moreover, the efficient technology-based approach provides enriching learning experiences for different targets of users, especially young people, attracted by a powerful interaction with computers. Hence, learning is enacted accordingly to a knowledge-as-a-process view (Lau & Tsui, 2009), following predetermined educational paths, as well as providing an interactive and entertaining experience. In fact, the results of several researches have showed that students involved in learning process, carrying on unique and personalized experiences through the placement of activities and tasks that can support their own interests, acquire more information with respect to a traditional lesson (Paraskeva et al., 2010). Regarding Calabrian Magna Graecia, virtual tours, virtual reality and stereoscopic Virtual Theatre are capable of arising interest and curiosity in users, as well as providing the opportunity to play with the virtual findings as in a videogame. In fact, these technologies exploit the entertaining factor, keeping the player motivated and spreading contents according to an Edutainment perspective.

REFERENCES

Adamo, A., Bertacchini, P. A., Bilotta, E., Pantano, P., & Tavernise, A. (2010). Connecting Art And Science For Education: Learning by an Advanced Virtual Theatre with "Talking Heads". *Leonardo*, *43*(5), 442–448. doi:10.1162/LEON_a_00036

Alfano, I., & Pantano, E. (2010). Advanced Technologies for promotion of cultural heritage: the case of Bronzes of Riace. *Journal of Next Generation Information Technology*, *1*(1), 39–46. doi:10.4156/jnit.vol1.issue1.4

Arnold, D. B., & Geser, G. (2007). *Research agenda for the applications of ICT to cultural heritage*. EPOCH.

Bertacchini, P. A., Bilotta, E., Cronin, M., Pantano, P., & Tavernise, A. (2007). 3D Modelling of Theatrical Greek Masks for an Innovative Promotion of Cultural Heritage. In Posluschny, A., Lambers, K., & Herzog, I. (Eds.), *Layers of Perception. Belrin: CAA2007*.

Bertacchini, P. A., Bilotta, E., Di Bianco, E., Di Blasi, G., & Pantano, P. (2006). Virtual Museum Net. *Lecture Notes in Computer Science*, *3942*, 1321–1330. doi:10.1007/11736639_165

Bilotta, E., Gabriele, L., Servidio, R., & Tavernise, A. (2009). Edutainment Robotics as learning tool. *Lecture Notes in Computer Science*, *5940*, 25–35. doi:10.1007/978-3-642-11245-4_3

Bilotta, E., Pantano, P., & Tavernise, A. (2010). Using an Edutainment Virtual Theatre for a Constructivist Learning. In *Proceedings of the 18th International Conference on Computers in Education (ICCE 2010) - "New paradigms in learning: Robotics, play, and digital arts"*. Putrajaya – Malaysia: Asia Pacific Society for Computers in Education.

Bird, S. A. (2005). Language Learning Edutainment: Mixing Motives in Digital Resources. *RELC*, *36*(3), 311–339. doi:10.1177/0033688205060053

Cai, Y., Lu, B., Zheng, J., & Li, L. (2006). Immersive protein gaming for bio edutainment. *Simulation & Gaming*, *37*, 466–476. doi:10.1177/1046878106293677

Chen, C.-M. (2009). Personalized E-learning system with self-regulated learning assisted mechanism for promoting learning performance. *Expert Systems with Applications*, *36*(5), 8816–8829. doi:10.1016/j.eswa.2008.11.026

Cutrì, G., Naccarato, G., & Pantano, E. (2008). Mobile Cultural Heritage: The case study of Locri. *Lecture Notes in Computer Science*, *5093*, 410–420. doi:10.1007/978-3-540-69736-7_44

Davis, F. (1989). Perceived usefulness, perceived ease of use and user acceptance of information technology. *Management Information Systems Quarterly*, *13*, 319–340. doi:10.2307/249008

de Freitas, S. (2006). Using games and simulations for supporting learning. *Learning, Media and Technology*, *31*(4), 343–358. doi:10.1080/17439880601021967

de Freitas, S., & Neumann, T. (2009). The use of 'exploratory learning' for supporting immersive learning in virtual environments. *Computers & Education*, *52*(2), 343–352. doi:10.1016/j.compedu.2008.09.010

Drettakis, G., Roussou, M., Asselot, M., Reche, A., Olivier, A., Tsingos, N., & Tecchia, F. (2005). Participatory Design and Evaluation of a Real-World Virtual Environment for Architecture and Urban Planning. In *Proceedings of IEEE Virtual Reality*. Bonn, Germany: MIT Press Cambridge.

European Commission (2007). *Access to and preservation of cultural heritage-Fact sheets of 25 research projects funded under the Sixth Framework Programme for Research and Technological Development (FP6)*. Luxemburg: Imprimé par OIL.

Ghiani, G., Paternò, F., Santoro, C., & Spano, L. D. (2009). UbiCicero: a Location-Aware, Multi-Device Museum Guide. *Interacting with Computers*, *21*(4), 288–303. doi:10.1016/j.intcom.2009.06.001

Guidi, G., Frischer, B., Russo, M., Spinetti, A., Crosso, L., & Micoli, L. L. (2006). Three-dimensional acquisition of large and detailed cultural heritage objects. *Machine Vision and Applications*, *17*, 349–360. doi:10.1007/s00138-006-0029-z

Hamalainen, R. (2008). Designing and evaluating collaboration in a virtual game environment for vocational learning. *Computers & Education*, *50*(1), 98–109. doi:10.1016/j.compedu.2006.04.001

Hansmann, U., Merk, L., Kahn, P., Nicklous, M., Stober, T., & Shelness, N. (2003). *Pervasive computing: the mobile world*. Berlin: Springer.

Hossian, L., & de Silva, A. (2009). Exploring user acceptance of technology using social networks. *The Journal of High Technology Management Research*, *20*, 1–18. doi:10.1016/j.hitech.2009.02.005

Hsu, C., & Lin, J. (2008). Acceptance of blog usage: The roles of technology acceptance, social influence and knowledge sharing motivation. *Information & Management*, *45*, 65–74. doi:10.1016/j.im.2007.11.001

Kiili, K. (2005). Digital game-based learning: Towards an experiential gaming model. *The Internet and Higher Education*, *8*(1), 13–24. doi:10.1016/j.iheduc.2004.12.001

Knipfer, K., Mayer, E., Zahn, C., Schwan, S., & Hesse, W. (2009). Computer Support for Knowledge Communication in Science Exhibitions: Novel Perspectives from Research on Collaborative Learning. *Educational Research Review*, *4*(3), 196–209. doi:10.1016/j.edurev.2009.06.002

Kotler, N., & Kotler, P. (2000). Can museums be all things to all people? Missions, goals, and marketing's role. *Museum Management and Curatorship*, *18*(3), 271–287. doi:10.1080/09647770000301803

Kuznik, L. (2009). Learning in Virtual Worlds. *US-China Education Review*, *6*(9), 43–51.

Kwon, O., & Wen, Y. (2010). An empirical study of the factors affecting social network service use. *Computers in Human Behavior*, *26*, 254–263. doi:10.1016/j.chb.2009.04.011

Lau, A., & Tsui, E. (2009). Knowledge management perspective on e-learning effectiveness. *Knowledge-Based Systems*, *22*, 324–325. doi:10.1016/j.knosys.2009.02.014

Lester, J., Stone, B., & Stelling, G. (1999). Lifelike Pedagogical Agents for Mixed-Initiative Problem Solving in Constructivist Learning Environments. *User Modeling and User-Adapted Interaction*, *9*, 1–44. doi:10.1023/A:1008374607830

Mason, D. D. M., & McCarthy, C. (2006). The feeling of exclusion: Young peoples' perception of art galleries. *Museum Management and Curatorship*, *21*, 20–31.

Mohino, E., Gende, M., Brunini, C., & Heraiz, M. (2005). SiGOG: simulated GPS observation generator. *GPS Solutions*, *9*, 250–254. doi:10.1007/s10291-005-0001-9

Owston, R., Widerman, H., Sinitskaya Ronda, N., & Brown, C. (2009). Computer game development as a literacy activity. *Computers & Education*, *53*(3), 977–989. doi:10.1016/j.compedu.2009.05.015

Pan, Z., Cheok, A. D., Yang, H., Zhu, J., & Shi, J. (2006). Virtual reality and mixed reality for virtual learning environments. *Computer Graphics*, *30*, 20–28. doi:10.1016/j.cag.2005.10.004

Pantano, E., & Servidio, C. (in press). The role of pervasive environments for promotion of tourist destinations: the users' response. *Journal of Hospitality and Tourism Technology*.

Pantano, E., & Tavernise, A. (2009). Learning Cultural Heritage through Information and Communication Technologies: a case study. *International Journal of Information Communication Technologies and Human Development*, *1*(3), 68–87. doi:10.4018/jicthd.2009070104

Paraskeva, F., Mysirlaki, S., & Papagianni, A. (2010). Multiplayer online games as educational tools: Facing new challenges in learning. *Computers & Education*, *54*, 498–505. doi:10.1016/j.compedu.2009.09.001

Parry, R. (2005). Digital heritage and the rise of theory in museum computing. *Museum Management and Curatorship*, *20*, 333–348.

Parsons, S., Leonard, A., & Mitchell, P. (2006). Virtual environments for social skills training: comments from two adolescents with autistic spectrum disorder. *Computers & Education*, *47*, 186–206. doi:10.1016/j.compedu.2004.10.003

Petric, J., Ucelli, G., & Conti, G. (2003). Real Teaching and Learning through Virtual Reality. *International Journal of Architectural Computing*, *1*(1), 2–11. doi:10.1260/147807703322467289

Pianesi, F., Graziola, I., Zancanaro, M., & Goren-Bar, D. (2009). The motivational and control structure underlying the acceptance of adaptive museum guides-An empirical study. *Interacting with Computers*, *21*(3), 186–200. doi:10.1016/j.intcom.2009.04.002

Pieraccini, M., Guidi, G., & Atzeni, C. (2001). 3D digitizing of cultural heritage. *Journal of Cultural Heritage*, *2*, 63–70. doi:10.1016/S1296-2074(01)01108-6

Roussou, M. (2002). *Virtual Heritage: From the Research Lab to the Broad Public*. Oxford, UK: Archaeopress.

Scaife, M., & Rogers, Y. (2001). Informing the design of a virtual environment to support learning in children. *International Journal of Human-Computer Studies*, *55*, 115–143. doi:10.1006/ijhc.2001.0473

Slator, B.M., Juell, P., McClean, P.E., Saini-Eidukat B., Schwert, D.P., White, A. R., & Hill, C. (1999). Virtual environments for education. *Journal of Network and Computer Applications*, *22*([REMOVED HYPERLINK FIELD]3), 161-174.

Steventon, S., & Wright, A. (2006). *Intelligent Spaces. The Application of Pervasive ICT*. Berlin: Springer.

Styliadis, A., Akbaylar, I. I., Papadopoulou, D. A., Hasanagas, N. D., Roussa, S. A., & Sexidis, L. (2009). Metadata-based heritage sites modeling with e-learning functionality. *Journal of Cultural Heritage*, *10*, 296–312. doi:10.1016/j.culher.2008.08.014

Tait, M. G. (2005). Implementing geoportals: applications of distributed GIS. *Computers, Environment and Urban Systems*, *29*, 33–47.

Taylor, G., Brunsdom, C., Li, J., Olden, A., Steup, D., & Winter, M. (2006). GPS accuracy estimation using map matching techniques: Applied to vehicle positioning and odometer calibration. *Computers. Environment and Urban Studies*, *30*, 757–772. doi:10.1016/j.compenvurbsys.2006.02.006

Tonta, Y. (2008). Libraries and museums in the flat world: Are they becoming virtual destinations? *Library Collections, Acquisitions & Technical Services*, *31*(1), 1–9. doi:10.1016/j.lcats.2008.05.002

Tredinnick, L. (2006). Web 2.0 and Business: A pointer to the intranets of the future? *Business Information Review*, *23*(4), 228–234. doi:10.1177/0266382106072239

UNESCO. (2005). *Text of the Convention for the Safeguarding of Intangible Cultural Heritage*. Retrieved May 28, 2008, from http://www.unesco.org/culture/ich/index.php?pg=00006

Van Dijck, J. (2006). The science documentary as multimedia spectacle. *International Journal of Cultural Studies*, *9*(1), 5–24. doi:10.1177/1367877906061162

Walker, J. (1990). Through the Looking Glass. In Laurel, B. (Ed.), *The art of computer-human interface design* (pp. 213–245). Menlo Park, CA: Addison-Wesley.

Wilson, J. (2006). 3G to Web 2.0? Can Mobile Telephony Become an Architecture of Participation? *Converge: The International Journal of Research into New Media Technologies*, *12*(2), 229–242. doi:10.1177/1354856506066122

Worboys, M., & Duckham, M. (2004). *GIS: A computing Perspective*. Taiwan: CRC Press.

Wu, Z.-H., Liu, Y.-L., Chang, M., Chang, A., & Li, M. (2006). Developing Personalized Knowledge Navigation Service for Students Self-Learning based on Interpretive Structural Modeling. In *Proceedings of the Sixth IEEE International conference on Advanced Learning Technologies*. The Netherlands: IEEE Computer Series.

Zhou, H., & Benton, W. C. Jr. (2007). Supply chain practice and information sharing. *Journal of Operations Management*, *25*, 1348–1365. doi:10.1016/j.jom.2007.01.009

KEY TERMS AND DEFINITIONS

Constructivism: Psychological theory that affirms that knowledge and meaning are generated by the interaction between thought and experience.

Educational Technology: Particular technology which studies the improving of learning by creating, managing, and using technological tools and processes.

Edutainment: Term that is the result of the mixture of "education" and "entertainment"; it means a learning experience that is formative and amusing at the same time.

Global Positioning System (GPS): System which recognizes user's geographical position and allows him to easily identify the most convenient route. It allows also the management of thematic maps consisting of different superimposed layers (each layer is a spatial dataset containing a common feature type representing objects, events and phenomena from real world) and extracting interesting aspects of a fixed territory. In this way the user is able to virtually travel across the area, increasing or reducing the scale level of the representation and choosing the features more interesting to visualize.

Knowledge Transfer: A process which allows to transmit a part of knowledge to a target audience. Especially the advanced technologies are capable of improving the efficiency of this process, by providing new effective tools such as augmented reality, web 2.0, mobile devices.

Virtual Heritage: The use of three-dimensional computer modelling in order to digitally reconstruct monuments, buildings, and finds that can be visualized by interfaces and allow a specific level of immersion and/or interaction to the user.

Virtual Tour: It is the 3D reconstruction of a location (existing or existed in the past and lost nowadays). It usually consists of a sequence of videos and can provide multimedia elements such as sound effects, music, narration, and text. A convincing immersive experience is given by virtual tours that allow the "exploration" of the computer environment as in real life.

Virtual Navigation Systems (VNS): Desktop application to stimulate visits for a wide range of environment. The capabilities of a desktop VNS can be also combined with the high performance of mobile devices. It can provide more exciting features such as allowing the user to have a real and virtual information combined depending on its location.

Chapter 14
A Case Study of Information Technology Education and Economic Development in Rural Nigeria

Adekunle Okunoye
Xavier University, USA

Nancy Bertaux
Xavier University, USA

Abiodun O. Bada
The George Washington University, USA

Elaine Crable
Xavier University, USA

James Brodzinski
St. Xavier University, USA

ABSTRACT

This essay presents a case study of Information Technology (IT) education as a contributor to economic and human development in rural Nigeria. The case of Summit Computers suggests that for developing countries to benefit from advances in IT, the following factors are of great importance and can be enhanced by IT education initiatives: convenience, affordability, emphasis on participation and empowerment of local users, encouragement of entrepreneurship, and building awareness among potential users. Additionally, careful attention should be given to how IT training can meet local employment and other needs are important factors in rural communities in developing countries such as Nigeria.

DOI: 10.4018/978-1-60960-497-4.ch014

INTRODUCTION

It is widely recognized that IT has great potential to increase efficiency and productivity, thus providing a positive force for economic growth in developing countries (UNDP, 2008). Often, rural areas present some of the greatest challenges for economic and human development in general, and for the extension of IT in particular (Tiwari, 2008; Ramirez, 2007). This essay presents a case study of Information Technology (IT) education as a contributor to economic and human development in rural Nigeria. The case of Summit Computers, in rural Osun State, Nigeria, summarized below, suggests that for rural areas in developing countries to benefit from advances in IT, a number of factors of great importance can be enhanced by IT education initiatives. These factors include convenience and affordability of IT, emphasis on participation and empowerment of local users, encouragement of entrepreneurship, and awareness building and skill development of potential users. Additionally, the case emphasizes the need for careful attention to the relationship of IT training to local employment in rural communities in developing countries such as Nigeria.

Many scholars have investigated how IT can contribute to economic and human development in the developing world (UNDP, 2008). Far fewer have focused on how to provide IT solutions to socio-economic development problems in rural areas (Richards, 2004; Avgerou, 1998; Kuriyan, Ray, & Toyama, 2008; Madon, 2000; Hollifield & Donnermeyer, 2003). Even fewer have discussed the role of empowerment in this context (Dawson & Newman 2002; Strover, Chapman, & Waters 2004). Most scholars emphasize that information technologies is crucial in the efforts of rural communities to attract and retain business and adapt to the new realities of a globalized world economy. For example, some African universities, such the University of Botswana, have focused on developing eLearning programs as a solution to educating both urban and rural populations in order to achieve an educated population and workforce that is digitally literate and prepared for a global economy (Ulys et al, 2004). Finally, the issue of IT education in rural areas of developing countries has received little attention from researchers, who have focused more on IT content and access to IT infrastructure (Grabill, 2003; O'Neil, 2002; Warschauer, 2003).

While the use of information and communication technology (ICT) is now worldwide, it is far from evenly distributed across the globe (WDI, 2008). The digital divide is not only between western industrialized countries and developing countries, but is present within many countries such as Nigeria, between rural and urban dwellers, and even within areas in urban centers (Mulama, 2009; Comfort et. al., 2003; Kvasny & Keil, 2002; Kvasny & Truex, 2001). The divide within countries between educated, employed, successful urban dwellers and the rest of the population can create tension and a "culture of discontent" that governments and other policy makers find difficult to ameliorate (Abraham 2009).

Many organizations seek to create and support local entities aimed at making ICT more accessible, for applications ranging from business to consumer to citizen (Roman & Colle, 2003). In seeking to explain the speed and pattern of Internet technology adoption, scholars have pointed to a decline in the importance of having a high proportion of English speakers, noting that economic and social factors are now the keys to understanding Internet technology diffusion. A cross-country study involving 21 countries found that both economic ability to access the Internet, and social factors including human capital/literacy, political stability, urbanization, and usage of other electronic media were important in explaining diffusion of Internet technology (Liu and San, 2006). Other research that has focused on efforts to successfully bring IT access to rural locations has also identified a number of relevant social and economic factors, including participation by local people in the design of such projects (Puri and Sahay, 2007).

Sometimes, projects that attempt to bring IT access to rural areas can have problems sustaining themselves once start-up funding is complete, as even modest user fees may stifle demand among rural dwellers with low cash incomes (Tiwari, 2008).

Even where rural people can clearly benefit from using IT, there is widespread lack of IT education (even among those with sufficient general education) to take advantage of these benefits. There are thus a number of specific issues affecting IT education in rural areas. Rural areas typically have low population density, meaning low demand level for IT education relative to concentrated urban areas (Hollifield & Donnermeyer, 2003). Profit-driven entrepreneurs are not likely to site an IT education center in such areas. In many countries, including Nigeria, IT education is largely private sector driven as IT education is not part of the curriculum at primary and post-primary public institutions, or even at many post-secondary government institutions[1]. Further, IT instructors with adequate skills are mobile and far more likely to live in urban areas. For all these reasons, IT education is not readily available in rural areas.

The purpose of this study was to investigate the concept of IT education centers in rural areas. The study specifically sought to address the following questions:

- Why is IT education not readily available in rural areas in developing countries compared to the urban centers?
- How is IT education contributing to human and rural community development?
- What are the factors that contribute to the unstainability of IT education in rural areas?

In this essay, we examine the concept of IT education, with special reference to developing countries. We characterize rural areas in developing countries and summarize how human and rural development issues relate to IT. In the analysis of the case, we discuss how the efforts of the case organization contribute to human and rural development. We consider the motivations for establishing the IT education center, the challenges and opportunities and the lessons that could be learned from their experience. Finally, we make some recommendations for policy-makers and present ideas for future research and practice.

CHARACTERISTICS OF RURAL AREAS IN DEVELOPING COUNTRIES

Both the population and labor force of developing countries are far more rural than is the case in developed countries. In Africa as a whole, for example, about 70% of the population lives in rural areas, compared with 25% in North America. Even more revealing is the fact that 68% of Africa's labor force is employed in agriculture, compared to a mere 3% of North America's labor force (Todaro, 2006). Low productivity in agricultural production means that while 68% of Africa's labor force works in agriculture, agriculture accounts for just 20% of the continent's GNP, and the great majority of agricultural work is basic, subsistence agriculture.

Rural areas in developing countries are currently facing a number of particular challenges such as hunger, de-population, declining agriculture, lack of employment opportunities, social exclusion, poor infrastructure, and environmental degradation (O'Malley, 2003). Per capita food production in Africa has actually declined steadily since the 1970s, and since food imports have not made up for the decline, this has indeed meant that the typical African has experienced a decrease in food consumption in recent decades. One author has summarized the reasons for this decline as,

"insufficient and inappropriate innovation, cultivation of marginal and sensitive lands, severe deforestation and erosion, sporadic civil wars, and misguided (incentive-reducing) pricing and

marketing policies—all of which were exacerbated by the highest rate of population growth in the world" (Todaro, 2006, p. 6).

Economic coordination between the private and public sectors must be addressed in order to stop this overall decline. Much could be gained by focusing on overcoming the economic coordination failures within the supply chain in poor rural areas, especially Africa. Policy makers, analysts and researchers must give more attention to understanding the constraints causing coordination failures and focus on the opportunities and mechanisms for overcoming these challenges (Kydd & Dorward, 2004).

Policy makers also need to be aware that poor rural households depend heavily on natural resources such as energy sources, fodder, raw materials and water for drinking and irrigation. A report produced by the United Nations Development Programme, the United Nations Environment Programme, the World Bank, and the World Resources Institute pointed out that an asset-based rather than just income-based approach should be considered when evaluating poverty (WRI, 2005). These natural resources are common to all and need to be managed so that the poor may access them, leading to increased income and migration out of these severe poverty levels (Narain, et al., 2008).

Nigeria, in which our case organization is located, is typical of many developing countries in a number of aspects. It is mostly rural (84% of the population resides in rural areas), yet there has been a dramatic trend of people migrating to urban areas, which has resulted in the growth of at least 24 cities with populations in excess of 100,000. Its economic reliance on exports of primary products (in the case of Nigeria, this has been largely oil exports since the 1970's) has led to a serious neglect of the agricultural sector. The stagnation in rural areas and the rural-to-urban exodus has led to a severe unemployment problem in urban areas, with both rural and urban areas experiencing widespread poverty (Todaro, 2006). Overall, Nigeria's public spending on education as evidence of its commitment to education has been ranked near the bottom when compared to other countries (UNDP, 2003). At the same time, an economic elite has emerged, causing large disparities in income and standards of living, on both a national as well as regional basis. Thus, the per capita income in a relatively rich state, Old Bendel (Delta and Edo state) is five times greater than that of a poorer state, Kaduna; similar disparities also exist in adult literacy rates (Todaro, 2006).

By all accounts, there is truly a crisis in rural Africa. In the effort to formulate responses to the urgent problems facing developing regions such as in Africa, it is understandable but unfortunate that the possible contributions of IT have been largely overlooked. We now turn to a consideration of the connections between economic, rural, and human development and IT.

IT EDUCATION

Basic Concept

Reichgelt et al. (2004) define IT as an academic discipline that focuses on meeting the needs of users in an organizational and societal context through the selection, creation, application, integration and administration of computing technologies. The Society for Information Technology Education (Lunt et al., 2003; Reichgelt et al., 2004) listed numerous skills and capabilities for information technology graduates, including the ability to

- Use and apply current technical concepts and practices in the core information technologies,
- Analyze, identify and define the requirements that must be satisfied to address problems or opportunities faced by organizations or individuals, and
- Effectively design IT-based solutions and integrate them into the user environment

taking into account user-centric design and interface.

Thus, any kind of training and education that sought to provide people with those skills and capabilities (see Lunt et al., 2003 for details) could be considered as IT education.

IT Education in Developing Countries

The challenges that have been identified (Barata et al., 2001; Darley, 2001; Morales-Gomez, & Melesse, 1998; Moyo, 1996) as the main constraints in IT development also apply to IT education (Bada et al., 2003; Okunoye et al., 2003; Shakya & Rauniar, 2002). Major challenges focus on expertise, infrastructure, funds, policies and regulations, literacy level, and income distribution. Like other development issues that differentiate urban and rural centers, provision of IT education is particularly problematic in rural areas of developing countries. The inequalities not only include access to the internet but also knowledge of search strategies, quality of technical connections, social support, and the ability to evaluate the quality of information and diversity of its uses (DiMaggio et al., 2001; Uys et al., 2004). This can be attributed to the elite nature of information technology as well as the basic literacy prerequisites that elude many residents of rural areas in developing countries.

IT education centers are usually concentrated in urban centers and universities, where most of the white-collar jobs are also available. This reflects the impression of IT as a business tool in many developing countries. It is only recently that IT has begun to be utilized by governments and public organizations in developing countries, and thus to have a greater direct impact on the local people and society in general. In a real sense, there are no genuine incentives for private investors to locate IT education centers in rural areas of developing countries. Although one could argue that education is a public good and that it is the responsibility of government to ensure its provision at all levels, IT education is not considered to be part of the standard curriculum in many developing countries and its provision is left to the elite in the society.

Nevertheless, many rural communities have begun to embrace IT and its applications. The influence of the Internet cannot be overemphasized in this regard, and globalization and internationalization of labor have also contributed to a greater awareness of the wide potentials of IT. Many immigrants that live abroad encourage family members in their home country to have an e-mail account as an inexpensive means of communication, and use IT for money transfers and other purposes.

Internet access is relatively new in Africa when compared to the rest of the world. In 1994 South Africa was one of the first African countries to begin building an infrastructure supporting the internet. Today, all African countries are connected to the Internet; however the internet remains out of the reach for most Africans. It is still too expensive and has inadequate connection ability to really serve the general population in Africa. One avenue policy makers can take to help build better connections is to take advantage of the mobile cellular growth in the country. They need to examine how best to utilize the mobile technology quickly moving forward to reduce the digital divide within Africa, but at the same time understand that huge investments in telecommunication infrastructure is still desperately needed (Ahmed, 2007).

While access must precede adoption and use, according to Hollifield & Donnermeyer (2003), it is only economically viable for service providers (IT education and others) to invest in the infrastructure required to provide access (and IT education) if they can expect a return on that investment in a reasonable period of time. The low population density of rural areas and other factors makes it difficult to recover those investment costs and thus discourage typical investors from providing IT education in rural areas. Since not-for-profit

IT education centers are rare, residents of rural areas are generally denied the expected benefits that accompany IT knowledge. Focusing on literacy alone is a major challenge for these areas so digital literacy is pushed aside in importance and funding (Ulys et al., 2004).

The Concept of Sustainability

Mursu, Sorinyan, and Korpela (2003) define IT as sustainable when it is appropriate, usable and useful, and when possible changes within an organization or society implementing the technology are far-reaching and humane, and further improvements to the technology can be conducted smoothly. Sustainability, therefore, focuses on the long term viability of a project and, in relation to IT projects, how the benefits and accompanying change from such projects will last beyond the its immediate conclusion and well into the future (Kanungo, 2004). Within the information systems domain such concerns have brought the concept of sustainability to the center of discussion and various authors have suggested different means through which the sustainability of IS-related projects can be viewed and achieved (Kimaro & Nhampossa, 2004; Oyomno, 1996; Srinath & Braa, 2005).

One of such approaches identified by Oyomno (1996) suggests that sustainability should be viewed from multiple perspectives that include the level of demand for the technology, availability of local technological capacity to sustain its beneficial use, and the appropriateness of such technology. According to Oyomno, demand is the measure of the extent to which the use of the technology supports the fulfillment of the *raison d'être* of an organization or supports certain fundamental goals within a society. Technological capacity, on the other hand, is defined as the entrepreneurial, technical, institutional, political, socio-cultural and managerial resources that are available in the organization or its environment to support effective use of the technology. Lastly, the author defines appropriate technology as how cost-effective, affordable (financial and human) and suitable the technology is to the requirements of the activity for which it used.

Based on the above work, Mursu, Sorinyan, and Korpela (2004) developed an extended sustainability framework and a checklist to guide IS developers and academics in analyzing the sustainability of IS projects. An important part of the work by Mursu, et al., is the incorporation of the concept of self-reliance or empowerment as developed by Pellegrini (1980), who suggests that the desirability and appropriateness of a given technology rests in its ability to empower its users. Consequently, Mursu, Sorinyan, and Korpela (2004) suggested an expanded framework that argues that the sustainability of an IS project is determined to some extent by the network of activities and resources supporting the project, and to a greater extent, by the participation and empowerment of the local user community in the project. In subsequent sections of this essay, concepts and ideas from these sustainability frameworks will be employed to highlight some of the major motivating factors that were found to be influential in our case study organization.

HUMAN AND RURAL DEVELOPMENT AND IT

The vitality of rural areas is crucial in determining how successful developing countries are in their efforts to promote human development. The United Nations' Human Development Report clearly outlines the connection between rural and human development, and states that human development is concerned with the creation of an environment in which people can fully develop their potential and lead lives that are productive and creative (UNDP, 2003). When agricultural productivity is rising, this increases households' nutrition and income levels. Rising incomes allow households to invest more in children's education and health, and also make increased public investments in areas such as health, education, transportation, communication, and other infrastructure more likely. Rising incomes and education levels also lead to declines in fertility and further investments in higher productivity agricultural and manufacturing techniques, constituting a positive upward spiral.

Conversely, when agricultural productivity is stagnant or even falling, as is the case in Africa generally, the spiral can be downward: low productivity in agriculture combined with high population growth lead to deforestation, erosion, soil depletion, and water scarcity, which in turn lead to further declines in productivity. Resulting low incomes and malnutrition inhibit efforts to increase education, health and other public services and investments. Fertility and infant mortality remain high, while investments in agricultural and manufacturing productivity remain low (UNDP, 2008).

In light of these proble0ms of the utmost gravity, is it not frivolous to talk of the need for IT and IT education? When people are lacking the very basics in terms of clean water, food, shelter, and health care, how can we advocate for funding for IT? As the United Nations Development Programme (UNDP, 2008) points out, we cannot ignore technology since technology will play a crucial role in raising productivity, thereby changing the downward spiral presently operative in Africa into a positive, upward spiral. We already have the technological knowledge to address many problems in the areas of hunger, water supplies, disease, and sanitation, but the developing world awaits new technological innovations to address problems such as HIV/AIDS, malaria, and the need for low-cost energy. The majority of technological advances today, including the dissemination of new knowledge, are intimately linked with access to, education in, and use of IT. Without significant progress in the IT area, spreading existing technological knowledge and creating new technology will be difficult or impossible. Thus, this is a task that cannot wait, in spite of the many other urgent issues facing Africa today.

As the Nobel Prize winning economist Gunnar Myrdal has said, “It is in the agricultural sector that the battle for long-term economic development will be won or lost” (Todaro, 2006, p. 6). IT and IT education have an important role to play in this battle for human and rural development. People usually have to move to urban areas to receive IT training. This limits the training to the few that have stronger financial resources. Significant proportions of the rural dwellers are thus alienated. Part of the needs of local people is to have IT training opportunity at affordable price within their local area so as to be able to compete with the urban people on jobs that require IT skills. Without the kind of initiative described in the paper, these rural people may not be able to get the IT training and are therefore unable to contribute to the social and economic development and family support in their area. This argument is, we believe, further strengthened by the case study presented below.

CASE STUDY METHODOLOGY AND DATA ANALYSIS

In this essay we use a single case study to investigate IT education in a developing country and how IT education is contributing to human and rural development. Our rationale is that the study of IT education in the context of rural economic and human development is in its infancy, so that a series of case studies in a variety of geographic areas will be crucial in the effort to design studies to address this issue at a more general level. We intend this study as a localized set of observations that is both interesting for those considering the specifics of economic development in western Nigeria, and for those who will contrast this case study with other efforts to introduce IT education to rural areas in developing countries in general (Benbasat, Goldstein, & Mead, 1987; Dul & Hak,

2008;). Case study methodology has been noted as particularly suitable for guiding investigations where theory is at an early or formative stage, as well as for practice-based problems where the experiences of the actors are important and the context of action is critical (Dul & Hak, 2008; Eisenhardt, 1989).

In support of case study analysis, Yin (2003) considered the crucial role of investigating a contemporary phenomenon or event in its real-life context, especially when the boundaries between the phenomenon and the context are not clearly evident. In a case study, the researcher does not, nor cannot, control or manipulate the situation; also, the researcher must take care to minimize the extent to which interpretation introduces bias and affects the outcome of the research. Another limitation of case study analysis is that a particular case study does not necessarily provide the basis for a generalized result; however, it should be a sound basis for informed knowledge.

In selecting the case organization, we identified a rural organization in a developing country with IT education as the main activity, and of course one for which we could gather information. Summit Computers was selected due to its status as a rural organization in Africa with a significant track record, and also due to our ability to arrange visits and interviews in this location. The study primarily used semi-structured interviews, short-duration, on-site observations, and narratives of the main actors in the case organization. Apart from the information provided by the case organization on the developmental influence of IT education, the authors independently investigated these influences to achieve triangulation of data and insights.

During the summer of 2004, one of the authors visited Summit Computers to interview the founder, Chief Muraina Oyelami. The interview was recorded and later transcribed. Interview notes were taken and physical structure and infrastructure of summit computers were recorded. During the visit, the researcher also interacted with the students in order to provide an opportunity for triangulation of the interview claims. Historical documents and public documents about Summit Computers were also examined. All these sources of data were analyzed, and the case description presented below was based on the synthesis of these various sources. A draft of the case of Summit Computers was presented to a conference in the summer of 2005. In the summer of 2007, one of the authors again returned to Nigeria to get further updates on Summit Computers. The project's concept remains the same, with the only substantive change being the introduction of a Cyber Café by the founder. The author again visited Summit Computers in the summer of 2010 and found that its branches in the neighboring village, especially the one at Ada, had grown significantly. Ada is very close to Iree, the location of Osun State Polytechnics; thus the Cyber Café is receiving patronage of the polytechnics students. Since 2007, many government services are now available online, including passport processing, University's entrance examination forms and results, visa processing etc. Many local residents who would normally have had to travel to cities to access the Internet for these services now visit Summit Computers and Cyber Café, initially to receive basic training on how to use the Internet and then to connect to the government and other services.

THE CASE ORGANIZATION: SUMMIT COMPUTERS

Summit Computers Ltd was established in 1997. It is located in Iragbiji, Osun State, Nigeria. Iragbiji is a rural community in southern part of Nigeria. Agriculture is the predominant occupation of the dwellers. Iragbiji is about 10km from Osogbo, the capital city of the Osun State. The population of Iragbiji is estimated to be about 30,000 according to 1991 Census. Summit Computers was established by a traditional chief within the community, Chief Muraina Oyelami. According to Chief Oyelami, Summit Computers was established to create IT awareness in the rural populace thereby discouraging the influx of rural dwellers to the urban areas, and thus is intended as his contribution to community development. At Summit Computers, the courses include general information technology studies, data processing, programming languages, Internet, and application packages (including PowerPoint, MS Word, Front Page, PageMaker, Excel, and CAD). The primary goal is to expose the community to the science of computing and information technology in general. The main motivation behind the establishment of Summit Computers came from Chief Oyelami's own personal penchant for IT.

Summit Computers offers personalized attention, based on individual needs. This has resulted in different levels of training for individuals (including children's programs, introduction to computer studies, secondary school leaver programs, and university graduate and mature or working class/civil servant programs). To date, Summit Computers has produced more than 500 graduates. Some of the graduates returned for more advanced study in information technology, and others obtained employment in their respective areas of interest. Most graduate with diplomas while a few (especially, the desktop publishing students) graduate at certificate level. Some graduates reported they became gainfully employed specifically because of their computer literacy. This confirms the idea that graduates stand a better chance of securing a job if they are IT literate, regardless of their discipline.

At Summit Computers, the students are trained using a syllabus designed in collaboration with Obafemi Awolowo University[2] Consultancy Services (UNIFECS). UNIFECS acts as an external regulating and examining body, managing all the examination scripts and grades and directly awarding the diplomas and certificates. This approach is being used to legitimize the training and provide security and industry recognition for the graduates, and shows the importance of connections between the formal and informal sectors. As one observer has noted, "Employers are always suspicious to see a diploma from a computer school in a remote and rural area" (Okunoye et al., 2005, p. 319).

The major problem confronting Summit Computers is in the area of financing. The founder related that he had been subsidizing the training and providing most of the equipment: "We have not made any profit in our seven years of operation and the finance issue is really affecting our growth" (Okunoye et al. 2005, p. 319). Summit Computers started with a few used Intel 386 processor-based systems but their inventory of computers has been agumented with many 486- and Pentium-processor-based systems. However, the need now is for additional systems with faster processors to accommodate new applications. Summit Computers is not supported by government funds in any form. Chief Oyelami believes that the Summit Computers model can be replicated elsewhere (contingent on funding, of course).

Among the success factors of Summit Computers is the consistency in the delivery of the training programs, despite the challenges posed by the environment. Summit Computers has been able to design new programs in response to upgrades in application software. This is essential to the marketability of graduates. Another success factor is the low tuition which allows the local farmers to be able to pay for the training of their children. Summit Computers has a positive impact on the community. For example, some of the current training staff at Summit Computers are alumni of the program. Other graduates have established their own businesses, including business centers and cyber cafés. In the future, Summit Computers hopes to go beyond provision of IT education and venture into assembling low cost, affordable computers for rural dwellers. This will encourage many to seek IT education and thus promote the main objectives of Summit Computers. They also plan to move to a large and permanent location to accommodate the growing demand for IT skills. Summit Computers has recently established new locations in neighboring towns, and as an educational support for the students, each training location now has a cyber café where students can browse the Internet for a nominal fee, designed to merely cover costs.

CASE ANALYSIS: MOTIVATING FACTORS AND DRIVERS OF IT EDUCATION IN RURAL AREAS

We next discuss notable factors in the case that might be indicators of sustainability in rural IT education programs.

Awareness and Exposure to Technology

In the case described above, the founder of the organization was exposed to western education and culture. Through his job and other interests, he often travels to western industrialized countries, is aware of current trends in information technology, and is also able to use many applications. While most founders of urban IT education centers have been specifically trained or have worked in IT in western industrialized countries, their rural counterparts may have only acquired IT skills through interactions with their international colleagues and by virtue of their primary job functions. In our case, the founder is an experienced artist who collaborates with many international artists and attends exhibitions, workshops and training sessions abroad. He secured most of his IT knowledge through his contacts and in the course of doing his job, which lies outside of IT but nevertheless utilizes IT.

Technical Ability

Awareness and exposure to technology is complemented by the development of technical capability. As expressed in the case, in rural areas there is a lack of IT expertise and thus it may be difficult to get qualified IT trainers. The cost of hiring and retraining personnel would defeat the purpose of low-cost education provision. Summit Computers addresses this issue by hiring the program's alumni. In our case, the founder of Summit Computers is highly knowledgeable in IT use and applications, has a clear understanding of hardware and software selection, and is able to complete some basic installation himself. He renders these services free of charge to the training center, thus reducing the out-of-pocket costs incurred by the organization.

"Son of the Soil" Factor

In the Nigerian context (and we believe this is likely to be the case numerous developing countries), a founder of a rural IT education center is likely to be well received in their community if they are actually resident in that community. Such an individual will typically be perceived as understanding the real needs of the people and thus will experience a higher level of trust. Ideally, organizers of rural IT education centers will be close to the community, and will be well known prior to starting the center. The community will then perceive the training center as their own and will work together to ensure its success. In our case, we believe the local origins of the founder have been extremely helpful, as illustrated by the following quote:

...Many of our graduates have voluntarily returned to render their services to the institute, some of them that we hired are glad with what we have to pay. It is all in the community spirit and expression of their support to what we are doing. (Oyelami, 2004, p. 1)

The founder of Summit Computer is also a high chief in the village. He commands respect and has been praised for his commitment to the village. He is close to all the local leaders and his intentions were credible to the people. If an external person would attempt IT education in a rural area, it is less likely to be well received even when it is completely free (as in the case of some NGO's).

Community Orientation and Service Spirit

Related to the 'son of the soil' factor is the community orientation. Without the genuine commitment to serve the community, there was no economic justification to situate IT education centers in the communities described in this case. The founder would have primarily established his center in the state capitol, which was about 10 kilometers away, before considering his rural village. As he expressed during the interview, he considers Summit Computers his contribution to community development and not a profit-making venture. Even though the use of this center might extend to other areas, it was originally planned to serve the people in the immediate rural community.

Financial Capability

As we have already noted, the founder of this rural-based IT education center started by investing his own personal equipment, efforts, and financial resources, providing services without financial remuneration. This individual is operating with charitable motives, since he does not receive income from the center but instead lives on revenue from other sources (e.g., from sales of artwork). If any immediate returns had been expected from his investment in the IT education center, the center would have been forced to charge market fees and, in the process, discourage people from attending. This would have likely led the center to fold without achieving its objectives. This case thus illustrates the difficulties associated with rural-based IT education centers that are conceived as profit-making ventures.

People's Need, Local Content and Service Driven

Many government and other service providers now require consumers to have the ability to use computer systems and to have access to the Internet. Beneficiaries of these services are now encouraged to acquire the knowledge necessary to operate computer systems and access other associated computer services. This need for knowledge is driving their interest in information technology education and thus increasing the patronage of the local IT education center. As the awareness of IT is growing local residents are increasingly interested in IT-based services. The presence of a local IT education center provides an opportunity that would have hitherto required migration to the urban center. Local IT centers now have expanding customer bases as a result of the increased availability of services and local content on the Internet.

CASE ANALYSIS: MAJOR CHALLENGES AND OPPORTUNITIES

Infrastructure Development

Based on this study, we note that providers of IT education and related services in rural areas may not be able to afford to build the base information infrastructure for a whole community and thus, the growth of IT education will still depend on individual providers operating on a small scale. As noted, unless providers are willing to donate time and equipment, there may be no IT education activities at all in rural areas. With the rural population in developing countries, the development of an information infrastructure would be extremely helpful in enabling the provision of IT education.

Support

The case we have presented shows that the IT education centers in rural areas can succeed with the efforts of committed and visionary, non-profit oriented members of the community. They may have to operate without financial support from government, even excluding tax-exempt status, and without support from international donor agencies.

We believe that our case demonstrates that support from both the national government and international donor agencies could assist in expansion of these IT education centers and also encourage interested people in other communities to start IT education centers. Having said this, we want to reiterate that the success of IT education initiative still relies on the ingenuity of local people, and that aid from governments and outside donor agencies should not prevent local people from being the main actors. This case suggests that a 'champion' may be required for IT education in rural areas. Without such a person who has many of the characteristics we describe above, provision of rural IT education in developing countries could be challenging. Additional research needs to be completed on how the number of resources of potential champions could be augmented.

Continuous Development and Skills Upgrade

Scholars examining IT education in developing countries in the 1990s (Odedra et al., 1993) noted that a dilemma exists with respect to human resource development when people in developing countries acquire IT education only to lose their skills shortly after training due to the lack of opportunities to apply and upgrade these skills. Some important aspects of IT training need constant application and updating to remain current and useful. This poses a major challenge to beneficiaries of IT education in rural areas of developing countries.

In our case, we did observe this problem. The graduates of the rural-based IT education center in Iragbiji, Nigeria are in danger of soon losing their skills. There are only a few places they can work within their community that would require application of IT skills. It is unfortunately the case that there are relatively few local jobs where their newly-acquired IT skills are needed and utilized. As the center's founder explains:

... Some of our graduates are gainfully employed on the merit of the knowledge. Most of these jobs are within the capital city. The demands for the skills in rural areas are low and that is affecting the demand for our programme (Oyelami, 2004, p. 1).

Also, since access to IT infrastructure in individual homes is almost non-existent, graduates are not able to apply and practice their skills at home.

Even graduates who are able to utilize their IT education often are unable to acquire new skills since they depend largely on their local center and limited training opportunities in their workplace. The problem of how to enable IT graduates to continuously apply and upgrade their skills is of course highly connected to the infrastructure problems discussed earlier. Increasing access to IT infrastructure will greatly aid the effort to utilize and improve skills. In this regard, the center in Iragbiji has begun to evaluate the feasibility of establishing a computer assembly plant where affordable computers and other devices could be produced. Bringing the actual production of IT equipment into rural areas could help extend the accessibility of IT infrastructure and provide further skills to local people, as well as address the problem of the affordability of IT equipment.

RECOMMENDATIONS AND CONCLUSION

Our findings from the case offer some relevant ideas for policies related to IT and development.

These ideas involve encouraging four groups of influential stakeholders to take a proactive role in developing an IT infrastructure that is accessible to residents in rural areas.

- *Elites*. We propose that policy makers encourage elites that reside in rural areas to use their expertise and exposure to modern IT technologies to establish training schools in their communities. Usually, these elites can afford to acquire their own personal computers, which they could use to train small groups in their local areas.
- *Local government administrators*. Members of the rural community with IT experience can also be encouraged to provide their expertise at low cost to train rural residents, provided the local government administrators provide the equipment and other necessary resources. The participation of the local government in such a capacity could be to provide the formal structural component, which would in turn provide the context in which individual IT education centers would operate (Cancian, 1981; Kuriyan, Ray, & Toyama, 2008; Rogers, 1995).
- *Local educational administrators*. Although our focus in this essay is on the nontraditional provision of IT education, we recognize that the leveraging of the local education infrastructure that is already available in rural areas could greatly assist such efforts. Space in local education centers (primary and secondary schools, libraries, etc.) could be made available at little or no cost to local people who are willing to set up an IT education center. While it may not be feasible to make such programs totally free, such IT education in rural areas must remain affordable to the intended beneficiaries.
- *Government Communication Officials*. Since this bottom-up approach can be effective in the delivery of IT education, we suggest that governments disseminate the success stories of existing rural based IT education centers to other rural areas. In such communication efforts, emphasis should be placed on the developmental implications and the benefits to the host communities. A local initiative can be more easily handled and managed than an initiative that was designed by an agency of central government (partially due to the many layers of bureaucracy typically present). There is also evidence that such initiatives can be considered alien to local people and thus reduce the chance of success. For example, the Mobile Internet Unit program of the federal government of Nigeria, which was meant to provide Internet connectivity to rural community, has not been very effective or popular, due to its over-centralized approach. Similarly, the Computer- for all- Schools program in Nigeria has yet to place computers to many schools in rural areas. We conclude that a centralized approach to IT education may be inappropriate in Nigeria and likely in other developing countries with similar characteristics.

This essay has also shown that IT education can be an essential factor in human development in rural areas. Emphasis has been placed on access to ICT, connectivity to the Internet, distance learning, and adoption of IT in previous research on IT and development. All these are not possible without adequate IT education, to which policy makers have paid inadequate attention. Multilateral and bilateral agencies have supported connectivity and access through various initiatives, yet very little has been done on IT education. Rural areas that want to take full advantage of IT for development purposes must find a way to support local initiatives that establish sustainable IT education centers. All the motivating factors and the challenges we presented need to be adequately considered. While IT has been long established as a driver of socioeconomic change and development, this study emphasizes the contribution of IT education to rural and human development, as it is focused on a case that differs from conventional thinking that associates IT education with urban dwellers.

An issue which requires further investigation is the possible relationship between rural IT education and rural de-population. There are legitimate concerns that rural people receiving such education may be motivated to relocate to urban areas where there would be a better IT infrastructure; but on the other hand, increasing the IT education, skills and infrastructure in rural areas may allow rural people who would have migrated away to stay in the area.

On the question of sustainable IT education in rural areas, our findings suggest that locally sponsored IT education centers could be a valuable approach to rural IT education provision. The analysis of the case suggests that for developing countries to benefit from advances in IT, awareness among the real users, convenience, affordability and consideration of how IT training can meet local needs and employment are important factors. IT education that considers the needs of local businesses and organizations could attract the participation of local people (as witnessed in this case study where alumni of the center became trainers themselves in the center) and thus increase the demand and sustainability of such centers. The government could encourage the provision of IT education and support the promoters of such efforts without making IT education provision their primary responsibility (except when it is part of the formal school curriculum).

As the world economy continues to be knowledge driven and with the influence of IT and globalization on these changes, we conclude that sustainability and development of rural communities in developing countries can be highly affected by the level of IT education. This study suggests that entrepreneurship, participation and empowerment of local users are important factors that enhance the sustainability of IT education in rural communities.

While we have offered some interesting findings, this study is not without limitations that need to be considered when interpreting and applying the recommendations. First, the study is based on one case in one country, and thus we cannot apply the findings to rural communities in general. However, a single case study is an appropriate methodology for exploratory purposes and where there is limited knowledge concerning the phenomena under investigation. We expect other studies of organizations with similar aims, in similar contexts, would reveal more about the representativeness of this case, as well as the reasons for and implications of similarities and differences uncovered by analyses of a number of cases. Also, we were unable to include the perspective of the other actors in the communities (the students, local government officials, etc.) in the current case presentation, and plan to do so in future endeavors. We believe the areas explored in this essay present important opportunities for future research, and that such research promises to make significant contributions to knowledge relevant to rural and human development in developing countries.

REFERENCES

Abraham, B. P. (2009). Preparing for the challenge of electronic globalization. *Decision.*, *36*(1), 30–32.

Ahmed, A. (2007). Open access towards bridging the digital divide-policies and strategies for developing countries. *Information Technology for Development.*, *13*(4), 337–361. doi:10.1002/itdj.20067

Avgerou, C. (1998). How can IT enable economic growth in developing countries? *Information Technology for Development*, *8*(1), 15–29. doi:10.1080/02681102.1998.9525288

Bada, A., Okunoye, A., Aniebonam, M., & Owei, V. (2003). Introducing information systems (IS) education in Nigerian higher institutions of learning: A context, content and process framework. *In Proceedings of the Nigeria Computer Society, 14*(1).

Barata, K., Kutzner, F., & Wamukoya, J. (2001). Records, computers, resources: A difficult equation for sub-Saharan Africa. *Information Management Journal, 35*(1), 34–42.

Benbasat, I., Goldstein, D., & Mead, M. (1987). The case research strategy in studies of information systems. *Management Information Systems Quarterly, 11*(3), 369–386. doi:10.2307/248684

Cancian, F. (1981). Community of reference in rural stratification research. *Rural Sociology, 46*, 626–645.

Castell, M. (1996). *The Rise of the Network Society*. Cambridge, MA: Blackwell.

Comfort, K., L. Goje, & K. Funmilola. (2003). Relevance and priorities of ICT for women in rural communities: A case study from Nigeria. *Bulletin of the American Society for Information Science and Technology 29, 1*(6), 24-25.

Darley, W. (2001). The internet and emerging e-commerce: Challenge and implications for management in sub-Saharan Africa. *Journal of Global Information Technology Management, 4*(4), 4–18.

Dawson, R., & Newman, I. (2002). Empowerment in IT education. *Journal of Information Technology Education, 1*(2).

DiMaggio, P., Hargittai, E., Neuman, W. R., & Robinson, J. P. (2001). Social implications of the internet. *Annual Review of Sociology, 27*, 307–336. doi:10.1146/annurev.soc.27.1.307

Dul, J., & Hak, T. (2008). *Case Study Methodology in Business Research*. Amsterdam: Butterworth-Heinemann/Elsevier.

Eisenhardt, K. M. (1989). Building theories from case study research. *Academy of Management Review, 14*(4), 532–550. doi:10.2307/258557

Grabill, J. (2003). Community computing and citizen productivity. *Computers and Composition, 20*, 131–150.

Hollified, C., & Donnermeyer, J. (2003). Creating demand: Influencing information technology diffusion in rural communities. *Government Information Quarterly, 20*, 135–150. doi:10.1016/S0740-624X(03)00035-2

Kanungo, S. (2004). Sustainable benefits of IT investments: From Concepts to implementation. In *Proceedings of the 10th Americas Conference on Information Systems*, New York, August, 925-933.

Kaushik, P., & Singh, N. (2003). Information technology and broad-based development: Preliminary lessons from North India. *World Development, 32*(4), 591–607. doi:10.1016/j.worlddev.2003.11.002

Kimaro, H., & Nhampossa, J. (2004). *The challenges of sustainability of health information systems in developing countries*. In the Proceedings of the 12th European Conference on Information Systems, Turku, Finland.

Kuriyan, R., Ray, I., & Toyama, K. (2008). Information and communication technologies for development: The bottom of the pyramid model in practice. *The Information Society, 24*(2), 93–104. doi:10.1080/01972240701883948

Kvasny, L., & Keil, M. (2002). *The challenges in redressing the digital divide: A tale of two cities*. In Proceedings of the International Conference on Information Systems, Barcelona, Spain, December, 15-18. Available at http://aisel.isworld.org/pdf.asp?Vpath=ICIS/2002&PDFpath=02RIP24.pdf; last accessed August 2003.

Kvasny, L., & Truex, D. (2001). Defining away the digital divide: the influence of institutions on popular representations of technology. In B. Fitzgerald, N. Russo, J. DeGross (Eds.), *Realigning Research and Practice in Information Systems Development: The Social and Organizational Perspective,* 399-415, New York: Kluwer Academic Publishers.

Kydd, J., & Dorward, A. (2004). Implications of market and coordination failures for rural development in least developed countries. *Journal of International Development, 16*, 951–970. doi:10.1002/jid.1157

Liu, Meng-chun, & San, Gee (2006). Social learning and digital divides: A case study of internet technology diffusion. *Kyklos, 59*(2), 307–321.

Lunt, B., Reichgelt, H., Ashford, T., Phelps, A., Slazinski, E., & Willis, C. (2003). *What is the new discipline of information technology? Where does it fit?* Session ETD 343, 2003 CIEC Conference, Tucson, Arizona. Available at http://faculty.csuci.edu/william.wolfe/csuci/create/asp/What_Is_IT.pdf.

Madon, S. (2000). The internet and socio-economic development: Exploring the interaction. *Information Technology & People, 13*(2), 85–101. doi:10.1108/09593840010339835

Mbarika, V., Jensen, M., & Meso, P. (2002). Cyberspace across sub-Saharan Africa: From technological desert towards emergent sustainable growth. *Communications of the ACM, 45*(12), 17–21.

Morales-Gomez, D., & Melesse, M. (1998). Utilizing information and communication technologies for development: The social dimensions. *Information Technology for Development, 8*(1), 3–14. doi:10.1080/02681102.1998.9525287

Moyo, L. M. (1996). Information technology strategies for Africa's survival in the twenty-first century: IT all pervasive. *Information Technology for Development, 7*(1), 17–21. doi:10.1080/02681102.1996.9627211

Mulama, J. (2009). *Technology-Africa: A rural-urban digital divide challenges women.* IPS. Available at http://ipsnews.net/news.asp?idnews=36563; last accessed April 22, 2009.

Mursu, A., Sorinyan, A., & Korpela, M. (2003). *ICT for development: sustainable systems for local needs*. Proceedings of IFIP WG 8.2 & 9.4 Joint Conference, Athens, Greece.

Mursu, A., Sorinyan, A., & Korpela, M. (2004). *A generic framework for analyzing the sustainability of information systems.* In Proceedings of the 10th Americas Conference on Information Systems, New York, 934-941.

Narain, U., Gupta, S., & van't Veld, K. (2008). Poverty and the environment: Exploring the relationship between household incomes, private assets, and natural assets. *Land Economics, 84*(1), 148–167.

O'Malley, M. (2003). *Sustainable rural development. Sustainable Ireland.* Available at www.sustainable.ie/resources/community/art03.htm [Accessed Thursday, September 30, 2004]

O'Neil, D. (2002). Assessing community informatics: A review of methodological approaches for evaluating community networks and community technology centers. *Internet Research: Electronic Networking Applications and Policy, 12*(1), 76–102. doi:10.1108/10662240210415844

Odedra-Straub, M., Lawrie, M., Bennett, M., & Goodman, S. (1993). International perspectives: Sub-Saharan Africa: A technological desert. *Communications of the ACM, 36*(2), 25–29. doi:10.1145/151220.151222

Okunoye, A., Bada, A., Pick, J., & Adewumi, S. (2003). *Call for more information systems education in developing countries: Perspectives from information systems researchers and practitioners*. In Palvia P. & Liu X. (eds.), Proceedings of the 4th Global Information Technology Management World Conference, 319, June 8-10, Calgary.

Oyelami, M. (2004, 2010) *Personal Communication and Interview Transcript*, June/July.

Oyomno, G. (1996). Sustainability of governmental use of micro-computer based information technology in Kenya. In Odedra-Straub, M. (Ed.), *Global Information Technology and Socio-Economic Development*. Nashua, USA: Ivy League Publishing.

Pellegrini, U. (1980). The problem of appropriate technology. In Criteria for Selecting Appropriate Technology under Different Cultural, Technical and Social Conditions, Roveda, D. (ed.), *Proceedings of the IFAC Symposium*, Pergamon Press, 1-5.

Pick, J., & Azari, R. (2008). Global digital divide: Influence of socioeconomic, governmental, and accessibility factors on information technology. *Information Technology for Development*, *14*(2), 91–115. doi:10.1002/itdj.20095

Puri, S. K., & Sahay, S. (2007). Role of ICTs in participatory development: An Indian experience. *Information Technology for Development*, *13*(2), 133–160. doi:10.1002/itdj.20058

Reichgelt, H., Lunt, B., Ashford, T., Phelps, A., Slazinski, E., & Willis, C. (2004). A comparison of baccalaureate programs in information technology with baccalaureate programs in computer science and information systems. *Journal of Information Technology Education*, *3*, 19–34. Available at http://jite.org/documents/Vol3/v3p019-034-098.pdf.

Richards, C. (2004). Information technology and rural development. *Progress in Development Studies*, *4*(3), 230–244. doi:10.1191/1464993404ps087oa

Rogers, E. (1995). *Diffusion of Innovation*. New York: The Free Press.

Roman, R., & Colle, R. (2003). Content creation for ICT development projects: Integrating normative approaches and community demand. *Information Technology for Development*, *10*(2), 85–94. doi:10.1002/itdj.1590100204

Shakya, S., & Rauniar, D. (2002). Information technology education in Nepal: An inner perspective. *Electronic Journal of Information Systems in Developing Countries*, *8*(5), 1–11.

Srinath, U., & Braa, J. (2005). *Training and capacity building to sustain healthcare information systems at a local level in India*. Proceedings of the 8th IFIP WG 9.4 International Working Conference, Abuja, Nigeria, 493-504.

Strover, S., Chapman, G., & Waters, J. (2004). Beyond community networking and CTCs: Access, development, and public policy. *Telecommunications Policy*, *28*, 465–485. doi:10.1016/j.telpol.2004.05.008

Tiwari, M. (2008). ICTs and poverty reduction: user perspective study of rural Madhya Pradesh, India. *European Journal of Development Research*, *20*(3), 448–461. doi:10.1080/09578810802245600

Todaro, M. (2006). *Economic Development*. Reading, MA: Addison-Wesley.

Ulys, P., Nleya, P., & Molelu, G. (2004). *Technological Innovation and Management Strategies for Higher Education in Africa: Harmonizing Reality and Idealism*. Omaha, NE: Education Media.

United Nations Development Programme (UNDP). (2003). *Human Development Report 2003*. New York: Oxford University Press.

United Nations Development Programme (UNDP). (2008). *UNDP Annual Report 2008*. Available at http://www.undp.org/publications/annualreport2008/. Accessed 30 December 2008.

Warschauer, M. (2003). Social capital and access. *Universal Access Information Society*, *2*, 315–330. doi:10.1007/s10209-002-0040-8

World Development Indicators (WDI). (2008). *Table 5.11, Information Age*. Available at http://siteresources.worldbank.org/DATASTATISTICS/Resources/WDI08_section5_intro.pdf Accessed 30 December 2008.

WRI. (2005). *The Wealth of the Poor: Managing Ecosystems to Fight Poverty*. Washington, D. C.: World Resources Institute.

Yin, R. K. (2003). *Case Study Research: Design and Methods* (3rd ed.). Thousand Oaks, CA: Sage.

ENDNOTES

1. Our focus in this article is the provision of IT education outside the formal educational systems like University, Colleges and Polytechnics
2. A national university in the same state as Iragbiji

Chapter 15
Strategic Metamorphoses of ICT Sector for Human Development in India

Meeta Mathur
University of Rajasthan, India

Sangeeta Sharma
University of Rajasthan, India

ABSTRACT

As Indian economy gets integrated to the global economy and strives to improve in terms of human development indicators, a special role exists for information and communication technologies (ICT) in this process. The strategic metamorphoses and the resultant expansion of ICT linked telecommunication services in India have favorably influenced the effort to accelerate the pace of human development by enabling equality in access to information, creation of employment, improving the quality of life, better livelihood opportunities in rural areas, growth of agriculture, impetus to business development, environmental management and many more. After the initiation of economic planning in India, telecom services were assumed to be natural monopoly and were provided by one entity without competition. The government launched ambitious ICT infrastructure initiatives, radically changing its communication policy framework. The resultant growth of ICT services in India has led to significant improvement in human development levels. It has led to a reduction in information asymmetry between the rich and the poor, improvement in telecom density and ICT accessibility in rural areas, fostering inclusive growth, providing better access to market information to people in remote and rural areas, facilitating technological leapfrogging, enhancing business networking and offering new opportunities from the perspective of human development.

DOI: 10.4018/978-1-60960-497-4.ch015

INTRODUCTION

The ICT is renaissance period for Telecommunications reforms in India and has opened new vistas for socio-economic development. There is a policy-shift for the Government to converge this sector into the core physical infrastructure so as to meet the global challenges with prudence. It is the cornerstone upon which creation of knowledge based society will depend for its prosperity and future economic development. Recognizing its importance for the human development as a whole, the telecom services are also referred to as social overheads, thus indicating its enormous potential for creating an egalitarian society. In information age where everyone would like to keep abreast with the latest information of interest ICT is an important media for providing access to information. Today India has a vibrant and competitive communications sector. Development of communications is the national priority for the government. This has been reflected in strategic and technological metamorphoses of the Indian telecommunications sector to promote the universal accessibility of a wide range of modern telecom services provided by multiple service providers in a well – regulated competitive environment stimulating economic growth, encouraging fair competition and social cohesion. It is based on the fundamental assumption that with the metamorphoses of communication sector mobilization of people can be enhanced thereby resulting into enhanced economic growth. Thus this convergence will have compounding impact on the economy to respond to the emerging demands. The facilitation of human development in terms of generating better avenues for growth and also providing quality services though ICT is attainable by incorporating the genuine expectations into the policy-frames. The pertinent thrust would be to identify the mechanics of achieving human development by creating a viable environment for this sector to take firm roots in India. In recent years, high financial returns on telecom services and the benefits they imply in harnessing critical concerns of human development have been increasingly recognized. The importance of rural telecom for any country in general and in particular for India, whose 74% of population still lives in villages and agriculture provides livelihood to 65% of population, is significant and well-established. This paper offers insight into the importance of increased rural coverage from the point view of economic development.

Today Indian telecommunications industry has the higher growth rate in the world. India has the second largest telecom network in the world after China. Presently, the country has one of the cheapest mobile communication services available anywhere in the world. Telecom sector has continued to grow even during the period of economic contraction in India. Teledensity has shown a phenomenal increase from a mere 1% in 1991 to 16% in 2006 and is currently 34%. This increase reflects the potential of sector to further grow to provide leverage to economic sustenance. However unprecedented growth of Indian ICT and telecom sector is attributable mainly to the factors like strong macro economic fundamentals; modernized policy environment; consumer affordability which have been addressed later.

The ICT along with telecommunications is now outreaching people with different aims and objectives. Through Web portals in different areas such as e-education, e- commerce, e-banking, e-taxation, e- travel and tourism, lives of common person is witnessing a new change. Such a change is capacitating people with more focused vision about setting right priorities of development. Apart from priorities, people's participation at grass-root level is also mobilized thus connecting various components of the system into a more composite system of effective governance. This paper deals with the changing patterns of strategic decisions that have contributed to the steady transformation of the entire telecommunication sector This aims at examining the transformation of ICT linked telecommunications sector in India, the visibility of its impacts, and policy initiatives of the government towards this sector. It also focuses on the changing rural telecom scenario in India and the benefits it has brought to the village economy. It explores and highlights the areas where India still has a huge growth potential. It is based on the assumption that strategic modifications brought out in this sector have facilitated a more inclusive growth.

Review of Literature

There are some studies regarding the restructuring of ICT and telecommunication sector in various developed and developing countries and their performance evaluation has been authenticated by various studies. Empirical studies have shown that both quantity (lines per 100 people) and quality (availability of digital communication links) of telecommunications are significant factors for economic development. The human side of ICT has been explored recently with finer focuses of understanding various dimensions of this interface. The growth of modern technology has led to better management of human resources. (Mehdi Khosrow-Pour, 2006; Shimizu, Carvalho and Laurindo, 2006). Feltenstein and Ha have examined the relationship between the provision of infrastructure and private output in 16 sectors (Feltenstein and Ha, 1995). Their study concludes that investment in communication and electricity generally reduces the cost of sector production. Donald has critically examined major economic and public policy issues in telecommunications industry focusing on the complex interaction among the forces of technological change, competition, and regulation (Donald, 1997). A similar discourse by Williams where he has assessed the relationship between telecom and regional economic development and discussed how and why telecommunications can be used as a policy instrument in the economic growth of rural areas is also important to understand this relationship (Williams 1997). Adding to the welfare dimension Leff has analyzed the welfare effects of investment in telecommunications facilities in developing countries (Leff, 1984). In another study Nair has analyzed the development of telecommunications in India enquiring into the policy of the government towards this sector (Nair, 1995). Saran has presented the scenario of rural telecom in India. His study makes a detailed assessment of technology, operational features, and network topology (Saran, 1999). Studies have explored the importance of telecommunications as a powerful tool for equitable distribution of national investment (See Indersan, 1993). By allowing easy acquisition and transfer of information among economic units, and by facilitating rapid two-way communications over distance, telecommunications helps in the coordination of economic activity (Nandi 2002). Malecki and Moriset advocate that IT is shaping the economic spaces (Malecki & Moriset, 2007). However Stahl refers to the effects of ICT on the patterns of human interaction (Stahl, 2008). Saith, Vijayabasker and Gayatri have put together the optimistic voices of techno idealists, critical social science perspectives on technology and a range of empirical material on the impact of ICT on the lives of people (A. Saith, M. Vijayabasker and V. Gayatri,2008). Sawarajiva, Rohan and Zainudeen in the volume brings together scholars, practitioners policy makers to address the problem of expanding ICT connectivity in emerging Asia (Sawarajiva, Rohan and A. Zainudeen, 2008). A plethora of literature is available on the telecommunication and ICT but this paper pointedly views the strategic maturity that has been registered by this sector in India. The inevitability of sustaining this sector in modern era cannot be subdued, for this being the focal axis of socio-economic transformations. Mostly available literature does not address to the issue of assessing impact of technology on human dynamics in India. This paper however examines the transformations in terms of the strategic interventions that have affected the process of creating a more inclusive society in Indian context. (Figure 1)

Figure 1. Factors accelerating telecom growth in India (Source: Authors, 2009)

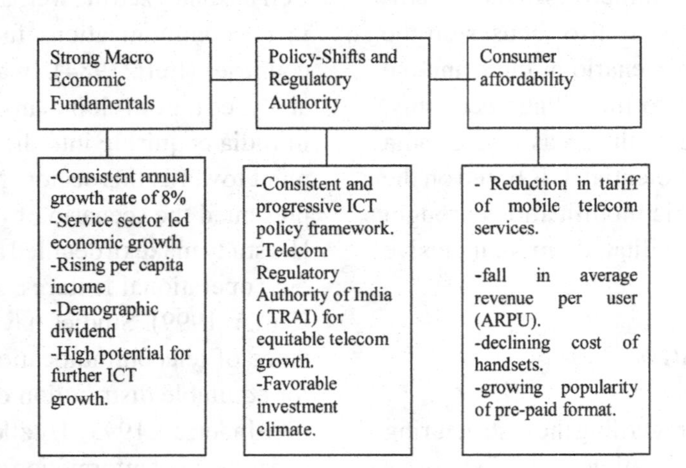

- **Growth Infusing Factors**: The figure below depicts those factors, which have increased the pace of socio-economic transformations *in lieu* of the global synchronizations. The connectivity at the global level is mobilizing people to get informed in their respective fields in India. It will be pertinent here to acknowledge the growth factors for the diversification of this sector. The telecommunication sector of India since its inception has been transformed from manual to the digital communication, indicating the fact that even being a developing economy with many hampering overtones; the issue of communications has been at the pinnacle of attaining modernization for the policy makers. The factors precisely have been categorized into three broader categories with further justifications.

The above schema depicts the factors as mentioned in the preceding paragraph. The phenomenal ICT linked telecom growth is attributable to macro economic fundamentals such as high annual economic growth rate, dominant service sector, and higher levels of income and immense scope of further ICT growth due to current lower teledensity in comparison to other growing Asian countries. Furthermore, a strategic shift towards modernized policy framework along with a regulatory authority has contributed to the creation of congenial environment for the progressive policies to be carried out effectively. The increasing consumer affordability has also significantly led to the penetration into the remote and rural areas of economy. This has created possibilities of remaining attuned with the changing scenario world over, even in those areas which otherwise were marginalized earlier due to lack of having connectivity. The precise reasons for this are cost effective handsets, availability of varied payment options and extending low cost wireless services to the consumers. The factors cannot be viewed in isolation and thereby the compounding effect on the process of transformation can only be understandable by considering the resultant effect collectively. These resultatnt effects have been converged into strategic metamorphoses, which continue to put leverage for incorporating the policy changes.

- **Strategic Metamorphoses and Human Development Impacts:** India since independence has witnessed many changes in socio-economic arena. Initially its adherence to the regulatory form of governance metamorphosed into a liberal economy to respond to the emerging global challenges. It will be apposite to discuss how Indian ICT sector has responded to the necessity of revisiting its existent policies to introduce renewed policy measures. The following specific strategic shifts have been identified and have been subsequently discussed.

1. Transformation of Telecom Market Structure:

The State run monopoly has now invited multiple players to create a competent market for this sector. The preliminary years of Indian independence had observed the controlled system of telecommunication sector with government being the sole player due to certain politico-administrative compulsions. India adopted the Nehruvian model of growth after 1950, patterned somewhat on the Soviet model and involved strict regulation in most areas of the economy. The government made it a state run monopoly under the department of telecommunications (DoT). Monopoly led to inefficiencies and sluggish telecom growth. The weakest area of DoT was the almost non-existence of after sales services and its lacking a professional touch. There was pressure on DoT to be more efficient and involve more players in the area of service provision. Towards the end of 1980s India witnessed winds of change leading to New Economic Policy 1991 with sweeping measures of deregulation and privatization aiming at growth, efficiency and equity through competitiveness. Wide-ranging reforms were initiated in telecom sector as well. The period from 1991 onwards has been a period of revolution in Indian telecom sector. Some factors which forced for the transformation of telecom sector include - traffic congestion during peak hours, high degree of unsatisfied demand evidenced in long waiting lists, unreliable services, frequent faults and long delays in repairing failures. The first wind of liberalization in telecom sector began in 1980s when private sector was allowed in telecom equipment manufacturing. Given the rapid pace of technological innovations in communications worldwide, the reform process was carried out in India on a continuous basis. The ICT and Telecom sector's transformation is an eloquent and successful example of economic infrastructure becoming market focused and customer centric.

2. Improved Telecommunication- Density and Coverage:

Throughout its history, telecom services in India have operated within a paradigm in which demand has exceeded supply. With eroding monopoly and increased competition, this paradigm has been reversed and supply now exceeds demand. Penetration, coverage and access to ICT in general and telecom in particular have dramatically improved, especially in mobile sector. Telecommunications has gone from a total five million telephone lines in 1991 to five million telephones every month. Over the past ten years India has seen a compound annual growth in the subscriber base of more than 35%. A spectacular increase in number of subscribers from mere18.68 millions in March 1998 to over 374 millions in November 2008 indicates the impact of strategic shift. The main driver of this growth has been the expansion of mobile communication services. Wireless services have grown 6.5 millions in March 2002 to a staggering mark of 335 millions in November 2008. There has been a notable increase in access to Internet, mainly in urban areas. Sharp rise in Internet penetration is exclusively on account of expansion in business sector and greater use among upper middle-income consumers of the economy. Tele-density (telephones per 100 persons) has shown a remarkable growth. The total teledensity has reached to 34% in December 2008 compared to 8.9% in March 2005 and 1.3% only in March 1996. Urban density levels have shown much faster rate of growth than rural density levels. (Figures 2 and 3)

Figure 2. (Source: Authors 2009, based on secondary data)

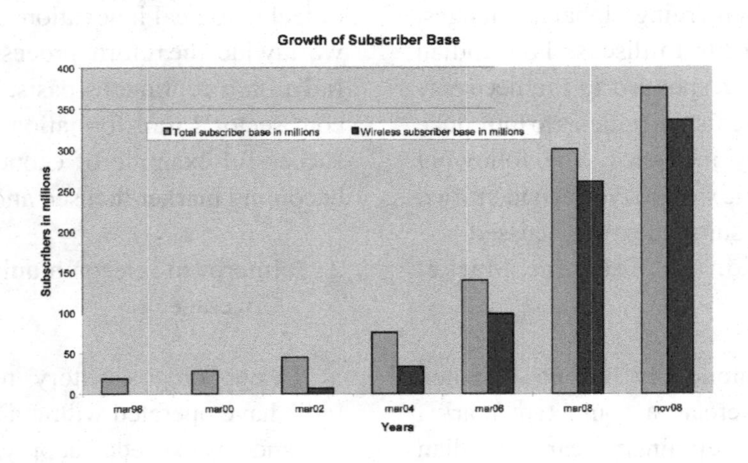

Figure 3. (Source: Authors 2009, based on secondary data)

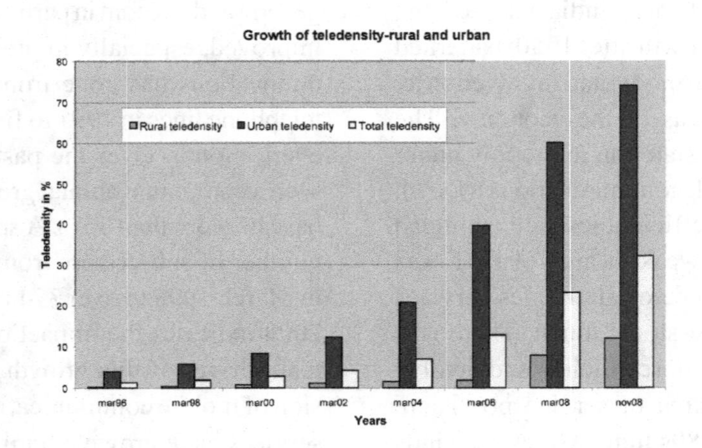

Privatization and liberalization measures have led significantly to quantitative augmentation of telecom services and a marked improvement in the quality of services provided. Private participation was necessary to meet the challenge of large investment requirements in telecom sector. In addition, the waiting list for telephone connections in most areas has been virtually wiped out. With digitization, telephone networks have become, in effect, computer networks, creating more opportunities to provide value-added services such as online data base, E-mail, video conferencing etc. using telecom network.

3. Modernization of the Policy and Regulatory Environment:

In policy terms, the Indian telecom sector has grappled with three legacies or periods: pre-liberalization, winds of liberalization and a period of revolution after 1990s. While telecom services, made a start in India almost simultaneously with the advanced countries of the world, its development was very slow initially. The terrible presence of unsatisfied demand, deteriorating network capability and service quality fuelled demand for change in the Department Of Telecommunications for a new, more competitive and liberalized structure. Recognizing the fact that efficient and advanced telecom holds the key to globalization, consistent efforts were made by the DoT for upgrading and expanding telecom network.

First telegraph, later telephones and telex were the traditional telecom services provided by the Indian government. From 1980s private sector entry into telecom equipment manufacture and supply began. Measures taken included broad based licensing, liberal import of technology and foreign collaboration. 'Mission: Better communication' was launched in 1984 wherein India was making a serious and comprehensive effort at the national level to reform telecom sector. Pilot projects were encouraged by the government in order to have access to new technologies, new systems both basic as well as value added services. A framework for the entry of local and international private sector was evolved. Development of small digital exchanged by the Centre for Development of Telematics (C-DOT) set up by the government revolutionized the telecom scenario in rural areas. National Telecom Policy- 1994 and later New Telecom Policy- 1999 were announced by the govt. to foster telecom growth. Limit to foreign direct investment in telecommunications was raised from 49% to 74%. In 1997, Telecom Regulatory Authority of India (TRAI), charged with promoting non-discriminatory competition, enabling private sector participation and promoting universal access, was created. TRAI was entrusted with the responsibility of a regulator while DOT was left with the task of policy making. Bharat Sanchar Nigam Limited (BSNL) was formed in 2001 to provide telecom services in competition with private sector service-providers. The BSNL is now an important player in ICT sector and is providing quality services to its customers.

4. Propelling Employment Generation:

An important determinant of the prosperity of a nation is the quality of its human resource. In the long run a nation's economic growth depends on its ability to attract, develop and retain the talented people who drive its economy. As telecom technologies continue to evolve in India, it presents opportunities for creation of new businesses and new jobs. India's identity as national and international hub and developer of talent has been reinforced, with India being at the cutting edge of deployment of new telecom technologies. It is the breeding ground for development of the new ICT services made possible by communication capabilities. Indian telecom sector is expected to create 150,000 jobs in the year 2009, the figure being based on the estimates of hiring consultants. The growth has been attributed to the launch of operations in new areas such as 3G and WiMAX. It is expected that the number of skilled and experienced telecom personnel required would exceed the available workforce in India. Mobile telephone growth is promoting a better enterprenual culture and supporting employment generation through proliferation of kiosks. Global system for mobile communications (GSM) has unleashed a new class of entrepreneurs who might otherwise have been unemployed. There is a nationwide network of dealers, vendors, GSM accessory sellers and ubiquitous operators. Manpower requirements for ICT growth do not stop with engineers and technicians. This sector has immense potential to absorb well-trained personnel in the areas of financial planning, law, accountancy, consultancy services, computer science and many more.

5. Fostering Inclusive Growth:

No country can register humane growth by marginalizing larger section of its society. The ICT has helped in assimilating the marginalized sections into mainstream to a larger extent thus controlling the pestilent effects of alienations.

Telecom accessibility has brought the population of distant and far-flung areas into the national mainstream. In general, rural telecom appeared to command lower order of priority. There was a great disparity between urban and rural economies, with rural becoming increasingly marginalized. Electronic communications, by removing friction of distance and reducing locational constraints, has led to convergence between cores and peripheries. While urban telecom network is relatively inexpensive to establish and comparatively easier to install and maintain, rural telecom network, on the other hand, are much more expensive to establish and difficult to maintain. These factors led to extremely low development of telecom in rural and remote areas of India till 1990s. Progressive policy initiatives and increased telecom density has gradually brought them into mainstream of development. The divide in the distribution of telecom services in urban and rural areas referred to as the digital divide has shown some reduction. Studies have proved that benefits generated by rural telecom are ecologically friendly. Given the relationship between telecom expansion and growth, there is scope for further narrowing down of digital divide. Comprehensive deployment of Universal Service Obligation on a larger scale and sharing of telecom infrastructure can help in achieving Digital Balance. A study of fertilizer movement in India showed that out of all vehicle trips made by farmers to fertilizer distribution points the proportion of failed trips, due to non-availability of fertilizer, was 10% to 25% and in some locations was more than 50%. This indicates the human dynamics in quantitative terms.

6. Enhanced Accessibility To Market Information in Rural Areas:

As telephones mobiles and other digital devices get cheaper and widely accessible across the nation, the life of rural people has definitely undergone a tremendous alteration providing the needed impetus for growth. Today, institutional demand for telecommunications is continuously on rise. Rural telecom is playing a crucial role in order to meet the rising demand of farm sector, small industries, irrigation, water supply, bank credit and meeting out marketing needs. There have been large economic savings and increased productivity in remote rural areas through substitution of low priced and reliable telephone usage in place of more expensive mode of traveling in order to have access to markets. Farmers have immensely benefited through better information flow on agriculture prices, products and attributes. Extension of ICT services into India's rural economy has rejuvenated the market functioning. Well-functioning markets have facilitated the commercialization and diversification of farming and they have a crucial role in efficiently bringing food and agriculture products to domestic and international consumers.

Recently govt. has extended a one-time subsidy support to operators from Universal Service Fund to set up shared towers and sites in remote and rural areas. Government has also proposed to provide broadband connectivity in all gram panchayats, public health care centers, government high schools to achieve goals like E-education, telemedicine, E-governance etc. by the end of year 2012.

Telecom growth in rural India has given a big push to the export of agricultural products in international market. To stay internationally competitive, farmers must resort to increased specialization and react to shift in consumer demand. In order for this to happen, the information infrastructure has to be extended to the rural areas. Greater investment on rural telecom coverage proved critical to rural development process.. Cropping decisions of the farmers are today influenced by international prices of exportable agricultural products. It has been found that 85% of use of rural telephones is accounted for by agriculture and trade

7. Improvement in the Quality Of Life:

There are far reaching human development impacts of providing telecom accessibility. The ultimate aim of human development is to improve the quality of life and the spread of ICT is a major determining factor in ensuring it through better access to health and education. Telecommunications has played a major role in improving health care in remote areas from the monitoring and control of diseases to improving availability of better health services in rural and desert areas by allowing local clinics to consult experts in urban hospitals. It has helped in alleviating poverty and income disparities by creating new job opportunities for workers living in geographically far flung areas and for those who lack the transportation to distant work places thereby not able to compete for certain jobs. When it comes to environmental concerns, advanced communication is again important. Remote sensing satellite and other measuring instruments are indispensable tools monitoring earth's atmosphere, agricultural crops, and pollution levels. In optimizing the use of human resources, through the creation of 'telework' opportunities, telecommunications is counteracting the social displacement and disorder seen in today's industrial society. Increased telecom coverage has contributed towards arresting the migration of population to urban areas. There have been increased employment opportunities for rural women looking for attractive employment and income source from the expansion of business in rural areas. Community telephones have been used to call up for medical assistance in emergencies, and to provide warnings in case of natural disasters or epidemics. During rainy seasons and floods in remote areas of a country like India, with uncertainty of monsoon, the telephones have become a lifeline to keep up the contact necessary to administer governmental services, manage development activities and reduce the sense of isolation. The ICT and telecommunications systems are blurring the separation between the home and the workplace. The information brought into home through high speed phone lines and Internet has increased the number and kinds of activities that can occur within the confines of a residence. Growth of telecom also has its impact on equality in access to information and communication. Extension of telecom network has equalizing effect by permitting people outside of the upper classes to gain access to what had earlier been the exclusive preserve of upper class groups.

8. Technological Leapfrogging:

As the sophistication in ICT is progressing all over the world and the wired technology getting converted into wireless, there has been a significant increase in capital expenditure in Indian telecommunications sector in trying to bring the network up to the modern business requirement. When India opened up mobile communication in 1994, it leapfrogged all legacy technologies to build one of the world's modern mobile phone networks. This played a major role in fulfilling the concept of Personal Communication Network (PCN). The Integrated Services Digital Network (ISDN) with built-in-capacity to offer high-speed data transfer and Global Mobile Personal Communications (GMPCS) policies have augmented the liberalized investment climate of Indian industry, agriculture and rapidly expanding service sector. Technological innovations in telecom have proved to be a cost effective measures of providing reliable public telephones and mobile services in hilly, forest, tribal and desert areas of the country. The growth and emergence of digital electronic technologies in fixed line and mobile telephony has virtually sounded the death knell to the fixed line technology itself. By offering a viable techno-economic alternative, mobile services have helped in improving telecom penetration bypassing shortages of fixed lines.

Increased competition has put pressure on service providers to be more cost efficient. The reduction in tariffs for airtime, national long distance, international long distance, and handset prices has driven demand. With a tariff of Rs.

18.00 per call when telecom revolution began in the country in early 1990s, it has fallen to 40 paise a call and incoming becoming free. Comprehensive spectrum policy and third generation (3G) mobile devices with capability of access to mobile data and voice will make India outpace any other nation in telecom technology.

9. Business Networking:

Since the transaction structure of the world economy is becoming integrated, an international market is being created and competitive telecom sector has propelled India in optimally exploiting international business opportunities. This has led to the development of various switching and transmission technologies in the field of telecom. In the era of globalization, multinational companies (MNCs) need access to global networks to link their offices, manufacturing facilities and resources worldwide. Growth of advanced communications in India since 1990 has made it an attractive destination for multinationals. Competitive telecom sector has created enabling environment for generating exports and attracting foreign investment. Export of products like garments, handicrafts, gems and jewellery etc. are particularly sensitive to the availability of efficient telecom network.

Unless telecom infrastructure is adequate, it is impossible to build facilities like electronic- mail which do not only confer competitive advantage but which are central to the electronic- way of doing business. Large corporations, in particular, are catalyzing the development of new networks that provide them with inexpensive and flexible communications.

The secondary and tertiary sectors of the Indian economy are today, the most intensive users of telecommunications. With economic growth, these sectors have rapidly expanded and innovations in ICT have brought spectacular gains to these sectors.. In India service sector is expanding at a fast rate contributing more than 50% to GDP. Advanced communication is very crucial to service industries like banking, retailing, transportation, etc. where information is vital to their activities. A reduction in the cost of these services through improved telecommunications has enhanced India's international competitiveness. In financial markets funds transfer between banks and their branches takes only a few seconds today, compared to week's time that was typical of Indian banks earlier. There exists a substantial degree of cross elasticity between telecommunications and transportation and increasing shift in favor of the former has resulted in substantial gains.

Telephone companies have taken up crucial role as packagers and popularizers of new information services. Geographical coverage of mobile network has rapidly expanded from urban into more high cost rural areas due to market competition, decreasing equipment costs and the expanded demand. Cellular telephones are no longer considered luxury items but rather they have become the preferred de-facto basic services for many low income consumers due to lower prices, calling party pays and pre-paid plans. There has been a notable increase in access to Internet mainly in urban areas. The main driver of improved overall telecom penetration has been the private sector investment unleashed by liberalization initiatives.

- **Visible Impacts on Human Dynamics**: The ICT linked telecommunications sector invariably affects the human dynamics of the society. The effectiveness of this sector is also indicative of the prime distinction between developing and developed economies. All dimensions of communications viz. infrastructure extension, outreaching function, financial viability, connectivity collectively map out mobilization of people in any society. Higher is the rate of mobilization more balancing of socio-economic dynamics is attainable thus leading to the formation of equitable society. It will be pertinent to analyze the multifold impacts of strategic development in this particular arena. This in particular has resulted into,

- Inclusion of marginalized sections of society into mainstream of social configuration.
- Articulation of public opinion to be assimilated at the politico-administrative fronts.
- Integration of the enviable public demands and expectations into policy-frames.
- Incorporating the climate change effects on humanity especially in the developing countries where it has vulnerable populations at high risk of facing consequences.
- Conversion of feeble public voices into cumulative impact making voices capable of providing right direction to sustain development in right direction.
- Building up of right perspective and vision through consolidation of creative and innovative ideas into convertible program with sharp contours.
- Nurturing self-reliant economy by bringing out equilibrium.
- Construction of a composite social structure with unified purpose of relegating purposeful humane services.

Visibility of impacts does have invisible underpinnings shaping the thought process of people at cognitive level. These are latent but are also important in the process of societal reengineering needed for the social prognosis, as it propagates the power of words. The content of communication forms the basis of reorienting the mindsets. The obsolescence of idle thinking can be converted into vibrant and resilient thinking with the help of connectivity transmuted though ICT and telecommunication. The technological advancement in this sector has made it possible to connect with people at the idea level. In India it has helped in dampening the excessive bureaucratization by empowering common person. The unheard voices are now converting into collective voices sufficient to affect the entire governmental system. The survival of effective system of governance is possible if people get mobilized to participate in the process of decision making. The impacts are tangible and predictive therefore can be converted into progressive policy documents.

Policy Trajectory- Future Focus

The transformation of Indian ICT and telecommunication sector is still in its infancy and politico-administrative indifference has eclipsed its fuller growth. Nevertheless ICT and telecommunication sector would always remain the epicenter of development in India. The opportunity to tap the full potential of this sector to improvise human capital in India is underutilized. Hence it will be important to look for plausible solutions to enrich the quality of human lives. A suggestive trajectory on the bases of following solutions can be prepared.

1. **Techno-Bureaucrats**: A change in the Indian bureaucracy is visible with increasing numbers of Technocrats opting for public services. Thus a generic transformation of conventional system of bureaucracy into a more technically advanced system of bureaucracy is in progress. They, in particular, have very special skills to understand the intricacies of technical dimensions.
2. **Enhancing Rural Tele-density**: Although Indian telecom sector has the unique distinction of being the fastest growing telecommunication sector in the world; there is sufficient evidence to suggest that the persistent digital divide between rural-urban sectors has not shown any significant reduction. Reducing the gap between Rural and Urban teledensity can develop an egalitarian society. It is ironical to see that rural teledensity is only 13% as against urban teledensity of 74%. A concerted policy action is required so that

the benefits of telecom revolution percolate down to the vast majority of rural citizens of India to the maximum extent. There is a challenging task ahead to cover the villages with telecom facilities. WiMax could be the answer to connect the 250,000 villages and rural areas in India.

3. **Strengthening Infrastructure**: Telecom operators do not find the much-needed infrastructure required for highly capital-intensive telecom industry. There are bottlenecks such as inadequate and erratic power supply, lack of remote and desert area connectivity, poor security, vandalization etc. which makes it difficult to have inroads into many potential market segments as the service providers have to incur a huge initial fixed cost. There is a further need of simplifying licensing processes for rural telecom operators. In order to strengthen the feasibility for the operators to deliver the services efficiently these hurdles need to be overcome.
4. **Improving International Competitiveness**: Tele-density in India is still low when comparison is made internationally. The digital access index (DAI) introduced by the international telecommunications union in 2003 is an inclusive index that measures the overall ability of the individuals in a country to access and use telecommunications and information and communications technology in general. This index includes: infrastructure, affordability, knowledge, quality and usage. The values range from 0 to 1; the closer it is to 1 better it is. Out of the 178 countries for which DAI are computed, India stands at 119. India needs to improve in terms of quality, affordability and teledensity (infrastructure variable) to better its score in Digital Access Index.
5. **Removing Bureaucratic Blockades**: Despite the technocrats entering into civil services systemic blockades continue to overpower their skills. It is the vestiges of bureaucratic control, which is seen as a stumbling block. The ultimate goal is to provide 'universal' telecommunications services. An integrated public policy has to be evolved to determine a vision and chart out a course of action. This is a prerequisite for building up an effective and competitive ICT network in India.
6. **Augmenting Expenditure on Research and Development**: Research and development is an important component for successful global competition. In Europe, U.S.A. and China the national government are spurring research and development activities to foster the Information Technology industry. To be able to penetrate into highly competitive export markets, there has to be a spurt of innovation in the Indian telecom industry to produce competitive products of high quality. Industry-academia-government partnership in research activities should be promoted.
7. **Promoting Domestic Telecom Equipment Industry**: In a country like India, which has massive unemployment problem, there is a need to initiate policy measures that can promote the growth of telecom equipment manufacturing industry to create more employment opportunities. This will also help the growth of other industries through forward and backward linkages.
8. **Better Spectrum Management:** Scarcity of spectrum- the airways that carry cell phone signals - is a major issue of concern for telecom operators. In a regulated sector, where telecom growth is dependent on government's release of spectrum, entrenched telecompanies have been using their clout to derail greater competitiveness and reforms. Efficiency and transparency in spectrum management is required. High cost

of international bandwidth should also be taken care of along with creating congenial environment.

CONCLUSION

The above analysis points out the importance of ICT in reducing the gaps across the social strata in India where divide is easily perceptible. The process of development has multifarious dimensions with sequential consequences. Hence for holistic and inclusive growth, the identification of tangible and intangible factors is indispensable as this can translate the seemingly distant objectives into reality. There can be difference of opinion on the notional contents of development and the ways of attaining it but no argument can stand against the fact that ultimately it has to aim at improving the quality of human life. The deprivations can lead to imbalanced growth causing other societal repercussions. Of many alternatives, having connectivity is the prime mover in the direction of development. The technological progression in the field of communications has given extra leverage to this aspect through advancements in ICTs linked telecommunication especially in the developing countries. India over the years has responded to the demands of improvisation in this sector strategically. As the part of ongoing process formulating progressive policies to extract maximum potential from telecommunications system can further facilitate reaching out to every person. This sector should not be considered in isolation and must be integrated in national and regional plans. For entire ICT sector, policy needs to be coordinated with the policies of other economic and social infrastructure such as transport, electrification, health and education etc. This requires working out of precise logistics at the formulation and implementation level respectively so that pragmatic policies can yield the desirable results.

The discussion makes it clear that a more comprehensive plan will not only enhance the rate of economic growth but would also mobilize weaker sections of the society. The relationship between ICT and human development is exponential which, in turn, is capable of changing the eugenics with finer focus for the directed growth towards the betterment of the society. The physical development without assimilating human element will only aggravate the miseries. Therefore it is necessary to incorporate this dimension through innovative policies that can address to the various facets with equanimity. In the times of turbulence facelessness of humanity need to be reverted back to the humane epicenter. In this respect the role of ICT sector is crucial. The strengthening of this sector can increase the rate of assimilation of different sections of the society without discriminations by empowering them and thus changing the human dynamics.

The affordability and tremendous potential for ICT growth in India benefiting socio-economic sectors of the economy has offered a great opportunity to outgrow several stages of development. Through innovations, right choice of technology and careful strategic moves in ICT sector, India is poised to take a giant leap forward.

REFERENCES

Brynjolfsson, E., & Hitt, L. (1995). Information Technology as a Factor of Production: The role of Differences among Firms. *Economics of Innovation and New Technology*, *3*, 183–199. doi:10.1080/10438599500000002

Elsadig M.A. (2008). ICT and Human Capital intensities Effects on Malaysian productivity Growth. *International Research Journal of Finance and Economics,* (13), 152-161.

Feltenstein, A., & Ha, J. (1995). The Role of Infrastructure in Mexican Economic Reforms. *The World Bank Economic Review*, *9*(2), 287–304. doi:10.1093/wber/9.2.287

Government Of India. (1996). *India Infrastructure Report.* Department of Economic Affairs, Ministry of Finance, Government of India.

Indiresan, M. (1993). Telecommunications: Social and Economic Impact. *Telecommunications* (pp. 22-25).

Khosrow-pour, M. (2006). *Advanced Topics in information resources Management*. Hershey, PA: IGI Global.

Leff, N. H. (1984). Externalities, Information Costs And Social Benefit-Cost Analysis For Economic Development; An Example From Telecommunications. *Economic Development and Cultural Change*, *32*, 255–276. doi:10.1086/451385

Nair, K. R. G. (1995). Telecommunications in India. *Productivity*, *36*(2), 209–214.

Rama, R. T., Venkat, R., Bhatnagar, S. C., & Satyanarayan, J. (2004). *E-Assessment Frameworks*. Retrieved on Apri 17, 2009 from http://egovt.mit.gov.in.

Saith, A., & Vijaybharkar, M. (2005). *ICTs and Indian Economic Development*. New Delhi: Sage Publication.

Saran, P. S. (1999). Rural Telecommunications. *Telecommunications*. DoT. India.

Saxena, K. K., & Satyananda, S. (1997). Infrastructure and Economic Development: Some Empirical Evidence. *The Indian Economic Journal*, *47*(2).

Schware, R. (2005). *E Development: From Excitement to Effectiveness*. USA: World Bank Publications.

Shah, A. (1997). Telecommunication: Monopoly vs Competition. *India Development Report* (pp. 239-240). Indira Gandhi Institute of Development Research.

Telecom Regulatory Authority of India. (2008). *Annual Report*. Ministry of Communications Information Technology, Government of India.

United Nations. (2008). *Human Development Reports 2007/2008*. UNDP.

Wade, P. (2001). *Information and Communication Technologies and Rural Development.* OECD Publishing.

World Bank. (1994). *World Development Report: Infrastructure*. NewYork: Oxford University Press.

This work was previously published in Information Communication Technologies and Human Development, Volume 1, Issue 4, edited by S. Chhabra and H. Rahman, pp. 16-29, copyright 2009 by IGI Publishing (an imprint of IGI Global)

Chapter 16

Users' Perception of Internet Characteristics in the Academic Environment

Abdullah Almobarraz
Imam University, Saudi Arabia

ABSTRACT

This paper examines the characteristics of internet that motivate faculty members of Imam Muhammad Bin Saud University (IMSU) in Saudi Arabia to utilize the Internet in their research and instructional activities. The framework of the study was the attributes of innovations offered by Rogers. A modified instrument was adopted to collect the data and measure the attributes. The result revealed that the majority of IMSU faulty members used the Internet for research and academic activities twice a month or less, indicating a low Internet adoption rate. Multiple regression analysis showed that all attributes of innovation individually predicted Internet adoption. The combination of all attributes indicated the model could predict Internet adoption among faculty.

INTRODUCTION

Adopting new innovations is one of the major areas in information technology that has been researched extensively in order to determine the primary factors influencing people to accept technologies and implement them in their activities.

DOI: 10.4018/978-1-60960-497-4.ch016

Among different types of technology, the Internet as a research tool has remained the most valuable source of information. In academic environment practically, researchers from different disciplines have become aware of the potential benefits of the Internet not only as a research tool but also as a communication medium. Multiple communication applications provided by the Internet inspire scholars and professionals to keep in contact with each other regularly and exchange information in a short period of time. To conduct their research studies, scholars and university faculty members have access to a wide variety of services, including information sources, electronic mail, file transfer, interest group membership, interactive collaboration, and multimedia displays (Cohen, n.d.) Moreover, Web 2.0 applications have added new ways for faculty members to obtain and organize information. These applications include but are not limited to blogs, wikis, Webcasts, podcasts, RSS feeds, social networks, tags, and AJAX. Hence, the diffusion of Internet adoption can be considered as the most important event of the late 20th century (Vadillo, Bárcena, & Matute, 2006).

In spite of the increasing contents of scientific information published on the Internet, Saudi Arabia as a developing country was late in connecting the Internet to the public compared to other countries. As a result, IMSU was provided with Internet services at a later period, and the faculty, until a short time ago, was unable to access the Internet through the university. Therefore, it is important to study the effect of this delay in adopting the Internet for academic and research purposes and to understand Internet characteristic encouraging or preventing faculty members from the adoption.

Butler and Sellbom (2002) state many factors and predictors affect users' decisions and the rate of adoption, including an innovation's characteristics and economic, sociological, organizational, and psychological variables. The current study focuses on innovation attributes as they appear in diffusion of innovation theory created by Rogers to determine Internet characteristics impacting IMSU faculty members' decision to adopt or reject the Internet in their research activities.

Two core research questions can be formulated to explore the situation concerning this issue including:

1. To what extent do faculty members at IMSU adopt the Internet for academic purposes?
2. Do the attributes of innovations: relative advantage, compatibility, results demonstrability, ease of use, image, visibility, voluntariness, trialability, as perceived by faculty members predict their Internet adoption?

Previous Work

The region or the country where people reside plays a major role in accepting new innovations. For example, a comparative study of home computer adoption in the United States, Sweden, and India showed that Indian households are still behind those of the United States and Sweden (Shih & Venkatesh, 2003). The reason for the difference in use might be attributed to the various infrastructural and cultural factors in countries' communities. Cultural differences around the world result in both divergent attitudes toward technology and culturally distinctive ways of implementing and utilizing technologies (Tully, 1998).

Among other institutions, universities sector particularly between countries differs in the abilities and support regarding technology tools provided to their faculty members. Research revels that the type of support to faculty members has a big influence on their attitude toward accepting the technology. In the United States, Dewald and Silvius (2005) surveyed business faculty members to assess their satisfaction with Web information compared with subscription database usage. The survey measured five factors of user satisfaction: content, accuracy, format, ease of use, and timeline. The study reported significantly higher levels of Web usage than subscription databases usage; however, faculty members were not satisfied with free Web information sources for their own professional research. In another study, Al-Asmari (2005) investigated the use of the Internet by teachers in Saudi Arabian. The result confirmed the existence of a positive correlation between teachers' level of use of the Internet and independent variables including computer and Internet expertise, place of accessing the Internet, perception of the advantages of the Internet, and computer and Internet experience. Nasir Uddin (2003) measured the level of Internet usage for information and communication needs by faculty members of the University of Rajshahi, Bangladesh, related to five categories of Internet activities: e-mailing, browsing, downloading, using newsgroups, and recreation. The major finding of the study revealed that the Internet is not popular among faculty members mainly because of the high cost of communication systems in the country. Results also showed that academic rank was a significant predictor to identify the level of Internet use and the priority of information needs.

Methodology

This exploratory study surveyed full-time faculty members of IMSU in Riyadh to examine their Internet adoption for academic and research purposes. The population included all faculty members who had PhD degrees. Lecturers with master's degrees, teaching assistants, teachers with bachelor's degrees, and staff were excluded. Also, faculty members who had administrative responsibilities without teaching or giving lectures were not incorporated in the study.

The instrument used was adopted from a general purpose scale created by Moore and Benbasat (1991) who aimed to measures individual's perceptions regarding the use of a technological innovation. The framework of their instrument was based on Rogers (2003) five attributes of an innovation: relative advantage, compatibility, complexity, observability, and trialability. Based on reviewing related literature, Moore and Benbasat modified the attributes and developed a scale to measure the following eight attributes:

1. Relative advantage defined as the degree to which an innovation is considered a better than an alternative innovation. The greater the degree an individual perceives the advantages of an innovation to be, the more rapid the innovation's rate of adoption will be (Rogers, 2003).
2. Compatibility which is the degree of the consistency of the innovation with the existing values, past experience, and needs for potential adopters. If an idea is inconsistent with the values of a society, it will not be adopted in the same rapidity as if it is compatible (Rogers, 2003).
3. Ease of use which is defined as the degree to which an individual believes that using a particular system will be free of physical and mental effort (Davis, 1989).
4. Result demonstrability is the tangibility of the results of using the innovation, including their Observability and Communicability (Moore & Benbasat, 1991, p.203).
5. Visibility is the degree to which others can see that an innovation is being used (Benham & Raymond, 1996).
6. Trialability is the degree to which an innovation may be experimented with on a limited basis. The trial provides individuals with less uncertainly and gives them the opportunity to learn and practice by doing (Rogers, 2003).
7. Image is "the degree to which use of an innovation is perceived to enhance one's image or status in one's social system." (Moore & Benbasat, 1991, p.195).
8. Voluntariness is "the degree to which use of innovation is perceived as being voluntary or of free will" (Moore & Benbasat, 1991, p.195).

DATA ANALYSIS

Distribution of the Rate of Adoption

The rate of adoption was determined by providing faculty members with five options to report their frequency of Internet use for academic and research purposes: I do not use it, rarely (once a month), sometimes (twice a month), often (once a week), and constantly (once or more a day).

Table 1. Distribution of the rate of Internet adoption

Rate	Frequency	Percent	Cumulative percent
None	34	9.9	9.9
Rarely	48	14	23.8
Sometimes	106	30.8	54.7
Often	100	29.1	83.7
Constantly	56	16.3	100

Among all participants, the highest number (30%) reported that they used the Internet twice a month, 29.1% of faculty members used it once a week, 16.3% used it once or more a day, 14% used it once a month, and 9.9% did not use the Internet at all (see Table 1). Although a small number of participants did not use the Internet, the results indicated a low Internet adoption rate among users.

FACULTY MEMBERS' PERCEPTION OF INTERNET ATTRIBUTES

A multiple regression analysis was conducted to determine the relationship between Internet adoption and each one of the following attribute: voluntariness, relative advantage, compatibility, images, ease of use, result demonstrability, visibility, and trialability. The multiple regression model used each attribute as an independent variable and Internet adoption as the dependent variable. This technique provided comprehension of the most influential predictors of the decision to adopt the Internet as perceived by faculty members.

Before analyzing each predictor variable individually, a multiple regression analysis was applied to the entire model of innovation attributes to account for the variance in the dependent variable, Internet adoption. As shown in Table 2, the analysis of the combined variables indicated that the entire model was statistically significant in predicting Internet adoption at the level of .05 (p-value = .001 < .05). The R square value for the model was .332, which means that 33.2% of the variance in Internet adoption by faculty members was explained by the eight predictors together.

AFFECT OF VOLUNTARINESS ATTRIBUTE

The results of multiple regression analysis indicated a significant statistical relationship between Internet adoption and the independent variable of voluntariness. As Table 3 shows, voluntariness explains 3% of variance in predicting the adoption of Internet by faculty members. The *p* value

Table 2. Regression Analysis of the Eight Attributes Perceived by Faculty Members

	Sum of Squares	*df*	Mean Square	*F*	*P*	R^2
Regression	94.597	29	3.262	4.791	.001	.332
Residual	190.642	280	.681			
Total	285.239	309				

Table 3. Voluntariness attribute perceived by faculty Members

	Sum of Squares	*df*	Mean Square	*F*	*p*	R^2
Regression	8.535	2	4.268	4.735	.009	.030
Residual	276.704	307	.901			
Total	285.239	309				

= .009 < .05, meaning that the voluntariness is a statistically significant predictor of the dependent variable adoption.

AFFECT OF RELATIVE ADVANTAGE ATTRIBUTE

Multiple regression analysis using relative advantage variable as the predictor was also conducted. The analysis indicated that relative advantage is a statistically significant predictor for the Internet adoption at the .05 level (p-value = .001 < .05) (see Table 4). The R square value in this model was .126, meaning that 12.6% of the variance in predicting Internet adoption is explained by relative advantage. Therefore, faculty members who perceived the Internet as more advantageous were more likely to adopt it in their academic activities.

AFFECT OF COMPATIBILITY ATTRIBUTE

Multiple regression analysis showed that the independent variable of compatibility has a statistically significant relationship with the dependent variable at the .05 level as shown in Table 5 (p-value = .001 < .05). The R square value in the model was .135. This indicates that the compatibility variable explained only 13.5% of the variance in Internet adoption.

AFFECT OF IMAGE ATTRIBUTE

Multiple regression analysis of composite scores of image showed this variable as a significant statistical predictor for Internet adoption at the .05 level (p-value = .001 < .05). The R square value for image was .089, which means that 8.9% of the variance in predicting Internet adoption is explained by the image variable (see Table 6). Consequently, faculty members who perceived the Internet as a factor to enhance their status in the university were more likely to adopt the Internet.

AFFECT OF EASE OF USE ATTRIBUTE

The analysis of composite scores in this model indicated that ease of use is a significant statistical predictor of Internet adoption among faculty members at the .05 level (p-value = .001 < .05). The R square value was .084, as shown in Table 7.

Table 4. Relative Advantage Attribute Perceived by Faculty Members

	Sum of Squares	*df*	Mean Square	*F*	*p*	R^2
Regression	36.066	7	5.152	6.245	.001	.126
Residual	249.172	302	.825			
Total	285.239	309				

Table 5. Compatibility Attribute Perceived by Faculty Members

	Sum of Squares	*df*	Mean Square	*F*	*p*	R^2
Regression	38.501	4	9.625	11.898	.001	.135
Residual	246.738	305	.809			
Total	285.239	309				

Table 6. Image attribute perceived by faculty Members

	Sum of Squares	*df*	Mean Square	*F*	*p*	R^2
Regression	25.417	4	6.354	7.459	.001	.089
Residual	259.822	305	.852			
Total	285.239	309				

This shows that ease of use explained only 8.4% of the variance in Internet adoption, which can be an indicator that faculty members who believed that using the Internet is free of difficulty are more likely to adopt it.

AFFECT OF RESULTS DEMONSTRABILITY ATTRIBUTE

Multiple regression analysis revealed a significant correlation between result demonstrability and faculty members' Internet adoption at the .05 level (p-value = .001 < .05). As shown in Table 8, the R square value for this variable was .073, meaning that 7.3% of the variance in Internet adoption is explained by this variable.

AFFECT OF VISIBILITY ATTRIBUTE

Table 9 indicates that visibility is also a statistically significant predictor for faculty members' Internet adoption ($p = .006 < .05$). Also statistically significant, the R square value suggests that only 4% of variance in predicting Internet adoption is explained by this factor. The frequency to which faculty members saw the Internet as accessible inside or outside the university influenced their Internet adoption can be inferred from this result.

AFFECT OF TRIALABILITY ATTRIBUTE

Similar to other variables in the model, Table 10 shows trialability is a statistically significant predictor of Internet adoption ($p = .006 < .05$). The R square value was equal to that of visibility ($R^2 = .040$). This

Table 7. Ease of Use Attribute Perceived by Faculty Members

	Sum of Squares	*df*	Mean Square	*F*	*p*	R^2
Regression	23.854	5	4.771	5.549	.001	.084
Residual	261.384	304	.860			
Total	285.239	309				

Table 8. Result Demonstrability Perceived by Faculty Members

	Sum of Squares	*df*	Mean Square	*F*	*p*	R^2
Regression	20.888	2	10.444	7.459	.001	.073
Residual	264.351	307	.861			
Total	285.239	309				

Table 9. Visibility perceived by faculty

	Sum of Squares	*df*	Mean Square	*F*	*p*	R^2
Regression	11.480	3	3.827	7.459	.006	.040
Residual	273.759	306	.895			
Total	285.239	309				

Table 10. Trialability Perceived by Faculty Members

	Sum of Squares	*df*	Mean Square	*F*	*p*	R^2
Regression	11.296	2	5.648	6.330	.002	.040
Residual	273.942	307	.892			
Total	285.239	309				

means that the trialability variable explains 4% of variance of predicting Internet adoption.

OTHER IMPACTS ON INTERNET ADOPTION

One open-ended question was included in the questionnaire to identify the most common barriers preventing faculty members from using the Internet for research and academic activities. While some responses confirmed some of the obstacles found in previous studies, other responses pointed out new obstacles. Out of all faculty members included in this study, 60 (17%) members answered the open-ended question.

Quality of Internet Connection

The low quality of Internet connection was reported as the most common barrier by 49 faculty members. This included slow speed and frequent disconnection during browsing. One respondent stated that "the recurrent disconnection makes me discourage my students to use the Internet for course assignments." Another respondent complained that "the Internet is supposed to reduce the time spent to find information, yet with continuous interruption, I sometimes find it easier to use the library to meet my information needs."

English Proficiency

Faculty members who reported the English language as a major barrier to using the Internet confirmed the results found in similar studies (Al-Salih, 2004; Al-Salem, 2005). The demographic analysis shows that 70% of the participants were at or below the average level in English proficiency. This common barrier might be due to the rareness of academic and scholarly Arabic Web sites. Two faculty members stated they had problems with a lack of adequate Arabic search engines to locate needed information, especially classified and specialized resources. Some popular search engines, such as Google, provide Arabic translation for retrieved Web sites, which might solve part of this problem. However, one respondent said, "The need of translation software built cooperatively with Arabic specialists in each discipline arises today, especially with the inaccuracy of the translation of foreign search engines."

Filtering System

All incoming Web traffic to Saudi Arabia passes through a proxy system to filter forbidden Internet contents. Blocked Web sites include those that contain content in violation of Islamic tradition and national regulations as well as pornographic sites. Faculty members complained that inaccuracy of a filtering system resulted in overblocking of unrelated contents. In fact, one faculty member asserted, "Even some academic and research sites are blocked.... I do not request to unblock those sites because it wastes my time to do so."

Internet Access Points

The availability of Internet access throughout the university was addressed as another common barrier to the diffusion of the Internet. Forty-five faculty members reported a lack of enough access points in the university. Surprisingly, the result indicated that the female campus was not connected to the Internet. One female faculty member said, "The Internet is not provided for our campus even in faculty offices." Another female faculty member commented, "By not connecting our campus to the Internet, the university seems to not encourage faculty use of the Internet."

Although Internet access in the male campus was much better, some responses indicated that not each faculty member had access to the Internet. "I am not provided with Internet access in my office, so I have to go to the library sometimes to browse the Internet," one faculty member stated.

The lack of Internet in the classrooms was also mentioned as a barrier to adopting the Internet for instructional and research purposes for both faculty members and students. As one faculty member complained, "Students need to be taken to the computer lab if the instructor wants to show them Web sites related to their assignment."

Cost of the Internet

The cost of Internet access has been an issue since the service was implemented in Saudi Arabia. Some faculty members reported the cost as one of the barriers to Internet diffusion among faculty members. One faculty member said "I decided to switch to DSL connection because of the bad connection of dial-up, but I found it unaffordable to pay for DSL."

The expensive cost of the Internet is not only a barrier because of the connection expenses, but also because of funds needed to purchase resources available through the Internet. One said, "Most academic databases in my field are not free on the Internet, and the library does not subscribe to them." This result is in accordance with other studies that considered cost as a barrier to adopting the Internet in Saudi Arabia. (Al-Fulih, 2002; Al-Kahbra, 2003).

Resistance of Technology

One of the listed issues against diffusion of the Internet was the university administration's awareness of the importance of the Internet in teaching and learning. "The university is still unaware of the valuable information on the Internet, which resulted in not providing the classrooms with computers and Internet," one faculty member said. The lack of knowledge about the valuable resources on the Internet also includes some faculty members who still have negative views toward electronic resources. As one faculty member said, "Some professors ask students to obtain information from books, not from the Internet or even other electronic resources."

CONCLUSION

Based on the findings of this study, the following recommendations are presented for enhancing IMSU of faculty members' adoption of the Internet:

1. All attributes of innovations examined in this study were found to be statistically significant predictors of Internet adoption, so it is recommended the IMSU administra-

tion concentrate on the factors that enhance aspects of each one of the attributes to allow all faculty members to utilize Internet resources in their research and teaching activities. The strategy should be planned based on the needs and skills of faculty members through conducting relevant research for this purpose.

2. IMSU administration needs to develop a new strategy plan to integrate Internet applications into the academic environment. This might include providing each faculty member with a computer and Internet access in his or her office. The findings disclosed that female faculty members are not provided with Internet connection. Thus, their campus needs more attention to give equal service for both genders.
3. Many of faculty members did not receive adequate technology training opportunities. To overcome this barrier, training programs should be held on a regular basis to instruct faculty members on the use of different Internet applications and services. The library may take the responsibility for arranging such activities and provides locations for training. Individual training upon request by faculty members is another way to diffuse Internet adoption. Individual training is critical to address individual faculty members' unique needs. During training, increasing the trialability should be taken into consideration. The perceived attribute of trialability was not statistically significant predictor of Internet adoption. Therefore, faculty members should be given the opportunity to try out different Internet applications, especially applications that support research and teaching activities.
4. Faulty members' English proficiency needs to improve since the majority of Internet content is written in English. The findings of this study revealed that the English level of most faculty members is average or below. This skill can be improved by encouraging faculty members to enroll in English courses. Additionally, IMSU could offer faculty members scholarships to English speaking countries, especially for faculty members who received their degrees from Saudi Arabian or Arabic universities.
5. Increasing the awareness of the importance of the Internet in teaching and research is recommended. Many methods are available to spread awareness among faculty members. One method is to demonstrate successful experiments and projects implemented in similar environments, particularly in developed countries. Another method is to provide more computer labs through IMSU and to connect classrooms with the Internet.
6. Faculty members should be supported and encouraged to employ Web-based instruction in class activities and assignments. This can be accomplished by creating an interactive Web site for classes where students can access class assignments, supplemental information related to study topic, and communicate with each other through discussion forums. Entire courses can also be taught using appropriate software such as WenCT.

REFERENCES

Al-Asmari, A. M. (2005). *The use of the Internet among EFL teachers at the colleges of technology in Saudi Arabia*. Unpublished doctoral dissertation, Ohio State University, Ohio.

Al-Fulih, K. (2002). *Attributes of the Internet perceived by Saudi Arabian faculty as predictors for their Internet adoption for academic purposes*. Unpublished doctoral dissertation, Ohio University.

Al-Khabra, Y. (2003). *Barriers in using the Internet in academic education in the perspective of Education College faculty at King Saud University in Riyadh.* Unpublished doctoral dissertation, King Saud University, Riyadh.

Al-Salem, S. A. (2005). *The impact of the Internet on Saudi Arabian EFL females' self-image and social attitudes.* Unpublished doctoral dissertation, Indiana University of Pennsylvania.

Al-Salih, Y. N. (2004). *Graduate students' information needs from electronic information resources in Saudi Arabia.* Unpublished doctoral dissertation, Florida State University, Florida.

Benham, H., & Raymond, B. (1996). Information technology adoption: Evidence from a voice mail introduction. *Computer Personnel, 17*(1), 3–25. doi:10.1145/227005.227006

Butler, D., & Sellbom, M. (2002). Barriers to adopting technology for teaching and learning. *Educase Quarterly, 25*(2), 22–28.

Cohen, L. (n.d.). *Conducting research on the Internet.* Retrieved October 20, 2007, from http://www.Internettutorials.net/research.html

Davis, F. (1989). Perceived usefulness, perceived ease of use, and user acceptance of information technology. *MIS Quarterly, 13*(3), 318–340. doi:10.2307/249008

Dewald, N. H., & Silvius, M. A. (2005). Business faculty research: Satisfaction with the Web versus library databases. *Portal: Libraries and the Academy, 5*(3), 313–328. doi:10.1353/pla.2005.0040

Moore, G. C., & Benbasat, I. (1991). Development of an instrument to measure the perceptions of adopting an information technology innovation. *Information Systems Research, 2*(3), 192–222. doi:10.1287/isre.2.3.192

Nasir Uddin, M. (2003). Internet use by university academics: A bipartite study of information and communications needs. *Online Information Review, 27*(4), 225–237. doi:10.1108/14684520310489014

Rogers, E. M. (2003). *Diffusion of innovations.* New York; London: Free Press.

Shih, C., & Venkatesh, A. (2003). *A comparative study of home computer adoption and use in three countries: US, Sweden, and India.* Centre for Research on Information Technology and Organizations. Retrieved December 05, 2007, from www.crito.uci.edu/noah/paper/MISPaperforWeb.pdf

Tully, P. (1998). Cross-cultural issues affecting information technology use in logistics. In C. Ess & F. Sudweeks (Eds.), *Proceedings, cultural attitudes towards technology and communication* (pp. 317-320). Australia: University of Sydney.

Vadillo, M. A., Bárcena, R., & Matute, H. (2006). The Internet as a research tool in the study of associative learning: An example from overshadowing. *Behavioural Processes, 73*(1), 36–40. doi:10.1016/j.beproc.2006.01.014

This work was previously published in Information Communication Technologies and Human Development, Volume 1, Issue 4, edited by S. Chhabra and H. Rahman, pp. 30-39, copyright 2009 by IGI Publishing (an imprint of IGI Global)

Chapter 17
Technology-Related Trust Issues in Inter-Organizational Business Relations

Muneesh Kumar
University of Delhi South Campus, India

Mamta Sareen
University of Delhi, India

ABSTRACT

The emergence of inter-organizational system has facilitated easy and fast flow of information among the trading partners. This has affected the business relations among the trading parties involved. Though the inter-organizational systems have helped a lot in improving the business relations, the vulnerability and the virtual environment of such systems raise the issues of trust that may affect the long-term business relations. This chapter makes an attempt to empirically examine the relationship between the levels of assurance with regard to deployment and implementation of relevant technology tools in addressing the identified technology-related trust issues and ultimately enhancing the perceived level of trust in inter-organizational business relations. The empirical evidence presented in this chapter is based on a survey of 106 Indian companies using inter-organizational systems for managing their business relations.

INTRODUCTION

The traditional manner of conducting business has changed with the advent of information technology and the focus of efforts to improve performance and relations among organizations has shifted from the organizational level to the inter-organizational level (den Hengst and Sol). Inter-organizational systems have been effective in improving efficiency and reducing costs (Swatman & Swatman, 1992; Subramani, 2004). They also help in reducing environmental uncertainty by facilitating communication and providing information (Henderson, 1990; Scala & McGrath, 1993) and encouraging closer relations with sup-

DOI: 10.4018/978-1-60960-497-4.ch017

pliers and customers (Cash & Konsynski, 1985; Swatman & Swatman, 1992; West, 1994; Siau, 2003). The importance of inter-organizational information sharing is that it provides benefits to organizations including economies of scale, lower overhead and reduced risks (Alexander, 1995). Numerous studies have been conducted to examine the impact of IT in inter-organizational relationships (Christiaanse & Huigen, 1997; Christiaanse & Venkatraman, 2002). Drawing on Malone et al (1987), Riggins and Rhee (1998), technology can enhance existing relationships between trading partners by promoting closer integration and increasing degree of interdependence between the trading partners. Furthermore, Bakos and Brynjolfsson (1993) proposed that IT use in business exchanges leads to closer cooperative relationships. Mukhopadhyay, Kekre and Kalathur (1995) asserted that technology leads to improved information sharing between trading partners. Although, the use of effective technology helps in sharing accurate and relevant information among the trading partners in less time (Anderson & Weitz, 1989; Malone, Yates & Benjamin, 1987), thereby greatly influencing their relationships, yet lack of trust in technology aided environment is often cited as a major hurdle in the growth of inter-organizational systems.

The role of inter-organizational trust has often been recognized as most effective in business relations. Trust in the physical trading environment is often built by various factors like face-to-face interactions, physical evidences, etc. The faceless environment offered by technology in electronic transactions, at times, is unable to provide such evidences, thereby generating reluctance on the part of the trading parties. Inter-organizational trust in an electronic business environment may relatively improve if the old adage "trust needs touch" (Olson & Olson, 2002) is addressed. In the virtual environment offered by inter-organizational systems, 'touch' needs to be replaced by effective evidences offered by technology. The issue is whether and how technology can be used to provide various 'trust building cues' to the trading parties, thereby improving the performance and relations in inter-organizational systems. The present chapter makes a modest attempt to address this issue. It also attempts to identify certain technology-related trust issues, which can be addressed through effective deployment and implementation of relevant technology tools, and attempts to relate them with the levels of trust in various Indian companies involved in inter-organizational business relationships.

LITERATURE REVIEW

Trust is the basis of commerce and it is essential for the growth of commerce that the trading partners trust each other and also the trading environment in which they operate. In the context of electronic commerce, trust may be regarded as a judgment made by the user, based on general experience learned from being a customer/seller and from the perception of a particular merchant. It is a belief or expectation that the promise by the merchant can be relied upon and the other party will not take advantage of the one's vulnerability. Literature on trust in inter-organizational e-commerce has shown it as a social element (Mayer et al, 1995), as related to supplier performance (Zaheer et al, 1998), as related to favorable economic outcomes (Ba & Pavlou, 2001) and related to perceived benefits (Ratnasingam, 2003). Institutional trust has often been cited as fundamental in building and retaining inter-organizational relationships (Pavlou, 2002). Keen et al (1999) has explicitly pointed out the importance of trust for the potential growth of e-commerce. Various other studies have also shown that lack of trust as one of the biggest concerns in e-commerce transactions (Cox, 1999; Levin, 2000; Westin and Maurici, 1998). Without trust, development of e-commerce cannot reach its full potential (Cheskin & Sapient, 1999).

Giddens (Giddens, 1984) and Luhmann (Luhmann, 1989) emphasized that system trust plays a key role in reference to inter-organizational relationships. Thus, inter-organizational trust was related as trust in between two abstract systems. Loose and Sydow (1994) stated that although the first step towards trust building process of a company is the social system that consists of individuals that eventually develop the trust, but it is the system trust as trust in abstract systems, institutions or other impersonal structures that play an important role in inter-organizational relations.

Certain issues pertaining to technology have been cited to impact trust. Mahadevan and Venkatesh (2000) stated security, privacy and authentication as the key factors which need to be addressed by technology in inter-organizational systems. Marchany (2002) focused on lack of online security as one of the major factors for the absence of trust. The study emphasized that the eradication of trust may cause users to avoid use of Internet and revert back to traditional method of business.

Perhaps, one of the most interesting studies that relate technology with trust is the one done by Pauline Ratnasingam (2003). She emphasized that trust in inter-organizational systems is a function of trust in the transaction infrastructure and underlying control mechanism (technology trust), which deals with transaction integrity, authentication, confidentiality, non-repudiation and best business practices. Haiwook (2001) found that technological infrastructure in inter-organizational systems provides a better means of coordination.

Pavlou (2003) integrated trust and risk as variables of technology acceptance model into a research model (TAM). He posited that trust in e-commerce retailer affects consumer's perceived risk of the transaction. Further, ease of use and usefulness of the website also had a positive affect on consumer's intention to transact. Mary Anne Patton and Audon Jøsang (2004) identified certain technology-aided methods that help in developing trust in the inter-organizational environment. The presented methods comprised of privacy strategies, communicating trustworthiness through the web interface, self-regulation and trustmark seals, security strategies, payment intermediaries and insurance providers, reputation systems, and alternative dispute resolution. They asserted that use of such methods have the ability to influence trust in business relations. George Staikos, (2006) stated that in order to build trust in the virtual environment of computers, there is a need to build a system that users have full confidence in and are able to use without falling victim to fraud. The solution needs to be technologically fool proof and must cover the security of all aspects and parties involved. Any single weak link can lead to a complete loss of trust in the entire system.

Hexmoor et al (2008) examined the interplay of inter-personal and inter-organizational trust through a theoretic inter-organizational trust-based security model for a multi-agent system information-sharing community. They offered a soft security approach that presented benefits like robustness, scalability, and adaptability over traditional hard security mechanisms.

Most of the existing research as reviewed above has acknowledged that the issue of trust is an important impediment in business relations in inter-organizational systems. A number of issues that affect the trust have also been identified in some of these studies. However, no comprehensive effort seems to have been made to relate trust with various trust–related issues that needs to be addressed by appropriate practices regarding effective deployment and implementation of relevant technology tools. This chapter attempts to fill this research gap.

Trust and Technology Model for Business Relations

Figure 1 shows a 'trust and technology' model for business relations in inter-organizational systems as proposed by the authors. This model emphasizes that trust in inter-organizational systems can be influenced by addressing two major issues namely: (a) Technology–related trust issues and (b) Envi-

Figure 1. Trust and technology model

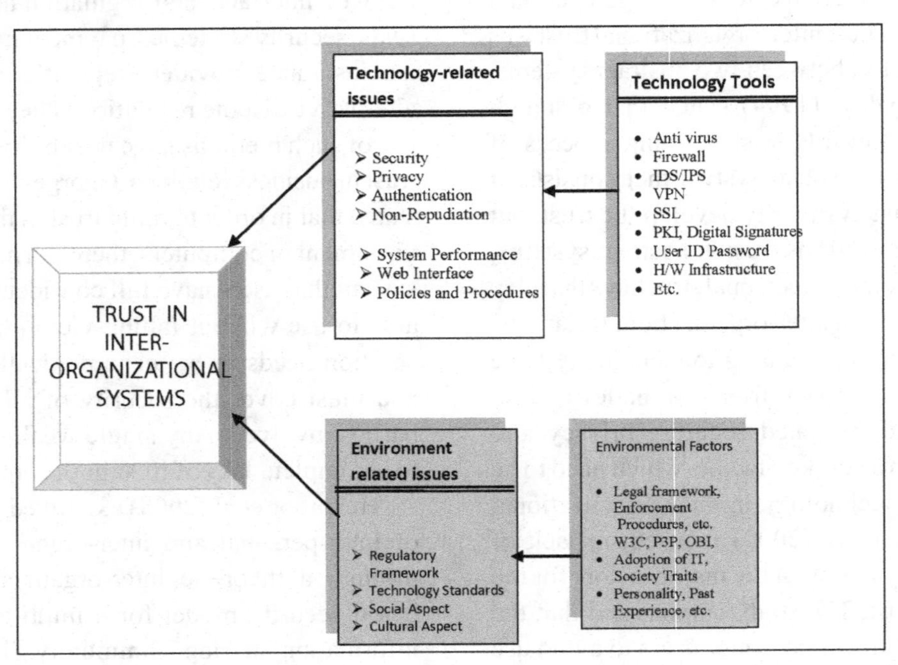

ronment related issues. The model also provides a conceptual framework for understanding the relationships between technology tools and the environmental factors on one side and the levels of trust on the other side in business relations.

The model identifies Security, Privacy, Authentication, Non-Repudiation, System Performance, web interface and technology-related policies and procedures as the technology-related trust issues which can be addressed by effective deployment and implementation of relevant technology tools like Anti-Virus, Firewalls, VPN, etc. The model also identifies environment-related trust issues such as regulatory framework, technology-standards, social aspect and cultural aspect that can be addressed by various environmental factors like legal framework, enforcement procedures, adoption of IT etc.

The scope of the present chapter has been purposely restricted to four important technology-related trust issues and the relevant technology tools in addressing these issues. The trust issues include: (a) Security; (b) Privacy; (c) Authentication; and (d) Non-Repudiation. This chapter makes an attempt to empirically validate the relation between the levels of assurance with regard to deployment and implementation of relevant technology tools in addressing the identified technology-related trust issues and ultimately enhancing the perceived level of trust in inter-organizational business relations.

Sample Selection and Methodology

This chapter is primarily based on a survey of the practices regarding deployment and implementation of technology tools followed by the companies that are using inter-organizational systems for business relations. The survey was carried out in the year 2007-2008. The collection of the primary data was done with the help of a structured interview-guide administered during personal interviews. A pilot survey of three companies preceded the main survey. During the pilot survey, the respondents were requested to offer suggestions regarding the design and content of the interview guide. The interview guide was modified in the light of the suggestions received before its use.

In India, no formal list of various organizations focusing on inter-organizational activities is available. Hence leading companies that are offering both the software and hardware solutions for inter-organizational systems were contacted to obtain the list of their customers. In this process, a list of 200 companies from all over India could be generated. These companies were contacted through post, e-mail, phone, etc. to seek their participation in the survey. However, only one hundred six (106) companies responded and agreed to participate in the survey. The companies, located in different cities like Delhi, Mumbai, Pune, Chennai, Hyderabad, Bangalore, Chandigarh, etc. constituted the sample. The respondents were of the rank of the chief executive officer, chief information officer, or the IT head of the company. The respondents were interviewed in person in order to offer clarifications, if needed. To the extent possible the information so gathered during the interview was also crosschecked by browsing their web sites.

The sample companies were asked to indicate their practices regarding the effective implementation of the deployed technology tools. Based on the practices (as indicated in Appendix A), the sample companies were classified on various levels of assurance with respect to the technology tools. Since, these technology tools help in addressing the technology-related issues, the level of assurance with respect to each of these issues was also calculated and the sample companies were then classified on that basis. For this purpose, various levels of assurance were converted into ratings (i.e. high level of assurance - 3, medium - 2 and low level of assurance -1). These ratings of different technology tools deployed by the sample company were added to arrive at the overall ratings of the issue for the sample company. Based on this overall rating, the sample company was classified into three broad categories. The cutoff for this classification was based on careful observation and assessment of the data. The assessment of level of assurance for security and privacy was done based on this method. And the assessment for level of assurance for authentication and non-repudiation was collectively based on the practices regarding the relevant technology tools and the policies and procedures followed by the sample companies. This assessment of levels of assurance regarding the technology-related trust issues was specifically done to test the following hypotheses:

H1: *High level of Assurance regarding Security has the potential to influence trust in inter-organizational systems.*

H2: *High level of Assurance regarding Privacy has the potential to influence trust in inter-organizational systems.*

H3: *High level of Assurance regarding Authentication has the potential to influence trust in inter-organizational systems.*

H4: *High level of Assurance regarding Non-Repudiation has the potential to influence trust in inter-organizational systems.*

Further, in order to relate the level of trust with the technology-related trust issues, the respondents were requested to indicate the perceived level of satisfaction of their trading partners with regard to these issues on a 3-point Likert scale. The respondents were requested to base their indicated level of satisfaction on the basis of the various enquires made by their trading partners with regards to these issues. The respondents were also asked to indicate their perceived level of satisfaction for the entire inter-organizational environment. This level was taken as indicator of level of trust of their trading partners in the inter-organizational system. Here, it is assumed that the levels of satisfaction as indicated by the respondents reflect the trust of trading partners.

Finally, regression analysis was used to relate the levels of trust with trust issues.

Hence, this paper empirically validates the relationships between the levels of trust and the trust issues in three broad steps. Step one identifies the levels of assurances with regards to effective implementation of relevant technology tools. Step 2 validates the relationship between technology-related trust issues and the relevant technology tools. Finally, step 3 attempts to relate the select trust issues with the levels of trust.

TECHNOLOGY TOOLS

Today, inter-organizational systems face challenges that lower the levels of trust. The key to building trust in inter-organizational systems is to implement the various technology- aided tools to ensure safe and trustworthy online transactions. As the cost of technology has fallen considerably due to both economies of scale and open-source initiatives, the cost of setting up and operating an infrastructure required for inter-organizational systems is increasingly becoming affordable. Although, there exist various technology tools that influence the overall trustworthiness of the system, the present paper has taken the most relevant and commonly used technologies into consideration that are now becoming almost mandatory for any inter-organizational system. These technology tools include Antivirus software solutions, IDS/IPS, Firewalls, VPN, SSL, IPSec, PKI/ Digital Signatures, etc.

However, these technology tools can be effective only when suitable practices are followed in their deployment and implementation. These practices are indicated for each of the technology tool in Appendix A. A company involved in inter-organizational business relationships need to put in place appropriate policies and procedures that help in fully exploiting the potential of these tools in addressing the trust issues. The degree of effectiveness of these technology tools is likely to be influenced by the extent to which these practices are followed. Table 1 presents the summary of the levels of assurance maintained by the sample companies with respect to each of the technology tools.

As can be observed from Table 1, most of the respondents were maintaining medium to high level of assurance with regard to deployment and implementation of most of the technology tools. Anti-virus and Firewall technologies were found to be most commonly used technologies. SSL was also found to be almost a mandatory technology for communication of information among inter-organizational systems. However, IPSec technology was found to be less in use mainly because of the high cost involved. This reinstates the importance of technology in inter-organizational systems. On further classification, it was found

Table 1. Level of assurance maintained by respondents with respect to deployed technology tools

Level of Assurance Technology Tools	High	Medium	Low	Total
Anti Virus	61(57%)	24(23%)	21(20%)	106(500%)
Firewall	67(63%)	21(20%)	18(17%)	106(100%)
IDS/IPS	36(34%)	16(15%)	54(51%)	106(100%)
VPN	32(30%)	56(53%)	18(17%)	106(100%)
SSL*	80(75%)	26(25%)		106(100%)
IPSec*	15(14%)	91(86%)		106(100%)
PKI	32(30%)	53(50%)	21(20%)	106(100%)

that often high level of assurance of technology tools is related to the size of the company, its annual turn over, its inter-organizational business venture and the nature of business of the organization. Respondents belonging to banks and IT sector, respondents with high annual turn over (more than 250 crores), and respondents with e-services and e-procurement aspect of inter-organizational systems were often found to be following practices relating to high level of assurance of various technology tools deployed. Since, this relation is not the focus of the present paper, the details of this analysis is purposely excluded.

These technology tools play an important part in addressing the identified technology-related trust issues; hence an attempt was made to relate the levels of assurance maintained by the sample companies regarding these tools with the trust issues.

Trust and Technology-Related Issues

As proposed in the model, the paper identified four main technology-related trust issue namely security, privacy, authentication and non-repudiation to have the potential to influence the levels of trust in inter-organizational business relationships. These issues and the relevant technology tools addressing them are discussed below.

Security: Security in inter-organizational systems, primarily, deals in protecting the integrity of the business network and its internal system and accomplishing transaction security between the trading partners. Since, in the inter-organizational systems, all the partners (as well as the employees) have on-line access to information generated by the organizations' internal systems, the players do not trust their counter parties to keep their information secure from espionage. The security failures in inter-organizational systems result in huge losses because of the high value of transactions being committed in this area. Hence, these security hazards need to be effectively addressed. There is a need to build a secure environment in which the trading parties have complete confidence and they are able to use it without falling victim to security risks.

Based on the study of the inputs from the respondents of the survey, seven technology tools were identified to be in common use for addressing the security related trust issue. These tools included antivirus solutions, firewalls, IDS/IPS, VPN, SSL, IPSec, and encryption mechanisms (using Public key infrastructure, digital certificates electronic signatures). For this purpose, level of assurance with respect to each technology tool was first converted into rating and overall ratings were computed as described earlier. These overall ratings were, then, used to classify companies into those with high level of assurance, medium level of assurance and low level of assurance. The companies with overall rating of 19 and above were classified as those with high level of assurance. The companies with overall rating between 10 and 18 were classified as those with medium level of assurance and those with overall rating less than 10 were classified as those with low level of assurance. The cutoff for this classification was based on careful observation and assessment of the data. Table 2 presents the distribution of sample companies for various level of assurance in respect of security related trust issue.

As may be observed from Table 2, almost one-third of the respondents were maintaining high level of assurance with regards to deployment of security related technology tools. These respondents were found to have effectively deployed technologies like anti-virus solutions, firewalls, IDS/IPS, VPN, PKI, SSL, etc. and were also fol-

Table 2. Level of assurance in respect of security related trust issue

Level of Assurance: Security				
	High	Medium	Low	Total
No of Companies	34	54	18	106
Percentage	32%	51%	17%	100%

lowing the relevant practices for their effective implementation. 17% of the respondents, however, maintained low level of assurance in this regard.

Further, information was also obtained from the responding companies regarding the security vulnerability of their infrastructure based on the frequency of security threats, their nature, their source, their impact on the system and the consequent action taken by the company. On the basis of information so obtained, the degree of vulnerability of the company's infrastructure was assessed. This degree of vulnerability was related with the level of assurance maintained by the company with respect to various security related technology tools. It was found that the respondents maintaining high level of assurance regarding these tools had a low degree of vulnerability. This would imply that degree of vulnerability is inversely related to the level of assurance with regard to the various technology tools deployed and implemented. A low degree of vulnerability towards security threats influences the credibility of the company and affects business relations among trading partners.

Privacy: Privacy concerns usually follow security concerns. The information flow in inter-organizational systems needs to be only accessed by authorized parties i.e., it should be kept private and confidential. The trading parties require an assurance that their sensitive data will not be exposed to any unauthorized party, and that it will not be shared in any way without their express authorization. Lack of privacy can lead to breach of trust.

Based on the survey results, five technology tools were identified to address the privacy aspect of inter-organizational business relations. These technology tools include IDS/IPS, VPN, SSL, IPSec and PKI. Further, an overall level of assurance with regard to privacy issue of each of the sample company was also assessed. For this purpose, level of assurance with respect to each technology tool was first converted into rating and overall ratings were computed as described earlier. These overall ratings were, then, used to classify companies into those with high level of assurance, medium level of assurance and low level of assurance. The companies with overall rating 14 and above were classified as those with high level of assurance. The companies with overall rating between 8 and 13 were classified as those with medium level of assurance and those with overall rating less than 8 were classified as those with low level of assurance. The cutoff for this classification was based on careful observation and assessment of the data. Table 3 presents the distribution of sample companies for various level of assurance in respect of privacy related trust issue.

As may be observed from the Table 3, three-fourth of the respondents were maintaining high to medium level of assurance with regards to privacy. On further enquires these respondents stated that such levels of assurance helps to build their credibility and maintain long term business relations with their trading partners. However, nearly one-fourth of the respondents were maintaining low levels in this regard. This would imply that these respondents might be compromising on their privacy regarding their inter-organizational business interactions, thereby affecting business relations among their trading partners.

Authentication: All the participants in a business transaction need to be confident of the identity of the trading parties. Since the transactions based on Inter-organizational systems are conducted in a virtual environment without face-to-face interactions, one needs to be sure

Table 3. Level of assurance in respect of privacy related trust issue

Level of Assurance: Privacy				
	High	Medium	Low	Total
No of Companies	34	50	22	106
Percentage	32%	47%	21%	100%

of the identity of the interacting party. Thus, the business interactions require authentication of the entities involved in the interactions. This issue attains greater importance in inter-organizational systems particularly due to fairly high value and frequency of transactions. The trading partners need to ascertain that the party is correctly associated with who they are pretending to be and has the required authorization for the action. There is a need for a process that verifies whether an organization or an individual exists, has a name, and is entitled to use that name. Technologies like VPN, SSL, PKI, Username and password id, etc. are widely used for authentication and authorization. Moreover, the use of digital certificates and USB smart card tokens are also becoming an important tool for authentication. For an effective authentication system, a number of practices need to be followed. These practices along with the basis of classification for assessing levels of assurance regarding this issue are indicated in Appendix B. Table 4 presents the distribution of sample companies for various level of assurance in respect of authentication related trust issue.

It was observed that more than 80% of the respondents were maintaining medium to low level of assurance with regards to authentication. These respondents were found to have laid policies and procedures for active and strong authentication and authorization system using various technologies like VPN and USB smart card with digital certificates. The respondents also felt that such levels of assurance is almost mandatory in the faceless environment of inter-organizational systems and helps to build their credibility and maintain long term business relations with their trading partners.

Table 4. level of assurance in respect of authentication related trust issue

Level of Assurance: Authentication				
	High	Medium	Low	Total
Number of Companies	39	51	16	106
Percentage	36%	50%	14%	100%

Non-Repudiation: If any transaction occurs, both the parties should be accountable for it, i.e. he/she should not be able to deny his/her entering into the transaction. Conversely, they should also be able to prove their non-participation in an exchange transaction where they have not actually been involved. The purpose of non-repudiation procedure is to generate verifiable proof or evidence for the transaction. For transactions in inter-organizational systems, non-repudiation procedures ensure that transacting parties cannot disown the transaction or cannot subsequently repudiate (reject) a transaction by exploiting virtual nature of electronic transactions. To protect and ensure digital trust, various technologies exist like PKI using Digital Signatures, which not only validate the sender, but also 'time stamp' the transaction, so it cannot be claimed subsequently that the transaction was not authorized or not carried out. Further, there is a need to adopt various policies and procedures like maintaining audit logs, time stamping the transactions, involving trusted third parties, enforcing digital signatures, etc. which help in ensuring non-repudiation of the transactions. Appendix B enlists the number of practices that needs to be followed in order to address the non-repudiation issue in an inter-organizational system. It also indicates the manner of assessment of level of assurance with regard to this issue.

As may be observed from Table 5, 71% of the respondents were maintaining high to medium level of assurance with regard to non-repudiation related trust issue. However, on informal discussions they asserted that if the non-repudiation issue is adequately addressed, then it helps in building credibility among their trading partners and assures them to commit themselves in long-term relations. The respondents asserted that most often the trading enquiries was related to the policies and procedures followed in addressing the non-repudiation issue and for this, they take adequate help from technology. This helps

Table 5. Level of assurance in respect of non-repudiation related trust issue

Level of Assurance: Non-Repudiation				
	High	Medium	Low	Total
Number of Companies	28	47	31	106
Percentage	27%	44%	29%	100%

to enhance the credibility of the organization and helps to build long-term business relations among the trading partners.

Trust and Technology-Related Issues

In order to understand the relationship between the perceived trust in inter-organizational systems and the levels of assurance with regard to technology-related issues, Spearman rank correlation was calculated between the level of assurance (as ascertained on the basis of technologies deployed and the related practices) with regard to each of technology-related issues and the perceived level of satisfaction as indicated by the respondent regarding each of these issues. Table 6 gives the correlation coefficients associated with each of the technology-related trust issues.

As may be observed from Table 6, the correlation coefficient is very high for each of the technology-related trust issue. This would imply that the perception about each of these trust issues is influenced to a great extent by the level of assurance maintained through effective deployment

Table 6. Correlation between perceived level of satisfaction and technology related trust issues

Technology Issues	Spearman rank Correlation Coefficients
Security	.910
Privacy	.840
Authentication	.780
Non-Repudiation	.954

Note: Correlation is significant at .01 level (2 sided)

and implementation of suitable technologies. A pertinent question, however, is whether the level of assurance for each of these technology-related issues influences the level of trust in the inter-organizational systems.

Level of Trust and Technology Related Trust Issues

In order to examine the relationship between the level of trust in the inter-organizational system and the level of assurance for the technology related trust issues, regression analysis was carried out. The regression analysis was carried on two sets of data.

First, the level of trust in the inter-organizational system was regressed on the perceived level of satisfaction (as indicated by the respondents) for each of the technology-related trust issues. This was done in order to ascertain the relationship between level of trust and the satisfaction/perceptions of the users as to what extent the technology related trust issues are being addressed.

Second, the level of trust was regressed on the levels of assurance in respect of the technology-related trust issues (as ascertained on the basis of practices relating to effective deployment and implementation of technology tools in addressing). This analysis aimed at finding out the relationship between the level of trust in inter-organizational systems and the relevant practices that address the technology-related issues.

Table 7 presents the results obtained from both the sets of data. In the first set of data, the adjusted R^2 obtained was 91%. This would imply that the level of satisfaction in respect of technology-related trust issues could collectively explain more than 90% of the variation in the trust ratings. Similarly, in the second set of data, adjusted R^2 obtained was 86%, indicating fairly high degree of relationship between the level of trust in the inter-organizational system and the levels of assurance in respect of different technology related trust issues.

Table 7. Regression results of technology-related trust issues

Model	Un-standardized Coefficients		Standardized Coefficients		
	B	Std. Error	Beta	t	Sig
Perceived Level of Satisfaction					
(Constant)	.250	.10		2.273	.021
Security	.486	.070	.486	6.955	.000
Privacy	.213	.059	.249	7.269	.000
Authentication	.123	.061	.241	3.663	.001
Non-Repudiation	.213	.055	.293	4.184	.001
Level of Assurance					
	B	**Std. Error**	**Beta**	**t**	**Sig**
(Constant)	.568	.148		3.809	.000
Security	.584	.197	.550	3.959	.001
Privacy	.234	.172	.022	3.281	.001
Authentication	.169	.154	.201	2.487	.010
Non-Repudiation	.480	.247	.467	3.578	.001

The results depicted in Table 6 and Table 7 provide evidence in support of all the four hypotheses stated earlier as all the four technology related trust issues have significant influence on level of trust. The regression results further indicate less significant role of authentication in influencing the level of trust as compared to the other technology-related issues.

Thus, our model emphasis and support the belief that the three major concerns of players in inter-organizational systems are (a) "How to be sure that the data has not been tampered"; (b) "How to ensure that no-one else see the confidential information exchange between two trading parties"; and (c) "How to ensure that the trading party would not repudiate the transactions". The parties in the inter-organizational systems need to be confident that the transactions details and other valuable commercial information are not accessible to anyone other than those involved in the transactions. They need to be assured that all confidential information regarding the business exchange is kept confidential and not leaked to any third party. The participants of inter-organizational systems need certainty that a transaction conducted over the virtual environment of inter-organizational systems is supported by adequate evidence so that none of the involved parties in the e-transactions are able to deny later that the transaction took place. If the above issues are carefully addressed then trust in inter-organizational systems is automatically built.

The inter-organizational business relationships can be improved significantly by maintaining high level of assurance with regard to each of technology related trust issues by following practices that ensure effective deployment and implementation of suitable technologies which may in turn influence business relations among the trading parties.

CONTRIBUTIONS

An important contribution of this paper is in terms of offering a 'Trust and Technology' model. The model relates the level of trust in the inter-organizational systems with various technology related trust issues and attempts to establish their relationship with various technology tools. This attempt is perhaps the first of its kind that focuses on technology tools in addressing the trust related issues that arise in inter-organizational systems. The empirical evidence presented by the paper makes the paper unique in its scope and approach. Since the model has been developed with reference to existing literature, its development is traceable in terms of identification of sources. However, it is not fully traceable in terms of the replicability of the development process. That means that it is very likely that other researchers may identify similar technology-related trust issues but their method of model construction might be different. Such a model involving trust and technology in inter-organizational systems does not seem to have been proposed by the previous researches and thus, it in itself is a unique contribution towards the growing body of literature.

As it seems, none of the previous researches have focused on the practices relating to various technology tools. Another main contribution of this study lies in the fact that it has identified various practices, polices and procedures regarding effective deployment and implementation of the relevant technology tools that help in enhancing the levels of trust in inter-organizational systems and thereby improve business relations. Further, it also offers a comprehensive list of various technology tools that help in addressing the technology-related trust issues identified in the model.

LIMITATIONS

Like any other study of this kind, this paper also suffers a number of limitations. The main limitation of this chapter can be imputed to the research's coverage. The authors tried to include as many trust-shaping factors as possible without trying to scope down the online trust problem to one particular aspect. This high-level aspect that resulted might therefore seem too general when viewed only from one discipline. Secondly, the survey should have been distributed to a larger population, giving a more accurate representation of the participant's perceptions of trust in inter-organizational systems. This limits the generaliability of our results to inter-organizational systems. Last but not the least, a potential limitation also comes from the some answers sought from the sample data. The empirical study was based on the satisfaction levels regarding various technology-related trust issues, indicated by the respondents as perceived by their trading partners. This may bias the results if the respondents have not given a true refection of their trading partner's perceptions. Ideally, the trading partners of the respondents should have been sought to answer their satisfaction levels.

SCOPE FOR FUTURE RESEARCH

The technology related trust issues proposed by the present chapter are merely representative of the notion of technology and are not exhaustive of the other potential drivers of inter-organizational systems. Despite the high validation of the proposed model, it may be possible to identify additional technology related factors that may influence trust in inter-organizational systems. Future research could explore other technology related constructs that may also predict trust and business relations among inter-organizational systems.

Having established the components of trust, future research could examine how a specific characteristic of technology and its related polices and procedures in inter-organizational system impact trust. For example how do Privacy policies, Security statements, return policies influence users' perceptions of predictability and or reliability? How patch-service provisions, vulnerability assessments, etc. individually affect the various trust related issues? Research could identify specific characteristics and perform experiments to test their influence on trust components to determine which characteristics have the greatest effect on users' perceptions of trust.

REFERENCES

Anderson, E., & Weitz, B. A. (1989). Determinants of continuity in conventional industrial channel dyads. *Marketing Science*, *8*, 310–323. doi:10.1287/mksc.8.4.310

Anderson, J., & Narus, J. (1990). A model of distribution firm and manufacturing firm working partnerships. *Journal of Marketing*, *54*(1), 42–59. doi:10.2307/1252172

Ba, S., & Pavlou, P. A. (2002). Evidence of the effect of trust building technology in electronic markets: Price premiums and buyer behavior. *Management Information Systems Quarterly*, *26*(3), 243–268. doi:10.2307/4132332

Bakos, J., & Brynjolfsson, E. (1993). From Vendors to Partners: Information Technologies and Incomplete Contracts in Buyer Seller Relationships. *Journal of Computing*, *3*(3), 301–329.

Cash, J. I., & Konsynski, B. R. (1985). IS redraws competitive boundaries. *Harvard Business Review*, (2): 134–142.

Christiaanse, E., & Huigen, J. (1997). Institutional Dimensions in Information Technology Implementation in Complex Network Settings. *European Journal of Information Systems*, *6*(2), 268–285. doi:10.1057/palgrave.ejis.3000258

Christiaanse, E., & Venkatraman, N. (2002). Beyond SABRE: An Empirical Test of Expertise Exploitation in Electronic Channels. *Management Information Systems Quarterly*, *26*(1), 15–38. doi:10.2307/4132339

Clemons, E., & Row, M. (1993). Limits to interfirm coordination through information technology: Results of a field study in consumer packaged goods distribution. *Journal of Management Information Systems. M E Sharpe Inc. Armonk. 10* (1), pp 73-89.

Cox, J. (1999). *Trust, Reciprocity, and Other-Regarding Preferences of Individuals and Groups. Department of Economics*, University of Arizona, mimeo.

Den Hengst, M. Hlupic, & V. Serrano, (2005). *Design of Inter-Organizational Systems: Collaboration & Modeling*. Proceedings of the 38th Annual Hawaii International Conference

Haiwook, C. (2001). *The Effects of Inter-organisational Information systems Infrastructure on Electronic Cooperation*: An Investigation of the "move to the Middle". PhD Abstract. Proquest Digital Dissertations. www.lib.umi.com/dissertations, Accessed 13.3.2003

Henderson, J. C. (1990). Plugging into strategic partnerships: The critical IS connection. *Sloan Management Review*, (Spring): 7–18.

Levin, C. (2000). Web Dropouts: Concerns About Online Privacy Send Some Consumers Off-Line, *PC Magazine*, http://www.zdnet.com/

Malone, R. W., Yates, J., & Benjamin, I. (1987). Electronic Markets and Electronic Hierarchies. *Communications of the ACM*, *30*(6), 484–497. doi:10.1145/214762.214766

Malone, T. W., Yates, J., & Benjamin, Y. R. I. (1987, June). Electronic Markets and Electronic Hierarchies. *Communications of the ACM*, *30*(6), 484–497. doi:10.1145/214762.214766

Marchany, R. C., & Tront, J. G. (2002). *E-Commerce Security Issues*. Proceedings of the 35th Hawaii International Conference on System Sciences, Hawaii.

Mayer, R. C., Davis, J. H., & Schoorman, F. D. (1995). An integrative model of organizational trust. *Academy of Management Review*, *20*(3), 709–734. doi:10.2307/258792

Mehadevan, B., & Venkatesh, N. S. (2000). *A framework for building on-line trust for business to business e-commerce.* In Proceedings of The IT Asia Millennium Conference,Bombay, India.

Mukhopadhyay, T., Kekre, S., & Kalathur, S. (2002). Business Value of Information Technology: A Study of Electronic Data Interchange. *Management Information Systems Quarterly*, (June): 137–157.

Olson & Olson. (2000). i2i trust in e-commerce. *Communications of the ACM, 43*(12), 41–44. doi:10.1145/355112.355121

Patton, M. A., & Jøsang, A. (2004). Technologies for Trust in Electronic Commerce. *Electronic Commerce Research, 4*(1), 9–21. doi:10.1023/B:ELEC.0000009279.89570.27

Ratnasingham, P., & Klein, S. (2001). *Perceived Benefits of Inter-Organizational Trust in Ecommerce Participation: A Case Study in the Telecommunication Industry*. Proceedings of the Seventh Americas Conference on Information Systems. Boston, Massachusetts, pp. 769-780.

Riggins, F. J., & Mukhopadhyay, T. (1999). Overcoming EDI adoption and Implementation Risks. *International Journal of Electronic Commerce, 3*(4), 103–113.

Riggins, F. J., & Rhee, H.-S. (1998). Toward a Unified View of Electronic Commerce. *Communications of the ACM, 40*(10), 88–95. doi:10.1145/286238.286252

Scala, S., & McGrath, R. (1993). Advantages and disadvantages of electronic data interchange: An industry perspective. *Information & Management, 25*, 85–91. doi:10.1016/0378-7206(93)90050-4

Siau, K. (2003). Interorganizational systems and competitive advantages - lessons from history. *Journal of Computer Information Systems, 44*, 33–40.

Staikos George. (2006). *Improving Internet Trust and Security*. KDE. http://www.w3.org/2005/Security/usability-ws/papers/33-staikos-improving-trust/

Studio Archetype and Cheskin(n.d.). *The Cheskin Research and Studio archetype/Sapient e Commerce Trust Study*,Retrieved from www.studioarchetype.com/cheskin/html/phase1.html

Subramani, M. (2004). How do suppliers benefit from information technology use in supply chain relationships? *Management Information Systems Quarterly, 28*, 45–74.

Subramani, M. E. Walden, (2000). *Economic Returns to Firms from Business-to-Business Electronic Initiatives: an Empirical Examination*. In Proceedings del International Conference on Information Systems (ICIS).

Swatman, P. M. C., & Swatman, P. A. (1992). EDI system integration: A definition and literature survey. *The Information Society, 8*, 169–205. doi:10.1080/01972243.1992.9960119

Wang, Y. D., & Emurian, H. H. (2005). An overview of online trust: Concepts, elements, and implications. *Computers in Human Behavior, 21*(1), 105–125. doi:10.1016/j.chb.2003.11.008

Wang, Y. D., & Emurian, H. H. (2007). Inducing Online Trust in E-Commerce: Empirical Investigations on Web Design Factors. In Khosrow-pour, M. (Ed.), *Utilizing and Managing Commerce and Services Online)*. Hershey, PA: Idea Group Publishing. doi:10.4018/9781591409328.ch005

West, L. J. (1994). Breaking down the barriers to EDI implementation. *TMA Journal, 14*(1), 10–15.

Westin, A., & Maurici, D. (1998). *E-commerce and privacy: What net users want* (p. 15). Price Waterhouse Coopers.

Zaheer, S., & Manrakhan, S. (2001). Concentration and dispersion in global industries: Remote electronic access and the location of economic activities. *Journal of International Business Studies*, *32*(4), 667–686. doi:10.1057/palgrave.jibs.8490989

ENDNOTE

* There are only two levels of assurance regarding SSL and IPSec technology i.e. either it exist or not. If the respondent is using these technology tools, then they are classified as maintaining high level of assurance.

APPENDIX A

Technology Tools	Practices Regarding Effective Deployment and Implementation
Anti-Virus Solutions	***a) Scanning of information:*** It can be done in two ways *(i)* ***Automatic scans*** which automatically scans specific files or directories and prompt at preset intervals to perform complete scans; and (ii) ***Manual scans*** which are carried out at user prompt as and when required. **b) Installing the *latest/ new version* of anti virus.** Installing the ***latest/ new version*** of anti-virus solutions ensures the protection against all old and the new viruses.
Level of Assurance	*High:* Automatic scanning with latest version *Low:* Manual scanning *Medium*: The remaining.
Firewall	**a) Selection of Firewall:** There are some rudimentary types of firewalls like packet filter and some elaborate firewalls like application layer and stateful multilayer. Application gateway firewall is a better option as it permits no traffic directly between networks, and also performs elaborate logging and auditing of traffic passing through it. **b) Placement of Firewall:** Firewalls should be placed in all layers of the system architecture. This helps in enhancing the security and decrease the vulnerability. **c) Creation of DMZ:** DMZ allows both present and potential trading partners to obtain the information that they need about the company without accessing the internal network. d)**Firewall Configurations:** There may exist various options like audit controls, content filtering, packet filtering, securing mail server, or real time monitoring, which the firewall may be configured to and which may decide the actual functionality of the firewall. The configurations spell out the level of implementation of a firewall and the extent of security it is imparting to the system or the network. e) **Documentation:** Firewall configuration gives a clear picture of the current strategy of the firewall configurations and the loopholes in the current configurations and where to incorporate changes if need arises. By documenting the configurations, one is able to check whether the configuration quality is correlated with other factors specifically, the operating system on which the firewall runs, the firewall's software version, and a new measure of rule-set complexity, etc. **f) Periodicity of log look ups:** Firewall logs provide summaries to the administrator about what kinds and amount of traffic passed through it, how many attempts were made to break into it, information on connection accounting active connections, multiple alerting capabilities, time of connection, destination, duration of connection, action taken, etc. The log files should be daily monitored as it helps to view in real time all traffic passing through the gateway so as to derive a useful summary of activity on the network. g) **Reviewing of Firewall configurations:** It is imperative that the rules concerning the configuration of every component in the firewall (Internet router, firewall, proxy server, and virus software) are properly implemented, fully documented and revisited for incorporating changes. This not only helps to be updated against the latest security threats but also helps to incorporate the changes in the business policies.
Level of Assurance	*High:* Following more than five of above stated practices *Low:* Following less than three of above stated practices *Medium*: The remaining.
IDS/IPS	**a) Subscription to latest IDS/IPS signatures**: The companies using IDS/IPS solutions should subscribe to their latest versions so that they get continuously updated by newer patches and signatures. This involves updating IDS/IPS software versions and signature release levels that enable the company to update signature files on their sensors as newer files become available. **b) Placement of IDS/IPS:** For effective implementation, it is essential to ensure the right placement of IDS/IPS. The best option is to place the IDS/IPS on every layer consisting of web servers, the application servers and the data repositories. This covers all the areas of vulnerabilities and ensures detection of any intrusion from all aspects. Placing the IDS in this location allows it to do its job on all traffic that gets through the edge firewall and provides an extra layer of protection. **c) Regular checking of IDS/IPS logs:** The IDS/IPS logs provide vital information regarding security incident identification and response. They also contain security content updates and outage notifications. These logs help in data analysis and to improve the accuracy of security event identification, and incident escalation and remediation.
Level of Assurance	*High:* Following all the three stated practices *Low:* No IDS/IPS *Medium*: The remaining.

continues on following page

Appendix A. continued

Technology Tools	Practices Regarding Effective Deployment and Implementation
VPN	**a) Type of VPN:** There are basically two types of VPN connections: **Remote-access and site-to-site VPN.** Remote-access Site-to-site VPNs allow companies to create dedicated, secure connection between locations across the open Internet. Site-to-site VPNs are either **Intranet based or Extranet based**. Extranet VPN's allow more secure connections with business partners, suppliers and customers for the purpose of B2B e-commerce. **b) Tunneling:** Most VPNs rely on tunnelling to create a private, secure network that reaches across the otherwise public and insecure Internet. VPN supports two types of tunneling: **Voluntary tunneling and compulsory tunneling.** In the voluntary tunneling, the VPN client manages connection setup. In the case of compulsory tunneling, the carrier network provider manages VPN connection setup. The tunneling functionality can be based on either **Layer 2 or Layer 3 tunneling protocol.** Layer 3 tunneling is more effective. **c) Hardware or Software Based:** Hardware based VPN systems are encrypting routers. They are secure and easy to use. Companies also use firewall based VPN which are more secure. Some software based VPNs are generally harder to manage than hardware based, but it is a better option for dial-up links. **d) Others:** For VPN to function successfully there are number of essential practices and polices are laid. They include: *Security, Preventing Denial of Service Attacks, Scalability and Flexibility.*
Level of Assurance	*High:* Following extranet based VPN with compulsory tunneling, offering hardware based VPN with either layer 2 or Layer 3 tunneling functionality making VPN installation to be secure, scalable, flexible and apt in preventing 'DOS' attacks. *Low:* Companies not using any VPN connections *Medium*: The remaining.
PKI	**a) User authentication:** Anonymity and role-play cannot be tolerated in B2B e-commerce and a strong verifiable identity is required. In face-to-face transactions, there is a high level of trust between the participants as it is easy to verify the identity of the participants. In virtual world, PKI offers strong user authentication methods. Digital signatures and Digital certificates provide a better and more efficient identity process. **b) Strength of Encryption:** The strength of encryption algorithm is the key to success of encryption. The strength of encryption mainly depends upon three factors. They are a) the method of encryption, b) the length of key and c) the type of algorithm used. The analysis revealed that only one-fourth of the respondents were using encryption methods of more than 1024 bits. More than half of the respondents were using key length of 128 bits to 512 bits for their encryption. **c) Manner of Encryption:** The data is not only vulnerable to attacks from "outside" the system/network (i.e. Internet), but also is exposed to "inner" network i.e. malicious users within the network. Hence, the best practice is that the confidential data should not only be encrypted when transmitting outside the network but also is encrypted when traveling within the company's network. **d) Variable level of Encryption:** The best practice for any company which had a distributed security i.e. access of the data from both outside and inside of the network, is to have a variable level of encryption. This means the level of encryption should vary with respect to the type of data, the type of person(s) accessing the data and the authority of accession of that data. Since encryption process takes some time, hence it invariably increases the response time. In order to overcome this, variable level of encryption should be applied depending upon the confidentiality and importance of the data. **e) Installing the encryption patches:** Almost all the companies of the encryption/PKI technology are constantly releasing various update patches so as to further secure and strengthen the encryption process. The best practice with regards to PKI is to update these patches as and when released in the systems. The analysis revealed that almost one-third of the sample companies were regular in installing encryption patches in their systems to strengthen their PKI infrastructure. However, the remaining respondents were not installing any encryption patches.
Level of Assurance	*High:* Following more than three of above stated practices. *Low:* Encrypting the data whenever required or when transmitting outside the system/network. *Medium*: The remaining

APPENDIX B

Authentication	a) There should be variable level of authorization depending upon the sensitivity of information shared. b) There should be an active, strong authorization system, which should prompt the users to forcibly change the password on a regular basis. c) Authentication services should be fully integrated into the enterprise-wide security policy and should be centrally managed through the graphical user interface. d) There should be tools to detect any frauds or attempted frauds in to the system. e) There should be limited trials given to authenticate the user. Failure to successfully log within the attempted trials should lock the user from further trials i.e. there should be an account lockout mechanism to block the user. f) All authentication sessions should be monitored and tracked through the Log. g) Firm background checks done on employees with access to sensitive information. h) All inactive accounts should automatically be disabled. i) The access controls of all those employees leaving the company should be immediately revoked.
Level of Assurance	*High:* Following more than seven of above stated practices. *Low:* Following less than four of above stated practices *Medium*: The remaining
Non-Repudiation	a) Maintain audit logs or logs for all the transactions conducted. b) Time stamp the transactions. c) Regularly perform audit checks on a daily, weekly or a predefined interval. d) The system should provide the information about the type of event, date and time of event and the transaction information. e) The logs should not only contain the successful attempts but also the unsuccessful attempts for access. f) The logs need to be archived to a secure log server. g) The audit logs should be backed up and retained for a pre-defined time interval. If possible, the archived logs should be sent to a trusted third party for the purpose of monitoring and the audit report should be duly communicated to the trading partners.
Level of Assurance	*High:* Following more than five of above stated practices. *Low:* Following less than three of above stated practices *Medium*: The remaining

Chapter 18
Enhancing Service Quality in Hospitals:
Mining Multiple Data Sources

Anirban Chakraborty
Lal Bahadur Shastri Institute of Management, India

Sonal G Rawat
Lal Bahadur Shastri Institute of Management, India

Susheel Chhabra
Lal Bahadur Shastri Institute of Management, India

ABSTRACT

Large organizations use multiple data sources, centralize processing in these organizations require analysis of huge database originating from various locations. Data mining association rules help perform exploration and analysis of large amounts of data to discover meaningful patterns which can facilitate effective decision-making. The objective of this article is to enhance service quality in a hospital using data mining. The improvement in service quality will help to create hygienic environment and enhance technical competence among staff members which will generate value to patients. A weighting model is proposed to identify valid rules among large number of forwarded rules from various data sources. This model is applied to rank the rules based on patient perceived service parameters in a hospital. Results show that this weighting model is efficient. The proposed model can be used effectively for determining the patient's perspective on hospital services like technical competence, reliability and hygiene conditions under a distributed environment.

INTRODUCTION

Big organizations such as hospitals with multiple branches deploy multiple data sources, putting all data together from different sources create a complex environment for processing huge databases for centralized processing. Data mining involves the exploration and analysis of large amounts of data in order to discover meaningful patterns. The association's rules help to process multiple data sources and also facilitate forwarding these

DOI: 10.4018/978-1-60960-497-4.ch018

rules to the centralized company databases. This provides a feasible way to deal with multiple data source problems. However, these forwarding rules may be too many for the centralized processing environment. Therefore, there is a need to discover high frequency rules that can play a major role in decision-making processes to improve the service quality in hospitals.

The objective of this article is to design a model that can be used to enhance customer perceptions on the service quality in a hospital under a distributed database environment. The centralized department needs to gather information (rules) from its various branches. But the massive number of forwarded rules is difficult to analyze. Hence, there is a need to find high frequency rules to facilitate analysis of large quantity of database. The article has shown mining multiple data sources (branches of hospital) in terms of patient's viewpoint on the service conditions and how these can be applied on optimized Xindong Synthesizing Model (Wu & Zhang, 2003). The performance comparison of this model along with our proposed model has also been depicted.

REVIEW OF LITERATURE

Data mining helps analysis of information (rules) that can mine useful patterns from large databases for decision makers. The discovered knowledge can be referred to as rules describing properties of the data, frequently occurring patterns, clustering of objects in the database which can be used to support various intelligent activities such as decision making, planning and problem solving (Jiawei, Kamber, & Kaufmann, 2007).

Let I ={ i_1, i_2 .i_3,….in }be a set of N distinct literals called items, and D be a set of transactions over I. Each transaction contains a set of items i1, i2, i3,….ik € I. A transaction has an associated unique identifier called TID (Transaction Identification Number). An association rule is an implication of the form A→B, where A, B ⊂ I, and A∩B = null set. A is called the antecedent of the rule, and B is called the consequent. A set of items (such as the antecedent or the consequent of a rule) is called an item set. Each item set has an associated statistical measure called support, denoted as supp. For an item set A⊂ I, supp(A) = s, if the fraction of transactions in D containing A equals to s. A rule A→B has a measure of strength called confidence (denoted as Conf) which is defined as the ratio supp(A∪B) / supp(A).

The problem of mining association rules is to generate all rules A→B that have both support and confidence greater than or equal to some user specified threshold, called minimum support (minsupp) and minimum confidence (minconf), respectively (Hand & Mannila, 2004). For regular associations, supp(AuB) ≥ minsupp, conf(A→B) = supp(A∪B) / supp(A) ≥ minconf.

Synthesizing rules is the process of putting all rules together and to produce valid rules from that. To mine transaction databases for large organizations that have multiple data sources, there are two possible ways.

i. Putting all data together from different sources to amass a centralized database for centralized processing, possibly using parallel and distributed mining techniques.
ii. Reusing all promising rules discovered from different data sources to form a large set of rules and then searching for valid rules that are useful at the organization level.

There are many methods and algorithms suggested for this second task. Apriori algorithm (Agrawal & Srikant, 2001) uses a two step technique to identify association rules, and a search space in Apriori consists of all items and possible itemsets. Apriori algorithm is an influential algorithm for mining frequent itemsets for Boolean association rules. The algorithm uses a Level-wise search, where k-itemsets (An itemset that contains k items is a k-itemset) are used to explore (k+1)-itemsets, to mine frequent itemsets from transactional database for Boolean association rules. In this approach first the set of frequent 1-itemsets is found. This is used to find the set of frequent 2-itemsets, which is used to fine frequent 3-itemsets and so on, until no more frequent k-itemsets can be found.

FP-tree-based frequent patterns mining method was developed by (Han, Pei, & Yin, 2000). This method is found efficient than the Apriori algorithm also an OPUS-based algorithm has been reported by Webb (2000) to reduce the searched space by focusing association rules mining with which the searched space consists of all possible items and item sets in a database.

However, existing work has focused on mining frequent item sets in data sets (Brin, Motwani, & Silverstein, 1997) and (Lee, Lee, & Chen, 2001) and few research efforts have been reported on post mining that gathers, analyzes, and synthesizes association rules from different data sources.

Turinsky and Grossman (2000) discussed two types of strategies while mining multiple databases. The first strategy is to leave the data in place, building local models and combining the models at a central site—In place Strategy. On the other extreme, when the amount of geographically distributed data is very small, it is possible to move all the data to a central site and build a single model there. They called this a Centralized Strategy.

Zhang and Wu (2004) have deprecated the idea of moving data to a central location and then mining which they called Mono mining because apart from the cost of moving huge data over a communication network, the strategy will effectively obliterate the local patterns at each site. They suggested that any business organization with multiple branches has two decision levels: headquarter level (global) and branch level (local) decisions.

Wu and Zhang (2005) discussed the classification of databases which is a pre requirement for multi database mining. If a large organization has different types of business with different meta data, the database within the company will have to be classified before data mining could be attempted. For example a multinational company with 20 branches, it is not possible that all branches may be dealing with the same stock; some may deal with food items, some with textiles, some with agricultural products and so on. They argue that the first step in multi database mining is to identify the databases that are most likely to be relevant to an application, without doing so the mining process will unnecessarily be lengthy, directionless and ineffective. They have proposed a measure of relevance and have given an algorithm for identifying relevant databases.

Due to size of large databases and the amount of intensive computation involved in association analysis, parallel data mining has been a crucial mechanism for large scale applications. Existing parallel data mining techniques (Skillicon & Wang, 2001) endeavor to scale up sequential techniques to parallel form. However, such algorithms does not make use of local rules and require more computing resources (Parthasarathy, Zaki, Ogihara, & Li, 2001) to distribute components across parallel processors.

Another related research effort is hierarchical meta learning (Padromidis & Stolfo, 2000) and (Ortega, Koppel, & Argamon, 2001) which has a similar goal of efficiently processing large amounts of data. Meta learning starts with a distributed database or partitions a database into disjoint subsets, concurrently runs a learning algorithm on each of the subsets and combines the predictions from classifiers learned from these subsets. The focus of meta learning is to combine the predictions from learned models from the partitioned data subsets in a parallel and distributed environment. However meta learning does not produce a global learning model from classifiers from different datasets.

Xindong and Shichao (2003) tried a technique for this synthesize problem and came out with a solution of normalizing the weights of data sources. To make use of discovered association rules from different data sources, they have proposed a synthesizing model. When data source for each of the mined association rule is clear they have constructed a weighting model to synthesize these rules. While calculating the weight of data sources they have multiplied the weight of rule with its frequency but this process of normalizing the weights of data sources increases the space complexity and is too difficult and time consuming. Moreover extra storage space is needed to store the intermediate results for further processing. The objective of the article will be to optimize the Xindong algorithm and use it for determining the service quality of Hospital sector and identify the patient perception on service quality. For enhancing the service quality of hospital, we have used the SERVQUAL scale for devising the survey questionnaire to understand the customer perceptions of service.

The SERVQUAL method from Valarie A. Zeithaml, A. Parasuraman, and Leonard L. Berry is a technique that can be used for performing a gap analysis of an organization's service quality performance against customer service quality needs. SERVQUAL is an empirically derived method that may be used by a services organization to improve service quality. The method involves the development of an understanding of the perceived service needs of target customers. These measured perceptions of service quality for the organization in question, are then compared against an organization that is "excellent". The resulting gap analysis may then be used as a driver for service quality improvement. It takes into account the perceptions of customers of the relative importance of service attributes. This allows an organization to prioritize. And to use its resources to improve the most critical service attributes. The data are collected via surveys of a sample of customers. In these surveys, these customers respond to a series of questions based around a number of key service dimensions like Reliability, security, responsiveness, assurance, empathy, customer-oriented qualities and so forth.

Based on the above analysis, the problem can be formulated as follows:

Given n data sources from a large organization, we are interested in

1. mining each of these data sources for local rules for each data source
2. synthesizing these local rules to find valid rules for the overall organization that would have been discovered from the union of all these data sources

There are various existing data mining algorithms that can be used to discover local rules for each data source. This work only focuses on synthesizing valid rules for the organization. The proposed method for synthesizing valid rules has been applied for patient perceived service quality in different branches of a hospital. The remaining part of this article is organized as follows: The next section describes the problem of synthesizing valid rules with a case study of patient perceived service quality in a hospital. After that we present both informal and formal description of algorithm respectively. Finally an analysis on performance is discussed.

PROBLEM DESCRIPTION

The patients in the branches (six in total) of the hospital were questioned on the service quality of the hospital using 12-item service quality scale, adapted from the SERVQUAL scale.

A survey questionnaire is constructed incorporating 12 service quality items (as presented in Table 1) for assessing the influence of each indicator on the performance of hospital services. The questionnaire is simple and easy to understand. It is a closed-ended questionnaire (Annexure-I) based on SERVQUAL scale. An exploratory factor analysis is conducted on these 12 variables in order to identify the key factors influencing the services of a hospital. Factor analysis is a statistical tool to determine a minimum number of unobservable common factors by studying the covariance among a set of observed variables. It is a data reduction procedure that identifies the underlying relationships that exist within a set of variables. The essential purpose of factor analysis is to describe, if possible, the covariance relationships among many variables in terms of a few core, but unobservable, random quantities called latent variables or factors. Such factors obtained after factor analysis are called rules. A correlation coefficient matrix is constructed for all the above mentioned twelve variables using the SPSS software (Malhotra, 2009). The data used for the coefficient matrix is obtained from the responses of the filled in questionnaire. It was observed that there exists high correlation among few sets of variables. This motivates us to conduct Factor Analysis and find out the group of variables representing a single underlying construct called factor/rule (Rastogi & Shim, 1999) which is responsible for the observed correlations. The result of mining the above datasets using the factors such as patient oriented, competence, tangibles and convenience is given below as four different association rules.

R1: Patient Oriented

The rule patient oriented comprised of variables like extent of prompt service, consistent courtesy and knowledge to answer patient's questions, convenient, operating hours, personal attention and understanding of specific patient needs. The rule is expressed as follows:

If V9 = yes **AND** V11 = yes **AND** V12 = yes **then**
Patient oriented = yes.

R2: Technical Competence

Variables like extent of interest in problem solving, right service, error free records, service-time guarantee and safety of transactions combined to define "competence". Then the rule is expressed as follows:

Table 1. Items of service quality scale (Conclusion, Level 1)

VARIABLE DESCRIPTION		
V1	Hospital infection control (HIC)	Ability to reduce or eliminate infection risks to patients, visitors, and service providers in the hospital
V2	Privacy (PRI)	The extent to which a hospital is able to maintain private record of patients
V3	Grievance handling time (GHT)	Time taken by hospital administration to solve any grievances of patient
V4	Continuity of care record (CCR)	Capacity of a hospital to maintain proper and detailed record of the patients' case history, number of visits made, and so forth.
V5	Waste disposal policy (WDP)	It denotes the policy of a hospital related to handling, storage, transportation, and disposal of hazardous materials.
V6	Waiting time (WT)	This variable indicates the total time spent by a patient for fixing appointment as well as consulting with the doctor.
V7	Access (ACS)	Ability of a hospital to admit patients for whom it can provide services with its available resources.
V8	Administrative staff's attitude (ASA)	The administrative staff's behavior toward patients, visitors, and practitioners.
V9	Facilities availability (FA)	Availability of specialized departments and facilities in the hospital like burn care, skin care, outdoor, and so forth.
V10	Practitioner's Experience (PRE)	The practitioner's experience and competence.
V11	Cost of medicine (COM)	Cost related to the medicines prescribed by the doctors in the hospital
V12	Hospital Staff attitude(HA)	Your hospital understand your specific needs

If V5 = yes **AND** V6 = yes **AND** V7 = yes **AND** V4 = yes **AND** V10 = yes then Competence = yes.

R3: Tangibles

The tangibles factor consisted of variables like extent of visual appeal of physical facilities and information conveyed by published materials. Then the rule is expressed as follows:

If Vl = yes **AND** V2 = yes **then** Tangibles = yes.

R4: Convenience

Variables like guidance signs and timeliness combined to define "convenience". Then the rule is expressed as follows:

If V3 = yes **AND** V8 = yes **then** Convenience = yes.

The Table 2 gives the local support and local confidence values for the rules Rl, R2, R3 and R4. Local support and local confidence define the support and confidence of the local rules respectively. The problem here is when these association rules are forwarded from different known data sources in the branches of a hospital to their headquarters; it requires a method to synthesize these association rules for knowing the valid rules among them. Figure 1 illustrates this model.

Table 2 gives the local support and confidence of rules at branch level. In this analysis we are considering three branches with data sources D1, D2, D3. The rules present in the data sources are shown in the column Rule.

The performance analysis bar diagram shown in Figure 1 plotted as rules (R_1,R_2,R_3,R_4) on the x-axis against global support value of rules obtained from Xindong method (shown as GSuppX(Rj)) and global support obtained in our proposed method((GSuppW(Rj)).

The higher support obtained for rule R_1 shows that Patient-oriented information like prompt service, consistent courtesy and knowledge to answer patient's questions, convenient operating hours, personal attention and understanding of specific patient needs are most important from a customer's perception, and hospitals can take further steps for improving their services in this direction.

Let D1,. D2, ..., Dm be m different data sources from the branches of a large company of similar size, and Si be the set of association rules from Di (i = 1,2, ..., m). Also, let W1 W2,W3, ..., Wm, be the weights of these data sources. Then for a given rule Ri, expressed as X → Y, its global support and global confidence are defined as follows:

Table 2. Lsupport and Lconfidence values of the rules

Data	Rules	Lsupport	Lconfidence
D1	R1	0.50	0.27
	R2	0.31	0.30
	R4	0.47	0.82
D2	R1	0.40	0.69
	R4	0.27	0.59
	R3	0.30	0.60
D3	R2	0.31	0.71
	R1	0.43	0.73

Figure 1. Performance analysis bar diagram

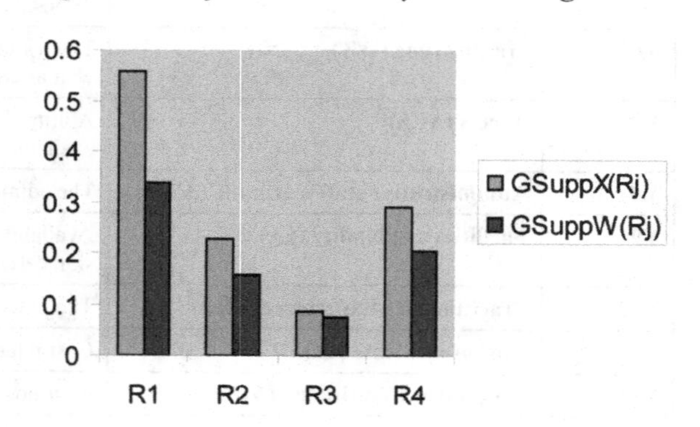

$$G\,\text{supp}(R_i) :\leftarrow \sum_{i=1}^{m} wD_i * L\,\text{supp}(R_i)$$

$$Gconf(R_i) :\leftarrow \sum_{i=1}^{m} wD_i * Lconfi(R_i)$$

where $Gsupp(R_i)$ is the support of R_i after synthesizing, $Lsupp_i(R_i)$ is the support of R_i in D_i and $Lconf_i(R_i)$ is the confidence of R_i in D_i, $i = 1, 2, ..., m$. The synthesis of rules in our model is generally straightforward once all weights are reasonably assigned. The weight of each rule is calculated by its frequency in the original data sources.

Thus this article proposes a uniform framework for synthesizing valid rules as follows:

- **Phase 1:** Computing weight of rules.
- **Phase 2:** Computing weight of data sources.
- **Phase 3:** Computing global support and global confidence.

ALGORITHM

This section in the first phase discusses the informal description of the synthesizing rules by weighting algorithm developed for ranking the rules based on patient perceived service qualities in a hospital. Next, it presents the formal description of the algorithm. Finally, it provides the impact of the algorithm.

DESCRIPTION (ALGORITHM, LEVEL 1)

In order to synthesize association rules from different data sources in the branches of a hospital, this method needs to determine the weight for each data source. In our opinion, if all data sources are of similar size, the weight of each data source can be determined by the rules discovered from it.

Let $D_1, D_2 ..., D_m$, be m different data sources in the branches of a hospital, Si the set of association rules from Di($i = 1, 2, ..., m$), and $S = \{S_1, S_2, S_3, S_4 .... Sm\}$. This method takes the frequency of a rule R_i in S to assign a rule weight wR_i.

The inter support relationship between a data source and its rules can be applied to assign the data sources a weight. If a data source supports a larger number of high-frequency rules, the weight of the data source should also be higher. Table 2 illustrates the above idea with the data. Let minsupp = 0.25. minconf = 0.42 (as per observation of previous data sets), and the following rules were mined from six different branches of the hospital. For the illustration purpose, we assume only three branches. The local support and confidence values were obtained by applying the definition of support and confidence (Agrawal & Srikant, 2001) using SPSS data miner.

Datasource DI: $S_1 = \{R_1, R_2, R_4\}$

R_1 with Lsupp = 0.50, Lconf = 0.27
R_2 with Lsupp = 0.31, Lconf = 0.30
R4 with Lsupp = 0.47, Lconf = 0.82

Datasource D2: $S_2 = \{R_1, R_4, R_3\}$

R_1 with Lsupp = 0.40, Lconf= 0.69 ;
R_3 with Lsupp = 0.30, Lconf = 0.60 ;
R_4 with Lsupp = 0.27, Lconf = 0.59 ;

Data source D_3:- $S_3 = \{R_2, R_1\}$

R_1 with Lsupp = 0.43, Lconf = 0.73 ;
R_2 with Lsupp = 0.31, Lconf = 0.71;

Thus S = {S1, S2, S3}. Here, the number of sources that contain $R_1 = 3, R_2 = 2, R_3 = 1$, and $R_4 = 2$.

We can use the frequency of a rule in S to assign a weight for rules. The weights are assigned as follows:

wR_1 = 3/ (3+2+1+2) = 3/8 = 0.375 ;
wR_2 = 2/(3+2+1+2) = 2/8 = 0..25 ;
wR_3 = 1/(3+2+1+2) = 1/8 = .125;
wR4= 2/(3+2+1+2) = 2/8 = 0.25 ;

We have seen that rule R_1 has the highest frequency and it has the highest weight; rule R3 has the lowest frequency and it has the lowest weight. Let $S = \{S_1, S_2, S_3, ..., Sm\}$, and R_1, R_2, ... Rn be all rules in S. Then, the weight of a rule Ri is defined as follows:

$$wR_i \frac{Num(R_i)}{\sum_{j=1}^{n} Num(R_j)}$$

where i = 1,2, ..., n ; and Num(R) is the number of data sources that contain rule R, or the frequency of R in S.

If a data source has a larger number of high-frequency rules, the weight of the data source should also be higher. If the rules from a data source are rarely present in other data sources, the data source would be assigned a lower weight.

To implement this argument, we can use the sum of the rule's weights divided by total number of data sources.

wD_1 = (.375 + 0.25 +0.25)/3 = 0.2917;
wD2 = (.375 + 0.125 + 0.25)/3 = 0.25;
wD3 = (0.375 +0.25)/3 = 0.2083;

Let $D_1, D_2, D_3, ...$, Dm be m different data sources in the branches of a hospital, S the set of association rules from D. (i = 1,2, ..., m), $S = \{S_1, S_2, ..., Sm\}$ and $R_1 R_2$,... Rn be all rules in S. Then, the weight is defined as follows:

$$wDi = \frac{\sum_{i=1}^{n} wRi}{m}$$

After all data sources have been assigned weights, it requires a synthesize process to evaluate these association rules. Hence this article introduces a simplified formula for computing global support and global confidence to replace the normalization process formula proposed by Xindong.

FOR RULE R_1: PATIENT ORIENTED

$Gsupp(R_1) = wD_1 * Lsupp_1(R_1) + wD_2 * Lsupp_2(R_1) + wD_3 * Lsupp_3(R_1)$
=0.2917* 0.50+ 0.25 * 0.40+ 0.2083 *0.43
=0.3354
$Gconf(R_1) = wD_1 * Lconf_1(R_1) + wD_2 * Lconf_2(Rl) + wD_3 * Lconf_3(Rl)$
=0.2917* 0.27+ 0.25 * 0.69 + 0.2083 * 0.73
=0.4033

For Rule R2: Technical Competence

$Gsupp(R_2) = wD_1 * Lsupp_1(R2) + wD_3 * Lsupp_3(R2)$
=0.2917* 0.31+ 0.2083 * 0.31
=0.155
$Gconf(R_2) = wD_1 * Lconf_1(R2) + wD3 * Lconf_3(R2)$
=0.2917* 0.30 + 0.2083 * 0.71 = 0.2354

For Rule R3: Tangibles

$Gsupp(R3) == wD_2 * Lsupp_2(R3)$
= 0.25 * 0.30
= 0.075
$Gconf(R_3) = wD2 * Lconf_2(R3)$
= 0.25 * 0.60 = 0.15

For rule R4: Convenience

$Gsupp(R_4) = wD_1 * Lsupp_1(R_4) + wD2 * Lsupp_2(R4)$
= 0.2917* 0.47+ 0.25 * 0.27
= 0.2045
$Gconf(R_4) = wD1 * Lconf_1(R_4) + wD_2 * Lconf_2(R_4)$
=0.2917* 0.82+ 0.25 * 0.59
= 0.3866

Thus the ranking of the rules by their global supports is R_1, R_4, R_2 and R_3. According to this ranking, we can select high-rank rules after the minimum support and minimum confidence. Table 3 gives the calculated value of Gsupport and Gconfidence for the rules using Xindong method. Table 4 gives the calculated value of Gsupport and Gconfidence for the same rules using the proposed method.

This procedure is transformed into an algorithm in the next section.

Table 3 is obtained by applying the frequency and weight of rules using the Xindong model to compare with our model under the same set of inputs and conditions.

Table 4 shows the global support and global confidence of rules from our proposed method. The global support and global confidence are obtained by multiplying the weight of data sources containing the rule and the Local support/confidence of the rules. The global supports are obtained at a centralized headquarter as a part of synthesization process to obtain the relative importance of the rules.

FORMAL DESCRIPTION (ALGORITHM, LEVEL 1)

```
Algorithm:- Synthesizing Rules By
Weighting method
Input:
S={S1,S2,....Sm}          : rule sets
;
Minsupp, minconf  : threshold values
;
Lsupp, Lconf              : local
support, local confidence
M                          : number
of data sources
N  : number of rules
Output:
R  : synthesized association rules
For each rule Ri in S do
Num(Ri)← the number of data sources
that contain rule Ri in S ;
```

$$wRi \leftarrow \frac{Num(Ri)}{\sum_{j=1}^{n} Num(Rj)}$$

```
For each data source do
```

$$wDi \leftarrow \frac{\sum^{n} wRi}{m}$$

```
3.. For each rule Ri in S do
```

$$GSupp(Ri) \leftarrow \sum_{i=1}^{m} wDi * LSupp_i(R)$$

$$Gconf(Ri) \leftarrow \sum_{i=1}^{m} wDi * Lconf_i(R)$$

```
4. Rank all rules in S by their
supports.

5. Output the high-rank rules in S
whose support and confidence are at
least minsupp and minconf respective-
ly.
```

IMPACT (ALGORITHM, LEVEL 1)

This synthesizing rules by weighting algorithm has been implemented in Java language (Jdk1.5) with MS-Access and it runs on Intel based Personal Computers. The method of ranking valid rules using synthesize by weighting in this algorithm has only less cost.

It does not involve any complicated formula computation. Hence the algorithm has less com-

Table 3. Gsupport and Gconfidence using Xindong method

Rule	Frequency	wR_i	$Gsupp^X(R_J)$	$Gconf^X(R_J)$
R1	3	0.375	0.5551	0.6621
R2	2	0.25	0.2271	0.4698
R3	1	0.125	0.0859	0.268
R4	2	0.25	0.2916	0.5096

Table 4. Gsupport and Gconfidence using proposed method

Rule	Frequency	wR_i	$Gsupp^X(R_J)$	$Gconf^X(R_J)$
R1	3	0.143	0.3354	0.4033
R2	2	0.429	0.155	0.2354
R3	1	0.286	0.075	0.15
R4	2	0.143	0.2045	0.3866

putation and space consumption than the Xindong method (Reference-2) and whose time complexity is greater than $O(n^2)$.The time complexity of the algorithm is $O(n^2)$, where n is the variable number on which the rules are defined.

PERFORMANCE ISSUES

To evaluate the effectiveness, there are many possible measures one can choose to determine, how good our approach works. Here, we defined one type of error that is known as average error to measure the effectiveness of our approach compared with Xindong approach.

The formula for computing average error is given by

$$AE(Ri) = \frac{G\,\text{supp}^X(R.) - G\,\text{supp}^W(R.)}{\substack{i = 1, 2...N \\ Freq(Ri.)}}$$

N: Total number of rules in the given set

Gsuppx(Ri): Golbal support of Ri in Xindong weighting model

Gsuppw(Ri): Golbal support of Ri in proposed weighting model

Freq(Ri): Frequency of Ri .

Table 5 summarizes the result and gives the performance comparison of our algorithm with Xindong algorithm. The bar diagram showing Gsupport of the rules determined by Xindong method and our proposed method is given in Figure 1.

Table 5. Results of performance analysis

Rule	$AE(R_J)$	$Gsupp^X(R_J)$	$Gsupp^W(R_J)$
R1	0.07	0.5551	0.3354
R2	0.04	0.2271	0.155
R3	0.01	0.0859	0.075
R4	0.04	0.2916	0.2045

The X-axis in graph represents the different rules in the data sources and the Y-axis represents the global support of these rules after synthesizing the rules.

The average error (shown as $AE(R_J)$) shown in Table 5 is the difference between the global support of a rule from Xindong method considered as standard and our proposed method divided by the frequency of a rule. In all cases we find that the average error obtained is negligible. However, for huge number of transactions, the space optimization done in our algorithm will definitely minimize the space requirements in a distributed environment where different branches forward information to headquarters.

CONCLUSION

Our proposed method designs a model that can analyze the customer perceptions on the service quality provided in a hospital under a distributed environment. For this we optimized the Xindong weighting model. Synthesizing rules by weighting model is presented in this article in order to rank the rules, which are getting from different branches of a hospital based on the patient perceived service qualities. The performance analysis of this algorithm is also done with the result of less computational complexity.

The main advantages of our method are simple calculations and low error amount. It avoids the normalization step used in the Xindong method for calculating weight of data sources. Large organizations like a hospital with different branches distributed in various locations can use this model for understanding the more important parameters based on customer perceptions for improving the service qualities.

The implementation of the model shows that factors like "facilities available in a hospital," "cost of medicine" and "hospital staff attitude" toward a patient are the higher parameters for determining service quality. This is shown from the higher values of global support of the "Patient Oriented" rule as shown in Table 4.

This article is inclined toward the human development in order to improve service quality in Hospitals. As a result from the four rules mentioned in the article "Patient oriented" gets the highest support and confidence among them. Rule Patient oriented includes variables like V9: Facilities availability (FA), V11: Cost of medicine (COM), V12: Hospitals attitude (HA)

Hospitals should focus on Availability of specialized departments and facilities like 24-hour cover in General Medicine, General Surgery, Intensive care, Master Health Checkup plans.

The article has revealed the importance of the relationship between the patient and the hospital and how this relationship affects the overall benefit of patient's treatments. Patients along with their families can sometimes feel overwhelmed by the medical terminology used by health care professionals. To be successful in obtaining the best medical care and minimize the accompanying stresses, it is necessary to understand the medical, financial and special education terminology also.

In general the synthesizing model can be applied in all such environments where there is a centralized headquarter and different branches which provide huge amounts of information, for example, analyzing data for a bank having many branches, in tour and travel industry for understanding customer expectation of services at different places in the country. Moreover other service parameters of the SERVQUAL scale like tangibility, empathy, courtesy, credibility can be applied in this direction using our model.

REFERENCES

Agrawal, R., & Srikant, R. (2001). Fast algorithms for mining association rules. *IEEE Transactions on Knowledge and Data Engineering*, *9*(3), 143–167.

Brin, S., Motwani, R., & Silverstein, C. (1997, May 13-15). Beyond market baskets: Generalizing association rules to correlations. In *Proceedings of the 1997 ACM SIGMOD International Conference on Management of Data,* Tucson, AZ (pp. 265-276). ACM Publishing.

Han, J., Pei, Y., & Yin, Y. (2000, May). Mining frequent patterns without candidate generation. In *Proceedings of the 2000 ACM SIGMOD International Conference on Management of Data*, Dallas, TX (pp. 1-12).

Han, Y.-J., & Yong, Q. (2006). Efficient improvement of FT-tree based frequent itemsets mining algorithms. *IEEE Transactions on Knowledge and Data Engineering, 3*, 374–377.

Hand, D., & Mannila, H. (2004). *Principles of data mining* (3rd ed.). Cambridge, MA: MIT Press.

Jiawei, H., Micheline, K., & Kaufmann, M. (2007). *Data mining: Concepts and techniques*. San Diego, CA: Academic Press.

Lee, G., Lee, K. L., & Chen, A. L. P. (2001). Efficient graph based algorithms for discovering and maintaining association rules in large databases. *Knowledge and Information Systems*, *3*, 338–355. doi:10.1007/PL00011672

Malhotra, N. K. (2009). Marketing research: An applied orientation using SPSS. Upper Saddle River, NJ: Prentice Hall.

Ortega, J., Koppel, M., & Argamon, S. (2001). Arbitrating among competing classifiers using learned referees. *Knowledge and Information Systems*, *4*, 470–490. doi:10.1007/PL00011679

Padromidis, A., & Stolfo, S. (2000). Meta learning in distributed data mining systems: Issues and approaches. In H. Kargupta & P. Chan (Eds.), *Advances in distributed and parallel knowledge discovery* (pp. 81-114). Cambridge, MA: MIT Press.

Parthasarathy, S., Zaki, M. J., Ogihara, M., & Li, W. (2001). Parallel data mining for association rules on shared memory systems. *Knowledge and Information Systems, 1*, 1–29. doi:10.1007/PL00011656

Rastogi, R., & Shim, K. (1999). Mining optimized support rules for numeric attributes. Proceedings ACM SIGMOID, Conference on Management of Data.

Sayal, M., & Scheuermann, P. (2001). Distributed web log mining using maximal large data sets. *Knowledge and Information Systems, 4*, 389–404. doi:10.1007/PL00011675

Skillicon, D. B., & Wang, Y. (2001). Parallel and sequential algorithms for data mining using inductive logic. *Knowledge and Information Systems, 4*, 405–421. doi:10.1007/PL00011676

Turinsky, A., & Grossman, R. (2000, August 20-23). *A framework for finding distributed data mining strategies that are intermediate between centralized strategies and in place strategies.* Paper presented at the ACMKDD Workshop on Distributed and Parallel Knowledge Discovery, Boston.

Webb, G. I. (2000, August 20-23). Efficient search for association rules. In *Proceedings of the 6th ACM SIGKDD International Conference on Knowledge Discovery and Data Mining,* Boston (pp. 99-107). ACM Publishing.

Wu, X. (2004, November 15-17). Knowledge discovery in multiple databases. In *Proceedings of the 16th IEEE International Conference on Tools with Artificial Intelligence,* Boca Raton, FL (pp. 2). IEEE.

Wu, X., & Zhang, C. (2005). Database classification for multi database mining. *Information Systems, 30*(1), 71–88. doi:10.1016/j.is.2003.10.001

Wu, X., & Zhang, S. (2003). Synthesizing high-frequency rules from different data sources. *IEEE Transactions on Knowledge and Data Engineering, 15*(2), 353–367. doi:10.1109/TKDE.2003.1185839

Zhang, S., & Wu, X. (2004). *Knowledge discovery in multiple databases*. New York: Springer.

ANNEXURE- I

Questionnaire

(For academic purpose only)

Name ____

Occupation

• Service Business Other, please specify ______________________

For each of the following statements, could you please tell me if you agree or disagree (YES/NO).

1. The procedure for admission in the hospital is not complicated and time consuming a. Yes b. No
2. The hospital has well qualified and efficient doctors in its panel. a. Yes b. No
3. The attitude of staff and administration toward patient is satisfying a. Yes b. No
4. Problems or grievances of patients are handled on time a. Yes b. No
5. Hospital does not charge any unfair claims from patients a. Yes b. No
6. Adequate support staff is present in the hospital a. Yes b. No
7. The hospital specializes in a particular treatment (a) Yes (b) No
8. Adequate surgical equipments and medicines are present in the hospital (a) Yes (b) No
9. The hospital has a proper waste disposal policy, that is, the policy of a hospital related to handling, storage, transportation, and disposal of hazardous materials (a) Yes (b) No
10. Cost related to the medicines prescribed by the doctors in the hospital is not very high (a) Yes (b) No
11. The hospital has a proper chemist and druggist store (a)Yes (b) No
12. Your hospital provides adequate security in terms of fire control, theft, controlling agitation (a)Yes (b) No
13. The hospital maintains detailed record of the patients' case history, number of visits made (a) Yes (b) No Hospital has proper arrangement of basic facilities like canteen, toilet, water and air conditions (a)Yes (b) No
14. Privacy of patients is given due importance by the hospital (a)Yes (b) No
15. The hospital's physical environment and ambience is satisfactory (a)Yes (b) No
16. The in patient food service quality is satisfying. (a)Yes (b) No
17. Doctors and nurses provide personalized care to patients. (a)Yes (b) No
18. The charges of rooms and service in the hospital are economical (a)Yes (b) No
19. There is separate department for infectious and fatal diseases. (a)Yes (b) No
20. The hospital has proper facilities for OPD and emergency circumstances. (a)Yes (b) No
21. Doctors associated with the hospital are available on call (a)Yes (b) No
22. The hospital provides for good arrangement of staying for acquaintances of admitted patients (a)Yes (b) No
23. The hospital has a proper enquiry department to handle queries of patients. (a)Yes (b) No
24. The surrounding of the hospital is healthy and hygienic and not any industrial area (a) Yes (b) No
25. There are separate pathology and radiology departments in the hospital (a) Yes (b) No

26. Hospital provides for corporate discounts (a) Yes (b) No
27. Hospital organizes for camps and subsidized health packages (a) Yes (b) No
28. A patient can conveniently fix an appointment for consulting with the doctor. (a) Yes (b) No

This work was previously published in Information Communication Technologies and Human Development, Volume 1, Issue 4, edited by S. Chhabra and H. Rahman, pp. 58-71, copyright 2009 by IGI Publishing (an imprint of IGI Global)

Compilation of References

Abraham, B. P. (2009). Preparing for the challenge of electronic globalization. *Decision.*, *36*(1), 30–32.

Abrams, Z. (2006). From Theory to practice: Intracultural CMC in the L2 classroom. In Duncate, L., & Arnold, N. (Eds.), *Calling to CALL* (pp. 181–209). San Marcos, TX: Computer Assisted Language Consortium.

Acevedo, M. (2005). Las TIC en la Cooperación al Desarrollo. In *La Sociedad de la Información en el Siglo XXI: Un Requisito para el desarrollo – Vol II: reflexiones y conocimiento compartido* (pp. 44-66). Madrid: State Secretariat for Telecommunications and the Information Society, Ministry of Industry, Turismo y Comercio, Spain. ISBN 84-96275-09-4.

Acevedo, M. (2007). Network capital: an expression of social capital in the Network Society. *The Journal of Community Informatics* [Online], *3(2).* Retrieved 18 Nov 2007 from <http://www.ci-journal.net/index.php/ciej/article/view/267/317>

ACM Special Interest group on Computer-Human Interaction Group. (1992). *ACM SIGCHI, Technical Report.* New York: ACM.

Adamo, A., Bertacchini, P. A., Bilotta, E., Pantano, P., & Tavernise, A. (2010). Connecting Art And Science For Education: Learning by an Advanced Virtual Theatre with "Talking Heads". *Leonardo*, *43*(5), 442–448. doi:10.1162/LEON_a_00036

Agrawal, R., & Srikant, R. (2001). Fast algorithms for mining association rules. *IEEE Transactions on Knowledge and Data Engineering*, *9*(3), 143–167.

AgriBazaar – Exchange for better price, project attempts to boost Malaysian agriculture by online trading. Retrieved December24, 2008 from http://www.agribazaar.com.my/.

Ahmed, A. (2007). Open access towards bridging the digital divide-policies and strategies for developing countries. *Information Technology for Development.*, *13*(4), 337–361. doi:10.1002/itdj.20067

Akshaya – Creating powerful e-Services to reach citizens, project addresses key issues in IT dissemination to masses. Retrieved December 24, 2008 from http://akshaya.kerala.nic.in

Al-Asmari, A. M. (2005). *The use of the Internet among EFL teachers at the colleges of technology in Saudi Arabia.* Unpublished doctoral dissertation, Ohio State University, Ohio.

Alfano, I., & Pantano, E. (2010). Advanced Technologies for promotion of cultural heritage: the case of Bronzes of Riace. *Journal of Next Generation Information Technology*, *1*(1), 39–46. doi:10.4156/jnit.vol1.issue1.4

Alfin (2006). *Bibliotecas por el aprendizaje permanente, Declaración de Toledo sobre la alfabetización informacional.* Retrieved December 30, 2007, from http://www.lectores.info/formacion/file.php/38/Modulos/Documentos/Dec_Toledo.pdf

Alfin (2006-2007). *Blog sobre alfabetización informacional.* Retrieved December 30, 2007, from http://www.alfinred.org/blog

Al-Fulih, K. (2002). *Attributes of the Internet perceived by Saudi Arabian faculty as predictors for their Internet adoption for academic purposes.* Unpublished doctoral dissertation, Ohio University.

Ali, M., & Bailur, S. (2007). The Challenge of Sustainability in ICT4D – Is Bricolage the Answer? *Proceedings of the 9th International Conference on Social Implications of Computers in Developing Countries*, Sao Paulo.

Aljaafreh, A., & Lantolf, J. P. (1994). Negative feedback as regulation and second language learning in the zone of proximal development. *Modern Language Journal*, *78*(4), 465–483. doi:10.2307/328585

Al-Khabra, Y. (2003). *Barriers in using the Internet in academic education in the perspective of Education College faculty at King Saud University in Riyadh.* Unpublished doctoral dissertation, King Saud University, Riyadh.

Al-Salem, S. A. (2005). *The impact of the Internet on Saudi Arabian EFL females' self-image and social attitudes.* Unpublished doctoral dissertation, Indiana University of Pennsylvania.

Al-Salih, Y. N. (2004). *Graduate students' information needs from electronic information resources in Saudi Arabia.* Unpublished doctoral dissertation, Florida State University, Florida.

Altman, D. (2002). Prospects for e-government in Latin America: Satisfaction with democracy, social accountability, and direct democracy. *International Review of Public Administration*, *7*(2), 5–20.

Ambrosi, A., Peugeot, V., & Pimienta, D. (Eds.). (2005). *Word Matters, Multicultural perspectives on information societies*, C&F Editions. Online version retrieved December 31, 2007, from http://www.vecam.org/article698.html?lang=en

Anderson, E., & Weitz, B. A. (1989). Determinants of continuity in conventional industrial channel dyads. *Marketing Science*, *8*, 310–323. doi:10.1287/mksc.8.4.310

Anderson, J., & Narus, J. (1990). A model of distribution firm and manufacturing firm working partnerships. *Journal of Marketing*, *54*(1), 42–59. doi:10.2307/1252172

Anheir, H., & Katz, H. (2005). Enfoques reticulares de la Sociedad Civil Global. In F. Holland, H. Anheir, M. Glasius, & M. Kaldor (Eds.), *Sociedad Civil Global 2004/2005* (pp. 221-238). Translated by José Luis González (original title: *Global Civil Society 2004-2005*). Barcelona: Icaria Editorial. ISBN: 84-7426-823-0.

Annan, K. (2000). *We the Peoples: The Role of the United Nations in the 21st Century. Millennium Report*. New York: United Nations Dept. of Public Information.

Anttiroiko, A.-V., & Mälkiä, M. (2007). *Encyclopedia of Digital Government*. Hershey, PA: Idea Group Inc.

Anttiroiko, A. (2002). Strategic Knowledge Management in Local Government. In Grönlund, Å. (Ed.), *Electronic Government: Design, Applications and Management* (pp. 268–298). Hershey, PA: IGI.

Anttiroiko, A. (2004). Towards Citizen-centered Local E-government: The Case of the City of Tampere. In M. Khosrow-Pour (Ed.), Annals of cases on information technology 6, 371-388. Hershey, PA: IGI.

Appadurai, A. (1996). *Modernity at Large. Cultural Dimensions of Globalization*. Minneapolis: University of Minnesota.

Arion, M., Numan, J. H., & Pitariu, H. (1994). Placing Trust in Human-Computer Interactions. In*Proceedings of 7th European Cognitive Ergonomics Conference*. (pp. 352-365).

Arnold, D. B., & Geser, G. (2007). *Research agenda for the applications of ICT to cultural heritage*. EPOCH.

Arquilla, J., & Ronfeldt, D. F. (1999). *The emergence of noopolitik: toward an American information strategy*. Santa Monica, California: Rand Corp.

Aucoin, P., & Heintzman, R. (2000). The dialectics of accountability for performance in public management reform. *International Review of Administrative Sciences*, *66*, 45–55. doi:10.1177/0020852300661005

Austin City Council. (2008). *Audit Report 2008: City of Austin's E-government Initiative*. Austin, Texas: Office of the City Auditor.

Avgerou, C. (1998). How can IT enable economic growth in developing countries? *Information Technology for Development*, *8*(1), 15–29. doi:10.1080/02681102.1998.9525288

Awad, E., & Ghaziri, H. (2004). *Knowledge Management*. Upper Saddle River, NJ: Prentice Hall.

Ba, S., & Pavlou, P. A. (2002). Evidence of the effect of trust building technology in electronic markets: Price premiums and buyer behavior. *Management Information Systems Quarterly*, *26*(3), 243–268. doi:10.2307/4132332

Bada, A., Okunoye, A., Aniebonam, M., & Owei, V. (2003). Introducing information systems (IS) education in Nigerian higher institutions of learning: A context, content and process framework. *In Proceedings of the Nigeria Computer Society, 14*(1).

Bagchi, K., Cerveny, R., Hart, P., & Peterson, M. (2003). The influence of national culture in information technology product adoption. In *Proceedings of the Ninth Americas Conference on Information Systems* (pp. 957-965).

Bajwa, F. R. (2007). Telecentre Technology: The application of free and open source software. APDIP eNote 19, 2007. Retrieved December 24, 2008 from http://www.apdip.net/ apdipenote/ 19.pdf

Bakos, J., & Brynjolfsson, E. (1993). From Vendors to Partners: Information Technologies and Incomplete Contracts in Buyer Seller Relationships. *Journal of Computing, 3*(3), 301–329.

Balaji, R., & Ravikumar, B. (2007). Community computing for development. Proceedings from ISED 2000: International Conference on ICT solutions for Socio-economic Development, (pp. 9–16).

Balaji, R., Neelanarayana, V., Ponraj, M., & Kailash, T. (2005). Establishment of Community Information Network in a Developing Nation. Proceedings from IEEE Tencon 2005: The International technical conference sponsored by IEEE Region, 10, 1- 6.

Banks, D.A. (2003, June). Collaborative Learning as a Vehicle for Learning about Collaboration. *Proceedings of Informing Science InSITE - "Where Parallels Intersect"*, (pp. 895-903), Pori, Finland.

Barata, K., Kutzner, F., & Wamukoya, J. (2001). Records, computers, resources: A difficult equation for sub-Saharan Africa. *Information Management Journal, 35*(1), 34–42.

Barker, A., Krull, G., & Mallinson, B. (2005, October 25-28). A Proposed Theoretical Model for M-Learning Adoption in Developing Countries. *mLearn 2005*, Cape Town, South Africa.

Bart Iakov, Y. Shankar Venkatesh, Sultan Fareena, &. Urban Glen L. (2005). *Are the Drivers and Role of Online Trust the same for all Web Sites and Consumers?*http://ebusiness.mit.edu

Bates, T. (1999, September). *Cultural and ethical issues in international distance education.* Paper presented at the *UBC/CREAD* conference, Vancouver, Canada. Available at: http:ilbate.o.cstudies.ubc.ca/pdfiCR~"AD.pdt.

Beahm, C. P., Rogers, P. C., & Liddle, S. W. (2006). Opportunities and Challenges of Utilizing Educational Technology in Developing Countries: The eCANDLE Foundation. In Mendez-Vilas, A., Solano Martin, A., Mesa Gonzalez, J. A., & Mesa Gonzalez, J. (Eds.), *Current developments in technology-assisted education* (pp. 1825–1831). Published by Formatex.

Behnken, E. (2005, June 20-22). The Innovation Process as a Collective Learning Process. *ICE 2005: 11th International Conference on Concurrent Enterprising*, Munich, Germany, Springer.

Belaal Mohammad Ahmad Ifhan. (2002).*Trust inducing model features for web sites*. http://eprints.uum.edu.

Bélanger, M. (2001). *Work-based distributed learning, Online-document.* Available from http://training.itcilo.org/actrav/library/english/publications/work-based_learning.doc

Benbasat, I., Goldstein, D., & Mead, M. (1987). The case research strategy in studies of information systems. *Management Information Systems Quarterly, 11*(3), 369–386. doi:10.2307/248684

Benham, H., & Raymond, B. (1996). Information technology adoption: Evidence from a voice mail introduction. *Computer Personnel, 17*(1), 3–25. doi:10.1145/227005.227006

Berg, E. C. (1999). The effects of trained peer response on ESL students' revision types and writing quality. *Journal of Second Language Writing, 8*(3), 215–241. doi:10.1016/S1060-3743(99)80115-5

Bergen Communiqué. (2005). *The European Higher Education Area – Achieving the goals*. Bergen, May 19–20.

Bertacchini, P. A., Bilotta, E., Di Bianco, E., Di Blasi, G., & Pantano, P. (2006). Virtual Museum Net. *Lecture Notes in Computer Science, 3942*, 1321–1330. doi:10.1007/11736639_165

Bertacchini, P. A., Bilotta, E., Cronin, M., Pantano, P., & Tavernise, A. (2007). 3D Modelling of Theatrical Greek Masks for an Innovative Promotion of Cultural Heritage. In Posluschny, A., Lambers, K., & Herzog, I. (Eds.), *Layers of Perception. Belrin: CAA2007.*

Bierens, H. J. (2004). *EasyReg International*. University Park, PA: Department of Economics, Pennsylvania State University.

Bilotta, E., Gabriele, L., Servidio, R., & Tavernise, A. (2009). Edutainment Robotics as learning tool. *Lecture Notes in Computer Science*, *5940*, 25–35. doi:10.1007/978-3-642-11245-4_3

Bilotta, E., Pantano, P., & Tavernise, A. (2010). Using an Edutainment Virtual Theatre for a Constructivist Learning. In *Proceedings of the 18th International Conference on Computers in Education (ICCE 2010) - "New paradigms in learning: Robotics, play, and digital arts"*. Putrajaya – Malaysia: Asia Pacific Society for Computers in Education.

Bird, S. A. (2005). Language Learning Edutainment: Mixing Motives in Digital Resources. *RELC*, *36*(3), 311–339. doi:10.1177/0033688205060053

Black, R. W. (2005). Access and affiliation: The literacy and composition practices of English language learners in an online fan fiction community. *Journal of Adolescent & Adult Literacy*, *49*(2), 118–128. doi:10.1598/JAAL.49.2.4

Black, R. W. (2006). Language, culture, and identity in online fan fiction. *E-learning*, *3*(2), 170–184. doi:10.2304/elea.2006.3.2.170

Black, R. W. (2008). *Adolescent and online fan fiction*. New York: Peter Lang.

Bober, M. J., & Dennen, V. P. (2001). Intersubjectivity: Facilitating knowledge construction in online environments. *Education Media International*, *38*(4), 241-250.

Bodycott, P., & Walker, A. (2000). Teaching abroad: Lessons learned about inter-cultural understanding for teachers in higher education. *Teaching in Higher Education*, *5*(1), 79–94. doi:10.1080/135625100114975

Boekema, F., Meeus, M., & Oerlemans, L. (2000). Learning, Innovation and Proximity: An Empirical Exploration of Patterns of Learning: a Case Study. In Boekema, F. (Ed.), *Knowledge, Innovation and Economic Growth: the Theory and Practice of Learning Regions* (pp. 137–164).

Bologna Declaration. (1999). *Joint declaration of the European Ministers of Higher Education*. Bologna, June 19.

Boud, D., Cohen, R., & Sampson, J. (2001). *Peer Learning in Higher Education: Learning from & With Each*. Routledge.

Boud, D., Keogh, R., & Walker, D. (1985). *Reflection: Turning Experience into Learning*. London: Kogan Page.

Bowditch J.L. & Buono (2005). *A Primer on Organizational Behavior*. 6th edition, Hoboken: John Willey & Sons.

Boyd, H., & Cowan, J. (1985). A case for self-assessment based on recent studies of student learning. *Assessment & Evaluation in Higher Education*, *10*(3), 225–235.

Braaksma, M. A. H., Rijlaarsdam, G., & Van den Bergh, H. (2002). Observational Learning and the Effects of Model-Observer Similarity. *Journal of Educational Psychology*, *94*(2), 405–415. doi:10.1037/0022-0663.94.2.405

Braaksma, M. A. H., Van den Bergh, H., Rijlaarsdam, G., & Couzijn, M. (2001). Effective learning activities in observation tasks when learning to write and read argumentative texts. *European Journal of Psychology of Education*, *1*, 33–48. doi:10.1007/BF03172993

Bradley, G. (2006). *Social and Community Informatics: Humans on the Net*. London: Routledge.

Bretschneider, S., Gant, J., & Ahn, M. (2003, October 9-11). A general model of e-government service adoption: Empirical exploration. *Public Management Research Conference*, Georgetown Public Policy Institute Washington, D.C. Retrieved May 10, 2004, from http://www.pmranet.org/conferences/georgetownpapers/Bretschneider.pdf

Bridges.org. (2001). *Comparison of e-Readiness assessment models*. Retrieved April 1, 2004 from http://www.bridges.org/e-Readiness/report.html

Brin, S., Motwani, R., & Silverstein, C. (1997, May 13-15). Beyond market baskets: Generalizing association rules to correlations. In *Proceedings of the 1997 ACM SIGMOD International Conference on Management of Data*, Tucson, AZ (pp. 265-276). ACM Publishing.

Broadfoot, P., James, M., McMeeking, S., Nuttal, D., & Stierer, S. (1988). *Records of achievement: report of the national evaluation of pilot schemes*. London: HMSO.

Brown, J. S., & Duguid, P. (2000). *The Social Life of Information*. Boston: Harvard Business School Press.

Brown, S. (1998). *Peer assessment in practice*. Birmingham: SEDA Administrator.

Brown, A., & Iwashita, N. (1996). Language background and item difficulty: The development of a computer-adaptive test of Japanese. *System*, *24*(2), 199–206. doi:10.1016/0346-251X(96)00004-8

Brown, S. (2002). Introduction: rethinking capacity development for today's challenges. In Browne, S. (ed.), *Developing capacity through technical cooperation* (pp. 1-14). London: Earthscan Publications (for UNDP). ISBN 0-185383-969-99.

Bruns (Jr.), E.L., & Takahashi-Welch, W. (2006). *Implementing Computer Supported Collaborative Learning in Less Developed Countries*. A course paper at the University of Texas at Austin.

Brynjolfsson, E., & Hitt, L. (1995). Information Technology as a Factor of Production: The role of Differences among Firms. *Economics of Innovation and New Technology*, *3*, 183–199. doi:10.1080/10438599500000002

Bui, T. X., Sankaran, S., & Sebastian, I. M. (2003). A framework for measuring national e-readiness. *International Journal of Electronic Business*, *1*(1), 3–22. doi:10.1504/IJEB.2003.002162

Bundesregierung (2002a). *Gesetz zur Gleichstellung behinderter Menschen (Behindertengleichstellungsgesetz – BGG)*. http://www.gesetze-im-internet.de/bundesrecht/bgg/gesamt.pdf, last visited 2010-09-24

Bundesregierung (2002a). Verordnung zur Schaffung barrierefreier Informationstechnik nach dem Behindertengleichstellungsgesetz (Barrierefreie Bundesrepublik Deutschland (2002). *Verordnung zur Schaffung barrierefreier Informationstechnik nach dem Behindertengleichstellungsgesetz (Barrierefreie Informationstechnik-Verordnung - BITV) vom 17*. Juli 2002

Burgess, J., & Chilvers, J. (2006). Upping the ante: a conceptual framework for designing and evaluating participatory technology assessment. *Science & Public Policy*, *33*(10), 713–728. doi:10.3152/147154306781778551

Burn, J., & Robins, G. (2003). Moving towards e-government: a case study of organizational change processes. *Logistics Information Management*, *16*(1), 25–35. doi:10.1108/09576050310453714

Burnett, R., & Marshall, P. D. (2003). *Web Theory – An Introduction*. London, New York: Routledge.

Butler, D., & Sellbom, M. (2002). Barriers to adopting technology for teaching and learning. *Educase Quarterly*, *25*(2), 22–28.

Buttle, F. (2004). *Customer Relationship Management: Concepts and Tools*. Oxford, England: Elsevier Publishing.

Cai, Y., Lu, B., Zheng, J., & Li, L. (2006). Immersive protein gaming for bio edutainment. *Simulation & Gaming*, *37*, 466–476. doi:10.1177/1046878106293677

Calvary, G.; Coutaz J.; Thevenin D.; Limbourg Q.; Bouillon L.; Vanderdonckt J.(n.d.). A unifying reference framework for multi-target user interfaces. *Interacting with Computers, 15*, pp. 289--308

Cancian, F. (1981). Community of reference in rural stratification research. *Rural Sociology*, *46*, 626–645.

Cao, M., Zhang, Q., & Seydel, J. (2005). B2C e-commerce web site quality: an empirical examination. *Industrial Management & Data Systems*, *105*(5), 645–661. doi:10.1108/02635570510600000

Carless, D. (2005). Prospects for the implementation of assessment for learning. *Assessment in Education*, *12*(1), 39–54. doi:10.1080/0969594042000333904

Cash, J. I., & Konsynski, B. R. (1985). IS redraws competitive boundaries. *Harvard Business Review*, (2): 134–142.

Cassell, J., & Bickmore, T. (2000, Dec.). External Manifestations of Trustworthiness in the Interface. *Communications of the ACM*, *43*(12), 50–56. doi:10.1145/355112.355123

Castell, M. (1996). *The Rise of the Network Society*. Cambridge, MA: Blackwell.

Castells, M. (1998a). *The rise of the Network Society (The Information Age: economy, society, culture; vol.1)*. Oxford: Blackwell Publishers. ISBN 0631221409.

Castells, M. (1998b). *End of millennium (The Information Age: economy, society, culture; vol.3)*. Oxford: Blackwell Publishers. ISBN 1-55786-872-7.

Castells, M. (2000). *Information Technology and Global Development.* [en línea] New York: UN Economic and Social Council (ECOSOC). Keynote address, ECOSOC High level segment July 2000. Retrieved 7 June 2001 from <http://www.un.org/esa/coordination/ecosoc/itforum/castells.pdf>

Cederlund, J., & Severinson Eklundh, K. (n.d). *JEdit: The logging text editor for Macintosh.* Stockholm, Sweden: IPLab, Department of Numerical Analysis and Computing Science, Royal Institute of Technology (KTH).

Celino, A., & Concilio, G. (2005, September 13). Open Content Systems for E-Governance: The Case of Environmental Planning. In Z. Irani, T. Elliman, & O.D. Sarikas (Eds.), *Proceedings of the eGovernment Workshop '05 (eGOV05)* (pp. 92-93), Brunel University, West London UB8 3PH, UK.

Chanchani, S., & Theivanathampillai, P. (2002). Typologies of culture. University of Otago, *Department of Accountancy and Business Law Working Papers Series, 04_10/02.* Dunedin: University of Otago.

Chang, C. Y., Sheu, J. P., & Chan, T. W. (2003). Concept and Design of Ad Hoc and Mobile classrooms. *Journal of Computer Assisted Learning, 19*(3), 336–346. doi:10.1046/j.0266-4909.00035.x

Chavan, A. (2004, November). Developing an Open Source Content Management Strategy for E-government. *Proceedings of the 42th Annual Conference on the Urban and Regional Information Systems Association*, Nevada.

Chen, C.-M. (2009). Personalized E-learning system with self-regulated learning assisted mechanism for promoting learning performance. *Expert Systems with Applications, 36*(5), 8816–8829. doi:10.1016/j.eswa.2008.11.026

Cheskin/Sapient Research and Studio Archetype/Sapient. (1999). *E-Commerce Trust Study*. http://www.sapient.com/cheskin.

Chong Ng, S.T. (2001, May). Taking e-learning education into the future – The global knowledge hall. *Sharing Knowledge and Experience in Implementing ICTs in Universities - Roundtable Papers* IAU/IAUP/EUA, Skagen Roundtable, Skagen, Denmark.

Chou, C. (2001). A model of learner-centered computer-mediated interaction for collaborative distance learning. In M. R. Simonson (Ed), *Annual Proceedings of Selected Research and Development [and) Practice Papers Presented at the National Convention of the Association for Educational Communications and Technology* (AECT) (pp. 74-80), Atlanta, Georgia.

Choucri, N., Maugis, V., Madnick, S., & Siegel, M. (2003). Global e-Readiness – for what? *MIT Sloan School of Management Research Paper 177.*

Christiaanse, E., & Huigen, J. (1997). Institutional Dimensions in Information Technology Implementation in Complex Network Settings. *European Journal of Information Systems, 6*(2), 268–285. doi:10.1057/palgrave.ejis.3000258

Christiaanse, E., & Venkatraman, N. (2002). Beyond SABRE: An Empirical Test of Expertise Exploitation in Electronic Channels. *Management Information Systems Quarterly, 26*(1), 15–38. doi:10.2307/4132339

Chutimaskul, W., & Chongsuphajaisiddhi, V. (2004, May 17-19). A Framework for Developing Local E-Government, In M.A. Wimmer (Ed.), *Proceedings of Knowledge Management in Electronic Government: 5th IFIP International Working Conference, KMGov 2004* (pp. 319-324), Krems, Austria.

CIA. (2003). *The World Fact book.* Retrieved April 1, 2004, from http://www.cia.gov/cia/publications/factbook

Clawson, R. A., & Choate, J. (1999). Explaining participation on a class newsgroup. *Social Science Computer Review, 17*, 455–459. doi:10.1177/089443939901700406

Clemons, E., & Row, M. (1993). Limits to inter-firm coordination through information technology: Results of a field study in consumer packaged goods distribution. *Journal of Management Information Systems. M E Sharpe Inc. Armonk. 10* (1), pp 73-89.

Cockburn, A., & McKenzie, B. (2001). What Do Web Users Do? An Empirical Analysis of Web Use. *International Journal of Human-Computer Studies, 54*(6), 903–922. doi:10.1006/ijhc.2001.0459

Cogburn, D. L. (2003). Globally-Distributed Collaborative Learning and Human Capacity Development in the Knowledge Economy. In Mulenga, D. (Ed.), *Globalization and Lifelong Education: Critical Perspectives*. New Jersey: Lawrence Erlbaum Associates.

Cohen, E. (2005). (Ed.) *Issues in informing science and information technology, 2*. Informing Science.

Cohen, L. (n.d.). *Conducting research on the Internet*. Retrieved October 20, 2007, from http://www.Internet-tutorials.net/research.html

Collignon, Benoît; Vanderdonckt, Jean; Calvary, Gaëlle(n.d.). Model-Driven Engineering of Multi-Target Plastic User Interfaces *Fourth International Conference on Autonomic and Autonomous Systems*, ICAS, 2008

Comfort, K., L. Goje, & K. Funmilola. (2003). Relevance and priorities of ICT for women in rural communities: A case study from Nigeria. *Bulletin of the American Society for Information Science and Technology 29, 1*(6), 24-25.

Commission of the European Communities (CEC). (2001). *Communication from the Commission: making a European area of lifelong learning a reality (COM (2001) 678, final of 21.11.01)*. Luxembourg: Office for Official Publications of the European Communities.

Commonwealth (2004). *Commonwealth local government handbook: Modernisation, Council Structures, Finance, E-Government, Local Democracy, Partnerships, Representation,*. Commonwealth Local Government Forum, Commonwealth Secretariat.

Connolly R., Bannister F. (Nov. 2007). Consumer Trust in Electronic Commerce: Social & Technical Antecedents. *Proceedings of World Academy of ScienceEngineering and Technology, 25*.

Coppola, N. W., Hiltz, S. R., & Rotter, N. G. (2002). Becoming a Virtual Professor: Pedagogical Roles and Asynchronous Learning Networks. *Journal of Management Information Systems, 18*(4), 169–189.

Cox, J. (1999). *Trust, Reciprocity, and Other-Regarding Preferences of Individuals and Groups. Department of Economics*, University of Arizona, mimeo.

CPSI. (2005). *Local Governance and ICTs Research Network for Africa. LOG-IN Africa, COUNTRY REPORT – SOUTH AFRICA, Produced for the LOG-IN Africa Project Planning, Compiled by: Centre for Public Service Innovation (CPSI) and the LINK Centre*. Nairobi, Kenya: University of the Witwatersrand.

CTG. (2000). *Putting information to work: Annual Report 2000*. Center for Technology in Government, University at Albany, State University of New York.

CTG. (2002). *New foundations: Annual Report 2002*. Center for Technology in Government, University at Albany, State University of New York.

CTG. (2003). *Annual Report 2003, Center for Technology in Government*. University at Albany, State University of New York.

Cutri, G., Naccarato, G., & Pantano, E. (2008). Mobile Cultural Heritage: The case study of Locri. *Lecture Notes in Computer Science, 5093*, 410–420. doi:10.1007/978-3-540-69736-7_44

Cyr, D., Kindra, G., & Dash, S. B. (2008). Website Design, Trust, Satisfaction and e-Loyalty: The Indian Experience. *Online Information Review, 32*(6), 773–790. doi:10.1108/14684520810923935

D'Haenens, L., Koeman, J., & Saeys, F. (2007). Digital citizenship among ethnic minority youths in the Netherlands and Flanders. *New Media & Society, 19*(2), 278–299. doi:10.1177/1461444807075013

Daniels, S. E., & Walker, G. B. (2001). *Working through environmental conflict—the collaborative learning approach*. Westport, CT: Praeger.

Darley, W. (2001). The internet and emerging e-commerce: Challenge and implications for management in sub-Saharan Africa. *Journal of Global Information Technology Management, 4*(4), 4–18.

Davenport, T. (1993). *Process Innovation: Reengineering Work through Information Technology*. Boston: Harvard Business School Press.

Davie, L., & Wells, R. (1991). Empowering the learner through computer-mediated communication. *American Journal of Distance Education, 5*, 15–23. doi:10.1080/08923649109526728

Davis, F. D. (1989). Perceived usefulness, perceived ease of use and user acceptance of information technology. *MIS Quaterly*, *13*(3), 319–340. doi:10.2307/249008

Davis, J., Mayer, R., & Shoorman, F. (1995). An integrated model of organizational trust. *Academy of Management Review*, *20*(3), 705–734.

Davis, F. (1989). Perceived usefulness, perceived ease of use and user acceptance of information technology. *Management Information Systems Quarterly*, *13*, 319–340. doi:10.2307/249008

Davis, F. (1989). Perceived usefulness, perceived ease of use, and user acceptance of information technology. *MIS Quarterly*, *13*(3), 318–340. doi:10.2307/249008

Dawson, R., & Newman, I. (2002). Empowerment in IT education. *Journal of Information Technology Education*, *1*(2).

de Freitas, S. (2006). Using games and simulations for supporting learning. *Learning, Media and Technology*, *31*(4), 343–358. doi:10.1080/17439880601021967

de Freitas, S., & Neumann, T. (2009). The use of 'exploratory learning' for supporting immersive learning in virtual environments. *Computers & Education*, *52*(2), 343–352. doi:10.1016/j.compedu.2008.09.010

de Juana-Espinosa, S. (2006). Empirical Study of the Municipalities' Motivation for Adopting Online Presence. In Al-Hakim, L. (Ed.), *Global E-government: Theory, Applications and Benchmarking* (pp. 261–279). Hershey, PA: IGI.

De Wit, B., & Meyer, R. (2004). *Strategy: Process, Content and Context*. London: Thomson Learning.

Deal, T. E., & Kennedy, A. A. (1982). *Corporate Cultures: The Rights and Rituals of Corporate Life*. Reading, MA: Addison-Wesley.

Dede, C. (2004). Enabling Distributed Learning Communities Via Emerging Technologies. *T.H.E. Journal.* Retrieved September 28, 2008 from http://www.thejournal.com/magazine/vault/A4963.cfm

Dehmel, A. (2006). Making a European area of lifelong learning a reality? Some reflections on the European Union's lifelong learning policies. *Comparative Education*, *42*(1), 49–62. doi:10.1080/03050060500515744

del Puerto, F., & Gamboa, E. (2010). The evaluation of computer-mediated technology by second language teachers: collaboration and interaction in CALL. *Educational Media International*, *46*(2), 137–152. doi:10.1080/09523980902933268

Den Hengst, M. Hlupic, & V. Serrano, (2005). *Design of Inter-Organizational Systems: Collaboration & Modeling*. Proceedings of the 38th Annual Hawaii International Conference

Denning, S. (2002). Technical Cooperation and Knowledge Networks. In S. Fukuda-Parr, S., C. Lopes, & K. Malik (Eds.), *Capacity for development: new solutions to old problems* (pp. 229-244). New York: Earthscan Publications. ISBN 1-85383-919-1.

Deutschland, B. (2006). Allgemeines Gleichbehandlungsgesetz der Bundesrepublik Deutschland vom 29.06.2006. http://gesetze-im-internet.de/agg/.

Dewald, N. H., & Silvius, M. A. (2005). Business faculty research: Satisfaction with the Web versus library databases. *Portal: Libraries and the Academy*, *5*(3), 313–328. doi:10.1353/pla.2005.0040

Dewapura, R. (2008, May). *Enabling Environment for e/m – Government: ICT Infrastructure & Interoperability*. A presentation at the Capacity-building Workshop on Back Office Management for e/m-Government in Asia and the Pacific Region, Shanghai, People's Republic of China. Available at http://unpan1.un.org/intradoc/groups/public/documents/UN/UNPAN030563.pdf

DIAS GmbH. (2010): Barrierefreie Informationstechnik-Verordnung-Test (BITV-Test. http://www.bitvtest.de, last visited 2010-09-24

Dijk, van J.A.G.M. (2005). *The Deepening Divide. Inequality in the Information Society*. London: Sage.

Diki-Kidiri, M. (2007). Comment assurer la présence d'une langue dans le cyberespace? *UNESCO*. Retrieved December 31, 2007, from http://unesdoc.unesco.org/images/0014/001497/149786F.pdf

DiMaggio, P., Hargittai, P., Neuman, E., Robinson, W. R., & John, P. (2001). Social implications of the Internet. *Annual Review of Sociology*, *72*, 307–336. doi:10.1146/annurev.soc.27.1.307

DiMaggio, P., Hargittai, E., Neuman, W. R., & Robinson, J. P. (2001). Social implications of the internet. *Annual Review of Sociology*, *27*, 307–336. doi:10.1146/annurev.soc.27.1.307

DINCERTCO.(2010): Barrierefreie Website. www.dincert-co.de/web/media_get.php?mediaid=9080&fileid=13930, last visited 2010-09-24

Doney, P. M., & Cannon, J. P. (1997, April). An Examination of the Nature of Trust in Buyer-Seller Relationships. *Journal of Marketing*, *61*, 35–51. doi:10.2307/1251829

Dowell, J., & Long, J. (1989). Towards a conception for an engineering discipline of human factors. *Ergonomics*, *32*, 1513–1535. doi:10.1080/00140138908966921

Drettakis, G., Roussou, M., Asselot, M., Reche, A., Olivier, A., Tsingos, N., & Tecchia, F. (2005). Participatory Design and Evaluation of a Real-World Virtual Environment for Architecture and Urban Planning. In *Proceedings of IEEE Virtual Reality*. Bonn, Germany: MIT Press Cambridge.

Dringus, L. P., & Terrell, S. (1999). The framework for DIRECTED online learning environments. *The Internet and Higher Education*, *2*(1), 55–67. doi:10.1016/S1096-7516(99)00009-3

Driscoll, M. P. (2000). *Psychology of learning instruction* (2nd ed.). Boston: Allyn and Bacon.

Dul, J., & Hak, T. (2008). *Case Study Methodology in Business Research*. Amsterdam: Butterworth-Heinemann/ Elsevier.

Dunleavy, P. (2006). *Digital era governance: IT corporations, the state and e-government*. Oxford: Oxford University Press.

Dutton, W. H., Kahin, B., O'Callaghan, R., & Wyckoff, A. W. (Eds.). (2005). *Transforming Enterprise: The Economic and Social Implications of Information Technology*. MIT Press. e-Asia 2007 (2007, February 6-8). Summary report, *e-Asia 2007*. Putrajaya International Convention Center, Malaysia.

Edmiston, K. D. (2003). State and local e-Government: Prospects and challenges. *American Review of Public Administration*, *33*(1), 20–45. doi:10.1177/0275074002250255

Edwards, A. D. (1988). The design of auditory interfaces for visually disabled users. In *Proceedings of the SIGCHI Conference on Human Factors in Computing Systems (Washington, D.C., United States, May 15 - 19, 1988)*. (pp.83-88). J. J. O'Hare, Ed. CHI '88. New York: ACM Press.

Egger, F N. (2003). From interactions to transactions: Designing the Trust Experience for Business-to-Consumer Electronic Commerce.

Egger, F. N. (2001). Affective Design of E-Commerce User Interfaces: How to maximize perceived trustworthiness. *Proceedings of The International Conference on Affective Human Factors Design*. London: Asean Academic Press.

Egorov, V., Jantassova, D., & Churchill, N. (2007). Developing pre-service English teachers' competencies for integration of technology in language classrooms in Kazakhstan. *Educational Media International*, *44*(3), 255–265. doi:10.1080/09523980701491732

E-GovReport. (2005). *Global E-Government Readiness Report 2004: Towards Access and Opportunity, I-WAYS - The Journal of E-Government Policy and Regulation, 28(1)*. IOS Press.

Eisenhardt, K. M. (1989). Building theories from case study research. *Academy of Management Review*, *14*(4), 532–550. doi:10.2307/258557

Elsadig M.A. (2008). ICT and Human Capital intensities Effects on Malaysian productivity Growth. *International Research Journal of Finance and Economics,* (13), 152-161.

Elsner, W. (2000). An Industrial Policy Agenda 2000 and Beyond: Experience, Theory and Policy. In Elsner, W., & Groenewegen, J. (Eds.), *Industrial Policies After 2000, Boston*. London: Dodrecht.

Elsner, W. (2003). Increasing Complexity in the "New" Economy and Co-ordination Requirements Beyond the "Market": Network Governance, Interactive Policy, and Sustainable Action. In Elsner, W., Frigato, P., & Steppacher, R. (Eds.), *Social Costs of the Global "New"*. Economy.

Elzen, B., Geels, F. W., & Green, K. (Eds.). (2004). *System Innovation and the Transition to Sustainability – Theory, Evidence and Policy*. Aldershot: Edward Elgar.

Englund Dimitrova, B. (2005). *Expertise and Explicitation in the Translation Process. (Benjamins Translation Library 64)*. Amsterdam: John Benjamins Publishing Company.

Englund Dimitrova, B. (2006). Segmentation of the writing process in translation: experts versus novices. In *Sullivan, K. P. H., & Lindgren, E. (2006). Computer keystroke logging: Methods and Applications* (pp. 189–201). Oxford, England: Elsevier.

Ess, C., & Sudweeks, F. (2005). Culture and computer-mediated communication: Toward new understandings. *Journal of Computer-Mediated Communication, 11*(1), Retrieved on July 17, 2006 from, Http://jcmc.indiana.edu/voll11/issue1/ess.html.

ETIC. (2005). *Estrategia Boliviana de TIC para el Desarrollo*. Retrieved February 12, 2008, from http://etic.bo

Etzkowitz, H. (2006). The new visible hand: an assisted linear model of science and innovation policy. *Science & Public Policy, 33*(5), 310–320. doi:10.3152/147154306781778911

Eubanks, V. (2007). Popular technology: exploring inequality in the information economy. *Science & Public Policy, 34*(2), 127–138. doi:10.3152/030234207X193592

European Commission (2000). Gleichbehandlung Behinderter in Beruf und Bildung: *Richtlinie des Rates 2000/78/EG vom 27. November 2000*, ABl. L 303 vom 2. Dezember 2000.

European Commission (2007). *Access to and preservation of cultural heritage-Fact sheets of 25 research projects funded under the Sixth Framework Programme for Research and Technological Development (FP6)*. Luxemburg: Imprimé par OIL.

European Union. (2005). *eAccessibility. COM(2005)425.* http://eur-lex.europa.eu/LexUriServ/LexUriServ.do?uri=COM:2005:0425:FIN:EN:PDF, last visited 2010-09-24

Fagan, G. H., Newman, D. R., McCusker, P., & Murray, M. (2006). *E-consultation: evaluating appropriate technologies and processes for citizens' participation in public policy.* Final Report, e-Consultation Research Project, UK.

Falchikov, N. (1998). Involving students in feedback and assessment. In Brown, S. (Ed.), *Peer assessment in practice* (pp. 9–23). Birmingham: SEDA Administrator.

Fang, Z. (2002). E-Government in digital era: Concept, practice, and development. *International Journal of the Computer, the Internet and Management, 10*(2), 1-22.

Farrell, G. (2008). *Survey of ICT and Education in Africa* (Vol. 1). Association for the Development of Education in Africa (ADEA). International Institute for Educational Planning, Paris, France

Federal Ministry of Finance Austria. (2010). *Accessible Documents*. http://formulare.bmf.gv.at/service/formulare/wai_formulare/_start.htm, last visited 2010-09-24

Feltenstein, A., & Ha, J. (1995). The Role of Infrastructure in Mexican Economic Reforms. *The World Bank Economic Review, 9*(2), 287–304. doi:10.1093/wber/9.2.287

Ferguson, M. (2005). Local E-Government in the United Kingdom. In Drüke, H. (Ed.), *Local Electronic Government: A Comparative Study* (pp. 156–196). Routledge.

Ferris, D. R. (2003). *Response to Student Writing: Implications for Second Language Students*. Mahwah, NJ: Lawrence Erlbaum Associates.

Fogg, B. J., & Nass, C. (1997). Effects of computers that flatter. *International Journal of Human-Computer Studies, 46*, 551–561. doi:10.1006/ijhc.1996.0104

Fogg B.J, Marshall J., Laraki O., Osipovich A., Varma C., Fang N., Paul J., Rangnekar A, Shon J., Swani P.& Treinen M. (2001). What Makes Web Sites Credible? A Report on a Large Quantitative Study. *ACM sigchi, 3* (1), 61-67.

Fougere, M., & Moullettes, A. (2007). The Construction of the Modern West and the Backward Rest: Studying the Discourse of Hofstede's Culture's Consequences. *Journal of Multicultural Discourses, 2*(1), 1–19. doi:10.2167/md051.0

Fountain, J. E. (2001). The virtual state: Transforming American government? *National Civic Review, 90*(3), 241–251. doi:10.1002/ncr.90305

Fountain, J. E. (2004). Digital government and public health. *Preventing chronic disease – Public Health Research. Practice, and Policy, 1*(4), 1–5.

Fox, W. F., & Gurley, T. (2006, May). *Will Consolidation Improve Sub-National Governments?* World Bank Policy Research Working Paper WPS3913, World Bank, Washington DC.

Fuchs, R. (1998). Little Engines that Did: Case Histories from the Global Telecentre Movement. IDRC, 1998. Retrieved December 24, 2008 from http://www.idrc.ca/ en/ ev- 10630-201- 1-DO_TOPIC.html.

Fukao, M. (1995). *Financial Integration, Corporate Governance, and the Performance of Multinational Companies*. Transaction Publishers.

Fukuda-Parr, S., & Hill, R. (2002). The Network Age: creating new models of technical cooperation. In S. Fukuda-Parr, S., C. Lopes, & K. Malik (Eds.), *Capacity for development: new solutions to old problems* (pp. 185-201). New York: Earthscan Publications. ISBN 1-85383-919-1.

Fukuyama, F. (1995). *Trust: The social virtues and the creation of prosperity*.New York: The Frees press.

FUNREDES. (2003-2005). *Multistakeholder partnership methodology*. Retrieved December 30, 2007, from http:// cmsi.funredes.org/inc/multistakeholder_en.htm

FUNREDES. (2005). *UNDP.DO supports Multi-stakeholder Partner Cheap*. Retrieved December 30, 2007, from http://funredes.org/undp.do

Gadomski, A. M. (1993, January/February) (Ed.). Toga: A Methodological and Conceptual Pattern for Modeling of Abstract Intelligent Agent. *Proceedings of the "First International Round-Table On Abstract Intelligent Agent* (pp. 25-27), Rome: Enea.

Garrett, N. (2009a). Technology in the service of language learning: Trends and new Issues. *Modern Language Journal, 93*, 697–718. doi:10.1111/j.1540-4781.2009.00968.x

Garrett, N. (2009b). Computer-Assisted Language Learning Trends and Issues Revisited: Integrating Innovation. *Modern Language Journal, 93*, 719–741. doi:10.1111/j.1540-4781.2009.00969.x

Garrety, K., & Badham, R. (2004). User-Centered Design and the Normative Politics of Technology. *Science, Technology & Human Values, 29*(2), 191–212. doi:10.1177/0162243903261946

Garrison, D. R., & Anderson, T. (2003). E-Learning in the 21st Century: A Framework for Research and Practice. London, UK: RoutledgeFalmer. Hamburg, I., Lindecke, C., & ten Thij, H. (2003, September 25-26). Social aspects of e-learning and blending learning methods. *A Proceedings of the 4th European Conference E-Comm-Line 2003* (pp. 11-15), Bucharest.

Gascó, M. (2007). *Civil Servants' Resistance toward E-Government Development*. In: Encyclopedia of Digital Government. Hershey, PA: Idea Group Inc.

Gee, X., Yamashiro, A., & Lee, J. (2000). Pre-class planning to scaffold students for online collaborative learning activities. *Journal of Educational Technology & Society, 3*(3).

Ghiani, G., Paternò, F., Santoro, C., & Spano, L. D. (2009). UbiCicero: a Location-Aware, Multi-Device Museum Guide. *Interacting with Computers, 21*(4), 288–303. doi:10.1016/j.intcom.2009.06.001

Gómez, R. (2001). *MISTICA: Re: Comunidades virtualizadas?* Retrieved December 31, 2007, from http://funredes.org/mistica/castellano/emec/pro/memoria6/0145.html

Goodfellow, R. (2001). Credit where it's due. In Murphy, D., Walker, R., & Webb, G. (Eds.), *Online Learning and Teaching with Technology: Case Studies, Experience and Practice* (pp. 73–80). Kogan Page.

Government Of India. (1996). *India Infrastructure Report*. Department of Economic Affairs, Ministry of Finance, Government of India.

Graafland-Essers, I., & Ettedgui, E. (2003). Benchmarking e-government in Europe and the US. *RAND, MR-1733-EC, 2003*. Retrieved May 10, 2004, from http://www.rand.org/publications/MR/MR1733/MR1733.pdf

Grabill, J. (2003). Community computing and citizen productivity. *Computers and Composition, 20*, 131–150.

Grameen Cyber Society. Reflecting on the pilot telecenter operation in rural bangladesh. Retrieved December 22, 2008 from http://siteresources. worldbank.org/ EDUCATION/Resources/ 278200-1126210664195/ 1636971-1126210694253/ Grameen_Final_ Report.pdf

Griffin, R. (2005). *Management* (8th ed.). Boston: Houghton Mifflin Company.

Grönlund, Å. (2003). Emerging electronic infrastructures: Exploring democratic components. *Social Science Computer Review, 21*(1), 55–72. doi:10.1177/0894439302238971

Gudykunst, W. B., Chua, E., & Gray, A. J. (1987). Cultural dissimilarities and uncertainty reduction processes. In McLaughlin, M. (Ed.), *Communication Yearbook* (*Vol. 10*, pp. 457–469). Beverly Hills, CA: Sage.

Guidi, G., Frischer, B., Russo, M., Spinetti, A., Crosso, L., & Micoli, L. L. (2006). Three-dimensional acquisition of large and detailed cultural heritage objects. *Machine Vision and Applications*, *17*, 349–360. doi:10.1007/s00138-006-0029-z

Gunawardena, C. N., Lowe, C. A., & Anderson, T. (1997). Analysis of a global online debate and the development of an interaction analysis model for examining social construction of knowledge in computer conferencing. *Journal of Educational Computing Research*, *17*(4), 397–431. doi:10.2190/7MQV-X9UJ-C7Q3-NRAG

Haiwook, C. (2001). *The Effects of Inter-organisational Information systems Infrastructure on Electronic Cooperation*: An Investigation of the "move to the Middle". PhD Abstract. Proquest Digital Dissertations. www.lib.umi.com/dissertations, Accessed 13.3.2003

Hamalainen, R. (2008). Designing and evaluating collaboration in a virtual game environment for vocational learning. *Computers & Education*, *50*(1), 98–109. doi:10.1016/j.compedu.2006.04.001

Han, Y.-J., & Yong, Q. (2006). Efficient improvement of FT-tree based frequent itemsets mining algorithms. *IEEE Transactions on Knowledge and Data Engineering*, *3*, 374–377.

Han, J., Pei, Y., & Yin, Y. (2000, May). Mining frequent patterns without candidate generation. In *Proceedings of the 2000 ACM SIGMOD International Conference on Management of Data*, Dallas, TX (pp. 1-12).

Hand, D., & Mannila, H. (2004). *Principles of data mining* (3rd ed.). Cambridge, MA: MIT Press.

Hansen, G. (2005). *Störquellen in Übersetzungsprozessen. Eine empirische Untersuchung von Zusammenhängen zwischen Profilen, Prozessen und Produkten*. Doctoral dissertation. Copenhagen: Copenhagen Business School.

Hanseth, O. (2002). From systems and tools to networks and infrastructures. Toward a theory of ICT solutions and its design methodology implications. Accessed October 25, 2008 from http://heim.ifi.uio.no/~oleha/Publications/ib_ISR_3rd_resubm2.html

Hansmann, U., Merk, L., Kahn, P., Nicklous, M., Stober, T., & Shelness, N. (2003). *Pervasive computing: the mobile world*. Berlin: Springer.

Hargittai, E. (1999). Weaving the Western web: Explaining differences in Internet connectivity among OECD countries. *Telecommunications Policy*, *23*(10/11).

Harridge-March, S. (2006). Can the building of trust overcome consumer perceived risk online? *Marketing Intelligence & Planning*, *24*(7), 746–761. doi:10.1108/02634500610711897

Harris, R. Telecentre 2.0: Beyond Piloting Telecentres. APDIP eNote 14, 2007. Retrieved December 24, 2008 from http://www.apdip.net/ apdipenote/ 14.pdf

Hassanein, K. S., & Head, M. M. (2004). *Building Online Trust through Socially Rich Web Interfaces*. http://dev.hil.unb.ca.

Hauge, H., & Ask, B. (2008). Qualifying University Staff in Developing Countries for e-Learning. *iLearning Forum 2008 Proceedings - European Institute for E-Learning (EIfEL)* (pp. 183-188), Paris.

Hauschildt, J. (2004). *Innovationsmanagement,* (3rd ed.), Munich.

Heeks, R. (2002). *Reinventing Government in the Information Age: International Practice in IT-Enabled Public Sector Reform* (2nd ed.). New York: Routledge.

Heeks, R. (2005). *Implementing and managing e-government*. London: Sage.

Heeks, R. (2008, June). ICT4D 2.0: The next phase of applying ICT for international development. *IEEE Computer* (pp.26-33), June 2008. IEEE Computer Society

Heinecke, W., Dawson, K., & Willis, J. (2001). Paradigms and Frames for R & D in Distance Education: Toward Collaborative Learning. *International Journal of Educational Telecommunications*, *7*(3), 293–322.

Henderson, J. C. (1990). Plugging into strategic partnerships: The critical IS connection. *Sloan Management Review*, (Spring): 7–18.

Herzberg, F. (1987). One more time: How do you motivate employees? *Harvard Business Review*, (January-February): 109–120.

Heskett, J., Jones, T., Earl, L., & Schlesinger, L. (1994). Putting the service-profit chain to work. *Harvard Business Review*, (March-April): 164–174.

Hiltz, R. (1990). Evaluating the virtual classroom. In Harasim, L. M., & Turnoff, M. (Eds.), *Online education: Perspectives on a new environment* (pp. 133–183). New York: Praeger.

Hofstede, G. (1980). *Culture's consequences: International differences in work-related values*. Beverly Hills, California: Sage Publications.

Hofstede, G. (1981). Culture and organizations. *International Studies of Management and Organization, 10*(4), 15–41.

Hofstede, G. (1983). National cultures in four dimensions – A research-based theory of cultural differences among nations. *International Studies of Management and Organization, 13*(1-2), 46–74.

Hofstede, G., & Bond, M. (1984). Hofstede's culture dimensions: An independent validation using Rokeach's value survey. *Journal of Cross-Cultural Psychology, 15*, 417–433. doi:10.1177/0022002184015004003

Hofstede, G. H. (2001). *Culture's consequences: Comparing values, behaviors, institutions, and organizations across nations*. Thousand Oaks, CA: Sage.

Hofstede, G. (2003). Cultural constraints in management theories. In Reddding, G., & Stening, B. W. (Eds.), *Cross-cultural management* (*Vol. II*, pp. 61–74). Cheltenham: Edward Elgar Publishing.

Hofstede, G. (1980). *Culture's consequences*. Beverly Hills, Ca: Sage.

Hofstede, G. (2004). *Geert Hofstede cultural dimensions*. Retrieved November 19, 2004, from http://www.geert-hofstede.com/hofstede_dimensions.php

Holden, S. H., Norris, D. F., & Fletcher, P. D. (2003). Electronic government at the grass roots: Contemporary evidence and future trends. In *Proceedings of the 36th Hawaii International Conference on System Sciences*. Big Island, Hawaii, January 06 - 09, 2003. Retrieved May 10, 2004, from http://csdl.computer.org/comp/proceedings/hicss/2003/1874/05/187450134c.pdf

Hollified, C., & Donnermeyer, J. (2003). Creating demand: Influencing information technology diffusion in rural communities. *Government Information Quarterly, 20*, 135–150. doi:10.1016/S0740-624X(03)00035-2

Holmen, H. (2002). *NGOs, networking, and problems of representation*. [Online]. Linköpings University and ICER, July 2002. Retrieved 18 Nov 2007 from <http://www.icer.it/docs/wp2002/holmen33-02.pdf>

Holmes, D. (2001). *EGov: eBusiness Strategies for Government*. London: Nicholas Brealey Publishing.

Hoogwout, M. (2003, September 1-5). Super Pilots, Subsidizing or Self-Organization: Stimulating E-Government Initiatives in Dutch Local Governments. In R. Traunmüller (Ed.), *Proceedings of Electronic Government: Second International Conference, EGOV 2003* (pp.85-90), Prague, Czech Republic: Springer.

Hossian, L., & de Silva, A. (2009). Exploring user acceptance of technology using social networks. *The Journal of High Technology Management Research, 20*, 1–18. doi:10.1016/j.hitech.2009.02.005

Hsu, C., & Lin, J. (2008). Acceptance of blog usage: The roles of technology acceptance, social influence and knowledge sharing motivation. *Information & Management, 45*, 65–74. doi:10.1016/j.im.2007.11.001

Hu, G. (2005). Using peer review with Chinese ESL student writers. *Language Teaching Research, 9*(3), 321–342. doi:10.1191/1362168805lr169oa

Hughes, M. (2006). *Change Management: A Critical Perspective*. London: CIPD Publications.

Hyland, F. (2000). ESL writers and feedback: giving more autonomy to students. *Language Teaching Research, 4*(1), 33–54.

Iahad, N., Dafoulas, G. A., Milankovic-Atkinson, M., & Murphy, A. (2005, January 3-6). E-learning in developing countries: suggesting a methodology for enabling computer-aided assessment. In *the proceedings of the 3rd ACS/IEEE International Conference on Computer Systems and Applications (AICCSA-05)* (pp. 847-852), Cairo, Egypt.

ICASO. (2002). *HIV/AIDS networking guide*. [Online]. Canada: International Council of Aids Service Organizations. Retrieved 7 April 2007 from <http://www.icaso.org/publications/NetworkingGuide_EN.pdf >

ICMA. (2005). *The Municipality Yearbook 2005. International City/County Management Association*. Washington, D.C.: ICMA.

Illustrated version available in Spanish (2004). Retrieved December 29, 2007, from http://funredes.org/mistica/castellano/ciberoteca/tematica/trabajando.pdf

Indiresan, M. (1993). Telecommunications: Social and Economic Impact. *Telecommunications* (pp. 22-25).

Informationstechnik-Verordnung – BITV. (2010). Retrieved from http://www.gesetze-im-internet.de/bundesrecht/bitv/gesamt.pdf, last visited 2010-09-24

Inter-American Development Bank. (2006). *Multiphase Program for the Strengthening of Chile's Digital Strategy.* Report # CH-L1001, Document of the Inter-American Development Bank, Washington D.C.

International Adult Literacy Service. (1998). Report for the Organisation for Economic Co-operation and Development (OECD). Retrieved from http://www.statcan.gc.ca/dli-ild/data-donnees/ftp/ials-eiaa-eng.htm, last visited 2010-09-24

International Perspectives on Surveillance: Technology and Management of Risk, 2004, *International Sociology*, vol. 19, No. 2, June, 131-254.

International Telecommunication Union. (2008). E-Services through post offices in bhutan, July 2008 http://www.itu.int/ITU-D/tech/RuralTelecom/UPU_Bhutan.pdf.

Ito, M., Okabe, D., & Matsuda, M. (Eds.). (2005). *Personal, Portable and Pedestrian: Mobile Phones in Japanese Life*. Cambridge, Ma: MIT Press.

ITU. (2003). *ITU Digital Access Index: World's First Global ICT Ranking*. Geneva: International Telecommunication Union.

Jakobi, A. P. & Rusconi, A. (2009). Lifelong learning in the Bologna process: European developments in higher education. *Compare: A Journal of Comparative and International Education*, 39 (1), 51 — 65

Jakobsen, A. L. (2003). Effects of Think Aloud on Translation Speed, Revision and Segmentation. In Alves, F. (Ed.), *Triangulating Translation* (pp. 69–95). Amsterdam, Philadelphia: John Benjamins Publishing Company.

Jansen, S., & Pimienta, D. (2007). *Perspectivas de la Cooperación Sur-Sur (CSS) en el marco de las Sociedades de los Saberes Compartidos: Visión desde el terreno.* Retrieved December 30, 2007, from http://funredes.org/mistica/castellano/ciberoteca/tematica/css-si-final.pdf

Jarvenpaa, S. L., & Tractinsky, N. (1999). Consumer Trust in an Internet Store: A Cross-Cultural Validation. *Journal of Computer-Mediated Communication, 5*(2).

Jassawalla, A. R., & Sahittal, H. C. (1999). Building collaborative new product teams. *Academy of Management Review, 13*(3), 50–60.

Jensen, A. (2001*). The Effects of Time on Cognitive Processes and Strategies in Translation. PhD dissertation.* Copenhagen Working Papers in LSP 2. Copenhagen: Copenhagen Business School.

Jiawei, H., Micheline, K., & Kaufmann, M. (2007). *Data mining: Concepts and techniques*. San Diego, CA: Academic Press.

Johns, S. K., Smith, M., & Strand, C. A. (2003). How culture affects the use of information technology. *Accounting Forum, 27*(1), 84–109. doi:10.1111/1467-6303.00097

Jones, S. E. (2006). *Against Technology: From the Luddites to Neo-Luddism*. New York: Routledge.

Kanuka, H., & Garrison, D. R. (2004). Cognitive presence in online learning. *Journal of Computing in Higher Education, 15*(2), 30–49. doi:10.1007/BF02940928

Kanungo, S. (2004). Sustainable benefits of IT investments: From Concepts to implementation. In *Proceedings of the 10th Americas Conference on Information Systems*, New York, August, 925-933.

Karacapilidis, N., Loukis, E., & Dimopoulos, S. (2005). Computer-supported G2G collaboration for public policy and decision-making. [Emerald Group Publishing Limited.]. *The Journal of Enterprise Information Management, 18*(5), 602–624. doi:10.1108/17410390510624034

Karacapilidis, N., Loukis, E., & Dimopoulos, S. (2005). Computer-supported G2G collaboration for public policy and decision-making. [Emerald Group Publishing Limited.]. *The Journal of Enterprise Information Management, 18*(5), 602–624. doi:10.1108/17410390510624034

Karl, G. (2007). Barrier-free accessibility: Not only talking people can be helped. In Zechner, A. (Ed.), *E-Government Guide Germany. Strategies, solutions, efficiency and impact*. Stuttgart: Fraunhofer IRB Verlag.

Katherine, R., & Ricardo, G. (2001). Comparing Approaches: Telecentre Evaluation Experiences in Asia and Latin America. The Electronic Journal of Information Systems in Developing Countries (EJISDC 2001), 4(3). Retrieved November 24, 2007, from http://www.ejisdc.org.

Kaufmann, D., Kraay, A., & Mastruzzi, M. (2008). *Governance matters VII: Governance indicators for 1996-2007*. World Bank Policy Research June 2008

Kaushik, P., & Singh, N. (2003). Information technology and broad-based development: Preliminary lessons from North India. *World Development, 32*(4), 591–607. doi:10.1016/j.worlddev.2003.11.002

Kayman, M. (2004). The state of English as a global language: Communicating culture. *Textual Practice, 18*(1), 1–22. doi:10.1080/0950236032000140131

Keegan, D. (2003). *The future of learning: From eLearning to mLearning. Hagen*. Germany: Femstudienforchung.

Keen, P. (1999). *Electronic Commerce Relationships: Trust by design*. Upper Saddle River, NJ: Prentice Hall.

Keen, P. G. W. (1997, April 21). Are you ready for 'Trust' Economy. *Computerworld, 31*(16), 80.

Kelegai, L., & Middleton, M. (2002). Information Technology Education in Papua New Guinea: Cultural, Economic and Political Influences. *Journal of Information Technology Education, 1*(1), 11–23.

Khosrow-pour, M. (2006). *Advanced Topics in information resources Management*. Hershey, PA: IGI Global.

Kiili, K. (2005). Digital game-based learning: Towards an experiential gaming model. *The Internet and Higher Education, 8*(1), 13–24. doi:10.1016/j.iheduc.2004.12.001

Kilduff, M., & Tsai, W. (2008). *Social Networks and Organizations*. London: SAGE Publications. (reprinted 2008, 1st published in 2003). ISBN 978-07619-6957-0.

KimJ., MoonJ.Y. (1998). Designing emotional usability in customer interface-trustworthiness of cyber banking system interface, interacting with computers, 10, *1-29*.

Kim, K.-J., & Bonk, C. J. (2002). Cross-cultural comparisons of online collaboration. *Journal of Computer-Mediated Communication, 8*(1). Retrieved April 1, 2004, from http://www.ascusc.org/jcmc/vol8/issue1/kimandbonk.html

Kimaro, H., & Nhampossa, J. (2004). *The challenges of sustainability of health information systems in developing countries*. In the Proceedings of the 12th European Conference on Information Systems, Turku, Finland.

King, A. B. (January 2000). *What Makes a Great Web Site?*http://goodmictices.com

Kirkman, G. S., Osorio, C. A., & Sachs, J. D. (2002). The network readiness index: Measuring the preparedness of nations for the networked world. In Dutta, S., Lanvin, B., & Paua, F. (Eds.), *The global information technology report 2001 – 2002: Readiness for the networked world* (pp. 10–29). New York: Oxford University Press.

Kleimann, B., & Wannemacher, K. (2004). *E-Learning an deutschen Hochschulen*. Hannover: HIS.

Knipfer, K., Mayer, E., Zahn, C., Schwan, S., & Hesse, W. (2009). Computer Support for Knowledge Communication in Science Exhibitions: Novel Perspectives from Research on Collaborative Learning. *Educational Research Review, 4*(3), 196–209. doi:10.1016/j.edurev.2009.06.002

Kolsaker, A. (2005. March 2-4). Third Way e-Government: The Case for Local Devolution. In M.H. Böhlen, J. Gamper, W. Polasek, & M.A. Wimmer (Eds.), *Proceedings of E-Government: Towards Electronic Democracy: International Conference, TCGOV 2005* (pp.70-80), Bolzano, Italy: Springer.

Korgaonkar., et al. (2006) as cited in Ganguly B. Dash S. B. Cyr D. Website characteristics, Trust and purchase intention in online stores: - An Empirical study in the Indian context. *Journal of Information Science and Technology, 6* (2), 22-44.

Kotler, N., & Kotler, P. (2000). Can museums be all things to all people? Missions, goals, and marketing's role. *Museum Management and Curatorship, 18*(3), 271–287. doi:10.1080/09647770000301803

Kovačić, Z. (2005). A brave new eWorld? An exploratory analysis of worldwide e-Government readiness, level of democracy, corruption and globalization. *International Journal of Electronic Government Research, 1*(3), 15–32. doi:10.4018/jegr.2005070102

Kuhlen, R. (2003, August 3). *Change of Paradigm in Knowledge Management Framework for the Collaborative Production and Exchange of Knowledge*. A paper presented in the Plenary Session, of the World Library and Information Congress: 69th IFLA General Conference and Council, Berlin.

Kuhn, N., Richter, S., & Naumann, S. (2007a). Improving Access to EGovernment Processes. In Khosrow-Pour, Mehdi (ed.). *Managing Worldwide Operations and Communications with Information Technology.* Proceedings of the 2007 Information Resources Management Association International Conference (IRMA 2007) Vancouver (British Columbia), Canada. Hershey, PA: IGI Global, pp. 1205-1206

Kuhn, N., Richter, S., & Naumann, S. (2007b). Improving Accessibility to Business Processes for Disabled People by Document Tagging. *Proceedings of the Ninth International Conference on Enterprise Information Systems (ICEIS 2007).* Funchal (Madeira), Portugal, pp. 286-289

Kumar, R. (2008, July). Convergence of ICT and Education. *Proceedings of World Academy of Science. Engineering & Technology, 30*, 557–569.

Kuriyan, R., Ray, I., & Toyama, K. (2008). Information and communication technologies for development: The bottom of the pyramid model in practice. *The Information Society, 24*(2), 93–104. doi:10.1080/01972240701883948

Kurzweil Technologies Inc. (2008). *National Foundation of the Blind: Reader software for a mobile phone.* http://www.knfbreader.com/index.php, last visited 2010-09-24

Kuznik, L. (2009). Learning in Virtual Worlds. *US-China Education Review, 6*(9), 43–51.

Kvasny, L., & Keil, M. (2002). *The challenges in redressing the digital divide: A tale of two cities*. In Proceedings of the International Conference on Information Systems, Barcelona, Spain, December, 15-18. Available at http://aisel.isworld.org/pdf.asp?Vpath=ICIS/2002&PDFpath=02RIP24.pdf; last accessed August 2003.

Kvasny, L., & Truex, D. (2001). Defining away the digital divide: the influence of institutions on popular representations of technology. In B. Fitzgerald, N. Russo, J. DeGross (Eds.), *Realigning Research and Practice in Information Systems Development: The Social and Organizational Perspective,* 399-415, New York: Kluwer Academic Publishers.

Kwon, O., & Wen, Y. (2010). An empirical study of the factors affecting social network service use. *Computers in Human Behavior, 26*, 254–263. doi:10.1016/j.chb.2009.04.011

Kydd, J., & Dorward, A. (2004). Implications of market and coordination failures for rural development in least developed countries. *Journal of International Development, 16*, 951–970. doi:10.1002/jid.1157

Labelle, R. (2005). *ICT Policy Formulation and e-Strategy Development: A Comprehensive Guidebook. United Nations Development Programme-Asia Pacific Development Information Programme (UNDP-APDIP) – 2005.* Reed Elsevier India Private Limited.

Ladner, Richard E. (2008). Access and empowerment. *ACM Trans. Access. Comput,. 1 (2),* (October 2008)

Lallana, E. C. (2004). *An Overview of ICT Policies and e-Strategies of Select Asian Economies. United Nations Development Programme-Asia Pacific Development Information Programme (UNDP-APDIP) – 2004.* Reed Elsevier India Private Limited.

Lallana, E.C. (2003, October). Comparative Analysis of ICT Polices and e-Strategies in Asia. *A presentation at the Asian Forum on ICT Policies and e-Strategies*, Kuala Lumpur: UNDP-APDIP.

Lalovic, K., Djukanovic, Z., & Zivkovic, J. (2004). Building the ICT fundament for local E-government in Serbia- Municipality of Loznica example. In M. Schrenk (Ed.), *Proceedings of the 9th International Symposium on Planning and IT*, Vienna, Austia.

Lam, W. S. E. (2000). Literacy and the design of self: A case study of a teenager writing on the Internet. *TESOL Quarterly, 34*, 457–482. doi:10.2307/3587739

Lane, C., & Plant, N. (1999). Community Computing in Rural Regeneration Networks, Cisc publication No 6, ISBN: 1860431666.

Lanzara, G. F., & Morner, M. (2004, 2-3 April). Making and Sharing Knowledge at Electronic Crossroads: the evolutionary ecology of open source. *Proceedings, The Fifth European Conference on Organizational Knowledge, Learning, and Capabilities*, Innskbruck.

Larsen-Freeman, D., & Long, M. (1991). *An introduction to second language acquisition research*. Harlow, Essex: Longman.

Lau, A., & Tsui, E. (2009). Knowledge management perspective on e-learning effectiveness. *Knowledge-Based Systems*, *22*, 324–325. doi:10.1016/j.knosys.2009.02.014

Lazaro, N., & Reinders, H. (2006). Technology in self-access: an evaluative framework. *PacCALL Journal*, *1*(2), 21–30.

Lea, M. (2000). Computer conferencing: new possibilities for writing and learning in Higher Education. In Lea, M., & Stierer, B. (Eds.), *Student writing in Higher Education: new contexts*. UK: Open University Press.

Lécouyer, Ch. (2005). *Making Silicon Valley: Innovation and the Growth of High Tech, 1930-1970*. Cambridge, MA: MIT Press.

Lee, M. K. O., & Turban, E. (2001). A Trust Model for Consumer Internet Shopping. *International Journal of Electronic Commerce*, *6*(1), 75–91.

Lee, G., Lee, K. L., & Chen, A. L. P. (2001). Efficient graph based algorithms for discovering and maintaining association rules in large databases. *Knowledge and Information Systems*, *3*, 338–355. doi:10.1007/PL00011672

Lee, J., Kim, J., & Moon, J. Y. (April 2000). What Makes Internet Users Visit Cyber Stores Again? Key Design Factors for Customer Loyalty. In *Proceedings of the Computer-Human Interaction Conference on Human Factors in Computing Systems*, The Hague, Netherlands, pp. 305-312.

Lefevere, A., & Bassnett, S. (1998). Where are we in Translation Studies? In Bassnett, S. & A. Lefevere: *Constructing Cultures. Essays on Literary Translation*. Clevedon: Multilingual Matters.

Leff, N. H. (1984). Externalities, Information Costs And Social Benefit-Cost Analysis For Economic Development; An Example From Telecommunications. *Economic Development and Cultural Change*, *32*, 255–276. doi:10.1086/451385

Leitner, C. (2003, July 7-8). E-government in Europe: the state of affairs. *Proceedings of the e-Government 2003 Conference*, Como.

Leki, I. (1990). Coaching from the margins: Issues in written response. In Kroll, B. (Ed.), *Second Language Writing* (pp. 57–68). Cambridge, UK: Cambridge University Press.

Lester, J., Stone, B., & Stelling, G. (1999). Lifelike Pedagogical Agents for Mixed-Initiative Problem Solving in Constructivist Learning Environments. *User Modeling and User-Adapted Interaction*, *9*, 1–44. doi:10.1023/A:1008374607830

Levin, C. (2000). Web Dropouts: Concerns About Online Privacy Send Some Consumers Off-Line, *PC Magazine*, http://www.zdnet.com/

LGAR. (2006). *Workforce Map of Local Government*. UK: Local Government Analysis and Research.

Lin, B., & Hsieh, C. (2001). Web-based Teaching and Learner Control: A Research Review. *Computers & Education*, *37*(3-4). doi:10.1016/S0360-1315(01)00060-4

Lin, B., & Hsieh, C. (2001). Web-based Teaching and Learner Control: A Research Review. *Computers & Education*, *37*(3-4), 377–386. doi:10.1016/S0360-1315(01)00060-4

Linders, L., & Goosens, (2004). Bruggen bouwen met virtuele middelen [Building bridges with virtual tools] In J. de Haan and O. Klumper (eds.) *Jaarboek ICT en samenleving: beleid in praktijk [ICT Yearbook and Society: Policy and Practice]* pp. 121-139. Amsterdam: Boom Publishers.

Lindgren, E. (2005). *Writing and Revising: Didactic and Methodological Implications of Keystroke Logging. (Skrifter från moderna språk, No. 18)*. Umeå, Sweden: Umeå University, Department of Modern Languages.

Lindgren, E., & Sullivan, K. P. H. (2003). Stimulated recall as a trigger for increasing noticing and language awareness in the L2 writing classroom: A case study of two young female writers. *Language Awareness, 12*, 172–186. doi:10.1080/09658410308667075

Lindgren, E., Stevenson, M., & Sullivan, K. P. H. (2008). Supporting the reflective language learner with computer keystroke logging. In Barber, B., & Zhang, F. (Eds.), *Handbook of Research on Computer Enhanced Language Acquisition and Learning* (pp. 189–204). Hershey, PA: IGI Global Inc.

Lindgren, E., Sullivan, K. P. H., Deutschmann, M., & Steinvall, A., (2009). Supporting learner reflection in the language translation class. *International journal of information technologies and human development*, 1(3),26-48.

Lingaard, G. (1999). *Does emotional appeal determine perceived usability of web sites?*www.cyberg.com

Liu, Meng-chun, & San, Gee (2006). Social learning and digital divides: A case study of internet technology diffusion. *Kyklos, 59*(2), 307–321.

Lobel, M., Neubauer, M., & Swedburg, R. (2005). Comparing how students collaborate to learn about the self and relationships in a real-time non- turn-taking online and turn-taking face-to-face environment. *Journal of Computer-Mediated Communication, 10*(4), 18. Available at http://jcmc.indiana.edu/vol10issue4/lobel.html.

Lockhart, C., & Ng, P. (1995). Analysing talk in ESL peer response groups: stances, functions and content. *Language Learning, 45*(4), 605–655. doi:10.1111/j.1467-1770.1995.tb00456.x

Lorenzo, M. (1999). Apuntes para una discusion sobre metodos de estudio del proceso de traduccion. In Gyde Hansen (Ed.), *Copenhagen Studies in Language, vol. 24: Probing the Process in Translation. Methods and Results* (pp. 21–42). Copenhagen: Samfundslitteratur.

Lorenzo, M. (2001). Combinación y contraste de métodeos de recogida y análisis de datos en el estudio del proceso de la traducción - Proyecto del grupo TRAP. Quaderns. *Revista de traducció, Barcelona, 6*, 33–38.

Lorenzo, M. (2002). ¿Es posible la traducción inversa? - Resultados de un experimento sobre traducción profesional a una lengua extranjera. In G. Hansen (Ed.), *Copenhagen Studies in Language, vol. 27: Empirical Translation Studies. Process and Product* (pp. 85–124). Copenhagen: Samfundslitteratur.

Lundquist, L. (2002*). L'anaphore associative: Etude contrastive et expérimentale de la traduction de l'anaphore associative du français en danois*. Romansk Forum XV, No. 16.

Lunt, B., Reichgelt, H., Ashford, T., Phelps, A., Slazinski, E., & Willis, C. (2003). *What is the new discipline of information technology? Where does it fit?* Session ETD 343, 2003 CIEC Conference, Tucson, Arizona. Available at http://faculty.csuci.edu/william.wolfe/csuci/create/asp/What_Is_IT.pdf.

Macdonald, J. (2003). Assessing online collaborative learning: process and product. [Elsevier Science.]. *Journal of Computers and Education, 40*(4), 377–391. doi:10.1016/S0360-1315(02)00168-9

MacFadyen, L. (2008). The perils of parsimony: "National culture" as red herring? *Proceedings of the Sixth International Conference on Cultural Attitudes towards Technology and Communication. Nîmes, France, June 2008*

Madon, S. (2000). The internet and socio-economic development: Exploring the interaction. *Information Technology & People, 13*(2), 85–101. doi:10.1108/09593840010339835

Maitland, C. F., & Bauer, J. M. (2001). National level culture and global diffusion: the case of the Internet. In Ess, C. (Ed.), *Culture, technology, communication: towards an intercultural global village* (pp. 87–128). Albany, NY: State University of New York Press.

Malhotra, N. K. (2009*)*. Marketing research: An applied orientation using SPSS. Upper Saddle River, NJ: Prentice Hall.

Malone, R. W., Yates, J., & Benjamin, I. (1987). Electronic Markets and Electronic Hierarchies. *Communications of the ACM, 30*(6), 484–497. doi:10.1145/214762.214766

Malone, T. W., Yates, J., & Benjamin, Y. R. I. (1987, June). Electronic Markets and Electronic Hierarchies. *Communications of the ACM, 30*(6), 484–497. doi:10.1145/214762.214766

Mangelsdorf, K. (1992). Peer response in the ESL classroom: What do the students think? *ELT Journal, 46*(3), 274–293. doi:10.1093/elt/46.3.274

Mansell, R., & Wehn, U. (1998). *Knowledge Societies: information technology for sustainable development.* New York: Oxford University Press (for United Nations Commission on Science and Technology for Development). ISBN 0198294107.

Mantha, R. W. (2001, May 4-6). Ulysses: Creating a Ubiquitous Computing Learning Environment. In *Proceedings of the Sharing Knowledge and Experience in Implementing ICTs in Universities EUA/IAU/IAUP Round Table*, Skagen.

Marchany, R. C., & Tront, J. G. (2002). *E-Commerce Security Issues*. Proceedings of the 35th Hawaii International Conference on System Sciences, Hawaii.

Markkula, M. (2006). Creating Favourable Conditions for Knowledge Society through Knowledge Management, e-Governance and e-Learning. A *Proceedings of FIG Workshop* (pp. 30-52), Budapest, Hungary.

Markus, B. (2008, June 11-13). *Thinking about e-Learning, A paper from the Proceedings of Sharing Good Practices: E-learning in Surveying, Geo-information Sciences and Land Administration FIG International Workshop 2008*, Enschede, The Netherlands.

Martin, J., & Cheong, P. (2008). Cultural Considerations of Online Pedagogy. In St.Amant, K., & Kelsey, S. (Eds.), *Computer-Mediated Communication across Cultures: International Interaction in Online Environments*. Hershey, PA: IGI Global.

Mason, D. D. M., & McCarthy, C. (2006). The feeling of exclusion: Young peoples' perception of art galleries. *Museum Management and Curatorship, 21*, 20–31.

Mataix, C., Moreno, A., & Acevedo, M. (2007). Estructuras en red: diseño y modelos para el Tercer Sector. Madrid: UNED (Spanish National Distance University), Fundación Luis Vives - *Module 8, 2007-2008 course on Strategic Management and Management Skills for Non-Profits Organizations.*

Mayer, R. C., Davis, J. H., & Schoorman, F. D. (1995). An integrative model of organizational trust. *Academy of Management Review, 20*(3), 709–734. doi:10.2307/258792

Mbarika, V., Jensen, M., & Meso, P. (2002). Cyberspace across sub-Saharan Africa: From technological desert towards emergent sustainable growth. *Communications of the ACM, 45*(12), 17–21.

McAlister, S., Ravenscroft, A., & Scanlon, E. (2004). Combining interaction and context design to support collaborative argumentation using a tool for synchronous CMC. *Journal of Computer Assisted Learning, 20*, 194–204. doi:10.1111/j.1365-2729.2004.00086.x

McIver, W. J., Jr. (2003). *The need for tools to support greater collaboration between transnational NGOs: implications for transnational civil society networking.* [Online]. State University of New York at Albany. Retrieved 20 Jul 2005 from <http://www.ssrc.org/programs/itic/publications/knowledge_report/memos/mcivermemo.pdf>

McMahon, K. (2005). *An exploration of the importance of website usability from a business perspective.* http://www.flowtheory.com/KTMDissertation.pdf

McNabb, D. E. (2006). *Knowledge Management in the Public Sector: A Blueprint for Innovation in Government.* M.E. Sharpe.

McSweeney, B. (2002). Hofstede's model of national cultural differences and their consequences: A triumph of faith – a failure of analysis. *Human Relations, 55*(1), 89–118.

MDL. (2006). *Local Government Structure and Efficiency. A report prepared for Local Government New Zealand.* New Zealand McKinlay Douglas Limited.

Mehadevan, B., & Venkatesh, N. S. (2000). *A framework for building on-line trust for business to business e-commerce.* In Proceedings of The IT Asia Millennium Conference,Bombay, India.

Mendonça, C. O., & Johnson, K. E. (1994). Peer review negotiations: Revision activities in ESL writing instruction. *TESOL Quarterly, 28*(4), 745–769. doi:10.2307/3587558

Miceli, T. (2006, December). Foreign language students' perceptions of a reflective approach to text correction. *Flinders University Languages Group Online Review, 3*(1). Retrieved October 3 2008 from http://ehlt.flinders.edu.au/deptlang/fulgor/volume3i1/papers/Miceli_v3i1.pdf

Millán, J. A. (2000). *La lengua que era un tesoro: el negocio del español y como nos quedamos sin el*. Retrieved December 31, 2007, from http://jamillan.com/tesoro.htm

Miller, E., & Findlay, M. (1996). *Australian thesaurus of educational descriptors*. Melbourne: Australian Council for Educational Research.

Miller, H. N. (2004). *Presentation by World Information Technology and Services Alliance (WITSA)*. WITSA President Harris N. Miller at the September 27, 2004 WITSA Steering Committee meeting in Bakubung, South Africa.

Min, H.-T. (2005). Training students to become successful peer reviewers. *System, 33*(2), 293–308. doi:10.1016/j.system.2004.11.003

Mindham, C. (1998). Peer assessment: report of a project involving group presentations and assessment by peers. In Brown, S. (Ed.), *Peer assessment in practice* (pp. 45–66). Birmingham: SEDA Administrator.

Misra, D. C. (2008). Emerging E-government Challenges: Past Imperfect, Present Tense but Future Promising. *A presentation at the 4th International Conference on E-government*, Indian Institute of Technology, Delhi.

MISTICA. (1999-2006). *Methodology and Social Impact of Information and Communication Technologies in America.* Retrieved December 29, 2007, from http://funredes.org/mistica

Misuraca, G. (2007). *E-Governance in Africa, from Theory to Action: A Handbook on ICTs for Local Governance*. IDRC.

Mohino, E., Gende, M., Brunini, C., & Heraiz, M. (2005). SiGOG: simulated GPS observation generator. *GPS Solutions, 9*, 250–254. doi:10.1007/s10291-005-0001-9

Moore, G. C., & Benbasat, I. (1991). Development of an instrument to measure the perceptions of adopting an information technology innovation. *Information Systems Research, 2*(3), 192–222. doi:10.1287/isre.2.3.192

Morales-Gomez, D., & Melesse, M. (1998). Utilizing information and communication technologies for development: The social dimensions. *Information Technology for Development, 8*(1), 3–14. doi:10.1080/02681102.1998.9525287

Moreno, A., Acevedo, M., & Mataix, C. (2006). *Redes 2.0 La articulación de las ONGD en España.* Madrid: Coordinadora de ONG para el Desarrollo-España (CONGDE).

Morgan, G., & Sturdy, A. (2000). *Beyond Organizational Change: Structure, Discourse and Power in UK Financial Services*. London: Macmillan.

Morse, K. (2003). Does one size fit all? Exploring asynchronous learning in a multicultural environment. *Journal of Asynchronous Learning Networks, 7*(1), 37–55.

Moyo, L. M. (1996). Information technology strategies for Africa's survival in the twenty-first century: IT all pervasive. *Information Technology for Development, 7*(1), 17–21. doi:10.1080/02681102.1996.9627211

Mukherjee, A., & Nath, P. (2003). A model of trust in online relationship banking. *International Journal of Bank Marketing, 1*(21), 5–15. doi:10.1108/02652320310457767

Mukhopadhyay, T., Kekre, S., & Kalathur, S. (2002). Business Value of Information Technology: A Study of Electronic Data Interchange. *Management Information Systems Quarterly*, (June): 137–157.

Mulama, J. (2009). *Technology-Africa: A rural-urban digital divide challenges women*. IPS. Available at http://ipsnews.net/news.asp?idnews=36563; last accessed April 22, 2009.

Muller, M. J., Wharton, C., McIver, W. J., & Laux, L. (1997). Toward an HCI research and practice agenda based on human needs and social responsibility. *Proceedings of the SIGCHI Conference on Human Factors in Computing Systems (Atlanta, Georgia, United States, March 22 - 27, 1997).* S. Pemberton, Ed. CHI '97, pp. 155-161. New York: ACM Press

Mursu, A., Sorinyan, A., & Korpela, M. (2003). *ICT for development: sustainable systems for local needs.* Proceedings of IFIP WG 8.2 & 9.4 Joint Conference, Athens, Greece.

Mursu, A., Sorinyan, A., & Korpela, M. (2004). *A generic framework for analyzing the sustainability of information systems.* In Proceedings of the 10th Americas Conference on Information Systems, New York, 934-941.

Mutula, S. M., & Wamukoya, J. (2007). *E-Government Readiness in East and Southern Africa*. In: Encyclopedia of Digital Government. Hershey, PA: Idea Group Inc.

Nah, S. H. (2006). *Breaking Barriers: The potential of Free and Open Source software for sustainable human development. UNDP-APDIP ICT4D Series* (pp. 3–16). Elsevier Publications.

Naidu, S., & Jasen, C. (2007). *Australia: ICT in Education, UNESCO Meta-survey on the Use of Technologies in Education*. Bangkok: UNESCO.

Nair, K. R. G. (1995). Telecommunications in India. *Productivity, 36*(2), 209–214.

Nante, J., & Glaser, E. (2005). *The Impact of Language and Culture on Perceived Website Usability*. http://www.rbcchair.com/chairerbc/fichiers/jetus.pdf

Narain, U., Gupta, S., & van't Veld, K. (2008). Poverty and the environment: Exploring the relationship between household incomes, private assets, and natural assets. *Land Economics, 84*(1), 148–167.

Nasir Uddin, M. (2003). Internet use by university academics: A bipartite study of information and communications needs. *Online Information Review, 27*(4), 225–237. doi:10.1108/14684520310489014

Nass, C., Moon, Y., & Carney, P. (1999). Are respondents polite to computers? Social desirability and direct responses to computers. *Journal of Applied Social Psychology, 29*, 1093–1110. doi:10.1111/j.1559-1816.1999.tb00142.x

Nath, V. (2000). *Knowledge Networking for Sustainable Development*. [Online]. Knownet Initiative, www.knownet.org. Retrieved 14 November 2001 from <http://www.cddc.vt.edu/knownet/articles/exchanges-ict.html>

Neilsen Normen group. (2000). *Trust: Design guidelines for e-commerce user experience*. Retrieved from www.nngroup.com

Nielsen, J. (1990). Navigation Through Hypertext. *Communications of the ACM, 22*(3), 296–310. doi:10.1145/77481.77483

Nielsen, J. (1993). *Usability Engineering*. New York: Academic Press.

Nielsen, J. (1999). *Designing Web Usability: the Practice of Simplicity*. New Riders.

Nielsen, J. (1999). *Trust or Bust: Communicating Trustworthiness in Web Design. Jacob Nielsen's Alertbox*. http://www.useit.com/alertbox/990307.html.

Nielsen, J. (August 19, 2001). *Did Poor Usability Kill E-Commerce*? Jacob Nielsen's Alertbox, http://www.useit.com/alertbox/20010819.html.

Nielsen, J. (May 12, 2002). *Top Ten Guidelines for Homepage Usability*. Jacob Nielsen's Alertbox. http://www.useit.com/alertbox/20020512.html

Nielsen, J. (May 16, 1999). *Who Commits the 'Top Ten Mistakes' of Web Design?* Jacob Nielsen's Alertbox, http://www.useit.com/alertbox/990516.html.

O'Malley, M. (2003). *Sustainable rural development. Sustainable Ireland*. Available at www.sustainable.ie/resources/community/art03.htm [Accessed Thursday, September 30, 2004]

O'Neil, D. (2002). Assessing community informatics: A review of methodological approaches for evaluating community networks and community technology centers. *Internet Research: Electronic Networking Applications and Policy, 12*(1), 76–102. doi:10.1108/10662240210415844

Odedra-Straub, M., Lawrie, M., Bennett, M., & Goodman, S. (1993). International perspectives: Sub-Saharan Africa: A technological desert. *Communications of the ACM, 36*(2), 25–29. doi:10.1145/151220.151222

OECD. (2007a). *OECD E-government Studies. Organisation for Economic Co-operation and Development*. Turkey: OECD Publishing.

OECD. (2007b). *OECD E-government Studies. Organisation for Economic Co-operation and Development*. Hungary: OECD Publishing.

Ogbonna, E., & Wilkinson, B. (2003). The false promise of organizational culture change: a case study of middle managers in grocery retailing. *Journal of Management Studies, 40*(5), 1151–1178. doi:10.1111/1467-6486.00375

Okot-Uma, W'O R. (2004). *Building cyberlaw capacity for eGovernance: Technology perspectives*. The Commonwealth Centre for e-Governance, London, United Kingdom.

Okunoye, A., Bada, A., Pick, J., & Adewumi, S. (2003). *Call for more information systems education in developing countries: Perspectives from information systems researchers and practitioners*. In Palvia P. & Liu X. (eds.), Proceedings of the 4th Global Information Technology Management World Conference, 319, June 8-10, Calgary.

Olaniran, B. A. (1994). Group performance and computer-mediated communication. *Management Communication Quarterly, 7*, 256–281. doi:10.1177/0893318994007003002

Olaniran, B. A. (1996). Social Skills Acquisition: A closer look at foreign students and factors influencing their level of social difficulty. *Communication Studies, 47*, 72–88.

Olaniran, B. A. (2006). Applying synchronous computer-mediated communication into course design: Some considerations and practical guides. *Campus-Wide Information Systems. The International Journal of Information & Learning Technology, 23*(3), 210–220.

Olaniran, B. A. (2009). Culture and Language Learning in Computer-Enhanced or Assisted Language Learning. *International Journal of Communication Technologies and Human Development, 1*(3), 49–67. doi:10.4018/jicthd.2009070103

Olaniran, B. A., Savage, G. T., & Sorenson, R. L. (1996). Experiential and experimental approaches to face-to-face and computer mediated communication in group discussion. *Communication Education, 45*, 244–259. doi:10.1080/03634529609379053

Olaniran, B. A., Stalcup, K., & Jensen, K. (2000). Incorporating Computer-mediated technology to strategically serve pedagogy. *Communication Teacher, 15*, 1–4.

Olaniran, B. (2006). Challenges to implementing e-learning and lesser developed countries. In Edmundson, A. L. (Ed.), *Globalized e-learning cultural challenges* (pp. 18–34). Hershey, PA: Idea Group, Inc.

Olaniran, B. A. (2001). The effects of computer-mediated communication on transculturalism. In Milhouse, V., Asante, M., & Nwosu, P. (Eds.), *Transcultural Realities* (pp. 83–105). Thousand Oaks, CA: Sage.

Olaniran, B. A. (2004). Computer-mediated communication as an instructional Learning tool: Course Evaluation with communication students. In P. Comeaux (Ed.), *Assessing Online Teaching & Learning*, 144-158.

Olaniran, B. A. (2007). Culture and communication challenges in virtual workspaces. In K. St-Amant (ed.), *Linguistic and cultural online communication issues in the global age* (pp. 79-92*)*. Hershey, PA: Information science reference (IGI Global).

Oliver, R., & Omari, A. (2001). Exploring Student Responses to Collaborating and Learning in a Web-Based Environment. *Journal of Computer Assisted Learning, 17*(1), 34–47. doi:10.1046/j.1365-2729.2001.00157.x

Olson & Olson. (2000). i2i trust in e-commerce. *Communications of the ACM, 43*(12), 41–44. doi:10.1145/355112.355121

Online, S. (2008). *Handy für Blinde liest gedruckte Texte vor.* http://www.spiegel.de/netzwelt/mobil/0,1518,532014,00.html, last visited 2010-09-24

Open Society Institute. (2007). *ICT for Local Government Handbook*. Budapest: Local Government and Public Service Reform Initiative. Open Society Institute.

Opera Software ASA. (2004). *New version of Opera Embeds ViaVoice from IBM.* http://www.opera.com/press/releases/2004/03/23/, last visited 2010-09-24

Ortega, J., Koppel, M., & Argamon, S. (2001). Arbitrating among competing classifiers using learned referees. *Knowledge and Information Systems, 4*, 470–490. doi:10.1007/PL00011679

Osman, G., & Herring, S. (2007). Interaction, facilitation, and deep learning in cross-cultural chat: A case study. *The Internet and Higher Education, 10*(2), 125–141. doi:10.1016/j.iheduc.2007.03.004

Owston, R., Widerman, H., Sinitskaya Ronda, N., & Brown, C. (2009). Computer game development as a literacy activity. *Computers & Education, 53*(3), 977–989. doi:10.1016/j.compedu.2009.05.015

Oyelami, M. (2004, 2010) *Personal Communication and Interview Transcript*, June/July.

Oyomno, G. (1996). Sustainability of governmental use of micro-computer based information technology in Kenya. In Odedra-Straub, M. (Ed.), *Global Information Technology and Socio-Economic Development*. Nashua, USA: Ivy League Publishing.

Pacific Council. (2002). *Roadmap for E-government in the Developing World: 10 Questions E-government leaders Should Ask Themselves. The Working Group report on E-government in the Developing World.* CA: Pacific Council on International Policy.

Padromidis, A., & Stolfo, S. (2000). Meta learning in distributed data mining systems: Issues and approaches. In H. Kargupta & P. Chan (Eds.), *Advances in distributed and parallel knowledge discovery* (pp. 81-114). Cambridge, MA: MIT Press.

Pan, Z., Cheok, A. D., Yang, H., Zhu, J., & Shi, J. (2006). Virtual reality and mixed reality for virtual learning environments. *Computer Graphics, 30*, 20–28. doi:10.1016/j.cag.2005.10.004

Panitz, T. (1997). Collaborative Versus Cooperative Learning: Comparing the Two Definitions Helps Understand the nature of Interactive learning. *Cooperative Learning and College Teaching, V8*(2).

Panitz, T. (1998). *Collaborative versus cooperative learning - a comparison of the two concepts that will help us understand the underlying nature of interactive learning.* Available at http://www.capecod. Net/

Pantano, E., & Servidio, C. (in press). The role of pervasive environments for promotion of tourist destinations: the users' response. *Journal of Hospitality and Tourism Technology*.

Pantano, E., & Tavernise, A. (2009). Learning Cultural Heritage through Information and Communication Technologies: a case study. *International Journal of Information Communication Technologies and Human Development, 1*(3), 68–87. doi:10.4018/jicthd.2009070104

Paraskeva, F., Mysirlaki, S., & Papagianni, A. (2010). Multiplayer online games as educational tools: Facing new challenges in learning. *Computers & Education, 54*, 498–505. doi:10.1016/j.compedu.2009.09.001

Pardo, T. A. (2000, October). Realizing the promise of digital government: It's more than building a web site. *IMP Magazine*.

Parry, R. (2005). Digital heritage and the rise of theory in museum computing. *Museum Management and Curatorship, 20*, 333–348.

Parsons, S., Leonard, A., & Mitchell, P. (2006). Virtual environments for social skills training: comments from two adolescents with autistic spectrum disorder. *Computers & Education, 47*, 186–206. doi:10.1016/j.compedu.2004.10.003

Parthasarathy, S., Zaki, M. J., Ogihara, M., & Li, W. (2001). Parallel data mining for association rules on shared memory systems. *Knowledge and Information Systems, 1*, 1–29. doi:10.1007/PL00011656

Patri, M. (2002). The influence of peer feedback on self- and peer-assessment of oral skills. *Language Testing, 19*(2), 109–133. doi:10.1191/0265532202lt224oa

Patsula, P. J. (2002). Practical guidelines for selecting media: An international perspective. *Usableword Monitor* (February 1). Available at: http:/uweb.txstate.edu/~-db15/edtc5335/docs/mediaselection_criteria.htm

Patton, M. A., & Jøsang, A. (2004). Technologies for Trust in Electronic Commerce. *Electronic Commerce Research, 4*(1), 9–21. doi:10.1023/B:ELEC.0000009279.89570.27

Pawar, U. S., Pal, J., & Toyama, K. (2006). Multiple mice for computers in education in developing countries. *Proc. IEEE/ACM ICTD* (pp. 64-71).

Pearlson, K., & Saunders, C. (2006). *Managing and Using Information Systems: A Strategic Approach* (3rd ed.). New York: John Wiley.

Pellegrini, U. (1980). The problem of appropriate technology. In Criteria for Selecting Appropriate Technology under Different Cultural, Technical and Social Conditions, Roveda, D. (ed.), *Proceedings of the IFAC Symposium*, Pergamon Press, 1-5.

Peppard, J., & Ronald, P. (1995). *The Essence of Business Process Re-Engineering*. Upper Saddle River, NJ: Prentice-Hall.

Perotti, E.C., & von Thadden, E-L. (2006, March). Corporate Governance and the Distribution of Wealth: A Political-Economy Perspective. *Journal of Institutional and Theoretical Economics JITE, 162*(1), 204-217(14). Mohr Siebeck.

Perry, D. (2003). *Handheld Computers (PDAs) in Schools*. Coventry, UK: Becta ICT Research, British Educational Communications and Technology Agency.

Peter, U. (2006): Accessible E-Government through Universal Design. In Anttiroiko, Ari-Veikko; Mäliä, Matti (2006) *Encyclopedia of Digital Government*, Hershey, PA: Idea Group Inc., pp. 16-19

Petric, J., Ucelli, G., & Conti, G. (2003). Real Teaching and Learning through Virtual Reality. *International Journal of Architectural Computing, 1*(1), 2–11. doi:10.1260/147807703322467289

Phan, M. (2007). *Customisable Checklist to evaluate Learning Management Systems regarding Speci_c Requirements in Viet Nam*. Diploma Thesis for the Department of Computer Science, Albert-Ludwigs-Universität Freiburg, Germany

Pianesi, F., Graziola, I., Zancanaro, M., & Goren-Bar, D. (2009). The motivational and control structure underlying the acceptance of adaptive museum guides-An empirical study. *Interacting with Computers, 21*(3), 186–200. doi:10.1016/j.intcom.2009.04.002

Pick, J., & Azari, R. (2008). Global digital divide: Influence of socioeconomic, governmental, and accessibility factors on information technology. *Information Technology for Development, 14*(2), 91–115. doi:10.1002/itdj.20095

Pickering, A. (2005). *Facilitating reflective learning: an example of practice in TESOL teacher education.* Retrieved May 29 2008 from http://www.llas.ac.uk/resources/goodpractice.aspx?resourceid=2395

Pieraccini, M., Guidi, G., & Atzeni, C. (2001). 3D digitizing of cultural heritage. *Journal of Cultural Heritage, 2*, 63–70. doi:10.1016/S1296-2074(01)01108-6

Pimienta, D. (1993, August). *Research Networks in Developing Countries: Not exactly the same story*! Paper presented at INET'93, San Francisco, CA. Retrieved December 30, 2007, from http://funredes.org/english/publicaciones/index.php3/docid/31

Pimienta, D. (2002). *The Digital Divide: the same division of resources*? Retrieved December 29, 2007, from http://funredes.org/mistica/english/cyberlibrary/thematic/eng_doc_wsis1.html

Pimienta, D. (2005, June). *Una vision desde la sociedad civil.* Presentation made at the Regional Preparatory Conference of Latin America and the Caribbean for the Second Phase of the World Summit on the Information Society, Rio de Janeiro, Brazil. http://www.riocmsi.gov.br/english/cmsi

Pimienta, D. (2006). Measuring linguistic diversity on the Internet. *UNESCO.* Retrieved December 31, 2007, from http://unesdoc.unesco.org/images/0014/001421/142186e.pdf

Pimienta, D. (2007). At the Boundaries of Ethics and Cultures: Virtual Communities as an Open Ended Process Carrying the Will for Social Change (the "MISTICA" experience). In Capurro, R. & al. (Eds.), *Localizing the Internet. Ethical Issues in Intercultural Perspective* (pp. 205-229). München: Fink Verlag. Online version retrieved December 31, 2007, from http://funredes.org/mistica/english/cyberlibrary/thematic/icie/

Pimienta, D., & al. (1993-2007). *Presentaciones de Funredes (y de sus asociados en proyectos).* Retrieved December 29, 2007, from http://www.funredes.org/presentation/

Pimienta, D., & Blanco, A. (2007). *The Hurdle Track from ICT to Human Development, first published in English at Global Knowledge Partnership Beyond Tunis: Flightplan.* Retrieved December 30, 2007, from http://www.globalknowledge.org/gkpbeyondtunis/INDEX.CFM?menuid=33&parentid=30

Pimienta, D., & Dhaussy, C. (1999). *Users Training: A Crucial but Ignored Issue in Remote Collaborative Environments.* Retrieved December 31, 2007, from http://www.isoc.org/inet99/proceedings/posters/157/index.htm

Pironti, J.P. (2006, May). Key Elements of a Threat and Vulnerability Management Program - Information Systems Audit and Control Association Member Journal, ISACA, 6.

Pitsis, T.S., Kornberger, M., & Clegg, S. (2004). The Art of Managing Relationships in Interorganizational Collaboration. *M@n@gement, 7*(3), 47-67. Special Issue: Practicing Collaboration.

Planet, D. (2004). *Digital Planet 2004: The Global Information Economy, a biannual report of the World Information Technology and Services Alliance*. VA, USA: WITSA.

Planet, D. (2008). *Digital Planet 2008: The Global Information Economy, Executive Summary of the report of the World Information Technology and Services Alliance*. VA, USA: WITSA.

Posner, G. J. (1996). *Field Experience: A Guide to Reflective Teaching*. White Plains, NY: Longman.

Prague Declaration. (2003). *Towards an Information Literate Society*. Retrieved January 31, 2008, from http://www.nclis.gov/libinter/infolitconf&meet/post-infolitconf&meet/PragueDeclaration.pdf

Proenza, F., Bastidas-Buch, R., & Montero, G. (2001). *Telecentres for Socio-Economic and rural Development in Latin America and the Caribbean*. Washington, D.C.

Puri, S. K., & Sahay, S. (2007). Role of ICTs in participatory development: An Indian experience. *Information Technology for Development*, *13*(2), 133–160. doi:10.1002/itdj.20058

Putnam, R. (2000). *Bowling alone, the collapse of and revival of civic America*. New York: Simon & Schuster.

Race, P. (1998). Practical pointers on peer assessment. In Brown, S. (Ed.), *Peer assessment in practice* (pp. 108–113). Birmingham: SEDA Administrator.

Rahman, H. (2005a). Distributed Learning Sequences for the Future Generation. In Howard, (Eds.), *Encyclopedia of Distance Learning* (pp. 669–673). Hershey, PA: Idea Group.

Rahman, H. (2005b). Virtual Networking: An essence of the Future Learners. In Howard, (Eds.), *Encyclopedia of Distance Learning* (pp. 1972–1976). Hershey, PA: Idea Group.

Rahman, H. (2008). An overview on Strategic ICT Implementations Toward Developing Knowledge Societies. In Rahman, H. (Ed.), *Developing Successful ICT Strategies: Competitive Advantages in a Global Knowledge-Driven Society* (pp. 1–39). Hershey, PA: IGI.

Rahman, H. (2007). Interactive Multimedia technologies for Distance Education Systems. In Tomei, L. (Ed.), *Online and Distance Learning: Concepts, Methodologies, Tools and Applications* (pp. 1157–1164). Information Science Reference.

Rahman, H. (2001, October 16-19*). Utilization of ICT Infrastructure for Collaborative Learning through Community Participation.* A paper presented at the IFUP2001, 4th International Forum on Urban Poverty, Marrakech, Morocco. http://www.ifup.org

Rahman, H. (2007). *Community-Based Information Networking in Developing Countries*. In: Encyclopedia of Digital Government. Hershey, PA: Idea Group Inc.

Rajaee, F. (2000). *Globalization on Trial. The Human Condition and the Information Civilization*. Ottawa: International Development Center.

Rajiv, C. S., & Jay, P. K. (2003, Summer). Incorporating Societal Concerns into Communication Technologies. *IEEE Technology and Society Magazine*, 31–33.

Rama, R. T., Venkat, R., Bhatnagar, S. C., & Satyanarayan, J. (2004). *E-Assessment Frameworks*. Retrieved on Apri 17, 2009 from http://egovt.mit.gov.in.

Ranganathan, C. (2003). Evaluating the options for Business-to-Business E-Exchanges. *Information Systems Management*, *20*(3), 22–28. doi:10.1201/1078/43205.20.3.20030601/43070.3

Ranganathan, C., & Ganapathy, S. (2002). Key dimensions of business-to consumer websites. *Information & Management*, *39*, 457–465. doi:10.1016/S0378-7206(01)00112-4

Ransdell, S. E. (1990). Using real-time replay of students' word processing to understand and promote better writing. *Behavior Research Methods, Instruments, & Computers*, *22*(2), 142–144.

Rastogi, R., & Shim, K. (1999). Mining optimized support rules for numeric attributes. Proceedings ACM SIGMOID, Conference on Management of Data.

Ratnasingham, P., & Klein, S. (2001). *Perceived Benefits of Inter-Organizational Trust in Ecommerce Participation: A Case Study in the Telecommunication Industry*. Proceedings of the Seventh Americas Conference on Information Systems.Boston, Massachusetts, pp. 769-780.

Reichgelt, H., Lunt, B., Ashford, T., Phelps, A., Slazinski, E., & Willis, C. (2004). A comparison of baccalaureate programs in information technology with baccalaureate programs in computer science and information systems. *Journal of Information Technology Education*, *3*, 19–34. Available at http://jite.org/documents/Vol3/v3p019-034-098.pdf.

Reinders, H., & Lazaro, N. (2007, April). Innovation in language support: The provision of technology in self access. *Computer Assisted Language Learning*, *20*(2), 117–130. doi:10.1080/09588220701331428

Report, V. (2001). The Role of Institutions in the Design of Communication Technologies, Telecommunications Policy Research Conference, Alexandria, Virginia Report No: TPRC-2001-086: Rajiv, C. S. & Jay, P. K, 19 - 21.

Ribbink, D., Allard, C. R., Liljander, V. V., & Treukens, S. (2004). Comfort your online customer: quality, trust and loyalty on the internet. *Managing*, *14*(6), 446–456.

Richards, C. (2004). Information technology and rural development. *Progress in Development Studies*, *4*(3), 230–244. doi:10.1191/1464993404ps087oa

Richter, S. (2003). Design and Implementation of a communication module for blind and visually impaired humans. *Diploma thesis.* Birkenfeld.

Ridley, C. E., & Nolting, O. F. (2003). *The Municipal Year Book. International City Managers' Association.* International City Management Association.

Riegelsberger J & Sasse, M.A.(2001). *Trust builders and Trustbusters: The Role of Trust Cues in Interfaces to e-Commerce Applications.*

Riggins, F. J., & Mukhopadhyay, T. (1999). Overcoming EDI adoption and Implementation Risks. *International Journal of Electronic Commerce*, *3*(4), 103–113.

Riggins, F. J., & Rhee, H.-S. (1998). Toward a Unified View of Electronic Commerce. *Communications of the ACM*, *40*(10), 88–95. doi:10.1145/286238.286252

Rischard, J. F. (2003). Integrating ICT in development programs. [Online]. *Keynote speech, Joint OECD/UN/World Bank Global Forum: Integrating ICT in Development Programmes*. Paris: OECD, 4-5 March 2003. Retrieved 14 Jun 2003 from www.oecd.org/dac/ictcd/docs/otherdocs/Forum_0303_summary.pdf

Ritter, L. (1998). Peer assessment: lessons and pitfalls. In Brown, S. (Ed.), *Peer assessment in practice* (pp. 79–86). Birmingham: SEDA Administrator.

Roach, K. D., & Olaniran, B. A. (2001). Intercultural willingness to communicate and communication anxiety in International Teaching Assistants. *Communication Research Reports*, *18*, 26–35.

Robbins, C. (2004). *Educational Multimedia for the South Pacific, Research Report for ICT Capacity Building at USP Project entitled "Maximising the Benefits of ICT/Multimedia in the South Pacific: Cultural Pedagogy and Usability Factors*. Fiji: Prepared for ICT Capacity Building at the University of the South Pacific Media Centre.

Roberts, T. S. (2004). *Online Collaborative Learning: Theory and Practice*. Information Science Publishing.

Roberts, R. (2005). Issues in Modeling Innovation Intense Environments: The Importance of the Historical and Cultural Context. *Technology Analysis and Strategic Management*, *17*(4), 477–495. doi:10.1080/09537320500357384

Robertson, J. (2004). *Successfully deploying a content management system*. Step Two Designs Pty Ltd, www.steptwo.com.au.

Robinson, K. K., & Crenshaw, E. M. (1999). *Cyber-space and post-industrial transformations: A cross-national analysis of Internet development*. Retrieved May 10, 2004, from http://www.soc.sbs.ohio-state.edu/emc/RobisonCrenshawCyber1a.pdf

Rogers, E. (1995). *Diffusion of Innovation*. New York: The Free Press.

Rogers, E. M. (2003). *Diffusion of innovations*. New York; London: Free Press.

Roman, R., & Colle, R. (2003). Content creation for ICT development projects: Integrating normative approaches and community demand. *Information Technology for Development*, *10*(2), 85–94. doi:10.1002/itdj.1590100204

Roschelle, J. (2003). Unlocking the learning value of wireless mobile devices. *Journal of Computer Assisted Learning*, *19*(3), 260–272. doi:10.1046/j.0266-4909.2003.00028.x

Rothe-Neves, R. (2003). The influence of working memory features on some formal aspects of translation performance. In Alves, F. (Ed.), *Triangulating Translation. Perspectives in Process Oriented Research* (pp. 97–119). Amsterdam: Benjamins.

Rotter, J. B. (1980). Impersonal trust, Trustworthiness and gullibility. *The American Psychologist*, *35*(1), 1–7. doi:10.1037/0003-066X.35.1.1

Roussou, M. (2002). *Virtual Heritage: From the Research Lab to the Broad Public*. Oxford, UK: Archaeopress.

Rovai, A. P. (2000). Building and sustaining community in learning network. *The Internet and Higher Education, 3*, 285–297. doi:10.1016/S1096-7516(01)00037-9

Roy, M. C., Dewit, C., & Auber, B. A. (2001). The *Impact of Interface Quality on Trust in Web Retailers*. http://www.cirano.qc.ca/pdf/publication/2001s-32.pdf

Sahana – Sahana FOSS disaster management system (2007). Retrieved December 22, 2008 from http://en.wikipedia.org/ wiki/Sahana_FOSS_ Disaster_Management_ System

Saith, A., & Vijaybharkar, M. (2005). *ICTs and Indian Economic Development*. New Delhi: Sage Publication.

Salaman, G., & Asch, D. (2003). *Strategy and Capability: Sustaining Organizational Change*. Oxford: Blackwell.

Salaman, G. (1997). Culturing Production. In Du Gay, P. (Ed.), *Production of Culture/Cultures of Production*. London: Sage.

Salmon, G. (2000). *E-moderating: the key to teaching and learning online*. London: Kogan Page.

Salmon, G. (2004). *E-tivities. Der Schlussel zu aktivizen online-lernen*. Zurich: Orell Fussili Verlag AG.

Samarajiva, R., & Zainudeen, A. (2008). (Eds.) *ICT Infrastructure in Emerging Asia: Policy and Regulatory Roadblocks*. SAGE Publications & IDRC.

Saran, P. S. (1999). Rural Telecommunications. *Telecommunications*. DoT. India.

Sarita, S (2006). eNRICH: Archiving and accessing local information, International Journal of Education and Development using Information and Communication Technology (IJEDICT), 2(1), 34-48.

Saxena, K. K., & Satyananda, S. (1997). Infrastructure and Economic Development: Some Empirical Evidence. *The Indian Economic Journal, 47*(2).

Sayal, M., & Scheuermann, P. (2001). Distributed web log mining using maximal large data sets. *Knowledge and Information Systems, 4*, 389–404. doi:10.1007/PL00011675

Sayo, P., Chacko, J. G., & Pradhan, G. (2004). (Eds.) *ICT Policies and e-Strategies in the Asia-Pacific: A critical assessment of the way forward*. United Nations Development Programme-Asia Pacific Development Information Programme (UNDP-APDIP) – 2004. Reed Elsevier India Private Limited

Scaife, M., & Rogers, Y. (2001). Informing the design of a virtual environment to support learning in children. *International Journal of Human-Computer Studies, 55*, 115–143. doi:10.1006/ijhc.2001.0473

Scala, S., & McGrath, R. (1993). Advantages and disadvantages of electronic data interchange: An industry perspective. *Information & Management, 25*, 85–91. doi:10.1016/0378-7206(93)90050-4

Schank, R. C., & Cleary, C. (1995). *Engines for education*. Hillsdale, NJ: Lawrence Erlbaum Associates.

Schellens, T., & Valcke, M. (2006). *Collaborative Learning in Synchronous Discussion Groups: What about the Impact of Cognitive Processing. Computers in Human behaviour*. Cape Town, South Africa: University of Cape Town.

Schlömer, T., Poppinga, B., Henze, N., & Boll, S. Gesture recognition with a Wii controller. Proceedings of the 2nd International Conference on Tangible and Embedded Interaction 2008, Bonn, Germany, February 18-20, 2008. ACM 2008, ISBN 978-1-60558-004-3

Schmidt, I. (2005). *Blended E-Learning*. Saarbrücken: VDM Verlag.

Schware, R. (2005). *E Development: From Excitement to Effectiveness*. USA: World Bank Publications.

Sen, A. (1999). *Development as Freedom*. London: Oxford University Press.

Shackleton, P., Fisher, J., & Dawson, L. (2004). E-Government Services: One Local Government's Approach. In Linger, H. (Eds.), *Constructing the Infrastructure for the Knowledge Economy: Methods and Tools, Theory and Practice*. Springer.

Shah, A. (1997). Telecommunication: Monopoly vs Competition. *India Development Report* (pp. 239-240). Indira Gandhi Institute of Development Research.

Shakya, S., & Rauniar, D. (2002). Information technology education in Nepal: An inner perspective. *Electronic Journal of Information Systems in Developing Countries, 8*(5), 1–11.

Shih, C., & Venkatesh, A. (2003). *A comparative study of home computer adoption and use in three countries: US, Sweden, and India.* Centre for Research on Information Technology and Organizations. Retrieved December 05, 2007, from www.crito.uci.edu/noah/paper/MISPaper-forWeb.pdf

Siau, K. (2003). Interorganizational systems and competitive advantages - lessons from history. *Journal of Computer Information Systems, 44*, 33–40.

Silvio, J. (2001, May). Building a typology for the comparative analysis of virtual universities worldwide. *Sharing Knowledge and Experience in Implementing ICTs in Universities Roundtable Papers (7), IAU/IAUP/EUA* Skagen Roundtable.

Siochrú, S. Ó., & Girard, B. (2005). *Community-based Networks and Innovative Technologies: New models to serve and empower the poor.* New York: UNDP, 'Making ICT Work for the Poor' Series.

Skillicon, D. B., & Wang, Y. (2001). Parallel and sequential algorithms for data mining using inductive logic. *Knowledge and Information Systems, 4*, 405–421. doi:10.1007/PL00011676

Slack, N., Chambers, S., Johnston, R., & Betts, A. (2006). *Operations and Process Management: Principles and Practice for Strategic Impact.* Essex, England: Pearson Education.

Slack, J. D., & Wise, J. M. (2002). Cultural studies and technology. In Livingstone, S., & Lievrouw, L. (Eds.), *Handbook of new media* (pp. 221–235). London: Sage.

Slator, B.M., Juell, P., McClean, P.E., Saini-Eidukat B., Schwert, D.P., White, A. R., & Hill, C. (1999). Virtual environments for education. *Journal of Network and Computer Applications, 22*([REMOVED HYPERLINK FIELD]3), 161-174.

Small, G. (2008). *iBrain: Surveying the Technological Alteration of the Modern Mind.* New York: HarperCollins.

Smith, B. (2004). Computer-mediated negotiated interaction and lexical acquisition. *Studies in Second Language Acquisition, 26*, 365–398. doi:10.1017/S027226310426301X

Smith, B., Alvarez-Torres, M., & Zhao, Y. (2003). Features of CMC technologies and their impact on language learners' online interaction. *Computers in Human Behavior, 19*, 703–729. doi:10.1016/S0747-5632(03)00011-6

Smith, P. B. (2002). Culture's consequences: Something old and something new. *Human Relations, 55*(1), 119–135.

Smith, T. J. Spiers, R. (2008). Perceptions of E-commerce Web sites across two generations. *Informing Science: the International journal of an emerging Transdiscipline,* 12, 159-179

Socitm (2003). *Managing e-government – a discussion paper.* A Socitm insight publication. Socitm. www.socitm.gov.uk

Sørensen, K. H. (2004). Cultural Politics of Technology: Combining Critical and Constructive Interventions. *Science, Technology & Human Values, 29*(2), 184–190. doi:10.1177/0162243903261944

Sorensen, E. K., & Takle, E. S. (2001). Collaborative Knowledge Building in Web-based Learning: Assessing the Quality of Dialogue. *The Proceedings of the 13th World Conference on Educational Multimedia, Hypermedia & Telecommunications (ED-MEDIA 2001).*

Sørensen, K. H., Aune, M., & Hatling, M. (2000). *Against Linearity – On the Cultural Appropriation of Science and Technology.* In: M. Dierkes, C. von Grote (Eds.) (2000). Between Understanding and Trust – The Public, Science and Technology. Amsterdam: Harwood.

Sørnes, J.-O., Stephens, K. K., Sætre, A. S., & Browning, L. D. (2004). The reflexivity between ICTs and business culture: Applying Hofstede's theory to compare Norway and the United States. *Informing Science Journal, 7.* Retrieved May 10, 2004, from http://inform.nu/Articles/Vol7/v7p001-030-211.pdf

Srinath, U., & Braa, J. (2005). *Training and capacity building to sustain healthcare information systems at a local level in India.* Proceedings of the 8th IFIP WG 9.4 International Working Conference, Abuja, Nigeria, 493-504.

Staikos George. (2006). *Improving Internet Trust and Security*. KDE. http://www.w3.org/2005/Security/usability-ws/papers/33-staikos-improving-trust/

Stallman, R. M. (2002). *Free Software, Free Society: Selected Essays of Richard M. Stallman, 2002*. GNU Press.

Steel, W. 1999). *Rules of Thumb for Web Designs*. http://www.mcst.edu/webhints.html.

Steventon, S., & Wright, A. (2006). *Intelligent Spaces. The Application of Pervasive ICT*. Berlin: Springer.

Strover, S., Chapman, G., & Waters, J. (2004). Beyond community networking and CTCs: Access, development, and public policy. *Telecommunications Policy, 28*, 465–485. doi:10.1016/j.telpol.2004.05.008

Studio Archetype and Cheskin(n.d.). *The Cheskin Research and Studio archetype/Sapient e Commerce Trust Study*,Retrieved from www.studioarchetype.com/cheskin/html/phase1.html

Styliadis, A., Akbaylar, I. I., Papadopoulou, D. A., Hasanagas, N. D., Roussa, S. A., & Sexidis, L. (2009). Metadata-based heritage sites modeling with e-learning functionality. *Journal of Cultural Heritage, 10*, 296–312. doi:10.1016/j.culher.2008.08.014

Subramani, M. (2004). How do suppliers benefit from information technology use in supply chain relationships? *Management Information Systems Quarterly, 28*, 45–74.

Subramani, M. E. Walden, (2000). *Economic Returns to Firms from Business-to-Business Electronic Initiatives: an Empirical Examination*. In Proceedings del International Conference on Information Systems (ICIS).

Subramanian, N., Balaji, R., & Ponraj, M. (2006). Model for Establishing Knowledge Platforms: A Case Study. *Proceedings from the International Conference on Digital Libraries, 2*, 559–566.

Subramanian, N., Balaji, R., & Neela Narayanan, N. (2007). Project management Approaches for developing Public Community Software Solutions. Proceedings from the Third International Conference on Project Management Leadership.

Sullivan, K. P. H., & Lindgren, E. (2002). Self-assessment in autonomous computer-aided second language writing. *ELT Journal, 56*(3), 258–265. doi:10.1093/elt/56.3.258

Sullivan, K. P. H., & Lindgren, E. (2006). *Computer keystroke logging: Methods and Applications*. Oxford, England: Elsevier.

Suthers, D. (2001). Towards a systematic study of representational guidance for collaborative learning discourse. *Journal of Universal Computer Sciences, 7*, 254–277.

Swatman, P. M. C., & Swatman, P. A. (1992). EDI system integration: A definition and literature survey. *The Information Society, 8*, 169–205. doi:10.1080/01972243.1992.9960119

Tait, M. G. (2005). Implementing geoportals: applications of distribuited GIS. *Computers, Environment and Urban Systems, 29*, 33–47.

Taylor, P. A., & Harris, J. L. (2005). *Digital Matters – Theory and culture of the matrix*. London, New York: Routledge.

Taylor, G., Brunsdom, C., Li, J., Olden, A., Steup, D., & Winter, M. (2006). GPS accuracy estimation using map matching techniques: Applied to vehicle positioning and odometer calibration. *Computers. Environment and Urban Studies, 30*, 757–772. doi:10.1016/j.compenvurbsys.2006.02.006

Teasley, S., & Wolinsky, S. (2001). Scientific collaborations at a distance. *Science, 292*(5525), 2254–2255. doi:10.1126/science.1061619

Telecom Regulatory Authority of India. (2008). *Annual Report*. Ministry of Communications Information Technology, Government of India.

Temple University Institute on Disabilities. (2010). *Wave – Web Accessibility Evaluation Tool*. Retrieved from http://wave.webaim.org/, last visited 2010-09-24

Teng, T. L., & Taveras, M. (2004-2005). Combining live video and audio broadcasting, synchronous chat, and asynchronous open forum discussions in distance education. *Journal of Educational Technology Systems, 33*(2), 121–129. doi:10.2190/XNPJ-5MQ6-WETU-D18D

Thomas, A., Howell, M. C., Patricia, C. K., & Angelo, B. (2001). *Collaborative Learning Techniques*. John Wiley & Sons Inc.

Thorne, S., Black, R., & Sykes, J. (2009). Second Language Use, Socialization, and Learning in Internet Interest Communities and Online Gaming. *Modern Language Journal*, *93*, 802–821. doi:10.1111/j.1540-4781.2009.00974.x

Thorne, S. (2008). Transcultural communication in open internet environments and massively multiplayer online games. In Magnan, S. (Ed.), *Mediating Discourse online* (pp. 305–327). Amsterdam: Benjamins.

Thornton, P., & Houser, C. (2004, March 23-25). Using Mobile Phones in Education. *In Proceedings of the IEEE International Workshop on Wireless and Mobile Technologies in Education (WMTE)* (p. 3), Taiwan.

Tiemann, M. (2006-09-19). History of the OSI: Open Source Initiative. Retrieved December 22, 2008 from http://www.opensource. org/history; Retrieved on 2008-12-17.

Tiffin, J., & Rajasingham, L. (1995). *In Search of the Virtual Class: Education in an Information Society*. New York: Routledge. doi:10.4324/9780203291184

Tiwari, M. (2008). ICTs and poverty reduction: user perspective study of rural Madhya Pradesh, India. *European Journal of Development Research*, *20*(3), 448–461. doi:10.1080/09578810802245600

Todaro, M. (2006). *Economic Development*. Reading, MA: Addison-Wesley.

Tonta, Y. (2008). Libraries and museums in the flat world: Are they becoming virtual destinations? *Library Collections, Acquisitions & Technical Services*, *31*(1), 1–9. doi:10.1016/j.lcats.2008.05.002

Toro, J. B. (2000). *Educación para la democracia.* Retrieved December 31, 2007, from http://funredes.org/funredes/html/castellano/publicaciones/educdemo.html

Traunmueller, R., & Wimmer, M. (2003, September 1-5). E-Government at a decisive moment: sketching a roadmap to excellence. *Proceedings of eGov-2003 International Conference: From E-Government to E-Governance, 2739*, Prague.

Tredinnick, L. (2006). Web 2.0 and Business: A pointer to the intranets of the future? *Business Information Review*, *23*(4), 228–234. doi:10.1177/0266382106072239

Tu, C.-H. (2004). *Online Collaborative Learning Communities*. Libraries Unltd Inc.

Tully, P. (1998). Cross-cultural issues affecting information technology use in logistics. In C. Ess & F. Sudweeks (Eds.), *Proceedings, cultural attitudes towards technology and communication* (pp. 317-320). Australia: University of Sydney.

Turinsky, A., & Grossman, R. (2000, August 20-23). *A framework for finding distributed data mining strategies that are intermediate between centralized strategies and in place strategies*. Paper presented at the ACMKDD Workshop on Distributed and Parallel Knowledge Discovery, Boston.

Turner, T. J. (2004). *Local Government E-Disclosure & Comparisons: Equipping Deliberative Democracy for the 21st Century*. University Press of America.

Udo, G. J., & Marquis Gerald, P. (2000). Effective Commercial Web Site Design: An Empirical Study, IEEE Engineering Management Society, 2000. *In Proceedings of the 2000 IEEE*, 313 – 318

UK Government. (2002). Www.localegov.gov.uk: The National Strategy for Local E-government, Office of the Deputy Prime Minister, Great Britain, Local Government Association (England and Wales), UK Online for Business, Office of the Deputy Prime Minister.

UK Government. (2003). Www.localegov.gov.uk One Year on: The National Strategy for Local E-government, Great Britain Office of the Deputy Prime Minister, Office of the Deputy Prime Minister, Local Government Association (England and Wales)., Great Britain, UK Online for Business, Office of the Deputy Prime Minister, 2003

Ulys, P., Nleya, P., & Molelu, G. (2004). *Technological Innovation and Management Strategies for Higher Education in Africa: Harmonizing Reality and Idealism*. Omaha, NE: Education Media.

UNDP. (2001). *Human Development Report 2001: making new technologies work for Human Development*. New York: Oxford University Press.

UNDP. (2000). *Optimizing efforts. A practical guide to NGO networking.* [Online]. Office to Combat Desertification and Drought (UNSO), UNDP. [Retrieved 3 Mar 2004 from http://www.energyandenvironment.undp.org/undp/indexAction.cfm?module=Library&action=GetFile&DocumentID=5256

UNESCAP. (2003). *Country Reports on Local government Systems*. Pakistan: United Nations Economic and Social Commission for Asia and Pacific.

UNESCAP. (1999). (Ed.) Local Government in Asia and the Pacific: A comparative Study of Fifteen Countries. Bangkok, (UN Economic and Social Commission for Asia and the Pacific), Thailand: UNESCAP.

UNESCO. (2001). *Monitoring Report on Education for All*. UNESCO.

UNESCO. (2005). *Text of the Convention for the Safeguarding of Intangible Cultural Heritage*. Retrieved May 28, 2008, from http://www.unesco.org/culture/ich/index.php?pg=00006

Unicode (1991-2007). *Unicode Works*. Retrieved December 31, 2007, from http://www.unicode.org

United Nations Development Programme (UNDP). (2003). *Human Development Report 2003*. New York: Oxford University Press.

United Nations Development Programme (UNDP). (2008). *UNDP Annual Report 2008*. Available at http://www.undp.org/publications/annualreport2008/. Accessed 30 December 2008.

United Nations, Division for Public Economics and Public Administration and American Society for Public Administration. (2001). *Benchmarking e-government: A global perspective --- Assessing the progress of the UN member states*. Retrieved May 10, 2004, from http://www.unpan.org/e-government/Benchmarking%20E-gov%202001.pdf

United Nations, International Telecommunication Union. (2003-2005). *World Summit on the Information Society*. Retrieved December 30, 2007, from http://www.itu.int/wsis

United Nations. (2000). *United Nations Information and Communication Technologies Task Force*. Retrieved December 29, 2007, from http://www.unicttaskforce.org

United Nations. (2007). Report by United Nations Development Account Project on Knowledge Networks through ICT Access Points for Disadvantage Communities (2007), March 2007 Assessment of the Status of the Implementation and Use of ICT Access points in Asia and the Pacific, 12. Usage of Action Apps. Retrieved December 22, 2008 from http://apc.org/ actionapps/english/ general/slices.shtml.

United Nations. (2008). *UN e-government survey 2008: From e-government to connected governance*. Retrieved August 25, 2008, from http://unpan1.un.org/intradoc/groups/public/documents/un/unpan028607.pdf

United Nations. (2008). *Human Development Reports 2007/2008*. UNDP.

United States of America. (1990). Public Law 101-336, 1990. *Text of the Americans with Disabilities Act, Public Law 336 of the 101st Congress,* enacted July 26, 1990.

Urban, G. L., Amyx, C., & Lorenzone, A. (2009). Online Trust: State of art, New frontiers and research potential. *Journal of Interactive Marketing*, *23*, 179–190. doi:10.1016/j.intmar.2009.03.001

US Government. (2006). FY 2005 Report to Congress on Implementation of the E-government Act of 2002, Office of Management and Budget, Executive Office of the President, United States.

USAID. (2006). *The Global Development Alliance: Public-Private Alliances for Transformational Development, U.* USA: S. Agency for International Development.

Use of FOSS in AgriBazaar. (2004). Reterived 24 November, 2007 from http://r0.unctad.org/ ecommerce/event_docs/ fossem/azzman_ agribazzar.pdf, 15.

Usun, S. (2004). Factors affecting the application of information and communication technologies (ICT) in distance education. *Turkish Online Journal of Distance Education,* 5(1). Available at: http://tojde.arladolu.edu.tr/tojde13/articles/us\m.html

Utsumi, T. (2005b). *Global University System with Globally Collaborative Innovation Network*. Finland: Global University System.

Utsumi, T. (2005a). Global E-Learning for Global Peace with Global University System. In Ruohotie, P. (Ed.), *Communication and Learning in the Multicultural World*. Finland: University of Tampere.

Vadillo, M. A., Bárcena, R., & Matute, H. (2006). The Internet as a research tool in the study of associative learning: An example from overshadowing. *Behavioural Processes, 73*(1), 36–40. doi:10.1016/j.beproc.2006.01.014

Valerie, P. (November 30, 1996). *Good Web Site Design*. Retrieved from www.geocites.com.

Van, D., Landay, J., & Hong, J. (2002). *The Design of Sites: Patterns, Principles, and Processes for Crafting a Customer-Centered Web Experience*. Reading, MA: Addison-Wesley.

Van Bavel, R., Punie, Y., & Tuomi, I. (2004, July). ICT-enabled changes in Social Capital. [Online]. *The IPTS Report,* 85. Retrieved 3 Feb 2006 from <http://www.jrc.es/home/report/english/articles/vol85/ict4e856.htm>

van Dam-Mieras, M. C. E. (Rietje) (2004, May 9-12). Learning In a Global Society. *The Hague Conference on Environment, Security and Sustainable Development*, Peace Palace, The Hague.

van den Berg, L., van der Meer, A., van Winden, W., & Woets, P. (2006). *E-governance in European and South African Cities: The Cases of Barcelona, Cape Town, Eindhoven, Johannesburg, Manchester, Tampere, The Hague*. Venice: Ashgate Publishing, Ltd.

van der Meulen, B. (2003, June). Integrating Technological and Societal Aspects of ICT in Foresight Exercises. *Technikfolgenabschätzung*, *2*(12), 66–74.

Van Dijck, J. (2006). The science documentary as multimedia spectacle. *International Journal of Cultural Studies*, *9*(1), 5–24. doi:10.1177/1367877906061162

Vanderheiden, Gregg C. (2008). Ubiquitous accessibility, common technology core, and micro-assistive technology. *ACM Trans. Access. Comput1, (2),* (October 2008)

Veiga, J. F., Floyd, S., & Dechant, K. (2001). Towards modeling the effects of national culture on IT implementation and acceptance. *Journal of Information Technology*, *16*(3), 145–158. doi:10.1080/02683960110063654

Venkatesh, M., & Small, R. V. (2003). *Learning-in-Community: Reflections on Practice*. Springer.

Villamil, O. S., & de Guerrero, M. C. M. (1996). Peer revision in the L2 classroom: social-cognitive activities, mediating strategies, and aspects of social behavior. *Journal of Second Language Writing*, *5*, 51–75. doi:10.1016/S1060-3743(96)90015-6

Virtual Community, M. I. S. T. I. C. A. (2002). *Working the Internet with a Social Vision*. Retrieved December 29, 2007, from http://www.funredes.org/mistica/english/cyberlibrary/thematic/eng_doc_olist2.html

Vonderwell, S. (2003). An examination of asynchronous communication experiences and perspectives of students in an online course: a case study. *The Internet and Higher Education*, *6*, 77–90. doi:10.1016/S1096-7516(02)00164-1

Vygotsky, L. S. (1978). *Mind in society: The development of higher psychological processes*. Cambridge, MA: Harvard University Press.

W3C (1994-2006). *Web Accessibility Initiative*. Retrieved December 30, 2007, from http://www.w3.org/WAI

W3C (2006). Guidelines and resources from the World Wide Web Consortium (W3C). Retrieved from www.w3.org/WAI/, last visited 2010-09-24

Wade, P. (2001). *Information and Communication Technologies and Rural Development.* OECD Publishing.

Walker, S. A. (2004). Socratic strategies and devil's advocacy in synchronous CMC debate. *Journal of Computer Learning*, *20*, 172–182. doi:10.1111/j.1365-2729.2004.00082.x

Walker, J. (1990). Through the Looking Glass. In Laurel, B. (Ed.), *The art of computer-human interface design* (pp. 213–245). Menlo Park, CA: Addison-Wesley.

Wang, E., & Barrett, S. Caldwell & Gavriel S. (2003). Usability comparison: similarity and differences between e-commerce and world wide web. *Journal of the Chinese Institute of Industrial Engineers*, *20*(3), 258–266. doi:10.1080/10170660309509234

Wang, Y. D., & Emurian, H. H. (2005). An overview of online trust: Concepts, elements, and implications. *Computers in Human Behavior*, *21*(1), 105–125. doi:10.1016/j.chb.2003.11.008

Wang, Y. D., & Emurian, H. H. (2007). Inducing Online Trust in E-Commerce: Empirical Investigations on Web Design Factors. In Khosrow-pour, M. (Ed.), *Utilizing and Managing Commerce and Services Online)*. Hershey, PA: Idea Group Publishing. doi:10.4018/9781591409328.ch005

Warschauer, M. (2003). Social capital and access. *Universal Access Information Society*, *2*, 315–330. doi:10.1007/s10209-002-0040-8

Wasserman A. I. (2001). *Principles for the Design of Web Applications*.

Wasukira, E., & Naigambi, W. (2002). *Report on the Usage of ICT in Local Governments in Uganda*. Canada: IICD.

Web Accessibility Initiative. (2010). *WAI Guidelines and Techniques*. Retrieved from http://www.w3.org/WAI/guid-tech.html, last visited 2010-09-24

WebAIM. (2010). *Accessible Site Certification*. Retrieved from http://webaim.org/services/certification/, last visited 2010-09-24

Webb, G. I. (2000, August 20-23). Efficient search for association rules. In *Proceedings of the 6th ACM SIGKDD International Conference on Knowledge Discovery and Data Mining,* Boston (pp. 99-107). ACM Publishing.

Weiss, J. W. (2001). *Organizational Behavior and Change: Managing Diversity, Cross-Cultural Dynamics, and Ethics*. Cincinnati, Ohio: South-Western.

Wellman, B. (2001). *Living networked in a wired world: the persistence and transformation of community. Report to the Law Commission of Canada*. Toronto: Wellman Associates.

Wellman, B. (1999). *Networks in the global village*. Boulder, Col.: Westview Press.

Wellman, B. (2001). Physical place and cyberplace: the rise of networked individualism. *International Journal of Urban and Regional Research*, No. 1.

West, L. J. (1994). Breaking down the barriers to EDI implementation. *TMA Journal, 14*(1), 10–15.

West, D. M. (2003). Achieving E-Government for All: Highlights from a National Survey. Commissioned by the Benton Foundation and the New York State Forum of the Rockefeller Institute of Government. (2003). Available at http://www.benton.org/archive/publibrary/egov/access2003.html?0#0, last visited 2010-09-24

West, D. M. (2008a). *Improving technology utilization in electronic government arount the world, 2008*. Governance Studies at Brookings. Retrieved August 24, 2008, from http://www.brookings.edu/~/media/Files/rc/reports/2008/0817_egovernment_west/0817_egovernment_west.pdf

West, D. M. (2008b). *State and federal electronic government in the United States, 2008*. Governance Studies at Brookings. Retrieved August 24, 2008, from http://www.brookings.edu/~/media/Files/rc/reports/2008/0826_egovernment_west/0826_egovernment_west.pdf.

Westin, A., & Maurici, D. (1998). *E-commerce and privacy: What net users want* (p. 15). Price Waterhouse Coopers.

Wikipedia. (2007). *Information Literacy*. Retrieved December 30, 2007, from http://en.wikipedia.org/wiki/Information_literacy

Wilson, J. (2006). 3G to Web 2.0? Can Mobile Telephony Become an Architecture of Participation? *Converge: The International Journal of Research into New Media Technologies, 12*(2), 229–242. doi:10.1177/1354856506066122

Wilson, D. (1999), *Rules of Thumb for Web Designs*. http://www.goodpractices.com

Winch, G. M., & Courtney, R. (2007). The Organization of Innovation Brokers: An International review. *Technology Analysis and Strategic Management, 19*(6), doi:10.1080/09537320701711223

WITSA (2008, May 28). *Global ICT Spending Tops $3.5 Trillion: Industry Experiences Subdued Spending Growth*, Press release, World Information Technology and Service Alliances (WITSA), Kuala Lumpur, Malaysia

WITSA. (2004). *WITSA Study: World IT Spending Rebounds Thanks Largely to Developing World*. World Information Technology and Services Alliance.

Woolgar, S. (Ed.). (2002). *Virtual Society, technology, cyberbole, reality*. Oxford University Press.

Worboys, M., & Duckham, M. (2004). *GIS: A computing Perspective*. Taiwan: CRC Press.

World Bank. (1994). *World Development Report: Infrastructure*. New York: Oxford University Press.

World Bank. (2007). Local Government Discretion and Accountability: A Local Governance Framework, Social Development Department in collaboration with the Finance, Economics and Urban Department (FEU) and the Social Protection Team (HDNSP), Report No. 40153, The World Bank, Washington DC.

World Development Indicators (WDI). (2008). *Table 5.11, Information Age*. Available at http://siteresources.worldbank.org/DATASTATISTICS/Resources/WDI08_section5_intro.pdf Accessed 30 December 2008.

WRI. (2005). *The Wealth of the Poor: Managing Ecosystems to Fight Poverty*. Washington, D. C.: World Resources Institute.

WSIS (2005). *Report of the WSIS Education*. Academia and Research Taskforce. Additional Readings Alexander, S. (2001). E-learning experiences. Education + Training, 43(4/5), 240-248.

Wu, X., & Zhang, C. (2005). Database classification for multi database mining. *Information Systems*, *30*(1), 71–88. doi:10.1016/j.is.2003.10.001

Wu, X., & Zhang, S. (2003). Synthesizing high-frequency rules from different data sources. *IEEE Transactions on Knowledge and Data Engineering*, *15*(2), 353–367. doi:10.1109/TKDE.2003.1185839

Wu, X. (2004, November 15-17). Knowledge discovery in multiple databases. In *Proceedings of the 16th IEEE International Conference on Tools with Artificial Intelligence,* Boca Raton, FL (pp. 2). IEEE.

Wu, Z.-H., Liu, Y.-L., Chang, M., Chang, A., & Li, M. (2006). Developing Personalized Knowledge Navigation Service for Students Self-Learning based on Interpretive Structural Modeling. In *Proceedings of the Sixth IEEE International conference on Advanced Learning Technologies*. The Netherlands: IEEE Computer Series.

Wynne, B. (1995). Technology Assessment and Reflexive Social Learning: Observations from the Risk Field. In Rip, A., Misa, T. J., & Schot, J. (Eds.), *Managing Technology in Society – The Approach of Constructive Technology Assessment*. London, New York: Pinter.

Wysocki, R. K., & DeMichiell, R. L. (1997). *Managing Across the Enterprise*. New York: John Wiley.

Yee, N., & Bailenson, J. (2007). The proteus effect: The effect of transformed self-representation on behavior. *Human Communication Research*, *33*, 271–290. doi:10.1111/j.1468-2958.2007.00299.x

Yen, B., Hu, P., & Wang, M. (2005). Towards Effective Web Site Designs: A Framework for Modeling, Design Evaluation and Enhancement. In *Proceedings. IEEE International Conference on e-Technology, e-Commerce and e-Service*, (pp. 716-721).

Yin, R. K. (2003). *Case Study Research: Design and Methods* (3rd ed.). Thousand Oaks, CA: Sage.

Yunus, M. (2008). *Un mundo sin pobreza.* Translated by Monserrat Asensio (original title: *Creating a World Without Poverty*). Barcelona: Editorial Paidós Ibérica. ISBN 84-493-128.3.

Zacher, L. W. (2001). *Between Risk and Trust – Values, Rules and Behaviour in the E-Society*. In: Innovations for an e-Society – Challenges for Technology Assessment, Berlin (conf. proc.), ITAS – VDI.

Zacher, L. W. (2007). *E-Government in the Information Society.* In: Encyclopedia of Digital Government, vol. II. Hershey – London – Melbourne – Singapore: Idea Group Inc.

Zaheer, S., & Manrakhan, S. (2001). Concentration and dispersion in global industries: Remote electronic access and the location of economic activities. *Journal of International Business Studies*, *32*(4), 667–686. doi:10.1057/palgrave.jibs.8490989

Zapata, G., & Sagarra, N. (2007). CALL on hold: The delayed benefits of an online workbook on L2 vocabulary learning. *Computer Assisted Language Learning*, *20*(2), 153–171. doi:10.1080/09588220701331352

Zhang, S., & Wu, X. (2004). *Knowledge discovery in multiple databases*. New York: Springer.

Zhao, H. (2008). *Who takes the floor: peer assessment or teacher assessment?: a longitudinal comparative study of peer- and teacher-assessment in a Chinese university EFL writing class*. Unpublished PhD thesis, University of Bristol, England.

Zheng, D., Li, N., & Zhao, Y. (2008). *Learning Chinese in Second Life Chinese language School*. Paper presented at CALICO Annual Conference. San Francisco, CA.

Zhou, H., & Benton, W. C. Jr. (2007). Supply chain practice and information sharing. *Journal of Operations Management*, *25*, 1348–1365. doi:10.1016/j.jom.2007.01.009

Zobl, H. (1983). Markedness and the projection problem. *Language Learning, 33*(3), 293–313. doi:10.1111/j.1467-1770.1983.tb00543.x

Zurita, G., & Nussbaum, M. (2004). Computer supported collaborative learning using wirelessly interconnected handheld computers. *Computers & Education, 42*(3), 289–314. doi:10.1016/j.compedu.2003.08.005

Zurita, L., & Bruce, B. C. (2005, June/July). Designing from the users side: reaching over the divide. *Submitted to the Computer Supported Collaborative Learning (CSCL) Conference 2005*, Taipei.

About the Contributors

Susheel Chhabra is Associate Professor of Information Technology at Lal Bahadur Shastri Institute of Management (Delhi, India) and is also acting as a programme coordinator, PGDM (MBA). His areas of research and consultancy include e-government, e-business, computer networks, and software engineering. He has published several research papers on international and national level journals. He has co-authored a textbook on human resource information systems, edited a special issue of International Journal of E-Government Research on strategic e-business model for government, and also co-authored the edited book Integrating E-Business Models for Government Solutions: Citizen-Centric Service Oriented Methodologies and Processes (IGI Global, USA). He is currently engaged in several consultancy and training assignments on social change for human development, e-governance, e-business, and ERP for ISID, NTPC, LBSRC, etc.

Hakikur Rahman (PhD) is the Project Coordinator of the Sustainable Development Networking Programme (SDNP) in Bangladesh, a global initiative of UNDP since December 1999. He also acts as the Secretary of South Asia Foundation Bangladesh Chapter. Before joining SDNP, he worked as the Director, Computer Division at Bangladesh Open University. He has written several books and many articles/papers on computer education for the informal sector and distance education. He is the Founder-Chairperson of Internet Society Bangladesh Chapter, Editor of the Monthly Computer Bichitra, Founder-Principal and Member Secretary of ICMS Computer College, Head Examiner (Computer) of the Bangladesh Technical Education Board, and Executive Director of BAERIN (Bangladesh Advanced Education Research and Information Network) Foundation. He is also involved in activities related to establishment of a IT based distance education university in Bangladesh.

* * *

Manuel Acevedo works since 2003 as an international consultant on ICT and Development (with government, civil society, multilateral agencies and business entities), researcher on development networks, and lecturer on ICT4D in Spanish universities. Recent engagements include working as senior advisor to the Argentinian National Programme for the Information Society Programme, strategic planning for UN Volunteers (UNV), co-production of a guide for ICT mainstreaming for the Spanish Secretariat for International Cooperation and evaluations for UNDP, UNV and civil society organizations. From 1994 to 2003 he worked with UNDP and UNV, where he set up in 2000-2003 the first e-Volunteer unit in an development agency, and was responsible for launching the programme UNITES and the UN Online Volunteer Service. He represented UNV during the first phase of the World Summit on the Information Society, and acted as co-chair of the Human Capacity Committee of the UN ICT Task Force during that period.

Minwir Al-Shammari is Professor of Management and Director of Graduate Studies at the University of Bahrain, College of Business Administration. He holds a PhD in Business Administration (Industrial Management) from University of Glasgow (UK, 1990) and MS in Industrial Management from Central Missouri State University (USA, 1986). He has been involved for about 20 years in teaching, research, training, and/or consultancy in the areas of operations management, knowledge management, supply chain management, management information systems, business process re-engineering, organization theory, organizational change and development, project management, spreadsheet modeling, management science, and research methodology. He is Editor-in-Chief of Journal of Supply Chain and Customer Relationship Management, IBIMA Publishing. He is the author of the premier reference source Customer Knowledge Management: People, Processes, and Technology, IGI-Global Publishing. He has published more than 30 research papers that have appeared in international refereed journals such as International Journal of Knowledge Management, Logistics Information Management, International Journal of Information Management, European Journal of Operational Research, Expert Systems with Applications, Journal of Computer Information Systems, International Journal of Information Management, International Journal of Information Communication Technologies and Human Development, International Journal of Operations and Production Management, Production and Inventory Management Journal, Business Process Management Journal, International Journal of Commerce and Management, International Journal of Computer Applications in Technology, Cross-Cultural Management, International Journal of Management, Leadership and Organization Development Journal, and Creativity and Innovation Management.

Abdullah Almobarraz was born in Riyadh, Saudi Arabia on 1963. He finished his undergraduate education in 1990 from 1990 King Saud University, Riyadh majoring in Information Science. In 2007, he got his PhD in Information Science from University of North Texas. Currently he is working as the Head of Information Studies Department, College of Computer and Information Sciences at Imam University in Riyadh.

Abiodun O. Bada is Assistant Professor of Information Systems in the Dept. of Engineering Management & Systems Engineering at The George Washington University, Washington DC, USA. Dr. Bada received his PhD in Information Systems from the London School of Economics, University of London, UK. His research interests include the application of resource-based and institutional theories to information systems phenomena and the implementation of IT in developing countries.

Nancy Bertaux is Professor of Economics at Xavier University, Cincinnati, Ohio, USA. She holds a PhD in Economics from The University of Michigan. She has published in a wide variety of scholarly books and journals on economic development, gender and diversity, and economic history issues, with a recent research focus on the role of information technology and entrepreneurship in economic development in Africa and South Asia.

James D. Brodzinski is Dean and Professor of Management in the Graham School of Management at Saint Xavier University in Chicago, Illinois. His PhD is from Ohio University in Human Information Systems. His research examines behavioral, perceptual, and information issues in organizations as well as occupational safety and health compliance programs. His work has appeared in various scholarly outlets including The Academy of Management Executive, Journal of Health and Human Resources Administration, HR Magazine, and Management Communication Quarterly.

Anirban Chakrabarty is Assistant Professor of Information Technology at Lal Bahadur Shastri Institute of Management, Delhi. He has published in journals and conferences at both National and International levels on data mining, knowledge management, and fraud detection issues. His research interest include Data mining, Object Oriented programming, Multimedia Technologies, Computer networks.

Elaine Crable is Professor of Management Information Systems at Xavier University, Cincinnati, Ohio, USA. She received her PhD from the University of Georgia with an MBA from Xavier University. Her research includes enterprise systems, social networking, pedagogical issues with online teaching and women's issues in developing countries.

Zlatko J. Kovačić is an associate professor in the School of Information and Social Sciences at the Open Polytechnic of New Zealand. Dr. Kovačić has a varied academic background and research interests, ranging from core interests relating to IT careers, learning and teaching, to e-commerce, e-learning, time series analysis and multivariate analysis. His current research is focused on financial time series analysis (stock exchanges in former Yugoslavia in particular), social and cultural aspects of e-government and on cognitive processes in distance education using computers and communications technologies. Dr. Kovačić is editor-in-chief of the academically peer refereed journal: Interdisciplinary Journal of Information, Knowledge, and Management (http://ijikm.org), editor of the Journal of Information Technology Education (http://jite.org) and senior associate editor of Informing Science: The International Journal of an Emerging Transdiscipline (http://www.inform.nu). He is member of the Informing Science Institute and International Association for Statistical Education.

Dr. Norbert Kuhn was born in 1957. He studied Computer Science and Mathematics at the Technical University Kaiserslautern (Germany) and graduated in 1986. Thereafter he was a staff member of the Computer Science department of the Technical University Kaiserslautern, from where he also received his PhD in 1991 with a thesis in Computer Algebra. From 1991 to 1995 he was a member of the German Research Centre for Artificial Intelligence (DFKI) and worked on projects in the fields of Multi-Agent systems and Office Automation. In 1995 he was appointed professor at the University of Applied Sciences Trier, (Germany). He is a member of the Institute for Software Systems in Business, Environment, and Administration. His current research areas cover E-Government applications, Document Analysis systems and Human Computer Interfaces.

Dr. Muneesh Kumar is a Professor, at Department of Financial Studies, University of Delhi (India). His responsibilities include teaching banking and information systems related courses to students of Masters in Finance and Control (MFC) programme and supervising research. He has published several articles in international journals and presented papers in several international conferences. He has also authored three books and co-edited three books. He is associated with the several expert committees appointed by Government of India such as expert committee for IT projects of India Post and Market Participation Committee of Pension Fund Regulatory and Development Authority (PFRDA).

Eva Lindgren, PhD, is a member of the Department of Language Studies at Umeå University, where she is a researcher and PhD-student supervisor. Her research interests include early language learning, writing, writing development and revision. She has published internationally in the areas of foreign language development, self-assessment, peer feedback, keystroke logging, revision and fluency. Together with Kirk Sullivan she has edited the book Computer Keystroke Logging and Writing: Methods and Applications. She is the Swedish country manager of the EU funded research project Early Language Learning in Europe (Project n°. 135632-LLP-2007-UK-KA1SCR), which takes a broad perspective on young learners' foreign language learning processes, including variables such as the teacher, language exposure and digital media.

Meeta Mathur is Assistant Professor in the Department of Economics at University of Rajasthan, India. She has specialized in International Economics. Her research areas include insurance, telecommunications and foreign trade. She has contributed to reputed national journals, national seminars and conferences. She has rich experience of teaching and research. Her excellent academic record has motivated many students to pursue their doctorate degree in her guidance. As an economist she is exploring socio-economic dimensions to understand the economic phenomena keeping in view the international perspective. She has been associated with many professional bodies and continue to work rigorously for the growth of the discipline.

Dr. Stefan Naumann was born in 1969 and studied computer science at the Universities of Kaiserslautern and Saarbrücken (Germany). Since 2008 he is a full professor at the University of Applied Sciences Trier, Location Environmental Campus Birkenfeld. He is a member of the Institute for Software Systems in Business, Environment, and Administration. His research interests are sustainable development in conjunction with online communities and the environmental and social impacts of information technology. Especially, he is engaged within the new research field "sustainability informatics". Within this context he investigates questions how non-professional IT users (like citizens) can be supported so that they are able to participate successfully e.g. in E-Government processes.

Adekunle Okunoye is Associate Professor of Management Information Systems at Xavier University, Cincinnati, Ohio, USA. He holds a PhD degree in Computer Science/Information Systems from University of Turku, Finland. Adekunle is a chartered information technology practitioner and member of the British Computer Society. He is also a member of Association for Information Systems. His research focuses on knowledge management, organizational implementation of IT and the resultant changes in organization, and IT & globalization. He has published in various journals, books and conference proceedings.

Bolanle A. Olaniran is a professor and interim Chair in the Department of Communication Studies at Texas Tech University, Lubbock, TX USA. He is internationally known scholar. His research includes Organization communication, Cross-cultural communication, Crisis Communication, and Communication technologies. He has authored several peer reviewed articles in discipline focus and interdisciplinary Journals (i.e., Regional, National, and International) and authored several edited book chapters in each of these areas. He edited book on e-learning. He also serves as consultant to organizations and Universities at local, national, international and government level. His works have gained recognition such as the American Communication Association's "Outstanding Scholarship in Communication field" among others.

Eleonora Pantano, is a Post doc research fellow at University of Calabria (Italy). She holds a PH-D degree in "Psychology of Programming and Artificial Intelligence". Her research interests are related to the applications of advanced technologies to retailing and tourism, with emphasis on the investigation of consumer behaviour in pervasive environments. She has been Assistant teacher of Integrated Marketing Communication, Engineering Faculty, University of Calabria; visiting lecturer at College of Business, University of Illinois (USA); visiting lecturer at Master in Business and Administration (MBA) Marketing Module at the Faculty of Economics & Business, University of Zagreb (HR). Furthermore, she is member of the Editorial Board of numerous international journals, guest editor of the special issue of Journal of Retailing and Consumer Services on Applications of New Technologies to Retailing, 17 (3); and of International Journal of Digital Content Technology and its Applications on Digital Contents Management for Improving Consumers Experience (in press). She was the Highly Commended Award winner of the 2008/2009 Emerald/EMRBI Business Research Award for Young Researchers.

Daniel Pimienta was born in Morocco, Applied Mathematics in Nice and hold a PhD in Computer Sciences. After creating a Software House specialized in APL, he joined IBM France (La Gaude Laboratory) and worked 12 years as Telecommunication System Architect and Planner. In 1988, he joined Union Latina, in Santo Domingo, as Scientific Advisor and Head of the REDALC project for creation of LA&C network. En 1993, he launched the Foundation Networks & Development (FUNREDES) and focused ICT4D. He is an active civil society actor in Information Society themes, especially the social impact of ICT, virtual communities and linguistic diversity. Member of several ICT4D related global groups such as Francophone virtual university, 3EL, GCNP, EUROLATIS, WINDS-LA, REDISTIC, APC, MAAYA, WSIS-AWARD, UN-GAID or Digital Solidarity Fund, he was given, in 2008, the Namur Award (IFIP WG9.2) for his comprehensive actions in the perspective of a positive social impact of ICT.

Balaji Rajendran received his BSc and MSc degrees in Computer Science from Madurai Kamaraj University, in 1998 and 2000 respectively. He is currently pursuing his PhD in Computer Science. He has worked as a lecturer in reputed institutions before moving to C-DAC (Centre for Development of Advanced Computing), formerly known as NCST, as a Scientist in 2001. Since then he has been working with C-DAC, executing various Research and Development Projects, funded by National and International agencies. He has research publications in International and National Conferences, and has also edited the proceedings of an International Conference. He has also received the PMP certification from PMI, USA, in 2005. His research interests are in the domain of Internet and Web Engineering with focus on Intelligent Systems, Security, and systems for Social Development.

Sonal G. Rawat is Assistant Professor of Information Technology at Lal Bahadur Shastri Institute of Management, Delhi. Her areas of interest are data analysis and algorithm, data structure, C++, Computer Graphics.

Stefan Richter was born in 1976 and studied computer science at the University of Applied Sciences in Birkenfeld. After his diploma thesis about an email client for blind people he worked for the SilverCreations AG which develops reading machines and workplaces for visually impaired people. He made significant contributions to the development of the "LiveReader" and the digital workplace camera named "Sceye" for document management and imaging. Since 2005 he is a member of the Institute for Software Systems in Business, Environment, and Administration and he works there in different research projects. His main research interests are accessibility for the visually impaired and blind people and document management. Now he investigates in the fields of ontology-based knowledge management and of E-Government research combining his experiences to make governmental forms accessible.

Dr. Mamta Sareen is an Associate Professor in Department of Computer Science, Kirori Mal College, University of Delhi, India. She has done her doctoral research in 'Trust and Technology in B2B e-commerce'. In addition, she is also pursuing research in areas like e-commerce, Internet banking, etc. She is teaching various courses on information technology like software engineering, management Information systems, Data Base Management Systems, etc. to various undergraduate and post graduate courses (B.Sc, MCA, MBA) of University of Delhi and I.P University, India.

M. P. Satija, is a Professor and Head of the Department of Library and Information Science, Guru Nanak Dev University, Amritsar, India. Having done his PhD on Ranganathan studies, he has been instrumental in interpreting and propagating Ranganathan's works and ideas to the new generation. As an author of more than a dozen books, 150 papers and 200 book reviews and many conference papers published in India and abroad. He has visited Germany, France, Finland, England, Nepal, the Netherlands, Belgium and Sri Lanka in connection with professional work. In 2005 he was invited for a year by the University of Kelaniya, Sri Lanka to serve as a Visiting Professor. There he was instrumental in instituting library and information science doctoral programme in the Department of Library of Information Science which is the first ever research degree programme in the country. Since April 2007 he is a member Advisory Board of the UDC Consortium, the Hague.

Michael Schmidt was born in 1975 and studied computer sciences at the University of Applied Science Birkenfeld and the University of Saarbrücken. After he finished his master thesis he joined the research group of Prof. Gollmer. In the research group he worked on a mathematical model of the e.coli bacteria. In 2007 he switched to the research group of Prof. Kuhn at the Institute for Software Systems in Business, Environment, and Administration. His main research interests are accessibility for the visually impaired and blind people and document management. He investigates in the field of E-Government research combining his experiences to make governmental forms accessible.

Sangeeta Sharma is a Professor of Public Administration in the Department of Public Administration at the University of Rajasthan, India. She has authored many books. Her articles have appeared in the Internationally reputed publications. Her research interests include exploring man-machine interface, development of conceptual constructs especially in the field of digital governance and ethics building. She is member of International Program Committees. Many students are pursuing Doctorate degree in her guidance. Her works on ethics and e-governance have been internationally acknowledged. She is also member of Editorial Board and has contributed in providing qualitative writings. The articulation of ideas as lucid narrations is highly appreciated.

Anders Steinvall is Senior Lecturer in English at Umeå University. His particular research interests lie in the field of cognitive linguistics and its application to lexical semantics, discourse analysis and second language learning.

Kirk P H Sullivan obtained his PhD from the University of Southampton, England and his EdD from the University of Bristol, England. He is currently Professor of Linguistics and member of the Department of Language Studies, Umeå University, Sweden. Prior to moving to Sweden he worked at Otago University, New Zealand. Together with Eva Lindgren he has edited the book Computer Keystroke Logging and Writing: Methods and Applications. Kirk's research interests lie at the intersection of linguistics and education.

Assunta Tavernise is a PhD in Psychology of Programming and Artificial Intelligence and collaborates with the Laboratory of Psychology at University of Calabria. Her research interests concern various scientific topics from an interdisciplinary point of view and comprise the following areas: - Educational Technology, - Virtual Worlds/Games, - Human Computer Interaction, - Edutainment, - Virtual Agents, - ICT for Cultural Heritage. At the moment, she is working on the constructivist approach to educational virtual worlds as learning environments, carrying on laboratories with students from grammar school to University. Moreover, she is carrying out studies on non verbal communication of virtual agents for the realization of didactic tutors, guides, and virtual shopper assistants. She has worked in national and international projects, among which "Virtual Museum Net of Magna Graecia" (ROP 2000–06, www.virtualmg.net) and "Connecting European Culture through New Technology - NETConnect" (Culture 2000 European Programme, http://www.netconnect- project.eu/ and http://netconnect-project.eu/index.aspx).

Neelanarayanan Venkataraman received his BSc degree in Physics from Madurai Kamaraj University, India, in 1993, his MSc degree in Computer Science from Madurai Kamaraj University, India, in 1995 and currently pursuing PhD in Computer Science at IT University of Copenhagen, Denmark. After receiving his Masters degree, he had taught Computer Science subjects in various colleges affliated to Madurai Kamaraj University and also in Madurai Kamaraj University. In 2001, he moved to Centre for Advanced Computing (C-DAC), formerly known as National Centre for Software Technology (NCST), where he was a Scientist. In 2007 he joined IT University of Copenhagen with Danish Government Scholarship for his PhD. He is a Life Member of ISTE, CSTM certified by STQC, ISMS LA certified by IRCA, and PMP certified by PMI, USA.

Huahui Zhao obtained her doctoral degree in Applied Linguistics at the University of Bristol, United Kingdom. She is currently working as a postdoctoral researcher in online peer collaboration in the Department of Language Studies, Umeå University, Sweden. Her main research interests lie in computer-enhanced learning and teaching, classroom-based language assessment, collaborative learning, and education research methodology.

Index

Symbols

A

B

C

D

F

G

H

R

S

U

V

W

X

Y

Z